Advances in Global Change Research

Volume 81

This book series is indexed in SCOPUS.

Advances in Global Change Research

Aims and Scopes

This book series is aimed at addressing a range of environmental issues through state-of-the-art and/or interdisciplinary approaches. The books in the series can either be monographs or edited volumes based, for example, on the outcomes of conferences and workshops, or by invitation of experts. The topics that the series can consider publishing include, but are not limited to:

Physical and biological elements of earth system science, in particular

- Climate change
- Biodiversity
- Sea-level rise
- Paleo-climates and paleo-environments

Social aspects of global change

- Environmentally-triggered migrations
- Environmental change and health
- Food security and water availability
- Access to essential resources in a changing world

Economic and policy aspects of global change

- Economic impacts
- Cost-benefit analyses
- Environmental governance
- Energy transition

Methodologies for addressing environmental issues

- Planetary data analysis
- Earth observations from space
- Proxy data analyses
- Numerical modeling
- Statistical analyses

Solutions to global environmental problems

- Sustainability
- Ecosystem services
- Climate services
- Technological/engineering solutions

Books in the series should be at least 200 pages in length, and include a table of contents.

Images supplied in colour will be reproduced in both the print and electronic versions of the book at no cost to the author/editor.

Manuscripts should be provided as a Word document and will be professionally typeset at no cost to the authors/editors.

All contributions to an edited volume should undergo standard peer review to ensure high scientific quality, while monographs should also be reviewed by at least two experts in the field.

Manuscripts that have undergone successful review should then be prepared according to the Publisher's guidelines manuscripts: https://www.springer.com/gp/authors-editors/book-authors-editors/book-manuscript-guidelines

Majda Khalil Suleiman · Shabbir Ahmad Shahid
Editors

Fostering Arid Lands Agriculture in the Face of Climate Change

Mitigation and Adaptation Synergy

Springer

Editors
Majda Khalil Suleiman
Food Security Program
Environment and Life Sciences Research Centre
Kuwait Institute for Scientific Research
Safat, Kuwait

Shabbir Ahmad Shahid
Food Security Program
Environment and Life Sciences Research Centre
Kuwait Institute for Scientific Research
Safat, Kuwait

ISSN 1574-0919 ISSN 2215-1621 (electronic)
Advances in Global Change Research
ISBN 978-3-031-98489-1 ISBN 978-3-031-98490-7 (eBook)
https://doi.org/10.1007/978-3-031-98490-7

This Springer imprint is published by the registered company Springer Nature Switzerland AG
The registered company address is: Gewerbestrasse 11, 6330 Cham, Switzerland

Dr. Mahdi Abdal

We would like to dedicate this book to Dr. Mahdi Abdal, Ex-Manager of Arid Land Agriculture Greenery Department (AAGD), and Research Scientist at Environment and Life Sciences Research Center of Kuwait Institute for Scientific Research, Kuwait. Dr. Abdal's impressive research career and dedication to advance arid lands agriculture research for food security inspired us to believe that he deserves dedication of this unique book. Dr. Mahdi's visionary approach to intensify crop production through climate smart agriculture practices within national boundaries to bridge gap between local production and food import was well received and appreciated by the scientific community of arid lands countries in general and the Gulf Cooperation Council countries in specific. We have no hesitancy to state that Dr. Abdal planted the seed of arid lands agriculture knowledge and nurtured it during his long professional career. Dr. Mahdi left behind his scientific footprints which are broadly followed by professionals, researchers and academia to foster agriculture for food security.

Foreword

Arid land countries are distributed in different geographical locations including but not necessarily limited to Asia, Australia, sub-Saharan Africa continents, etc. In these countries, the agriculture and food security mainly depend on the availability of water for crop production. Generally, the arid lands are considered water scarce due to low precipitation and high evapotranspiration, which determines the aridity index. The aridity index determines the level of aridity, that is used to classify hyperarid, arid, semi-arid and dry sub-humid regions. A general consensus among the scientists exists that the climate change in terms of increased greenhouse gas emission and induced warming is likely to impact these countries more than the cold humid region countries. Thus, the agriculture and food security in arid lands countries is on stake, and this may restrict achieving the 17 United Nations Sustainable Development Goals especially SDG Number-2 (end hunger, achieve food security and improved nutrition and promote sustainable agriculture). The SDG-2 has various targets to be achieved by 2030. According to Target 2.3, by 2030 double the agricultural productivity and the incomes of small-scale food producers, and Target 2.4 by 2030 ensure sustainable food production systems and implement resilient agricultural practices that progressively improve land and soil quality. The Intergovernmental Panel on Climate Change (IPCC) based on many studies covering a wide range of regions and crops has predicted "*negative impacts of climate change over the 21 century on crop yields more than positive impacts*". The recent (2023) IPCC Assessment Report number 6 has warned that the 1970–2020 was the warmest multi-century period in more than 100,000 years, when the temperature has risen from 0.2 to 1.0 °C. The countries highly vulnerable to rising temperatures will be those geographically located in arid region (hyperarid, arid, semiarid), where heat, prolonged drought, land degradation, and water scarcity will be the major constraints to sustain food security to the level to feed the escalating population.

Considering the aspects of climate change and the challenge to achieve SDG-2 in arid lands countries, we plan to centralize latest scientific information on various aspects of climate change and mitigation and adaptation synergy to improve agriculture production for food security. The book will be a stand-alone document for use by multi-stakeholders including but not necessarily limited to professionals and researchers, students, academia, policy, and decision-makers to aid in developing science-based decisions on natural resources uses and management for food security. To justify this, we have used a holistic and multi-disciplinary approach to cover a wider range of subjects to complete the theme of the book. The manuscripts are being contributed by a high level of foreign qualified expertise (restoration ecologist, soil scientist, plant eco-physiographist, water specialist, agronomist, entomologist, botanist, economist, forage expert, microbiologist, forester and range expert, modeller, livestock, poultry, aquaculture, animal biosciences, climate change and food security experts, etc.) whose work is well recognized internationally. Twenty cross-sectoral manuscripts are distributed into five themes covering wide-ranging aspects of the book. The five themes are: (i) Natural Resources Management and Governance; (ii) Agriculture and Food Security; (iii) Mitigation and Adaptation Synergy; (iv) Poultry and Aquaculture; and (v) Agricultural Pests and Soil Microbial Population.

The book is considered as a landmark and likely to set the precedence to understand and manage the arid lands for food security and natural resources services. It is believed that the book will be inspiration to diversified stakeholders who will take benefits and will be source of information for many years to come.

Abdullah Al-Enezi
Environment and Life Sciences
Research Center
Kuwait Institute for Scientific Research
Safat, Kuwait

Preface

Arid lands cover a massive portion of the Earth, roughly one-third to over 40% of the world's land surface, but dominate in Asia, sub-Sahran Africa, Middle East and North Africa, Australia and countries of Central Asia. These countries are facing aridity and climate-induced warming problems, leading to increase drylands and threatening crops intensification to meet the food demand of escalating population. The drylands are dominant in developing countries occupying more than 70% area. The global population was 0.98 billion in 1,800 which is predicted to be 10.35 billion by 2100. Cumulatively the hyperarid, arid, and semi-arid regions occupy 35.8 per cent of the Earth's land area. Of the global dryland area, approximately 11% is used as cropland and 30% as pasture, and about 2.8 billion people are inhabited in the dryland regions. Climate change is recognized as the main culprit for temporal expansion of drylands and prolonged drought phenomenon in dryland counties. The current level of greenhouse gasses in the atmosphere is 410 parts per million that has passed the safe limit of 350 ppm. The period between 1970 and 2020 is recorded as the warmest multi-century period in more than 100,000 years. The recent droughts of Africa and Asia intensely emphasized the need for vital solutions in these zones and the intricacy of the obstacles to their rational development.

The water and arable land scarce countries meet their food demand through heavy imports, which is threatened during epidemics like corona and political conflicts. The IPCC projection of the impact of rising temperature scenarios revealed that there are more chances of crop yield decrease than increase. Considering these evidences, there is an urgent need to unlock the causes and expression of aridity to develop a holistic approach for a possibility of delimitation of the occurrence of world's arid and semi-arid zones and to improve crop production for food security. It is believed that, agriculture with healthy and productive soils, sustainable production systems that save on water and inputs and are less harmful to the environment are pre-requisite to enhance food security, sequester carbon and in building resilience.

Considering the unprecedented escalation of population in arid lands countries and the impact of climate change on crop production systems for food security, we took the initiative to address the constraints to crop production for food security and improve the farmer livelihood by publishing an objective oriented and arid lands tailor made book as guide to foster agriculture for food security. This unique book consists of twenty multi-sectoral contributory manuscripts from eminent scientists, distributed into five themes covering comprehensive problems and solutions to foster agriculture in arid lands countries in the face of climate change. Part I (*Natural Resources Management and Governance*) presents global distribution of arid lands and agriculture, soil governance for sustainable agricultural services, climate-driven dynamics of water resources, and impacts on grazing lands and animal sector. Part II (*Agriculture and Food Security*) covers aspects like sustaining food security *vis-à-vis* food import, trade-offs and local production, innovative technologies to foster agriculture in the context of climate change, soil salinity dynamics and impacts on food security, open field and protected environment agriculture, challenges of feeding escalation population, revitalizing indigenous plants for food security, nature-based agroforestry solutions to mitigate climate change for food security, and economic impacts for policy development. Part III (*Mitigation and Adaptation Synergy*) addresses gaps and challenges in current mitigation and adaptation progress, mitigation and adaptation synergy, sustainable development goal 2 "End hunger" and climate change-induced variations in ecosystems structure. Part IV (*Poultry and Aquaculture*) broadly presents reproductive performance of poultry chicken in rising temperature and potential of in-land shrimp farming. Part V (*Agricultural Pests and Soil Microbial Population*) widely presents soil microbial dynamics and shifts and dynamics of agricultural pests and implication for pesticide use.

The editors have followed *double-blind peer review* procedure from internationally renowned scientists and edited the book, the contents of the manuscripts, and the ideas expressed are the exclusive responsibility of the authors.

Safat, Kuwait

Majda Khalil Suleiman
Shabbir Ahmad Shahid

Acknowledgements

I have been privileged to share the responsibility of editing this eminent book with Dr. Shabbir A. Shahid who has published multiple topics agriculture and food security books with international publisher Springer. We would like to gratefully and sincerely thank the eminent scientists for their contribution of manuscripts. We are thankful to Dr. Sameer Al-Zenki, Ex-Acting Executive Director Environment and Life Sciences Research Center (ELSRC), Kuwait Institute for Scientific Research for his support and approval to proceed for book preparation and publication. We wish to thank Dr. Afaf Al-Nasser (Acting Executive Director and DD/OD-ELSRC) and Dr. Abdullah Al-Enezi (Acting STD-ELSRC) for their interest and encouragements in publishing this unique book on fostering agriculture in arid lands countries for food security.

We also thank the international reviewers, whose expertise and valuable comments have certainly improved the final version of the manuscripts. Springer management, especially Margaret Deignan, deserves thanks for accepting the publication as a part of book series *Advances in Global Change Research* and expedition for signing the agreement and publication of book. We would like to thank Ms. Saritha Sarath for formatting the manuscripts and Ms. Shruthi Aravind Project Management Office for checking the manuscripts for similarity index and Artificial Intelligence Writing Report.

Safat, Kuwait

Majda Khalil Suleiman
Shabbir Ahmad Shahid

Reviewers

Asad Sarwar Qureshi International Center for Biosaline Agriculture, Dubai, United Arab Emirates

Asis Mukherjee Department of Agrometeorology Bidhan Chandra Krishi Vishwavidyalaya (BCKV) West Bengal, India

Azeem Khalid PMAS Arid Agriculture Rawalpindi, Pakistan

Davinder Kumar Grover Punjab Agricultural University, Ludhiana, India

Hafiz Tanvir Ahmad Institute of Soil and Water Resources, The Islamia University of Bahawalpur, Pakistan

Henda Mahmoudi International Center for Biosaline Agriculture, Dubai, United Arab Emirates

Ijaz Rasool Noorka College of Agriculture, University of Sargodha, Pakistan

In Kwon Jang National Institute of Fisheries Science, Republic of Korea

Jagdish Dagar Indian Council for Agriculture Research, India

Khalil ur Rahman Butt His Highness Private Department, Ruler of Dubai, United Arab Emirates

Khaliq ur Rahman Arshad Soil Salinity Research Institute, Government of Punjab Pakistan

Mirza Abdul Qayyum Institute of Plant Protection, Muhammed Nawaz Sharif

University of Agriculture, Multan, Pakistan

Mohammed Shahid International Center for Biosaline Agriculture, Dubai, United Arab Emirates

Muazzam Sabir College of Agriculture, University of Sargodha, Pakistan

Redouane Choukr-Allah Agricultural Innovation and Technologies Transfer Center, University Mohammed VI Polytechnic, Ben Guerir, Morocco

Samsul Huda University of Sydney, Australia

S. C. Kotur ICAR-Indian Institute of Horticultural Research, Bengaluru, Karnataka

Shaukat Ali Bhatti Ohio State University, United States of America

Zahoor Aslam Nuclear Institute for Agriculture and Biology, Faisalabad, Pakistan

Contents

Part I Natural Resources Management and Governance

1 **Global Distribution of Arid Lands: Opportunities and Constraints for Agriculture and Food Security** 3
Majda Khalil Suleiman

2 **Soil Governance in the Arid Lands Countries for Sustainable Agriculture Services and Food Security** 25
Shabbir Ahmad Shahid

3 **Climate-Driven Changes and Impact on Arid Lands Water Resources for Agriculture Production and Food Security** 45
Hussain Abdullah, Adel Al-Haddad, and Abdallah Abusam

4 **Impact of Climate Change on Grazing Lands and Response of Arid Lands Ecosystems to Livestock Industry** 67
Mariam E. Al-Bahouh and Samira Mohammed

Part II Agriculture and Food Security

5 **Sustaining Food Security in Arid Lands Countries-Augmenting Domestic Agricultural Production Through Corporate Farming and Food Imports: A Trade-Off** 87
Majda Khalil Suleiman and Shabbir Ahmad Shahid

6 **Potential Technological Innovations to Advance Arid Lands Agriculture in the Context of Climate Change** 109
Shabbir Ahmad Shahid and Majda Khalil Suleiman

7 **Soil Salinity Dynamics in Arid Lands and Impact on Agriculture for Food Security** 131
Shabbir Ahmad Shahid, Amal Jaafer Alkandari, and Fahad Meshari Alnajdi

8 Open-Field and Controlled Environment Agriculture Systems for Food and Environmental Security in Arid Regions: Integration of Emerging Technologies 155
Binson M. Thomas and Narayana R. Bhat

9 The Challenge of Feeding Escalating Population in Arid Land Countries 181
Abul Kalam Samsul Huda and Shabbir Ahmad Shahid

10 Revitalizing Indigenous and Native Plants for Food Security of Arid Lands Countries in Changing Climate Conditions 195
Tareq A. Madouh and Merlin K. Davidson

11 Agroforestry: A Nature Based Solution to Mitigate Climate Change Impacts and to Improve Crop Production Systems for Food Security in Arid Regions 229
M. Anisul Islam

12 Economic Impacts Attributable to Climate Change on Arid Lands Agriculture and Food Security—Policy Development 251
Adil Naseeb, Ahmed Al-Khayat, Shaima Al-Shamali, Nora Abdulmalik, May Al-Asfour, Abdullah F. Al-Jaber, and Mariam S. Behbehani

Part III Mitigation and Adaptation Synergy

13 Current Mitigation and Adaptation Progress: Gaps and Challenges in Arid Lands Agriculture 273
Majda Khalil Suleiman and Shabbir Ahmad Shahid

14 Mitigation and Adaptation Synergy in Arid Lands Agriculture for Food Security 287
Majda Khalil Suleiman

15 Mitigation and Adaptation in Arid Lands Agriculture to Achieve Sustainable Development Goal 2–End Hunger 307
Habibah Al-Menaie and Merlene Ann Babu

16 Climate Change-Induced Variation in Ecosystem Structure in Arid Lands Countries 339
Amal Jaafer Alkandari, Mohammad A Alsayegh, and Ali A Abdulraheem

Part IV Poultry and Aquaculture

17 Effect of Rising Temperatures on Overall Health, Productive and Reproductive Performance of Chickens—A Review 371
Hanan Sulaiman Al-Khalaifah and Afaf Yaqoob Al-Nasser

18 Impact of Climate Change on Shrimp Farming in Arid Region Countries: Challenges, Opportunities and Threats 403
Sherain N. Al-Subiai

Part V Agricultural Pests and Soil Microbial Population

19 Soil Microbial Population Dynamics and Shifts in Arid Lands Due to Global Warming 425
Ali M. Quoreshi and Faizah Al-Enezi

20 Dynamics of Agricultural Pests in a Changing Climate and Implications for Pesticide Use: A Review 449
Mustapha F. A. Jallow

Index 471

Editors and Contributors

About the Editors

Majda Khalil Suleiman is Research Scientist at Food Security Program at the Environment and Life Sciences Research Center of Kuwait Institute for Scientific Research. She has over 35 years of research, development and extension experience in agriculture and urban landscape development and management. She earned her Ph.D. degree in the year 2016 *Restoration Ecology* from the University of Western Australia. She earned a B.Sc. degree in Botany/Plant Biology from Kuwait University in 1989 and started her career as Botanical Researcher at the Public Authority for Agriculture Affairs and Fish Resources of Kuwait. She is Desert Landscaper certified by Desert Botanical Garden Phoenix, Arizona, USA. She is Recipient of nine achievement awards from different organizations, including Kuwait University Poster Day (2018–2019), KNPC HSE performance award (2018–2019), five Scientific Achievement Awards graced by KISR (2009–2015), and two Awards of Excellence from *Euro-Mediterranean Institute for Sustainable Development* in 2014 and 2015. She is Member of the Convention on International Trade in Endangered Species (CITES) of Wild Fauna and Flora. She has led several research projects and was part of the team of many national projects covering wide range of topics, such as restoration of coastline through mangroves plantation, greenery master plan, and urban demonstration garden. She is Prolific Author/Co-Author of 120 scientific papers published in international refereed journals (58), conference proceedings, abstracts, and book chapters (62). In addition, she published eight books and 137 technical reports. In 2023, she edited a book "Terrestrial Environment and Ecosystems of Kuwait" which was published by Springer. Currently she is leading a Government Initiative Food Security project and coordinating three projects related to desert rehabilitation through Kuwait Environment Remediation Program (KERP) funded by Kuwait National Focal Point.

Shabbir Ahmad Shahid was embraced with prestigious Sir William Roberts award to pursue Ph.D. degree in Soil Science specialization in Salt-affected Soils at the University of Bangor, Wales, UK, completed in 1989. He has over 44 years' experience as Soil Scientist in Pakistan, the UK, Australia, the United Arab Emirates, and Kuwait. Currently, Dr. Shahid is Research Scientist at Food Security Program, Environment and Life Sciences Research Center of Kuwait Institute for Scientific Research. He led several projects in natural resources assessment and management. He was Technical Coordinator in multi-million-dollar national soil surveys of the State of Kuwait and Abu Dhabi Emirate and developed the soil survey action plan for the Northern Emirates of UAE and the Republic of Mauritania. Currently he is leading National Salinity Management of Kuwait project. Dr. Shahid's discovery of anhydrite rich soil is formally added in the US Keys to Soil Taxonomy as a subsurface diagnostic horizon, mineralogy class and subgroups in the Salids suborder of the order Aridisols. He is also Principal Author of *United Arab Emirates Keys to Soil Taxonomy* and *Kuwait Soil Taxonomy* books published by Springer in the years 2014 and 2022, respectively. In addition, Dr. Shahid is Creator and Co-Founder of the Emirates Soil Museum, Dubai, UAE. He is Prolific Author of over 225 scientific papers published in peer-reviewed scientific journals, book chapters, conference proceedings, and newsletters. He published nine books through professional publisher Springer. In 2023, he co-edited a book "Terrestrial Environment and Ecosystems of Kuwait" published by

Springer. Dr. Shahid presented keynotes at several international conferences. Dr. Shahid's Research Interest Score (RIS) is 3,852, h-index 28, i10-index 62 with 4,326 citations, and 222,705 reads globally. His RIS is higher than 98% of global ResearchGate members. He is Recipient of David A Jenkins Award (1995) graced by the Soil Mineralogy and Micromorphology, Society of Pakistan.

Contributors

Hussain Abdullah Water Research Center, Kuwait Institute for Scientific Research, Safat, Kuwait

Nora Abdulmalik Techno-Economic Division, Kuwait Institute for Scientific Research, Safat, Kuwait

Ali A Abdulraheem Food Security Program, Environment and Life Sciences Research Center, Kuwait Institute for Scientific Research, Safat, Kuwait

Abdallah Abusam Water Research Center, Kuwait Institute for Scientific Research, Safat, Kuwait

May Al-Asfour Techno-Economic Division, Kuwait Institute for Scientific Research, Safat, Kuwait

Mariam E. Al-Bahouh Food Security Program, Environment and Life Sciences Research Center, Kuwait Institute for Scientific Research, Safat, Kuwait

Faizah Al-Enezi Food Security Program, Environment and Life Sciences Research Center, Kuwait Institute for Scientific Research, Safat, Kuwait

Adel Al-Haddad Water Research Center, Kuwait Institute for Scientific Research, Safat, Kuwait

Abdullah F. Al-Jaber Techno-Economic Division, Kuwait Institute for Scientific Research, Safat, Kuwait

Amal Jaafer Alkandari Food Security Program, Environment and Life Sciences Research Center, Kuwait Institute for Scientific Research, Safat, Kuwait

Hanan Sulaiman Al-Khalaifah Food Security Program, Environment and Life Sciences Research Center, Kuwait Institute for Scientific Research, Safat, Kuwait

Ahmed Al-Khayat Techno-Economic Division, Kuwait Institute for Scientific Research, Safat, Kuwait

Habibah Al-Menaie Food Security Program, Environment and Life Sciences Research Center, Kuwait Institute for Scientific Research, Safat, Kuwait

Fahad Meshari Alnajdi Food Security Program, Environment and Life Sciences Research Center, Kuwait Institute for Scientific Research, Safat, Kuwait

Afaf Yaqoob Al-Nasser Food Security Program, Environment and Life Sciences Research Center, Kuwait Institute for Scientific Research, Safat, Kuwait

Mohammad A Alsayegh Food Security Program, Environment and Life Sciences Research Center, Kuwait Institute for Scientific Research, Safat, Kuwait

Shaima Al-Shamali Techno-Economic Division, Kuwait Institute for Scientific Research, Safat, Kuwait

Sherain N. Al-Subiai Coastal and Marine Resources Program, Environment and Life Sciences Research Center, Kuwait Institute for Scientific Research, Safat, Kuwait

M. Anisul Islam Food Security Program, Environment and Life Sciences Research Center, Kuwait Institute for Scientific Research, Safat, Kuwait

Merlene Ann Babu Food Security Program, Environment and Life Sciences Research Center, Kuwait Institute for Scientific Research, Safat, Kuwait

Mariam S. Behbehani Techno-Economic Division, Kuwait Institute for Scientific Research, Safat, Kuwait

Narayana R. Bhat Ex-Principal Research Scientist, Food Security Program, Environment and Life Sciences Research Center, Kuwait Institute for Scientific Research, Safat, Kuwait

Merlin K. Davidson Food Security Program, Environment and Life Sciences Research Center, Kuwait Institute for Scientific Research, Kuwait City, Kuwait

Abul Kalam Samsul Huda School of Science, National Vegetable Protected Cropping Centre, Western Sydney University, Sydney, Australia

Mustapha F. A. Jallow Food Security Program, Environment and Life Sciences Research Center, Kuwait Institute for Scientific Research, Safat, Kuwait

Tareq A. Madouh Food Security Program, Environment and Life Sciences Research Center, Kuwait Institute for Scientific Research, Kuwait City, Kuwait

Samira Mohammed Food Security Program, Environment and Life Sciences Research Center, Kuwait Institute for Scientific Research, Safat, Kuwait

Adil Naseeb Techno-Economic Division, Kuwait Institute for Scientific Research, Safat, Kuwait

Ali M. Quoreshi Food Security Program, Environment and Life Sciences Research Center, Kuwait Institute for Scientific Research, Safat, Kuwait

Shabbir Ahmad Shahid Food Security Program, Environment and Life Sciences Research Center, Kuwait Institute for Scientific Research, Safat, Kuwait

Majda Khalil Suleiman Food Security Program, Environment and Life Sciences Research Center, Kuwait Institute for Scientific Research, Safat, Kuwait

Binson M. Thomas Food Security Program, Environment and Life Sciences Research Center, Kuwait Institute for Scientific Research, Safat, Kuwait

Part I
Natural Resources Management and Governance

Chapter 1
Global Distribution of Arid Lands: Opportunities and Constraints for Agriculture and Food Security

Majda Khalil Suleiman

Abstract Arid lands are distributed throughout the globe, where aridity is climate phenomenon specific to water scarcity. Arid lands countries are mostly located in Asia, Africa, Australia and Middle East regions. Arid lands can be divided based on the climatic variations (mainly temperature and rainfall) and aridity index (Precipitation-P/Potential evapotranspiration-PET that is P/PET), a numerical indicator of aridity. In general, deserts have arid climates and constitutes about 35% of the land areas of the world and are typically characterized by rainfall scarcity, higher temperatures and evapotranspiration, prolonged droughts, lower humidity, and a general paucity of vegetation cover. Based on the aridity index (AI) four subtypes of drylands covering a total of 47.2% global land area are classified, hyper-arid (AI < 0.05), arid (0.05 < AI < 0.2), semiarid (0.2 < AI < 0.5), sub-humid (0.5 < AI < 0.65). Based on the rainfall, arid lands receive annual rainfall (0–300 mm) and semi-arid (300 to 600 mm). Arid regions occupy area of about 28 million km^2 and semi-arid about 46 million km^2, and is home to approximately 2.5 billion people and occupy 41% of the Earth's land surface. Sustaining crop production to feed escalating population in arid regions is challenge due to water scarcity, high temperature and low rainfall. Long term disruption of rainfall will have far reaching impacts on food security in arid regions especially in Africa. The Intergovernmental Panel on Climate Change scenarios of rising temperature has established that there are more chances of crop yield decrease than increase over the twenty-first century. Therefore, the governments have to develop soil and water conservation policies to sustain and intensify domestic crop production. To be sustainable in food security in the arid lands the governments must support farmers (training, provision of timely inputs and marketing of commodities to avoid wastage) that could have potential implications for their country's food security. In addition, deals have to be developed with net food exporter countries to bridge the gap between domestic food production and local population food demand.

M. K. Suleiman (✉)
Food Security Program, Environment and Life Sciences Research Center, Kuwait Institute for Scientific Research, Safat, Kuwait
e-mail: mkhalil@kisr.edu.kw

M. K. Suleiman and S. A. Shahid (eds.), *Fostering Arid Lands Agriculture in the Face of Climate Change*, Advances in Global Change Research 81,
https://doi.org/10.1007/978-3-031-98490-7_1

Keywords Arid lands · Aridity · Antarctica · Drylands · Drought · Aridity index · Food security · Meteorology

1.1 Introduction

Arid lands are distributed around the globe where aridity is the main issue leading to drought and impact on arid lands ecosystems and ecosystems services and the food security. About half of the countries are facing aridity problems. The intensity of bioclimatic aridity can be determined by measuring the relative quantities of water gained from precipitation (P) and combined losses of water from soil surface (evaporation) and through plants (transpiration) cumulatively called evapotranspiration (ET). Based on the historical records of increasing global temperature, and changes in rainfall patterns, aridity is forecasted to increase, and as a consequence, drylands will continue to degrade and expand to cover 56% of the Earth's land surface by 2100 (Huang et al., 2016a, 2016b, 2017), leading to reduction in eco-services over time (Nachtergaele & Petri, 2013).

The aridity increases as precipitation decreases and as evaporation increases. It is important to understand the causes and manifestation of aridity. Recently recognized global warming phenomenon has alarmed the international community to take keen interest to mitigate the impacts of climate change in terms of decreasing greenhouse gasses (GHGs) emission, the culprits of global warming, leading to increase the drylands and temporal expansion more than the last decades. Droughts are common in Asia and Africa; in these regions a drought management strategy needs to be developed on a priority basis to sustain crop production for food security.

Arid lands are distributed over half of the countries of the world, but dominate in Asia, sub-Saharan Africa, Middle East and North Africa (MENA), Australia and countries of the Central Asia etc. The aridity calls upon to unlock the causes and manifestation of aridity whether caused through natural or anthropogenic interventions, and to develop a holistic approach for a possibility of decreasing the aridity in drylands including arid and semi-arid zones, possibly through the development and adoption of climate smart agricultural (CSA) and natural resources management practices leading to improved crop production, improve ecosystem services and mitigating climate change. In the early fifties Meig's on the request of UNESCO prepared world map based on Thornthwaite's index, which led to publish a map at scale of 1:25,000,000 exhibiting the distribution of arid climatic zones. The base of this map was the values of aridity index (P/PET), as well as other related information including soil data, landscape relief and vegetation cover. This map was used by diversified disciplined scientists as a reference to develop research projects for the development of arid and semiarid regions. It is to be noted that the P/PET was preferred over the difference between P and PET (P–PET) which gives the quantity of available water to plants, instead of aridity index. This can confuse, e.g., a difference between P and PET (P-PET) could be 300 from different areas coming from 1000–700, 700–400, 400–100 mm. However, in contrast P/PET expresses the aridity in a better way. The aridity can be higher where precipitation is lower and environmental temperature induced ET is higher. Aridity increased as the ratio of P/PET decreases.

In this chapter emphases are given on the distribution of drylands (hyperarid, arid, semiarid, dry subhumid) on the globe, and the challenges, constraints and opportunities to food security.

1.2 Aridity Index, Geographic Extent and Characterization of World's Arid Zones

There are over 2.5 billion people living in arid and semi-arid (ASA) regions (Fig. 1.1). Globally, the ASA region occupy about 41% of the Earth's land surface (Gaur & Squirs, 2018). Arid and semiarid regions are alienated based on the annual rainfall, for example, where the rainfall is less than 300 mm/annum, the zone is classified as arid, however, an inter-annual variability between 50 and 100% may occur, whereas in the case of semiarid region the annual rainfall occur between 300 and 600 mm/annum with an inter-annual variability ranging between 25–50% (Barakat, 2009). The ASA regions face low water availability through hydrological imbalances (increased evapotranspiration and decreased precipitation) causing various durations droughts (Qader et al., 2021). Drylands are characterized based on the aridity index (AI), which is less than 0.65 (Middleton & Thomas, 1997; Schimel, 2010), and covering 66.7 Million km^2, or 45% of the Earth's land surface (Prăvălie, 2016), where the water scarcity can limit crop production and increase soil erosion. Dryland soils store 646 Pg of organic C to 2 m, the equivalent of 32% of the global soil organic C pool (Plaza et al., 2018).

1.2.1 The Aridity Classes

The AI commonly projects level of aridity under climate change scenarios, which ultimately leads to the projection of drylands expansion (Huang et al., 2016a, 2016b). The CGIAR-CSI aridity database (Trabucco & Zomer, 2009), defines drylands into four zones based on P/PET values. Four dryland zones are, i) hyper-arid with $AI < 0.03$, ii) arid ($0.03 < AI < 0.20$), iii) semi-arid ($0.20 < AI < 0.50$) and dry sub-humid ($0.5 < AI < 0.65$).

Based on the AI standards, Huang et al. (2016a); Ji et al. (2015) and Spinoni et al. (2015) have reported an increase of 4% drylands between 1948 to 2004, where semi-arid area increased the most, that is, about 7% (Huang et al., 2016b). The use of AI has long history to define drylands which has $AI < 0.65$ (UNEP, 1992). Globally drylands occupy different Arial coverage of Earth land surface, such as, hyper-arid (5.8%), arid (14.0%), semi-arid (16.1%) and sub-humid (8.9%). Of the global dryland area, approximately 11% (7.6 Million km^2), is used as cropland and 30% (20.0 Million

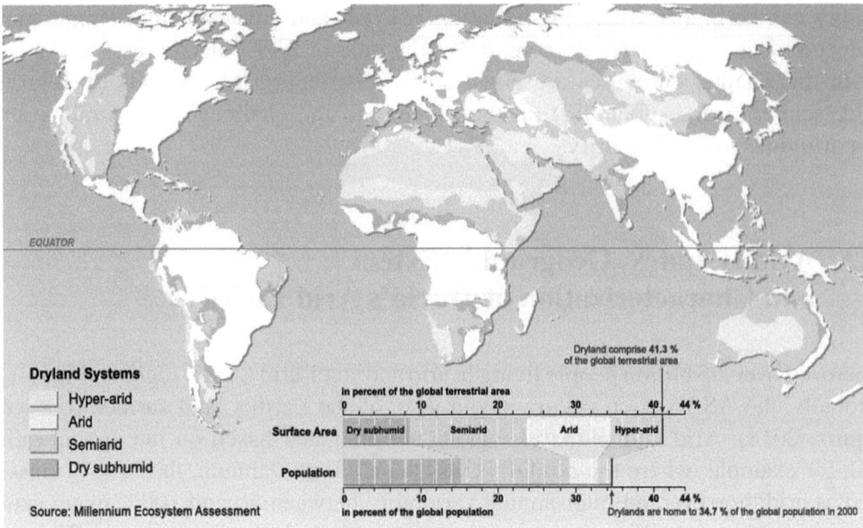

Fig. 1.1 Global map showing hyper-arid, arid, semiarid and dry subhumid areas. *Source* Millennium Ecosystem Assessment; Wikimedia commons, FAO, 2009a

km^2) as pastures (Plaza et al., 2018). Approximately 2.8 billion people (39% of the global population) is settled in the dryland regions. The review of literature revealed that drylands are dominant in developing countries (occupying more than 70% area), where water scarcity affected the desert vegetation, prolonged drought affects rainfed agriculture, food security causes famine, such as in poor African countries (AGRA, 2014), and their food demand is met through food supplements through UN aid policies. With the current scenario of temperature rise, decreasing rainfall trend through climate change induced global warming it is most likely that aridity (area under hyperarid, arid, semiarid, subhumid) will increase globally (Huang et al., 2016a, 2016b; 2017), and so is the case of drylands expansion, as has been the case over the last decades (Dai, 2012; Feng & Fu, 2013), with further expansion of 4% than the previous estimates (Prăvălie, 2016).The increase in drylands through climate change and associated warming impacts likely to affect the ecosystem services (Nachtergaele & Petri, 2013), thus affecting the livelihood of large population in the developing world located in the drylands (Reynolds et al., 2007a). These arid and semiarid zones are shown in Fig. 1.1. The arid region is covering about 28 million km^2 and semiarid zone 46 million km^2. The arid region is not limited to one region but distributed around the globe, where the most of the area of Australia is classified as arid region, where in number of years bush fires happened due to heat waves and destroyed hundreds of kilometers of forests and or bush.

1.2.1.1 Arid Region

Arid region includes the areas with low rainfall such as the desert area where perennial vegetation cover is highly degraded and scarce due to prolonged drought and unplanned grazing. The annual rainfall and interannual variability of rainfall is described in earlier section. In the arid region, agriculture is accomplished through irrigation systems where water is available to offset the crops requirement (Suleiman & Shahid, 2023a). This is mainly the groundwater (brackish/saline) in poor developing countries, and the groundwater and waste water treated at various levels (primary, secondary, tertiary and quaternary) in capital rich water scarce countries like countries in the Gulf Cooperation Council. The desalinated water is the main source of water for human consumption in the Gulf Cooperation Council countries.

1.2.1.2 Semiarid Zone

Semiarid areas are better than arid zone due to better rainfall relative to arid zone resulting into good vegetation cover and high capacity of grazing and agriculture farming. Mean annual rainfall and interannual rainfall variability is described in earlier section. Due to better rainfall than arid region open field agriculture is practiced on large scale, as well as grazing for meat production.

1.2.1.3 The Sub-Humid Zone

The sub-humid zone is better than the arid and semiarid zones in terms of rainfall and AI (AI $0.50 < P/PET < 0.65$), where interannual rainfall variability is $< 25\%$.

1.3 Geographic Distribution of Arid Lands

Drylands are home to over 40% of the population of both Africa and Asia, and to between 25–30% of the rest of the world's population, which in total is two billion people (Reynolds et al., 2007b), 90% of whom live in developing countries (UNEP, 2007). As explained earlier drylands based on the AI can be classified into four zones (hyper-arid, arid, semi-arid, and dry sub-humid). In the following section the position of major arid lands countries is described with respect to the impacts on the country climate and drought conditions.

1.3.1 Indian Subcontinent

The area of about 600,000 and 204,000 km^2 is covered by arid and semi-arid climate in India and Pakistan respectively. In Pakistan, the Indus Valley is mostly identified as arid zone. In these areas prolonged drought occurs during summer months except monsoon season where rainfall occurs to different levels and causing rivers outflows and floods.

1.3.2 Australia

In Australia over 80 percent of the continental area is categorized as arid or semiarid (Meigs, 1953). Australia has nowhere the extreme conditions of aridity (hyper-arid climates) that occur in other countries (Bahrain, Kuwait, Oman, Qatar, Saudi Arabia and United Arab Emirates) and countries of the sub-Sahara Africa. Gentilli (1971) noted that most of the Australia can be considered arid but only marginally, where a slight increase in rainfall can convert arid to semi-arid environment.

1.3.3 Sahara; Sahelian and Sudanese Zones

1.3.3.1 Sahara

In Sahara, the most common countries are Algeria, Egypt, Libya, Mali, Morocco, Mauritania, Niger, Sudan, Chad and Tunisia. These are located in the arid and semi-arid zones. Most of the area has a very uniform climate (less than 100 mm/year), it is irregular, thus drought is almost long-lasting. In Mauritania, Atar registers an average rain of 93 mm per annum and relative humidity between 20 and 40 per cent. High annual means temperature of the Sahara is generally ranging between 20° to 25° C.

1.3.3.2 Southern, South-Western Africa and Madagascar

The driest areas in the region are in the west, with coastal Namib Desert (2,000 km long), and the extreme south of Madagascar. The rainfall is between 150 and 300 mm and trees are the main natural vegetation. The work of Logan (1960, 1969) and the synthesis by Walter (1973) describe these regions in terms of their characteristics.

1.4 Major Soil Types and Their Characteristics in the Arid Land Countries

The soils in the dryland regions due to high temperature, low vegetation cover and limited rainfall are less developed, compared to soils in the humid regions. In general, arid lands being located in the drylands (aridity index < 0.65), present less developed soils (Entisols and Aridisols), have coarse textures with the dominance of sand over 90% (Omar & Shahid, 2013; Shahid, 2023a, 2023b; Shahid & Abdelfattah, 2008; Shahid & Omar, 2022; Soil Survey Staff, 2022), with low nutrients and water holding capacities (Shahid et al., 2014), higher bulk density compared to humid region soils (organic rich and low bulk density). In the arid lands regions, water scarcity is the central challenge for agriculture. The crop production is done through irrigating the farms using either groundwater (which is usually brackish) or treated sewage effluent (Khan et al., 2022; Suleiman & Shahid, 2023a). In terms of organic carbon contents in drylands, Plaza et al. (2018) has reported that as the aridity is increased the organic carbon is decreased, e.g., mean organic carbon in hyper-arid region is reported as 5 kg m^{-2} to a depth of 2 m, whereas, in dry subhumid regions a mean level of organic carbon is reported as 18 kg m^{-2} to a depth of 2 m is reported (Plaza et al., 2018). Similar to organic carbon, drylands soils on average store 73.2 ± 0.6 Pg of total nitrogen to 2 m depth, which is about 40% of the global soil nitrogen, whereas total phosphorous is higher in drylands compared to humid region soils (Plaza et al., 2018).

Among the twelve soil orders mapped in the world and reported in the US Soil taxonomy (Soil Survey Staff, 2022), Aridisols and Entisols are the dominant, following by other soil orders (Inceptions, mollisols etc.). The Histosols and Podzols are cold region soil orders, whereas Andisols are located in countries where volcanoes are common, and Gelisols in frozen soils. Review of literature has shown that the Soil Order Aridisols is not only dominantly occurring in arid and semi-arid regions but it can be located in sub-humid countries. Globally aridisols and entisols occupy 12% and 18% ice free land area, respectively. Aridisols are commonly used for range, wildlife and recreation. These soils have significant constraints for agriculture due to shortage of water, however, these soils are used for agriculture where irrigation water is available in sufficient quantities. The Entisols represent desert sandy terrain, either used for seasonal grazing or agriculture where either rainfall is significant or irrigation water is available. The Entisols are prone to high percolation losses of water and fertilizer nutrients, therefore, their water and nutrients retention properties must be improved by incorporating organic and inorganic amendments (Farrar et al., 2018) in to the poorly structured soils.

1.5 Extent of Salt-Affected Soils in the Arid Region

The scarcity of water, arable lands, and salinity problem in drylands regions countries raised special concerns for national and global food security systems. It is to be noted that where rainfall is over 1,000 mm per annum, the soil salinity does not exist, due to regular leaching of salts from the plants root zone and rain water is sufficient to grow crops, such as the case of humid regions. Understanding soil salinization through science based diagnostics can lead to sustainable management and revitalization of salt-affected soils (reversing salinity affects) for food production and developing national salinity management strategies globally (Shahid and Rahman, 2011), and especially in the drylands countries.

Historical perspectives of soil salinization have been the focus of few publications (Shahid et al., 2018). The recent estimates by FAO (2021) revealed globally 424 million hectares of topsoil (0–30 cm) and 833 million hectares of subsoil (30–100 cm) are salt-affected to various degrees, and are located dominantly in dryland countries. This estimate is based on 73% of land mapped for soil salinization. Another estimate shows 20% of all irrigated land is affected by soil salinization (Qadir et al., 2014). It has been reported that on daily basis 2,000 hectares (20 km^2) of irrigated land in arid and semiarid areas have been degraded due to soil salinity in over 75 countries (IWMI, 2014).

Based on the current population growth rate, it is projected that by 2100 the world population will reach 8.6 billion by 2023, 9.8 billion in 2050 and 11.2 billion in 2100 (UN, 2017; UN DESA 2017). Therefore, this suggests that to meet the rapidly growing population food demand, we should develop national plans based on consumption and production as highlighted in the UN Sustainable Development Goals (FAO, 2017). If we don't consider salinization as threat to food security, it will have negative effects on national economies, farmers livelihood and the functions and provision of ecosystem services (Ruto et al., 2018) especially the water scarce region (Drylands). However, if these marginal quality salinized soils are well managed through science based diagnostics and mitigation measures, the salinity can be reversed for substantially additional agricultural production (de Vos et al., 2016; Hirich et al., 2021; Rao et al., 2009). As stated above, that the current rate of soil salinization is greater than our interventions to tackle salinization. The business as usual will lead us to more salinized soils and the farm soil resources will be depleted fast, not allowing the future generations to meet their food demands. We must have a paradigm shift in our understanding and managing the salinization, through adoption of innovative and regenerative approaches to manage these resources to conserve the soils for higher production (Zaman et al., 2018).

1.6 Growing Water Scarcity in Arid Lands Countries

Water scarcity is a general phenomenon and the main issue for those living in drylands (Stringer et al., 2021). In these countries the water per capita is much smaller than humid region countries. It is projected that per capita water in drylands will be decreased especially in low income countries. Then the Middle East and North Africa (MENA) region will be the next stop for water scarcity (Baggio et al., 2021). There is agreement of scientists that in future water will become a significant issue to increase crop production for food security. This is the case especially in areas where maintenance and feasibility for storage of flood causing water are most important factors. Therefore, there is an urgent need to develop infrastructures for water harvesting for use where and when needed for crop production. In addition, to increase food production, knowledge on climate-adapted crops and multiple cropping, water harvesting structures, and livelihood diversification options are needed (Macchi et al., 2011).

Drylands in hot, tropical areas have already experiencing rising temperature than the global average, and temperatures are projected to increase by 2–4 °C by 2100. Climate change will have significant impact on water availability globally, especially in dryland countries, owing to unexpected trend of increase or decrease of precipitation and rainfall (Konapala et al., 2020). In water scarce dryland countries there is urgent need to understand the water dynamics and develop water governance and conservation strategies and synergies for sustainable and equitable uses on sustainable basis (EAD, 2009; EAD-ICBA, 2010; UAEMEW, 2010).

1.7 Trend of Population Growth in Arid Lands

As stated earlier that arid lands are considered one of the drylands, and according to UN-Habitat, compared to other ecological zones, population growth is much faster in drylands countries, that is around 18.5% faster growth. In general, this is the fact that population density (population per km^2) increases as the aridity decreases, e.g., 10 persons/km^2 in desert land countries compared to 71/km^2 in the dry sub-humid (rangeland) areas (UNEP, 2007). The densely populated drylands are in developing countries, such as the case of sub-Saharan Africa. However, dryland countries in the Gulf region are not densely populated, where significant number is expats, and populated in cities, rather in desert field.

1.8 State of Food Security and Nutrition in the Arid Lands Countries

There are three pillars of food security, (i) sufficient production and availability, (ii) economic affordability and access to available food, (iii) nutritional security (IFPRI, 1995). The countries located in drylands region in specific are in general in crises of food and nutrition securities. The reasons being, drylands geographic locations where there is low precipitation (P) and high potential evapotranspiration (PET) as described earlier. In addition, current climate change induced global warming is building up heats and prolonged drought has affected the crop production in general in dryland countries and sub-Saharan African (SSA) countries in specific, where both food and nutritional securities are significant issues. The repercussions of 2007/2008 food crises have been felt significantly in dryland poor countries compared to developing countries. The food crises are further exacerbated due to unprecedented hike in food prices, especially staple food (wheat, rice, maize etc.), as well as reduced food stocks.

1.9 Prospects and Challenge to Feed Escalating Population in Arid Lands Countries

There are multiple challenges in poor arid land countries, exceptions are in capital rich dryland countries where food trade with other countries keeps the countries sufficient food in stock. The challenges in local food production in poor drylands countries are multi facets. These may be, (i), poor economic resources for timely purchase of inputs (seeds, fertilizers, and pesticides), (ii) small scale subsistence farming, (iii) lack of agriculture related education, (iv) water scarcity and conservation constraints, (v) prolonged drought, poor water harvesting and storage infrastructures, (vi) loss of surface soil sediment due to water erosion, (vii) poor weed control, crops harvesting and timely transfer of the food products to nearby markets, (viii) significant food losses (globally 33% food is lost from farm to fork), and (ix) soil salinity problems. More recent data suggests up to 40% of food is lost or wasted along the value chain, exposing a major blind spot in global food systems (Goodwin & Lipinski, 2024). Provided, food loss is controlled, we can feed additional 2 billion mouths, where, currently over 1.5 billion peoples in poor African countries are sleeping hunger. The adoption of integrated approach (Water-Energy-Food-Ecosystem Nexus) may be the viable option to use the resources sustainably to improve food production (Suleiman & Shahid, 2023a) and optimize water footprints of crops in arid environment (Suleiman & Shahid, 2023b).

1.10 Possible Consequences of Drought in Semi-Arid and Arid Environments

Drought is a complex environmental phenomenon and refers to lack of rainfall for long period or reduced average-precipitation, increasing water scarcity in the arid lands region and that has devastating impacts on crop production, revitalization of grazing lands, and human society at large. Increasing temperature, low and infrequent rainfall (climate change induced global warming) leads to prolonged risks drought in arid lands countries. Patel and Patel (2024) worked for over a period of three decades on climate and associated land use changes and revealed risks of drought in dryland countries especially in the semi-arid region. They further recommended mitigation strategies for climate change impacts.

1.11 Impact of Extreme Weather Disasters on Crop Production

The global scientific community developed general consensus that if mitigation measures are not adopted at large scale on the planet, it is likely to face the consequences of extreme weather conditions. These conditions can be in the form of global warming (rising temperature), prolonged drought, shift of ecosystems including failure of certain crops to grow due to drought and high evapotranspiration, sea water intrusions to the coastal areas leading to inundate these areas, loss of vegetation and increase in soil salinization.

The IPCC (2023) has projected regional impacts on biophysical responses to the impact of climate change (temperature rise, reduction in precipitation, solar radiation, humidity, wind, and CO_2 enhancement) on growth and water retention in currently cultivated areas. The projection in the lights of increasing temperature has shown significant maize crop yield decline on the global landscape, however, there are few areas where positive signs are predicted where maize crop yield will increase especially in South African region. The IPCC (2014) based on numerous studies on a wide range of regions and crops, have reported negative impacts of climate change on crop yields more common than positive impacts, over the twenty-first century. Further insights have shown almost equal chances of crop yield increase or decrease over 2010–2029 (Fig. 1.2), however, after these two decades and 2090–2109 has indicated possibilities of less increase but higher decrease of crop yield over the twenty-first century.

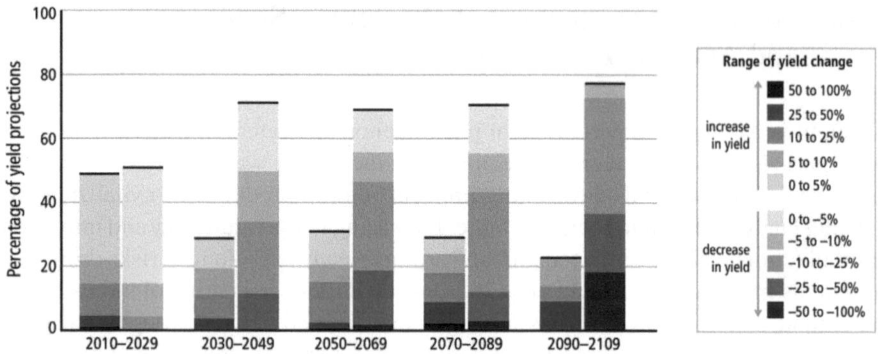

Fig. 1.2 Summary of projected changes in crop yields (mostly wheat, maize, rice and soy), due to climate change over the twenty-first century (Fig. 2.7 in IPCC, 2014: Climate Change 2014: Synthesis Report. Contribution of Working Groups I, II and III to the Fifth Assessment Report of the Intergovernmental Panel on Climate Change [Core Writing Team, R.K. Pachauri and L.A. Meyer (eds.)]. IPCC, Geneva, Switzerland, 151 pp

1.12 Agriculture Farming in Arid-Lands Countries

Agriculture farming in arid lands countries is mainly based on either irrigating the crops from multiple water sources (groundwater, rivers, treated sewage effluent), and rainfed agriculture where either other sources of water are not available or the landscape features (potowar) are not allowing to establish irrigation infrastructure (hilly terrain). Since the arid lands countries are located in arid environment, therefore, water scarcity is the main constraint for crop production for subsistence and commercial farming. Rainfed agriculture can be sustainable on long term basis provided environmental friendly and climate resilient technologies are adopted (Baig et al., 2013).

1.12.1 Potentials

Although the arid lands are constrained with water scarcity due to high aridity, high temperature, prolonged droughts, however, there are potentials of sustaining agriculture in arid lands countries. One of the greatest potential is to recycle the municipal waste water through various treatments (primary, secondary, tertiary and quaternary etc.) and use for agriculture (Khan et al., 2022). This is the fact that as the population will grow temporally, simultaneous to population growth the municipal waste water will be increased, and this should be considered as potential for crop production. Other potential source of water could be water harvesting and storing in small, medium and large-scale dams or small scale reservoirs constructed at farmers field scale.

1.12.2 *Constraints*

In most of the African countries the heavy rain causes heavy floods, remove productive soils and the fertilizer nutrients. Soil erosion "soil degradation" is the main threat to crop production and may lead up to 50% crop yield loss (Shahid, 2023b). There are multiple constraints in arid land agriculture. These are briefly listed below.

- Water scarcity
- High temperature
- Aridity index (Precipitation/Potential evapotranspiration)
- Groundwater and soil salinization
- Groundwater and soil sodification
- Lack of knowledge of farmers on modern technologies
- Lack of timely access to inputs (seeds, fertilizers, pesticides etc.)
- Marketing the product and gaining good price

1.13 Impacts of Climate Change

The climate change impact is not country-specific, but countries in the drylands are going to be exceedingly vulnerable to climate change, especially the poor arid lands countries. The impact of climate change will be especially on peoples living in the dry areas due to fragile and marginal nature of natural resources they are living (El-Beltagy and Madkour, 2012). In these countries the climate change will undermine the natural resources and the developing countries fail to achieve the UN SDG2 (end hunger), by 2030 (FAO, 2017). El-Beltagy & Madkour (2012) further highlighted that over 2,000 million hectares land is degraded, which has led to loss of agrobiodiversity and destruction of natural resources. In addition, exploitive use of natural resources due to the impact of climate change has set up further the climate change in motion. There are many impacts of climate change on arid lands agriculture, including but not necessarily limited to, increased temperature, unpredictable rainfall and flash floods, prolonged drought, shift of insects and pests, change in cropping patterns and soil salinization, all these factors are closely related to climate change impacts on arid lands agriculture. If we continue business as usual and do not take immediate action to mitigate climate change (cut greenhouse gasses emission), it is anticipated that crops, cropping system, rotations and soil biodiversity will undergo transformation. All IPCC climate change models revealed that the dry countries will become drier and more water stressed (IPCC, 2001, 2007a, 2007b). In addition to above water scarcity will increase both in arid and drought vulnerable areas and the areas of abundant rainfall.

Based on the current population growth, increased marginality of farm land resources, limitations to expand agriculture areas, FAO (2009b) predicts obligatory 70% food production increase by 2050. This can be only possible through optimization of inputs used (improved nutrients and water use efficiencies), improved cultural practices, reversing the farm land marginality (salinized farm land) and adoption of climate smart agricultural practices. Looking at the climate change impacts in a broader sense, it is likely that there may be positive impacts in colder countries where increased temperature will increase plants growth period, in hot climate regions, there may be negative impacts due to increased temperature and low rainfall leading to prolonged drought and failure of crops due to irrigation water scarcity, especially in rainfed agriculture area of sub-Sahara Africa. For instance, the increase in temperatures recorded in the last 25 years, has led to a decrease in rice production ranging from 10 to 20% in numerous countries of the world, and while daytime minimum temperatures rise, and nights are warmer, production will continue to fall (Peng et al. 2004). In some countries the climate trends from 1980 to 2008 had a large impact on average yields, and other agricultural factors (Lobell et al., 2011).

1.13.1 Mitigation Measures

As mentioned earlier, that if we continue business as usual and don't act to reverse climate change through adopting sustainable natural resources practices, increasing soil carbon sequestration through large scale forestation and using organic soil amendments, the problem of climate change will be aggravated. Therefore, we need to cut down the GHGs emission at significant level to control the rising temperature trend. Following should be considered on top priority basis to cope challenges of climate change, including but not necessarily limited to, (i) a paradigm shift in current agriculture practices including climate smart and regenerative agriculture, (ii) evolving high yielding drought resistant, salt-tolerant crops varieties, (iii) strengthening link between research and extension services for technology transfer at fast track mode, (iii) access to market to sell the product, timely and reasonable prices. Adopting these options at national and farm gate level enable the most affected farming communities of the arid lands areas to use their natural resources for sustainable uses, protect the environment and saves soils for generations to come.

1.14 Climate Smart and Regenerative Agriculture Practices

Recently Suleiman and Shahid (2023a) have enlightened few ideas on climate smart and regenerative agriculture practices. The FAO (2017) has recommended five principles to balance social, economic, and environmental considerations. These are briefly given here, (i) improve resource use efficiency, (ii) conserve and protect the natural resources, (iii) the agriculture that fail to livelihood improvement is unsustainable, (iv) enhancing resilience (peoples, ecosystem) is key to sustainable agriculture, and (v) requires responsible and effective governance mechanisms. Regenerative agriculture (RA) after climate smart agriculture term used globally is new approach. Broadly the RA seeks to work with natural systems to restore and enhance the biodiversity,

Table 1.1 Comparison of industrial and regenerative approach (Savills Research, 2021; *cf.* Suleiman & Shahid, 2023a)

Conventional-exhaustive or industrial approach	Regenerative agriculture approach
Extractive	Value added
Control nature	Works with nature
Higher chemical inputs	Lower chemical inputs
Monoculture	Diversity
Carbon emission	Carbon fixation
Moisture depletion	Moisture conservation
Soil degradation	Soil build up
Lower soil biology population	Higher soil biology population
High cost of production	Low cost of production

and soil fertility of farmed land (Savills Research, 2021). The RA concept once used, should conserve, protect and improve soil health through various means (conservation agriculture, crop rotation with legumes, soil mulching, no-to zero tillage, and integration of livestock with crop production).

Suleiman and Shahid (2023a) further emphasized on the integration of conventional and nature based climate smart and regenerative agriculture approaches including novel and innovative technologies to include into current agriculture production systems. Savills Research (2021) has highlighted that all activities in RA (Table 1.1) are sustainable when compared with conventional practices.

1.15 Precision Agriculture Techniques and Practices

The precision agriculture (PA) is defined in many aspects, but recently its definition is modified to accommodate wider aspect of agriculture. The PA includes collecting agriculture related data and taking actions to compile the recommendations for use in agriculture fields for the precise application of soil inputs (water, fertilizers, pesticides etc.) rationally where these are specifically to be added to avoid wrong application and saving the inputs. Shafi et al. (2019) have recently enlightened the scope of precision agriculture (PA) through Internet of Things (IoT).

1.16 Use of Remote Sensing in Agriculture in Arid Lands

Climate change is rapidly altering the landscape in many ways, e.g., salinization is increasing over the agricultural farming areas due to the use of marginal quality groundwater, coastal areas due to sea water intrusion and subsequent evaporation and land degradation due to wind and water erosion. The soil salinity in farming areas is taking high toll and many farms have been abandoned due to lack of fresh water and germplasm to grow in highly saline environment. Remote sensing technology offers practical benefits and rapid monitoring of landscapes for informed decisions at the national and global scales, however, if field surveys are also conducted simultaneous to RS, accurate and detailed information of farm land resources including soil properties, productivities and landscape change can be determined (Fadl et al., 2024). This provides guidelines to land use planners for informed decisions to sustain agriculture for food security, by allocating new areas for agriculture farming where sufficient water resources and arable lands are available to expand agriculture. Using the Normalized Difference Vegetation Index (NDVI), soil productivities can be predicted (Thenkabail, 2015) over a large area at various scales.

1.17 Prediction of Temporal Growth of Agriculture Farming Areas Using NDVI-Case of Kuwait

In Kuwait agriculture is practiced in three regions, Abdali in the north Kuwait (1945 farms), Sulaibiya in the central Kuwait (44 farms), Al-Wafra in the South Kuwait (1889 farms), and 35 farms scattered elsewhere in Kuwait, making a total of 3913 small, medium, and large vegetable and crop farms (Central Statistics Bureau, 2019). To predict the current extent of cultivated and non-cultivated areas, and temporal growth in Abdali and Al-Wafra, an RS-NDVI study was completed.

According to Kuwait Environment Protection Authority (KEPA) official website *emisk* (https://gisportal.emisk.org/arcgis/apps/MapJournal) the Al-Wafra agriculture farming area (cultivated and non-cultivated) is reported as 8 km^2 in 1973, which increased to 94, 171, and 205 km^2 in the years 1991, 2002, and 2016 respectively. These extents can misguide the potential users including policy makers as these areas represent both the cultivated and non-cultivated areas (bare area without farms). In order to determine the actual extent of area for both cultivated and non-cultivated categories we downloaded the Landsat images in the month of August for 3 years (2000, 2013, and 2022).

The boundaries of the entire farming areas in Abdali and Al-Wafar were established separately using the GIS tool based on the 2022 Landsat images (covered both by farms and empty areas between the farms) which come up as 298.2 and 276.8 km^2 in Abdali and Al-Wafra areas respectively. Following this determination and using Normalized Difference Vegetation Index (NDVI) the cultivated and non-cultivated areas were separated and determined. The results show that since 2000, the actual cultivated area is significantly increased, e.g., from 5.2 km^2 to 72.2 km^2 in Abdali and 4.5 km^2 to 74.5 km^2 in Al-Wafra (Suleiman & Shahid, 2023a) in the year of 2022.

1.18 National Level Policy Development to Intensify Crop Production for Food Security

It is apparent from the above descriptions that there will be food crises due to climate change impact in terms of water scarcity and global warming. The crop production for food security in arid lands countries will be highly vulnerable to climate change impact, therefore, the decision makers must make national level food security policies and trade-off with potential countries where natural resources (arable land and water resource) are abundant and surplus (Suleiman and Shahid, 2025). In addition, increase in local crop production, and development of national/regional level strategic food reserves should be on priority in poor arid lands countries. These reserves are to be used at the time of climate crises or political disturbances in the region. In addition, the government should establish research on systems and value chains that link farmers to local, regional and global markets.

1.19 The Role of Agriculture Extension Services to Boost Crop Production

It is generally recognized that there is a poor link between the research organizations and extension services in poor arid lands countries (especially sub-Saharan Africa countries) for technology transfer. Whereas, strengthening this link can play a critical role in providing information on the latest technologies to farmers for livelihood improvement and crops intensification, through increased farm productivities, that ultimately leads to improved food security, and adapt farming practices in response to the challenges of climate change. Timely transfer of technologies and financial assistance to farmers to purchase farming inputs is key to improved food security. In addition, the government should also buy their farm products to avoid post harvesting waste and losses (nearly 1/3rd of food produced). In order to achieve this, it is essential to strengthen the research-extension-farmer link.

1.20 Conclusions and Recommendations

It is concluded that arid lands countries are distributed in different regions including Asian, Africa, Australia, and Middle East and are constrained with arable lands and fresh water resources to produce the food to meet the demand of escalating population and to maintain food reserves for use at climatic and political crises. The agriculture in arid lands countries is also affected by rising temperature, soil salinization and erosion hazards. All these factors are affecting local crop production. These countries have the potential to maintain their agriculture industry to feed the escalating population, if proper infrastructures are developed to use treated waste water after suitable treatments. In addition, to avoid soil losses by water erosion, the government should help the farmers with low level technology and provide financial assistance to construct small dams to store water, conserve soil resources and to buy necessary inputs for crop production. In addition, governments should make food-import deals to meet the national food demand and to sustain a certain level of food reserved for emergency use. Above all, there is need to improve link between research and extension services for a timely provision of newly and innovatively developed technologies transfer to farming community, especially poor African countries.

Acknowledgements I would like to thank Wikimedia Commons (the free media repository) for allowing the free use of map shown in Fig. 1.1 under the Creative Commons CC0 License. I would also like to extend my thank to Vanessa Obiero of World Meteorological Organization (WMO)/ IPCC for granting permission to use Fig. 2.7 from IPCC (2014) in this chapter as Figure 1.2. IPCC is fully credited.

References

AGRA. (2014). *Africa agriculture status report: Climate change and smallholder agriculture in sub-Saharan Africa.* Alliance for a Green Revolution in Africa (AGRA). https://hdl.handle.net/10568/42343. Accessed on 10 Feb 2020.

Baggio, G., Qadir, M., & Smakhtin, V. (2021). Freshwater availability status across countries for human and ecosystem needs. *Science of the Total Environment, 792*, 148230. ISSN 0048–9697. ahttps://doi.org/10.1016/j.scitotenv.2021.148230.

Baig, M. B., Shahid, S. A., & Straquadine, G. S. (2013). Making rainfed agriculture sustainable through environmental friendly technologies in Pakistan: A review. *International Soil and Water Conservation Research, 1*(2), 36–52.

Barakat, H. N. (2009). Arid lands: Challenges and hopes. In *Earth system: History and natural variability*, Vol. III. EOLSS publishers/UNESCO.

Central Statistics Bureau (CSB). (2019). *State of Kuwait.*

Dai, A. (2012). Increasing drought under global warming in observations and models. *Nature Climate Change, 3*(1), 52–58. https://doi.org/10.1038/nclimate1633

de Vos, A., Bruning, B., van Straten, G., Oosterbaan, R., Rozema, J., & van Bodegom, P. (2016). *Crop salt tolerance under controlled field conditions in The Netherlands, Based on trials conducted at salt farm Texel.* Salt Farm Texel. https://edepot.wur.nl/409817. Last accessed 4 Jan 2025.

EAD. (2009). Abu Dhabi water resources master plan (p. 219). Environment Agency-Abu Dhabi, United Arab Emirates.

EAD-ICBA. (2010). A strategy for the use of recycled water–Abu Dhabi Emirate, Vol. 1 (Main Report) (p. 92). Joint publication of Environment Agency-Abu Dhabi, UAE and International Center for Biosaline Agriculture Dubai UAE.

El-Beltagy, A., & Madkour, M. (2012). Impact of climate change on arid lands agriculture. *Agriculture and Food Security, 1*, 3 (2012). https://doi.org/10.1186/2048-7010-1-3.

Fadl, M. E., Abdel Rahman, M. A. E., El-Desoky, A. I., & Sayed, Y. A. (2024). Assessing soil productivity potential in arid region using remote sensing vegetation indices. *Journal of Arid Environments, 222*, 105166. ISSN 0140-1963.

FAO. (2009a). *Global map of aridity. Spatial resolution of 10 arc minutes and temporal resolution of 1961–1990.* Food and Agriculture Organization of the United Nations (FAO). https://data.apps.fao.org/map/catalog/static/api/records/221072ae-2090-48a1-be6f-5a88f061431a. Last accessed on 5 Jan 2025.

FAO. (2009b). How to feed the world in 2050. In *Proceedings of a technical meeting of experts, Rome Italy* (2009)

FAO. (2017). Climate smart agriculture sourcebook. In *Overview-significant development, summary*, 2nd edn. ISBN: 978-92-5-109988-9

FAO. (2021). Global map of salt affected soils. Version 1.0. https://www.fao.org/soils-portal/data-hub/soil-maps-and-databases/global-map-of-salt-affected-soils/en/. Last accessed Dec 2021

Farrar, M., Gill, S., Shahid, S. A., & Babiker, K. (2018). Water saving innovation in urban landscaping and irrigated agriculture by using AustraBlend multimineral root zone conditioner. *Asian Journal of Science and Technology, 9*(12), 9092–9100.

Feng, S., & Fu, Q. (2013). Expansion of global drylands under a warming climate. *Atmospheric Chemistry and Physics, 13*(19), 10081–10094. https://doi.org/10.5194/acp-13-10081-2013

Gaur, M. K., & Squires, V. R. (2018) Geographic extent and characteristics of the World's arid zones and their peoples. In *Climate variability impacts on land use and livelihoods in drylands*. Cham, Switzerland: Springer.

Gentilli, J. (1971). Climates of Australia. In H. E. Lansberg (Ed.), *World survey of climatology* (Vol. 13, pp. 35–211). Elsevier.

Goodwin, L., & Lipinski, B. (2024). How much food does the world really waste? *What we know—and what we don't*. World Resources Institute. https://www.wri.org/insights/how-much-food-does-the-world-waste#:~:text=Around%20one%2Dthird%20of%20all,spot%20in%20global%20food%20systems.

Hirich, A., Choukr-Allah, R., Ezzaiar, R., Shahid, S. A., & Lyamani, A. (2021). Introduction of alternative crops as a solution to groundwater and soil salinization in the Laayoune area, South Morocco. *Euro-Mediterranean Journal for Environmental Integration, 6*, 52. https://doi.org/10.1007/s41207-021-00262-7.

Huang, J., Yu, H., Guan, X., Wang, G., & Guo, R. (2016a). Accelerated dryland expansion under climate change. *Nature Climate Change, 6*, 166–171. https://doi.org/10.1038/nclimate2837

Huang, J., Ji, M., Xie, Y., Wang, S., He, Y., & Ran, J. (2016b). Global semi-arid climate change over last 60 years. *Climate Dynamics, 46*, 1131–1150. https://doi.org/10.1007/s00382-015-2636-8

Huang, J., Yu, H., Dai, A., Wei, Y., & Kang, L. (2017). Drylands face potential threat under 2 °C global warming target. *Nature Climate Change, 7*(6), 417–422. https://doi.org/10.1038/nclimate3275

IFPRI (International Food Policy Research Institute). (1995). *Women: The key to food security*. http://www.ifpri.org/pubs/fpr/fpr21.pdf. Retrieved Dec 29 2008.

IPCC. (2001). Synthesis report. In *Contribution of working group I, II, and III to the third assessment report of the intergovernmental panel on climate change*. Cambridge University Press.

IPCC. (2007a). In M. L. Parry, O. F. Canziani, J. P. Palutikof, P. J. van der Linden, & C. E. Hanson (Eds.), *Assessing key vulnerabilities and the risk from climate change. Impacts, adaptation and vulnerability. Contribution of working group II to the fourth assessment report of the intergovernmental panel on climate change*. Cambridge University Press (IPCC/WGII/AR4).

IPCC. (2007b). In B. Metz, O. R. Davidson, P. R. Bosch, R. Dave, & L. A. Meyer (Eds.), *Agriculture. Mitigation of climate change. Contribution of working group III to the fourth assessment report of the Intergovernmental panel on climate Change*. Cambridge University Press (IPCC/WGIII/AR4).

IPCC. (2007c). Summary for policymakers. In M. L. Parry, O. F. Canziani, J. P. Palutiko, P. J. van der Linden, & C. E. Hanson (Eds.), *Climate Change 2007: Impacts, adaptation and vulnerability. Contribution of working group II to the fourth assessment report of the Intergovernmental Panel on Climate Change*. Cambridge University Press.

IPCC. (2014). *Climate Change 2014: Synthesis Report. Contribution of Working Groups I, II and III to the Fifth Assessment Report of the Intergovernmental Panel on Climate Change* [Core Writing Team, R.K. Pachauri and L.A. Meyer (eds.)] (p. 151). IPCC.

IPCC. (2023). Summary for policymakers. In H. Lee, & J. Romero (Eds.) *Climate Change 2023: Synthesis Report. Contribution of Working Groups I, II and III to the Sixth Assessment Report of the Intergovernmental Panel on Climate Change*, pp. 1–34. IPCC. https://doi.org/10.59327/IPCC/AR6-9789291691647.001.

IWMI. (2014). *World losing 2,000 hectares of irrigated farmland to salt-induced degradation Daily*. International Water Management Institute Press Release. https://www.iwmi.org/news/press-release-world-losing-2000-hectares-irrigated-farmland-salt-induced-degradation-daily/. Last accessed 4 Jan 2025.

Ji, M., Huang, J., Xie, Y., & Liu, J. (2015). Comparison of dryland climate change in observations and CMIP5 simulations. *Advances in Atmospheric Sciences, 32*(11), 1565–1574. https://doi.org/10.1007/s00376-015-4267-8

Khan, M. M., Siddiqi, S. A., Farooque, A. A., Iqbal, Q., Shahid, S. A., Akram, M. T., Rahman, S., & Al-Busaidi, W. (2022). Towards sustainable application of wastewater in agriculture: A review on reusability and risk assessment. *Agronomy, 12*, 1397. https://doi.org/10.3390/agronomy12061397

Konapala, G., Mishra, A. K., Wada, Y., & Mann, M. E. (2020). Climate change will affect global water availability through compounding changes in seasonal precipitation and evaporation. *Nature Communication, 11*, 3044. https://doi.org/10.1038/s41467-020-16757-w

Lobell, D. B., Schlenker, W., & Costa-Roberts, J. (2011). Climate trends and global crop production since 1980. *Science, 333*(6042), 616–620. https://doi.org/10.1126/science.1204531

Logan, R. F. (1960). *The central Namib Desert, South West Africa* (p. 758). National Academy of Sciences—National Research Council.

Logan, R. F. (1969). Geography of the central Namib Desert. In W. G. McGinnies & B. I. Goldman (Eds.), *Arid lands in perspective* (pp. 127–143). University of Arizona Press.

Macchi, M., Gurung, A. M., Hoermann, B., & Choudhury, D. (2011). *Climate variability and change in the Himalayas—community perceptions and responses*. ICIMOD.

Meigs, P. (1953). World Distribution of arid and semi-arid homoclimates. In UNESCO (Ed.), *Reviews of research on arid zone hydrology*, pp 203–210. UNESCO.

Middleton, N. J., & Thomas, D. S. G. (1997). *World Atlas of Desertification*, 2nd Ed (Arnold, 1997).

Nachtergaele, F., & Petri, M. (2013). Land degradation assessment in drylands. In *Mapping land use systems at global and regional scales for land degradation assessment analysis*. Version 1.1. FAO.

Omar, S. A. S., & Shahid, S. A. (2013). Reconnaissance soil survey for the State of Kuwait. In S. A. Shahid, F. K. Taha, & M. A. Abdelfattah (Eds), *Developments in soil classification, land use planning and policy implications: Innovative thinking of soil inventory for land use planning and management of land resources* (pp. 85–110). Springer. https://doi.org/10.1007/978-94-007-5332-7_3.

Patel, R., & Patel, A. (2024). Evaluating the impact of climate change on drought risk in semi-arid region using GIS technique. *Results in Engineering, 21*(1), 01957. ISSN 2590–1230, https://doi.org/10.1016/j.rineng.2024.101957.

Peng, S., Huang, J., Sheehy, J. E., Laza, R. C., Visperas, R. M., Zhong, X., Centeno, G. S., Khush, G. S., & Cassman, K. G. (2004). Rice yields decline with higher night temperature from global warming. *Proceedings National Academy of Sciences USA, 101*(27), 9971–9975. https://doi.org/10.1073/pnas.0403720101

Plaza, C., Zaccone, C., Sawicka, K., Méndez, A. M., Tarquis, A., Gascó, G., Heuvelink, G. B. M., Schuur, E. A. G., & Maestre, F. T. (2018). Soil resources and element stocks in drylands to face global issues. *Scientific Reports, 8*, 13788. https://doi.org/10.1038/s41598-018-32229-0

Prăvălie, R. (2016). Drylands extent and environmental issues. A global approach. *Earth-Science Reviews, 161*, 259–278. https://doi.org/10.1016/j.earscirev.2016.08.003

Qader, S. H., Dash, J., Alegana, V. A., Khwarahm, N. R., Tatem, A. J., & Atkinson, P. M. (2021). The role of Earth observation in achieving sustainable agricultural production in Arid and Semi-Arid Regions of the World. *Remote Sensing, 13*, 3382. https://doi.org/10.3390/rs13173382

Qadir, M., Quillerou, E., Nangia, V., Murtaza, G., Singh, M., Thomas, R. J., Drechsel, P., & Noble, A. D. (2014). Economics of salt-induced land degradation and restoration. *Natural Resources Forum, 38*, 282–295. https://doi.org/10.1111/1477-8947.12054.

Rao, N., Shahid, M., & Shahid, S. A. (2009). Alternative crops for diversifying production systems in the Arabian Peninsula. *Arabian Gulf Journal of Scientific Research, 27*(4), 195–203.

Reynolds, J. F., Smith, D. M. S., Lambin, E. F., Turner, B. L., Mortimore, M., Batterbury, S. P. J., Downing, T. E., Dowlatabadi, H., Fernandez, R. J., Herrick, J. E., Huber-Sannwald, E., Jiang, H., Leemans, R., Lynam, T., Maestre, F. T., Ayarza, M., & Walker, B. (2007a). Global desertification: Building a science for dryland development. *Science, 316*(5826), 847–851. https://doi.org/10.1126/science.1131634

Reynolds, J. F., Maestre, F. T., Kemp, P. R., Smith, M. S., & Lambin, E. (2007b). Natural and human dimensions of land degradation in drylands: Causes and consequences. In J. Canadell, D. Pataki, & L. Pitelka (Eds), *Terrestrial ecosystems in a changing world* (pp. 247–257). Springer. https://doi.org/10.1007/978-3-540-32730-1_20

Ruto, E., Gould, I., Bosworth, G., & Wright, I. (2018). The state, causes and impact of soil salinization: A global overview. SALFAR Work Package 3 Baseline January 2018 Report. Lincoln.

Savills Research. (2021). Regenerative agriculture. UK-Rural. www.savills.co.uk/insight-and-opinion

Schimel, D. S. (2010). Drylands in the earth system. *Science, 327*, 418–419.

Shafi, U., Mumtaz, R., García-Nieto, J., Hassan, S. A., Zaidi, S. A. R., & Iqbal, N. (2019). Precision agriculture techniques and practices: From considerations to applications. *Sensors, 19*, 3796. https://doi.org/10.3390/s19173796

Shahid, S. A. (2023a). Potential threats to soil functions and mitigation options for sustainable uses. In M. K. Suleiman, & S. A. Shahid (Eds.), *Terrestrial environment and ecosystems of Kuwait* (pp. 3–20). Springer. https://doi.org/10.1007/978-3-031-46262-7_1.

Shahid, S. A. (2023b). Desertification: A central problem to restoring ecosystems. In M. K. Suleiman, & S. A. Shahid (Eds.),*Terrestrial environment and ecosystems of Kuwait* (pp. 119–137). Springer. https://doi.org/10.1007/978-3-031-46262-7_5.

Shahid, S. A., & Abdelfattah, M. A. (2008) Soils of Abu Dhabi Emirate. In R. Perry (Ed), *Terrestrial environment of Abu Dhabi Emirate* (pp. 71–91).

Shahid, S. A., & Omar, S. A. S. (2022) *Kuwait Soil Taxonomy* (p. 149). Springer. https://doi.org/10.1007/978-3-030-95297-6.

Shahid, S. A., & Rehman, K. (2011). Soil salinity development, classification, assessment and management in irrigated agriculture. In M. Passarakli (Ed.), *Handbook of plant and crop stress* (3rd Edn, pp. 23–39). CRC.

Shahid, S,A., Zaman, M., & Lee, H. (2018). Soil salinity: Historical perspectives and world overview of the problem. In M. Zaman, S. A. Shahid, & H. Lee (Eds.), *Guideline for salinity assessment, mitigation and adaptation using nuclear and related techniques* (pp. 43–53). Springer. https://doi.org/10.1007/978-3-319-96190-3_2

Soil Survey Staff. (2022). *Keys to soil taxonomy*. USDA-NRCS.

Spinoni, J., Vogt, J., Naumann, G., Carrao, H., & Barbosa, P. (2015). Towards identifying areas at climatological risk of desertification using the Köppen-Geiger classification and FAO aridity index. *International Journal of Climatology, 35*, 2210–2222. https://doi.org/10.1002/joc.4124

Stringer, L. C., Mirzabaev, A., Benjaminsen, T. A., Harris, R. M. B., Jafari, M., Lissner, T. K., Stevens, N., & der Pahlen, C. T. (2021). Climate change impacts on water security in global drylands. *One Earth* (6), 851–864. ISSN 2590-3322, https://doi.org/10.1016/j.oneear.2021.05.010.

Suleiman, K. M. (2025). Sustaining food security in arid lands countries-national production vis-à-vis food import and foreign trade-offs for corporate farming. In M. K. Suleiman, & S. A. Shahid (Eds.), *Fostering arid land agriculture in the face of climate change: Mitigation and adaptation synergies*. Springer (In press).

Suleiman, M. K., & Shahid, S. A. (2023a). Prospective of agricultural farming in Kuwait and energy-food-water-climate nexus (Chap. 15). In M. K. Suleiman, & S. A. Shahid (Eds.),*Terrestrial environment and ecosystems of Kuwait* (pp. 363–391). Springer. https://doi.org/10.1007/978-3-031-46262_15.

Suleiman, M. K., & Shahid, S. A. (2023b). Agricultural water footprint of major crops grown in Kuwait compared to the world average: A review (Chap. 16). In M. K. Suleiman, & S. A. Shahid (Eds.), *Terrestrial environment and ecosystems of Kuwait* (pp. 393–414). Springer. https://doi.org/10.1007/978-3-031-46262_16.

Thenkabail, P. S. (Ed). (2015). Land resources monitoring, modeling, and mapping with remote sensing (p. 849). Taylor and Francis Inc. CRC Press. ISBN SBN 9781482217957-CAT# K22130.

Trabucco, A., & Zomer, R. J. (2009). Global aridity index (Global-Aridity) and Global potential evapo-transpiration (Global-PET) Geospatial Database. In *CGIAR consortium for spatial information*. http://www.cgiar-csi.org (2009).

UAEMEW. (2010). United Arab Emirates water conservation strategy (p. 215). UAE Ministry of Environment and Water.

UN DESA (United Nations Department of Economic and Social Affairs). (2025). https://www.un.org/development/desa/en/news/population/world-population-prospects-2017.html. Last accessed 5 Jan 2025.

UN. (2017). World population projected to reach 9.8 billion in 2050, and 11.12 billion in 2100. WWW Document.

UNEP. (1992). World atlas of desertification. United Nations Environment Programme.

UNEP. (2007). The global environment outlook 4. Environment for development. UNEP.

Walter, H. (1973). Die Vegetation der Erde. Stuttgart.

Wikimedia Commons. (2020). *Map of dry lands geomorphology*. https://commons.wikimedia.org/wiki/File:Drylandmap.jpg

Zaman, M., Shahid, S. A., & Heng, L. (Eds.) (2018). *Guideline for salinity assessment, mitigation and adaptation using nuclear and related techniques* (p. 164). Springer. https://doi.org/10.1007/978-3-319-96190-3.

Chapter 2
Soil Governance in the Arid Lands Countries for Sustainable Agriculture Services and Food Security

Shabbir Ahmad Shahid

Abstract Arid lands are distributed around the globe and are dominant in areas which have high temperature and limited rainfall. Based on the scientific assessment, soils can be categorized into different soil taxa. Each taxa presents suitability for different uses. A land can be unsuitable for agriculture production but may be vital for grazing or other potential uses, such as source of construction or industrial materials e.g., gravelly and gypsum rich soils respectively. Therefore, it makes sense to use the soils rationally based on science based information for informed decisions for sustainable soil governance. This suggests national systems to have an agenda for soil governance for better services for food security and ecosystem services without compromising the soil qualities. This chapter first outlines essential features of soil governance, potential authorities responsible for international and national level soil governance. The international review has revealed that there is no an overarching international body responsible for international soil governance, amid the creation of UN-FAO sponsored Global Soil Partnership, which educate and aware international communities the values of soils. Thus, it is evident that there is no central body at national or international levels designated for soil governance, but this responsibility is partially distributed to different national institutes e.g. land use authorities, municipalities, ministries etc. These authorities, unfortunately, occasionally allocate prime arable land for housing projects and thus these lands are sealed forever, and cannot be reversed for crop production and food security. In the author's opinion, such land allocation/governance rights to be given to national soil science societies for final approval of land for different projects, keeping in mind that no prime arable lands are used other than crop production and food security, and the housing projects and development of new cities to be planned on marginal quality lands, which will never be brought to crop production or food security projects, without amending the marginality level. The Sustainable Development Goal "land degradation neutrality" on its own is not a comprehensive soil policy, but it is forward looking planning

S. A. Shahid (✉)
Food Security Program, Environment and Life Sciences Research Center, Kuwait Institute for Scientific Research, Safat, Kuwait
e-mail: sshahid@kisr.edu.kw

M. K. Suleiman and S. A. Shahid (eds.), *Fostering Arid Lands Agriculture in the Face of Climate Change*, Advances in Global Change Research 81,
https://doi.org/10.1007/978-3-031-98490-7_2

element in particular starting point for guiding national policies and further work. It is therefore, essential to develop treaties on soil protection, to address gaps and shortcoming in current national and international soil governance, leading to the provision of long terms sustainable food security and ecosystem services.

Keywords Soil governance · Soil protection · Land uses · Food security · Ecosystem services · Rationale uses · Soil classification

2.1 Introduction

Arid lands are globally distributed where aridity is the main issue, and almost about half of the countries in the world are facing the aridity problems (low precipitation and high evapotranspiration). These countries are mainly distributed in Asia, Middle East, Africa regions and Australia. Naturally, where arid lands occur, the population is poor and have lack of resources to manage their soil optimally for crop production and food security at large (case of Africa and Asia). Soil governance is also an issue, where irrational use is leading to soil degradation, recently Sambo (2024) has addressed the soil governance phenomenon in Africa.

Soil governance refers to the policies, strategies, and the processes of decision-making employed by nation states and local governments regarding the use of *soil*. Currently there is no predominant international soil governance organization, but different organizations in different countries control the soil governance, in many cases without consultation with certified soil scientists or galaxy of soil scientists in national and international soil science societies. The absence of science based soil governance may lead to the allocation of prime arable land for housing projects to offset the escalating population growth and associated urbanization. Recently, Denmark has taken the initiative to develop an Act on land distribution and property trade for agricultural purposes (No. 1051 of 2023 www.retsinformation.dk). This Act contains rules regarding the consolidation of agricultural land and redistribution of agricultural land for purposes of agricultural development and improvement of land use planning. In the same context, it is necessary for all countries especially the poor countries in the arid lands to reassess their Acts pertaining to land use planning for agricultural development, to assure the prime arable lands are not allocated for other uses not relevant to food production.

Among different international organizations working with different countries, the UN-FAO sponsored Global Soil Partnership (GSP) educates and aware international communities the values of soils. Thus, it is evident that there is no central body at national or international levels designated for soil governance, but this responsibility is partially distributed to different national institutes e.g. land use authorities, municipalities and ministries etc. Who may in the absence of science based indigenous soil knowledge approve prime arable land for different uses, other than the food security.

Soil and land governance are two different aspects of using, managing and conserving terrestrial natural resources. Soil, in strict terms is the medium for plant growth, agriculture perspectives. From the purpose of soil taxonomy, soil is a natural body comprised of solids (minerals and organic matter), liquid, and gases that occur on the land surface, occupies space, and is characterized by one or both of the diagnostic horizons, or layers, that are distinguishable from the initial material as a result of additions, losses, transfers, and transformations of energy and matter or the ability to support rooted plants in a natural environment (Shahid & Abdelfattah, 2008).

Different scientists/potential soil users have viewed soil from different contexts, for example for a farmer soil is growing medium for plant, for a potter a source of mud to make pots, civil engineer considers as a source of construction material, or foundation to build houses and buildings. According to Minami (2009), soil is erroneously thought of as "little more than dirt" despite it being vital for the existence of humans and civilization. It is the soil that, created the earth's biosphere, and it could very easily become worn out in our lifetimes (Minami, 2009).

2.2 The Lower and Upper Boundary of Soil—Soil Classification Perspective

Soil consists of the horizons near the earth's surface that, in contrast to the underlying parent material, have been altered by the interactions of climate, relief, and living organisms over time (Jenny, 1941). The Soil Survey Staff (2022) has set up the upper limit of soil for soil classification purposes, the boundary between soil and either air, shallow water, live plants, or plant materials that have not begun to decompose. Areas are not considered to have soil if the surface is permanently covered by water too deep (typically more than about 2.5 m) for the growth of rooted plants. The lower boundary of soil for the practicality of soil survey is arbitrarily set at 200 cm. In areas where soil is thin, pedogenetically cemented horizons that are impermeable to roots, the soil extends as deep as the deepest cemented horizon, but not below 200 cm. For certain management goals, layers deeper than the lower boundary of the soil that is classified (200 cm) must also be described if they affect the content and movement of water and air or other interpretative concerns.

2.3 The Planet Earth, Eco-resources and Escalating Population Growth

The planet Earth was evolved nearly 4.54 billion years ago and the life on Earth could be mapped to 3,800 million years ago. In the Solar system, the Earth is the only planet to have an atmosphere containing free oxygen, oceans of liquid water on its surface, and, of course, *life*. Since the evolution of Earth and until the year 1971, the Earth used to provide surplus resources to live on sustainably. In the year 1971, human demand exceeded the capacity of Earth to provide resources (GFN, 2022).

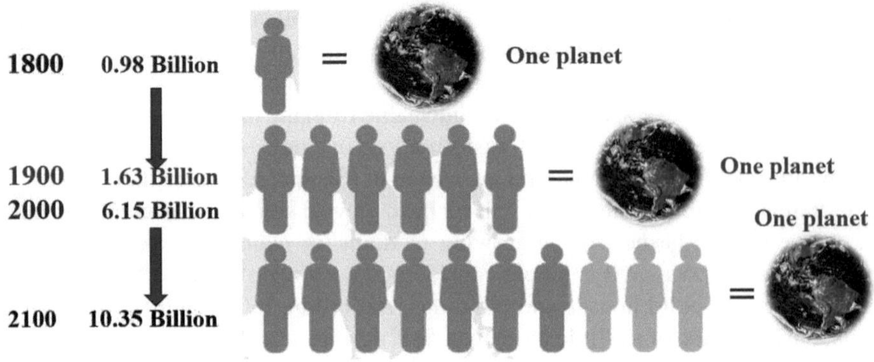

Fig. 2.1 Temporal population growth over three centuries

This shows that the Earth resources start shrinking rapidly and now the human needs 1.73 Earth planets to live on, and there is no virtual planet to import (Shahid & Suleiman, 2023). This shows that the human demand on nature is unsustainable and increasing, confirming that by taking more from our ecosystems and natural processes than can be replenished, we are jeopardizing our and new generations future. For example, in 2023, we consumed 100% Earth resources by 2 August 2023 (GFN, 2022). This suggests that the business as usual will lead us when we will require more than two Earth planets to meet our demand for food and eco-services for the current and escalating population estimated at 10.35 billion by 2100 (IPCC, 2023), which was 0.98 billion in 1800, 1.63 billion (1900) and 6.15 billion in 2000 (Fig. 2.1). After reviewing the population growth data (1950 and 2023) (UN, World Population Prospect, 2024) of five the most populated countries (China, India, USA, Indonesia, Pakistan) and the World, it is apparent that the population has increased multi-fold in all these countries (Table 2.1). Maximum increase is in Pakistan (5.9-fold increase), whereas, world average increase is over 200%. Given the current population growth trends and expected global population of over 9.3 billion by 2050 it seems a rather ambitious target to achieve food demand of escalating population. Whilst demand for food, feed and fiber may increase by 70% by 2050, 60% of the world's major ecosystems that help produce these resources and have already been degraded or are used unsustainably (Montanarella & Vargas, 2012).

2.4 Status of Urbanization

Urbanization is increasing at unprecedented levels globally and nationally. Looking at the temporal growth of urbanization, in the year 1970, about 36.6% global population was residing in urban build-up areas, and based on the level of urbanization in the past years, scientists made the projections that by 2050, over two-third population

Table 2.1 Temporal growth of population in five most populated countries and the work

Country↓ Year→	1950[a]	1975[a]	2000[a]	2023[a]	% increase (1923) over year 1950
Millions					
China	544.04	916.12	1,270	1,440	164.69
India	346.28	611.31	1,060	1,420	310.01
United States of America	154.20	219.13	281.4	343.48	122.75
Indonesia	68.80	131.84	216.08	282.19	310.16
Pakistan	35.85	69.46	154.88	247.50	590.38
World	2,490	4,070	6,170	8,090	224.89

Source UN, World Population Prospects (2024)
[a] Values as of 1 July of the indicated year

will live in urban areas mainly in Africa and Asia where population is also growing rapidly.

There are many factors behind this escalation and movement of population from rural to cities. The main factors include scarcity of essential basic resources, difficulty in accessing resources, lack of educational institutions, basic facilities, hospitals, unemployment, justice courts etc. The urbanization at high level can cause conflict for land use/land cover (Wang et al., 2023), usually taking productive lands for construction of houses and infrastructures, that can aggravate ecological services in urban areas. The search of literature revealed in China between 1995 and 2020, the level of urbanization expanded 9.1 times, with urban developed area escalation upto 16% between 1995 and 2005 (Wang et al., 2023), with China's urban population increasing from 29.04% in 1995 to 63.89% in 2020 (Ji & Chen, 2015; Li et al., 2018). This rapid urbanization has impacted noteworthy land use land cover (LULC) change. Such an extension of urbanization is likely to increase greenhouse gases emission and increase climate change impacts as well as pollute air and cause human health problems (Bardgett et al., 2021; Li et al., 2020a, 2020b; Sheng et al., 2017; Zhang et al., 2022).

2.5 Definition of Land

The UNCCD (2017) defines land as "*the terrestrial bioproductive system that comprises soil, vegetation, other biota, and the ecological and hydrological processes that operate within the system*". Alternatively, land is defined (CSD, 1996) as "*a delineable area of the Earth's terrestrial surface, encompassing all attributes of the biosphere immediately above or below this surface, including those of the near-surface climate, the soil and terrain forms, the surface hydrology (including shallow lakes, rivers, marshes, and swamps), the near-surface sedimentary layers and associated groundwater reserve, the plant and animal populations (biodiversity), the human settlement pattern and physical results of past and present human activity (terracing, water storage or drainage structures, roads, buildings*, etc.)".

The above definition makes it very clear that the term soil has limited coverage of the terrestrial environment, whereas the land has wider aspect of the terrestrial environment. The UNCCD (2017) has clearly indicted the trade-off of land for different uses in the context of climate change and Sustainable Development Goals. It states that "*our ability to manage trade-offs at a landscape scale will ultimately decide the future of land resources—soil, water, and biodiversity—and determine success or failure in delivering poverty reduction, food and water security, and climate change mitigation and adaptation. Indeed, integrated land and water management is recognized as an accelerator for achieving most of the Sustainable Development Goals*".

2.6 Momentous Pressures on Land Resources

The planet Earth on which the soil and land exists, is under stress from many aspects. There are number of effects on the planet, but the most severe are, (i) rapidly growing population, (ii) degradation of land resources (FAO-ITPS, 2015), (iii) water scarcity, (iv) global warming, (v) soil erosion is causing 60% crop yield loss, (vi) soil salinization is causing 27.3 billion $ annual loss (Qadir et al., 2014) due to decreased crop yields (20% irrigated lands are salinized), (vii) IPCC scenario confirms crops yield to be declined over 75% by 2100 (IPCC, 2023), (viii) 1/3rd loss and waste of food produced (FAO, 2011), (ix) increase in urbanization (Wang et al., 2023), (x) impact of climate change-rising temperature, low rainfall and subsequent decrease in aquifer recharge (Hughes et al., 2021), and (xi) increased drought (IPCC, 2023) etc. All are considered noteworthy threats to sustainable land resources uses. All these aspects limit the use of land resources and reduce the choices for variety of uses. This means that humanity is demanding more than the earth can currently provide. Currently, globally there is 1.1 global hectare (gha) eco-resources deficit per capita per year (Shahid & Suleiman, 2023). In this context, the achievement of UN-FAO, SDG 2 (End hunger by 2030) may be ambitious but questionable to achieve by 2030.

Food loss and *food waste* are two different elements in food supply chain. Food loss takes place during food production process at the farm levels after sowing, production, at harvesting and post harvesting and processing in the food supply chain (Partfitt et al., 2010) and at the market before the food reaches to the consumer. Whereas *food waste* (retailers and consumer behavior) is the phenomenon which occurs at sale of agriculture commodities when the consumer is buying and using the food (leftover food after eating in plates called waste at fork, unused cooked food, decaying of food before it is being used, due to poor storage) etc. The FAO (2011) reported food losses upto 70% of the total waste (losses and waste, farm and fork), and food waste at fork (30%). Provided, the food waste is reduced to zero, the food that is lost and wasted could feed 1.26 billion hungry people every year (FAO, 2011). The total food waste data (1.3 billion tons per year) is alarming, and if this high quantitates of food are not saved, then the land resources are stressed due to overexploitation for increased food production to meet the food demand of burgeoning population projected to be 10.9 billion by 2100.

When we discuss food losses at the global scale, peoples take the message that only food is lost, whereas, food production has strong nexus with other resources (energy-food-water-climate nexus) (Suleiman & Shahid, 2023a, 2023b), revealing that the resources (land, water, energy, inputs) used to produce food are also lost, as well as emission of greenhouse gasses during food production process. This is alarming situation, and the question is where do we stand currently to cope all the above impacts and what potential strategies we must adopt to mitigate the impacts to sustain our life with respect to improving soil health for food security and provision of ecosystem services in the broader context? On the positive side, FAO believes that sustainable soil management can lead to an average crop yield increase of 58% (FAO, 2022). Therefore, there is great challenge for us to adopt climate resilient soil management practices that enhance soil quality while producing the food, along with achieving land degradation neutrality. It is relevant at this stage that the soil users and conservators must consult the "*Status of the World's Soil Resources*" which offers a global perspective on the current state and trends of soils, their role in providing ecosystem services and the threats these soils face (FAO-ITPS, 2015). This will guide where do we stand currently and will lead the path for future to develop intervention to address the land degradation holistically.

2.6.1 Land Degradation and Sustainable Soil Management

The current trend of land degradation is faster than anytime recorded in the past. The main culprits of causing land degradation are, erosion (wind and water) and soil salinization, and to a less extent soil contamination with heavy metals and toxic compounds. The FAO-ITPS (2015) stated that the majority of the world's soil resources are in only *fair, poor or very poor conditions* and that conditions are getting worse in far more cases than they are improving. Currently, globally 33% land is degraded to various levels (FAO, 2022), and it seems the human lacks to understand the causes of land degradation to mitigate the degradation level (Fig. 2.2). FAO (2022) warned that if we do not deliver change now, degradation is on course to hit a high of 90% of all soils by 2050.

There are many definitions of sustainable soil management from different users point of view. However, to unify the definition, FAO (2022) defines sustainable soil management, if the supporting, provisioning, regulating, and cultural services provided by soil are maintained or enhanced without significantly impairing either the soil functions that enable those services or biodiversity.

Fig. 2.2 Resource degradation and interventions

2.6.2 Potential Interventions to Alleviate Land Degradation

Land degradation is a serious concern, when we target to achieve SDG-2 (end hunger) by 2030. The challenge is by 2050, agricultural production must increase by 60% globally, and by almost 100% in developing countries in order to meet food demand alone. Therefore, there are huge responsibilities on all of us to reverse land degradation and conserve the land for sustainable services including food security and ecosystem services. The potential interventions are (i) combating desertification (wind and water erosion) by protecting the soils and organized grazing (Shahid, 2023; Shahid & Omar, 2001; Shahid et al., 1999, 2003), (ii) reverse soil salinization using integrated soil reclamation programs (physical, chemical, hydrological, biological methods) (FAO, 2021; Shahid & Rahman, 2011; Shahid et al., 2008, 2018; Zaman et al., 2018), (iii) rehabilitation and restoration of degraded lands, (iv) phytoremediation of soils contaminated with chemical pollutants, petroleum hydrocarbons and heavy metals (Ashraf et al., 2019; Yan et al., 2020).

2.7 Soil Governance

Soil governance refers to the policies, strategies, and the processes of decision-making employed by nation states and local governments regarding the use of soil (Bonn Conference, 2011). Healthy soils in a community can achieve much more than food security and environmental harmony (Sambo, 2024). In general soil governance focuses on the use for agriculture for food security mostly in poor but most populated countries facing food insecurity. The UN-Global Soil Partnership (GSP) took the initiative to improve soil governance to assure healthy and productive soils are available for a food secured world, especially the food deficit arid lands countries, where there is arable land and water scarcities.

The mission of the GSP (FAO, 2022) is "*to improve soil governance to guarantee healthy and productive soils*". The GSP do so by making sure soils can provide the right set of essential ecosystem services and functions needed for food security, climate change mitigation and biodiversity conservation (FAO, 2022). The GSP not only focuses on food security but also the soil in the context of the provision of essential ecosystem services. The history and state of existing international soil governance (Bodle, 2022; Bodle et al., 2020; Boer et al., 2016) suggests that the case for addressing soil at the international level is not self-evident. Whereas, available legal frameworks for soil conservation at national and regional level seem not to be able to regulate the current use of soil resources in order to assure long-term sustainability (Montanarella & Vargas, 2012).

2.7.1 International Soil Governance Organizations and Treaties

The soil governance is not limited to at international level only, but it is important aspect at national levels so that the national policy makers use the soils rationally for maximum benefits and on sustainable manners. There are a number of international organizations whose one of the domains is international soil governance. Currently, there is no binding international law for soil governance (Bodle, 2022) to follow by all nations. However, if in any case such international body exists, it has different mechanism, other than national laws. Where treatise exist, these are applied to the countries who signed these treaties, and no binding to other non-signing countries. When we discuss international soil governance, it involves transitional entities/organizations in a cooperative manner, but the laws become so complex as these may not be relevant to one country. Following are the main international soil governance elements include the following treaties, institutions and non-binding instruments (Bodle, 2022), but none of these entities have full command at international or national levels to protect soil or provide a framework for soil governance:

- Agenda 2030 and the Sustainable Development Goals (SDGs)
- United Nations Convention to Combat Desertification (CCD)

- United Nations Convention on Biological Diversity (CBD)
- Ramsar Convention on wetlands
- Paris Agreement on climate change
- Soil Conservation Protocol to the Alpine Convention
- African Convention on the Conservation of Nature and Natural Resources (Maputo Convention)
- UN Food and Agricultural Organization (FAO)
- UN Environment Programme (UNEP) and the United Nations Environmental Assembly (UNEA)
- Intergovernmental Panel on Climate Change (IPCC)
- Intergovernmental Science-Policy Platform on Biodiversity and Ecosystem Services (IPBES).

2.7.2 Global Soil Partnership (GSP) and Soil Governance

Among all international organizations, UN-GSP took the initiative to highlight the importance of soil governance at national and global levels. To advance the agenda of soil governance, joint working group on soil legislation from FAO-GSP published peer reviewed journal "Soil Security". The FAO-GSP describes the necessity of soil governance "*Governing the soil requires international and national collaboration between governments, local authorities, industries and citizens to ensure implementation of coherent policies that encourage practices and methodologies that regulate the usage of the soil resource to avoid degradation and conflict between users* (https://www.fao.org/global-soil-partnership/areas-of-work/soil-governance/en/)". The mandate of GSP is to improve and maintain the health of at least 50 percent of the world's soils by 2030 and beyond (FAO, 2022). The FAO-GSP organized two webinars on 13 January 2021 and 9 March 2023 "*Enhancing soil governance: regional and national examples of soil legislation development*", where various topics related to soil governance were presented and deliberated.

The ***International Union of Soil Sciences*** (IUSS) is the global union of soil scientists. The purpose of the IUSS is to promote all branches of soil science and their applications, to promote contacts among scientists and other persons engaged in the study and the application of soil science, to stimulate scientific research and to further the application of such research (https://www.iuss.org/mission/).

The ***Intergovernmental Panel on Climate Change*****(IPCC)** was created to provide policymakers with regular scientific assessments on climate change, its implications and potential future risks, as well as to put forward adaptation and mitigation options (https://www.ipcc.ch/).

The ***United Nation's Food and Agriculture Organization*** works to achieve food security for all to make sure people have regular access to enough high-quality food to lead active, healthy lives. The FAO mandate is to improve nutrition, increase agricultural productivity, raise the standard of living in rural populations and contribute to global economic growth.

United Nations Convention to Combat Desertification (UNCCD) is the only legally binding framework set up to combat desertification and mitigate the effects of drought through national action programs that incorporate long-term strategies (https://www.unccd.int/convention/overview).

Convention on Biological Diversity (CBD), parties are obligated to inventory and monitor biodiversity, incorporate the concepts of conservation and sustainable development into national strategies and economic development, and preserve indigenous conservation practices (https://www.un.org/en/observances/biological-diversity-day/convention).

2.7.3 Soil Governance is a Neglected Element

It is evident from the above discussion that international soils are under severe anthropic and biotic stresses, leading to high level of degradation and becoming inferior in quality for sustainable agriculture for food security and ecosystem services. Comparing to other issues of global concerns (environment, climate change, biodiversity, water and pollutants, air), soil is merely being ignored to the level of other issues. However, in contrast, Schimel (2010) very wisely presented value of soil defining peoples culture and society more strongly than other environment variables.

Although, the global community celebrates 5 December as World Soil Day from the Rome based forum of FAO-Global Soil Partnership, but no international organization exists explicitly focusing on Global Soil Governance. As of today, no international soil governance treaty exists. Considering the high level of land degradation (FAO-ITPS, 2015) and need to increase food production to achieve the objectives of SDG 2 (End Hunger), the international policy makers and the organizations should make negotiations and agreed on one agenda and establish a forum dealing with international soil governance, which is transnational and signed by the interested parties at national level. Lal et al. (2021) after reviewing the 17 UN-SDGs has stated that there are 11 SDGs intricately related to soil health. The most relevant are 1–3 (ending poverty, zero hunger and good health and wellbeing).

2.7.4 National Soil Governance

National level soil governance for rational and sustainable uses of soil resources is implemented by national institutes, authorities, private and public sectors. Each country has its own national authorities designated for soil governance, and legislation for land uses, these may be following in different countries.

- Municipalities (urban development)
- Land use authorities (rational use of soil resources)

- Ministry of Agriculture or Public Authorities for agriculture (allocation of land for agriculture expansion and extension services)
- Agriculture departments (recommendations for new locations for agriculture)
- Directorate of soil surveys (provision of soil and land suitability maps).

2.7.4.1 Pre-requisites for National Soil Governance

In the author's opinion for a sustainable and rational uses of national soil resources, it is a fundamental pre-requisite to assess the soil suitability for agriculture through science based national soil surveys (Omar & Shahid, 2013). Such surveys provide soil and multiple maps addressing different themes. In many countries including Pakistan national level soil surveys have been completed in seventies of the twentieth centuries. At that time there was lack of technological advancements, such as the use of remote sensing (RS) and geographical information system (GIS), most of the soil maps were prepared using low level available technologies (black and white stereo photographs). The literature revealed few countries have conducted soil surveys during 2000 to 2024. However, a number of countries have assessed their soils in the twentieth century.

The most recent soil surveys are completed in the Gulf Cooperation Council (GCC) Counties, for example Kingdom of Saudi Arabia (MAW, 1985); Sultanate of Oman (MAF, 1990), State of Kuwait (KISR, 1999a, 1999b; Omar & Shahid, 2013; Shahid & Omar, 1999, 2022; Shahid et al., 2004), State of Qatar (Scheibert et al., 2005), Abu Dhabi emirate (Abdelfattah & Shahid, 2007; EAD, 2009a, 2009b; Shahid & Abdelfattah, 2008; Shahid & Omar, 2022; Shahid et al., 2013), Northern Emirates of UAE (EAD-MOEW, 2012), Dubai emirate (Dubai Municipality, 2005), New Zealand soil classification (Hewitt, 1992), Australian Soil classification (Isbell, 2016; Moore et al., 1983; Northcote, 1979), South Africa (Soil Classification Working Group, 1991), Canada (Haynes, 1998; Soil Classification Working Group, 1998), Russia (Fridland, 1988), China (Feng, 2001), and revised World Soil Map (FAO-UNESCO, 1990). These countries have science based soil information for soil governance at national level for informed decisions with respect to land use planning and policy development (Shahid et al., 2013).

Following to these soil surveys Shahid et al. (2014) in collaboration with eminent USDA-NRCS staff published United Arab Emirates Keys to Soil Taxonomy, and in 2022 published Kuwait Soil Taxonomy (Shahid & Omar, 2022), and established Emirates Soil Museum, hosted at Dubai based International Center for Biosaline Agriculture (ICBA), United Arab Emirates. In the recognition of the importance of national level soil surveys, the FAO has called on all countries to urgently improve their soil information and capacities by making stronger commitments towards sustainable soil management. In addition, in the author's opinion, centralizing national soils under one ceiling (soil museum) at different levels (province, division etc.) and publish National Soil Taxonomy books for education and future research is essential for sustainable soil management and uses.

2.7.4.2 Opportunities and Constraints in National Soil Governance

As discussed in the above section, that the soil governance either at international or national levels is in general infancy. The future of the global and national soil governance has optimistic future agenda, since diversified initiatives have been taken in the past. Therefore, there are great opportunities to build up framework for soil governance through legislations and implementation through potential public–private entities in a cooperative manner. For such a development, there is need to have science based information about global and national soils to predict soil suitability for diversified uses.

With the opportunities to set up soil governance mechanism at national level, there are constraints to implement soil governance in the form of overlap of mandates of different organizations, and especially, in arid lands countries, even the suitable soils are isolated for future expansion of irrigated agriculture. These countries are constrained with scarcity of water resources, prolonged drought, high temperature and marginal quality soils.

2.8 Laws of Soil Governance and Role of Soil Science Societies in National Soil Governance

Arid lands are prone to water scarcity and associated prolonged drought. The legislation of soil governance is poorly stipulated in government documents. Previous attempts for global soil governance have failed (Montanarella & Vargas, 2012). The soil governance is accomplished by various organizations. Governance of the soil differs from soil management. Soil management involves practices and techniques used to increase and maintain soil fertility, structure, and carbon sequestration, etc.

Specific laws of soil governance with respect to rational uses of soils (science based decisions) in most of the countries either do not exist or poorly defined. The existing laws are very general, and land is allocated for new infrastructures (housing schemes, new cities development, industrial growth etc.) in the absence of science based soil information (national soil and land use maps) and guidance from national soil science societies. This has led occupation of prime arable lands (highly suitable for crop production) by other uses, and these soils are permanently sealed, and can never be reversed for crop production to address food security issue. This occupation over a period of time is likely to reduce per capita share of crop land and may cause food security issue, knowing that population is growing unprecedentedly and arable land resources are shrinking at rapid rate. Currently crop land per person, has fallen below 0.25 ha (Montanarella & Vargas, 2012).

In this regard, national soil science societies can bridge the gap and guide the policy developers and decision makers based on the Indigenous Soil Knowledge (ISK) or based on the published soil survey report in the country, where to select lands for new infrastructures. This way the prime arable land can be set aside for future use for crop production, whenever, the government decides to expand agriculture, to meet food demand of its population. Based on the science based knowledge, new infrastructures can be developed on marginal lands which have no capability to bring into crop production due to either permanently unsuitable soil quality, remote areas or high cost is required to prepare for agriculture use. Currently, the national soil science societies have narrow or no soil governance agenda, and specifically aiming at e.g., Soil Science Society of Pakistan *(i) promote soil science research and its application, (ii) create facilities and provide a forum for close cooperation among the scientists working in soil science and allied disciplines, (iii) foster publication of a Soil Science Journal, Newsletter* (https://sss-pakistan.org/about-us/).

2.9 Conclusions and Recommendations

It has been concluded that soils are important natural resource essential for diversified uses. The soil governance has not yet been given due consideration either at national or international level, like other elements (environment, water, biodiversity, air etc.) of nature. Regardless of this neglect, the soils are essential part of our landscape and must be protected in their own line to meet the demand of food of escalating population, mitigate climate change impacts and other challenges related to crop production and ecosystem functions and services. There are international organizations addressing the soil agenda in their own domains, but no international soil governance legislation occurs to follow by the international communities for rational uses of soil resources. The National Soil Science Societies exist in most of the countries, their mandate is to organize soil related seminars, symposia, congress and conferences, publish soil science journals of national and international interests and communicate with international soil related organizations. Their role to guide the land use planners for rational use of soil resources is not seen in the forefront. This has led the use of prime arable lands for construction as a part of urbanization, and these soils are permanently sealed, which cannot be reversed for crop production and food security.

In this regard, national soil science societies can bridge the gap and guide the policy developers and decision makers based on the Indigenous Soil Knowledge (ISK) or based on the published soil survey report in the country, where to select lands for new infrastructures. This way the prime arable land can be set aside for future use for crop production, whenever, the government decides to expand agriculture, to meet food demand of its population. Based on the science based knowledge, new infrastructures can be developed on marginal lands which have no or low capability to bring into crop production due to either permanently unsuitable soil quality or high cost is required to prepare for agriculture use.

References

Abdelfattah, M. A., & Shahid, S. A. (2007). A comparative characterization and classification of soils in Abu Dhabi coastal area in relation to arid and semiarid conditions using USDA and FAO soil classification systems. *Arid Land Research and Management, 21*(3), 245–271. https://doi.org/10.1080/15324980701426314

Ashraf, S., Ali, Q., Zahir, Z. A., Ashraf, S., & Asghar, H. N. (2019). Phytoremediation: Environmentally sustainable way for reclamation of heavy metal polluted soils. *Ecotoxicology and Environmental Safety, 174*(13), 714–727. https://doi.org/10.1016/j.ecoenv.2019.02.068

Bardgett, R. D., Bullock, J. M., Lavorel, S., Manning, P., Schaffner, U., Ostle, N., Chomel, M., Durigan, G., Fry, E. L., Johnson, D., Lavallee, J. M., Le, G., Provost, S., Luo, S., Png, K., Sankaran, M., Hou, X., Zhou, L., Ma, L., Ren, W., Li, X., Ding, Y., Li, Y., & Shi, H. (2021). Combatting global grassland degradation. *Nature Reviews Earth and Environment, 2*(10), 720–735. https://doi.org/10.1038/s43017-021-00207-2

Bodle, R. (2022). International soil governance. *Soil Security, 6*, 100037. ISSN: 2667-0062. https://doi.org/10.1016/j.soisec.2022.100037. Accessed 1 Jan 2025.

Bodle, R., Stockhaus, H., Wolff, F., Scherf, C.-S., & Oberthür, S. (2020). *Improving international soil governance—Analysis and recommendations* (p. 62). Final report on behalf of German Environment Agency, Umweltbundesamt, Dessau-Roßlau Publisher.

Boer, B., Ginzky, H., & Heuser, I. L. (2016). International soil protection law: History, concepts and latest developments. In H. Ginzky, I. L. Heuser, T. Qin, O. C. Ruppel, & P. Wegerdt (Eds.), *International yearbook of soil law and policy* (pp. 49–72). Springer International Publishing. https://doi.org/10.1007/978-3-319-42508-5_7

Bonn Conference. (2011). *The water energy and food security nexus—Solutions for the green economy, 16–18 November 2011, Bonn, Nordrhein-Westfalen, Germany.* https://sdg.iisd.org/events/bonn-2011-conference/

CSD (Convention on Sustainable Development). (1996). *Progress report on Chapter 10 of Agenda 21, United Nations, New York, NY, USA.*

Dubai Municipality (DM). (2005). *Satellite imagery and thematic mapping project* (p. 221). Executed jointly by Global Scan Technologies (Dubai) and National Remote Sensing Agency.

EAD. (2009a). *Soil survey of Abu Dhabi Emirate: Extensive survey* (Vol. 1, pp. xx+506). Environment Agency-Abu Dhabi.

EAD. (2009b). *Soil survey of Abu Dhabi Emirate: Intensive survey* (Vol. 3, pp. xviii+435). Environment Agency-Abu Dhabi.

EAD-MOEW. (2012). *Soil survey of the Northern Emirates.* Soil report (Vol. 1, pp. xxvi+410). Environment Agency-Abu Dhabi.

FAO. (2011). *Global food losses and food waste—Extent, causes and prevention, Rome.*

FAO. (2021). *Global map of salt-affected soils version 1.* Retrieved December, 2021, from https://www.fao.org/soils-portal/data-hub/soil-maps-and-databases/global-map-of-salt-affected-soils/en/

FAO. (2022). *Global soil partnership 2012-2022—Sustainable soil management in action, Rome.* https://doi.org/10.4060/cc0921en

FAO-ITPS. (2015). *Status of the world's soil resources (SWSR—Technical summary)* (p. 77). Food and Agriculture Organization of the United Nations and Intergovernmental Technical Panel on Soils.

FAO-UNESCO. (1990). *Soil map of the world: Revised legend* (World soil report, No. 60). FAO.

Feng, L. I. (2001). *Chinese soil taxonomy* (p. 203). Science Press Beijing, Coordinated by Institute of Soil Science, Chinese Academy of Sciences.

Fridland, V. M. (Ed.). (1988). *Soil map of the Russian Soviet Federative Socialist Republic at Scale 1:2.5 million.* All Union Academy of Agricultural Science. Government Administration for Geodesy and Cartography (GUGN), 16 sheets.

GFN. (2022). *York university ecological footprint initiative and global footprint network (2022)*. National Footprint and Biocapacity Accounts, 2022 edition. Retrieved August 27, 2023, from https://data.footprintnework.org

Haynes, R. H. (Ed.). (1998). *The Canadian system of soil classification* (3rd ed.). Soil Classification Working Group, Research Branch, Agriculture and Agri-Food Canada, NRC Research Press.

Hewitt, A. E. (1992). *New Zealand soil classification* (DSPIR Land Resources Scientific Report No. 19).

Hughes, A., Mansour, M., Ward, R., Kieboom, N., Allen, S., Seccombe, D., Charlton, M., & Prudhomme, C. (2021). The impact of climate change on groundwater recharge: National-scale assessment for the British mainland. *Journal of Hydrology, 598*, 126336. ISSN: 00221694. https://doi.org/10.1016/j.jhydrol.2021.126336

IPCC. (2023). Intergovernmental panel on climate change report. In *Sixth assessment report*. Retrieved January 01, 2025, from https://www.ipcc.ch/report/sixth-assessment-report-cycle/

Isbell, R. F. (2016). *The Australian soil classification* (2nd ed.) CSIRO. https://www.clw.csiro.au/aclep/asc_re_on_line_V2/soilbgro.htm

Jenny, H. (1941). Factors of soil formation. In *A system of quantitative pedology*. United States of America Dover Publications, Inc.

Ji, X., & Chen, B. (2015). Assessing the energy-saving effect of urbanization in China based on stochastic impacts by regression on population, affluence and technology (STIRPAT) model. *Journal of Cleaner Production, 24*(7). https://doi.org/10.1016/j.jclepro.2015.12.002

KISR. (1999a). *Soil survey for the state of Kuwait: Reconnaissance survey* (Vol. II & III). Kuwait Institute for Scientific Research.

KISR. (1999b). *Soil survey for the state of Kuwait: Semi-detailed survey* (Vol. IV & V). Kuwait Institute for Scientific Research.

Lal, R., Bouma, J., Brevik, E., Dawson, L., Field, D. J., Glaser, B., Hatano, R., Hartemink, A. E., Kosaki, T., Lascelles, B., Monger, C., Muggler, C., Ndzana, G. M., Norra, S., Pan, X., Paradelo, R., Reyes-Sánchez, L. B., Sandén, T., Singh, B. R., Spiegel, H., Yanai, J., & Zhang, J. (2021). Soils and sustainable development goals of the United Nations: An International Union of Soil Sciences perspective. *Geoderma (Regional), 25*, e00398. ISSN: 2352-0094. https://doi.org/10.1016/j.geodrs.2021.e00398. Accessed October 2024.

Li, G., Sun, S., & Fan, C. (2018). The varying driving forces of urban land expansion in China: Insights from a spatial-temporal analysis. *Landscape and Urban Planning, 174*(2–4), 63–77. https://doi.org/10.1016/j.landurbplan.2018.03.004

Li, X., Zhou, Y., Zhao, M., & Zhao, X. (2020a). A harmonized global nighttime light dataset 1992–2018. *Scientific Data, 7*(1), 168. https://doi.org/10.1038/s41597-020-0510-y

Li, X., Gong, P., Zhou, Y., Wang, J., Bai, Y., Chen, B., Hu, T., Xiao, Y., Xu, B., Yang, J., Liu, X., Cai, W., Huang, H., Wu, T., Wang, X., Lin, P., Li, X., Chen, J., He, C., Li, X., Yu, L., Clinton, N., & Zhu, Z. (2020b). Mapping global urban boundaries from the global artificial impervious area (GAIA) data. *Environment Research Letters, 15*(9), 094044. https://doi.org/10.1088/1748-9326/ab9be3

MAF. (1990). *General soil map of the Sultanate of Oman*. Ministry of Agriculture and Fisheries, Directorate General of Agricultural Research.

MAW. (1985). *General soil map of the Kingdom of Saudi Arabia*. Ministry of Agriculture and Water.

Minami, K. (2009). Soil and humanity: Culture, civilization, livelihood and health. *Soil Science and Plant Nutrition, 55*(5), 603–615. https://doi.org/10.1111/j.1747-0765.2009.00401.x. Accessed 1 January 2025.

Montanarella, L., & Vargas, R. (2012). Global governance of soil resources as a necessary condition for sustainable development. *Current Opinion in Environmental Sustainability, 4*(5), 559–564. ISSN: 1877-3435. https://doi.org/10.1016/j.cosust.2012.06.007

Moore, A. W., Isbell, R. F., & Northcote, K. H. (1983). Classification of Australian soils. In *Soils: An Australian viewpoint* (pp. 253–256). Division of Soils CSIRO, CSIRO, Melbourne/Academic Press.

Northcote, K. H. (1979). *A factual key for the recognition of Australian soils* (4th ed.). Rellim Technical Publications.

Omar, S. A. S., & Shahid, S. A. (2013). Reconnaissance soil survey for the state of Kuwait. In S. A. Shahid, F. K. Taha, & M. A. Abdelfattah (Eds.), *Developments in soil classification, land use planning and policy implications: Innovative thinking of soil inventory for land use planning and management of land resources* (pp. 85–110). https://doi.org/10.1007/978-94-007-5332-7_3

Parfitt, J., Barthel, M., & Macnaughton, S. (2010). Food waste within food supply chains: Quantification and potential for change to 2050. *Philosophical Transactions of the Royal Society, 365*(1554), 3065–3081. https://doi.org/10.1098/rstb.2010.0126

Qadir, M., Quillerou, E., Nangia, V., Murtaza, G., Singh, M., Thomas, R. J., Drechsel, P., & Noble, A. D. (2014). Economics of salt-induced land degradation and restoration. *Natural Resources Forum, 38*, 282–295. https://doi.org/10.1111/1477-8947.12054

Retrieved October 20, 2024, from https://www.ipcc.ch/

Retrieved October 20, 2024, from https://www.un.org/en/observances/biological-diversity-day/convention

Retrieved October 20, 2024, from https://sss-pakistan.org/about-us/

Retrieved October 21, 2024, from https://www.fao.org/global-soil-partnership/areas-of-work/soil-governance/en/

Retrieved October 21, 2024, from https://www.iuss.org/mission/

Retrieved October 21, 2024, from https://www.unccd.int/convention/overview

Sambo, P. T. (2024). An African legal, cultural and religious perspective of sustainable soil governance. In H. Ginzky et al. (Eds.), *International yearbook of soil law and policy.* https://doi.org/10.1007/978-3-031-40609-6_13

Scheibert, C., Stietiya, M. H., Sommer, J., Abdalla, O. E. S., Schramm, H., & Memah Al, M. (2005). *The atlas of soils for the state of Qatar*. Ministry of Municipal Affairs and Agriculture.

Schimel, J. (2010). Soil and culture. *Nature Geoscience, 3*, 301. Retrieved June 22, 2022, from https://rdcu.be/cIbeZ

Shahid, S. A. (2023). Desertification: A central problem to restoring ecosystems. In M. K. Suleiman, & S. A. Shahid (Eds.), *Terrestrial environment and ecosystems of Kuwait* (Ch. 5, pp. 119–137). Springer. https://doi.org/10.1007/978-3-031-46262-7_5

Shahid, S. A., & Abdelfattah, M. A. (2008). Soils of Abu Dhabi Emirate. In R. Perry (Ed.), *Terrestrial environment of Abu Dhabi Emirate* (pp. 71–91). Environment Agency-Abu Dhabi.

Shahid, S. A., & Omar, S. A. S. (1999). *Order 1 soil survey of the demonstration farm sites with proposed management* (pp. viii + 144). Kuwait Institute for Scientific Research. ISBN: 0 957700369.

Shahid, S. A., & Omar, S. A. S. (2001). Causes and impacts of land degradation in the arid environment of Kuwait. In E.M. Bridges et al. (Eds.), *Response to land degradation* (pp. 76–77).

Shahid, S. A., & Omar, S. A. S. (2022). *Kuwait soil taxonomy* (p. 149). Springer. https://doi.org/10.1007/978-3-030-95297-6

Shahid, S. A., & Rahman, K. (2011). Soil salinity development, classification, assessment and management in irrigated agriculture. In M. Pessarakli (Ed.), *Handbook of plant and crop stress* (3rd ed., pp. 23–39). CRC Press.

Shahid, S. A., & Suleiman, M. K. (2023). Ecological footprint and biocapacity of Kuwait and proposed eco-resources management strategies: A review. In M. K. Suleiman, & S. A. Shahid (Eds.), *Terrestrial environment and ecosystems of Kuwait* (Ch. 4, pp. 393–414). Springer. https://doi.org/10.1007/978-3-031-46262-7_4

Shahid, S. A., Omar, S. A. S., & Al-Ghawas, S. (1999). Indicators of desertification in Kuwait and their possible management. *Desertification Control Bulletin, 34*, 61–66.

Shahid, S. A., Omar, S. A. S., Misak, R., & Rizq, H. A. (2003). Land resource stresses and degradation in the arid environment of Kuwait. In A. S. lsharhan, W. W. Wood, A. S. Goudie, A. Flower, & E. M. Abdellatif (Eds), *Desertification in the third millennium* (pp. 351–260). Swets and Zeitlinger Publishers. https://doi.org/10.1201/NOE9058095718.ch37

Shahid, S. A., Omar, S. A. S., Jamal, M. E., Shihab, A., & Abo-Rezq, H. (2004). Soil survey for farm planning in northern Kuwait. *Kuwait Journal of Science and Engineering, 31*(1), 43–57.

Shahid, S. A., Dakheel, A. H., Mufti, K. A., & Shabbir, G. (2008). Automated in-situ salinity logging in irrigated agriculture. *European Journal of Scientific Research, 26*(2), 288–297.

Shahid, S. A., Abdelfattah, M. A., Othman, Y., Kumar, A., Taha, F. K., Kelley, J. A., & Wilson, M. W. (2013). Innovative thinking for sustainable use of terrestrial resources in Abu Dhabi Emirate through scientific soil inventory and policy development. In S. A. Shahid, F. K. Taha, & M. A. Abdelfattah (Eds.), *Developments in soil classification, land use planning and policy implications—Innovative thinking of soil inventory for land use planning and management of land resources* (pp. 3–49). Springer. https://doi.org/10.1007/978-94-007-5332-7_1

Shahid, S. A., Abdelfattah, M. A., Wilson, M. A., Kelley, J. A., & Chiaretti, J. V. (2014). *United Arab Emirates keys to soil taxonomy* (p. 108). Springer. https://doi.org/10.1007/978-94-007-7420-9

Shahid, S. A., Zaman, M., & Heng, L. (2018). Salinity and sodicity adaptation and mitigation options. In M. Zaman, S. A. Shahid, & L. Heng (Eds.), *Guidelines for salinity assessment, mitigation and adaptation using nuclear and related techniques* (Ch. 3, pp. 55–88). Springer. https://doi.org/10.1007/978-3-319-96190-3_3

Sheng, J., Han, X., & Zhou, H. (2017). Spatially varying patterns of afforestation/reforestation and socio-economic factors in China: A geographically weighted regression approach. *Journal of Cleaner Production, 153*(5583), 362–371. https://doi.org/10.1016/j.jclepro.2016.06.055

Soil Classification Working Group. (1991). *Soil classification a taxonomic system for South Africa* (Memoirs on the Agricultural Natural Resources of South Africa No. 15) (p. 257). A Report on a Research Project Conducted under the Auspices of the Soil and Irrigation Research Institute, Department of Agricultural Development.

Soil Classification Working Group. (1998). *The Canadian system of soil classification* (3rd ed., p. 187). Agriculture and Agri-Food Canada Publication 1646.

Soil Survey Staff. (1999). *Soil taxonomy: A basic system of soil classification for making and interpreting soil surveys* (2nd ed., p. 436). Natural Resources Conservation Service. US Department of Agriculture Handbook.

Soil Survey Staff. (2022). *Keys to soil taxonomy* (13th ed.). USDA-NRCS.

Suleiman, M. K., & Shahid, S. A. (2023a). Prospective of agricultural farming in Kuwait and energy-food-water-climate nexus. In M. K. Suleiman, & S. A. Shahid (Eds.), *Terrestrial environment and ecosystems of Kuwait* (Ch. 15, pp. 363–391). Springer. https://doi.org/10.1007/978-3-031-46262_15

Suleiman, M. K., & Shahid, S. A. (2023b). Agricultural water footprint of major crops grown in Kuwait compared to the world average. In M. K. Suleiman, & S. A. Shahid (Eds.), *Terrestrial environment and ecosystems of Kuwait* (Ch. 16, pp. 393–414). Springer. https://doi.org/10.1007/978-3-031-46262_16

UN, World Population Prospects. (2024). *Processed by our world in data.* "Population, total—UN WPP". United Nations, "World Population Prospects". Retrieved October 20, 2024, from https://ourworldindata.org/population-growth

UNCCD. (2017). *Article 1 of the text of the convention.* http://www2.unccd.int/sites/default/files/relevant-links/2017-01/UNCCD_Convention_ENG_0.pdf

Wang, C., Liu, S., Zhou, S., Zhou, J., Jiang, S., Zhang, Y., Feng, T., Zhang, H., Zhao, Y., Lai, Z., Cui, S., & Xuegang Mao, X. (2023). Spatial-temporal patterns of urban expansion by land use/land cover transfer in China. *Ecological Indicators, 155*, 111009. ISSN: 1470-160X. https://doi.org/10.1016/j.ecolind.2023.111009

Yan, A., Wang, Y., Tan, S. N., Yusof, M. L. M., Ghosh, S., & Chen, Z. (2020). Phytoremediation: A promising approach for revegetation of heavy metal-polluted land. *Frontiers in Plant Sciences, Plant Biotechnology Section, 11*. https://doi.org/10.3389/fpls.2020.00359

Zaman, M., Shahid, S. A., & Heng, L. (2018). *Guideline in salinity assessment, mitigation and adaptation using nuclear and related techniques* (p. 164). Springer. https://doi.org/10.1007/978-3-319-96190-3

Zhang, X., Brandt, M., Tong, X., Ciais, P., Yue, Y., Xiao, X., Zhang, W., Wang, K., & Fensholt, R. (2022). A large but transient carbon sink from urbanization and rural depopulation in China. *Nature Sustainability, 5*(4), 1–8. https://doi.org/10.1038/s41893-021-00843-y

Chapter 3
Climate-Driven Changes and Impact on Arid Lands Water Resources for Agriculture Production and Food Security

Hussain Abdullah, Adel Al-Haddad, and Abdallah Abusam

Abstract Arid lands are geographically distributed on the globe. They usually receive lower rainfall than the cold humid regions. Therefore, an arid region is often considered water scarce, in which crop production and consequently food security is significantly affected. The impact of climate change (emission of greenhouse gases and subsequent warming, low and infrequent rainfall) will further affect the volume and quality of water resources in arid lands. The reduced aquifer recharge and recycling of groundwater for irrigation will increase the marginality of the groundwater for better crop production, thus the crop production may be improvised and affects the food security. In this chapter a comprehensive review of the water resources (conventional and non-conventional) of the arid region countries are amassed, the impact of climate change on water resources highlighted, recommendations are given for strategic integrated water management in the arid region countries to intensify crop production for food security. Water withdrawal by agriculture sector to total withdrawal in arid region countries is presented. Finally, developmental policies and legal framework are suggested to mitigate the impact of climate change for sustainable use of water resources for food security.

Keywords Arid regions · Water scarcity · Food security · Agricultural production · Water management

3.1 Introduction

The challenge of balancing water demand against supply in dryland region countries are enormous, due to their location in hyper-arid, arid and semi-arid climate zones. The arid regions are characterized by low aridity index (Precipitation/Potential

H. Abdullah (✉) · A. Al-Haddad · A. Abusam
Water Research Center, Kuwait Institute for Scientific Research, Safat, Kuwait
e-mail: hsafar@kisr.edu.kw

M. K. Suleiman and S. A. Shahid (eds.), *Fostering Arid Lands Agriculture in the Face of Climate Change*, Advances in Global Change Research 81,
https://doi.org/10.1007/978-3-031-98490-7_3

evapotranspiration P/PET), high temperatures and solar radiation, low rainfall and atmospheric humidity, and high wind velocity. The weather in the aid regions is characterized by long summer season and short winter season. Short periods (around one month) of Autumn and Spring seasons exist in these areas. The maximum temperature in the summer season is above 50 °C, while the maximum winter temperature is around 0 °C.

Due to arid conditions, the soils are dry, low in soil fertility leading to low crop production. The sandy soils are common in desert environments, which have low water and nutrients retention capacities, leading to high drainage of water (Gaur & Squires, 2018). Based on the average rainfall, the drylands are divided into hyper-arid, arid, semiarid and dry subhumid regions (Adeel et al., 2005). Adeel et al. (2005) presented relative distribution of different zones on global scale basis. Dryland comprises of four zones, occupying 41.3% of the global terrestrial area.

The arid region countries are located in South East, West and Central Asia, Africa, Australia, South Africa, East and West South America, and West and South of North America.

3.2 Water Resources in Arid Countries

The water resources in the arid regions are divided into conventional natural water resources and non-conventional water resources. The natural water resources include surface water and groundwaters. Great and temporary rivers, lakes and natural springs are forms of surface water. Great rivers, Nile (Egypt), Tigris and Euphrates (Iraq), Amazon (Brazil), Indus (Pakistan), Bengal (Bangladesh), China river (Republic of Chine), Darling River (south east Australia) and the Orange river (south Africa) are found in the arid regions zones.

The surface water is used for domestic, irrigation, industrial and for other purposes. Most of the natural lakes are fed from the main rivers. Some of these lakes under high temperatures might be dried (for example Urmia Lake located in the north west of Islamic Republic of Iran). These lakes are used for irrigation, fishing, transportation, drinking water for human and animal consumption. In general, when the water table meets the ground surface, it produces natural springs (for example Jordanian spring).

3.2.1 Groundwater

Groundwater aquifers can be confined or unconfined. Groundwater aquifers are highly porous and permeable and composed of sandstones and carbonate rocks (Todd & Mays, 2005). The high permeability of the aquifers is due to the presence of the primary porosity and secondary/karstification porosity (Todd & Mays, 2005). The unconfined aquifers are consisting of surface permeable aquifers underline by impermeable layers such as clay. On the other hand, the confined aquifers consist of high permeable zones surrounded by impermeable rocks above and below the porous media. The Nubian sandstone aquifer is one of the main aquifers located in Egypt, Sudan, Libya and Chad, producing potable water in the arid regions (Groundwater). Carbonate groundwater aquifers are found in Kingdom of Saudi Arabia (KSA), Jordan, Syria, Lebanon and other Gulf Corporation Council Countries (GCCC). The main source of the groundwater is rain water, surface water and runoff.

Groundwater salinity is expressed either as electrical conductivity (millisiemens/cm) or as Total Dissolved Solids (TDS) in milligrams/liter. Based on the TDS levels the groundwater can be fresh (<1,000 mg/L-TDS), brackish (1,000–10,000 mg/L-TDS), saline (10,000–100,000 mg/L-TDS) and brine (more than 100,000 mg/L-TDS) (Todd & Mays, 2005). Some countries in the arid regions might have one or more types of groundwater. In Kuwait, four types of groundwater (fresh, brackish, saline and brine groundwater) are recognized, while Egypt has mostly potable (fresh water) quality groundwater. The thickness of the groundwater aquifers might be shallow (30 m depth) and deeper aquifers also exist with thickness range between 300 and 500 m. The groundwater may be contaminated due to leaching of fertilizers, pesticides, animal waste in the agricultural farm areas; crude oil and natural gas in the industrial oil refinery areas; contamination with dissolved gases such as hydrogen sulfide, methane, ammonium, microbial parameters in the residential areas (Mukhopadhyay et al., 2019). In general, the contaminated groundwater should be treated before being used for irrigation, industrial processes and landscape activates. The cost of the groundwater remediation should be considered when groundwater reclamation projects are executed (Al-Haddad et al., 2019).

3.2.2 *Non-conventional Water Resources*

Non-conventional groundwaters include desalinated (seawater) and municipal waste water treated at various levels.

3.2.2.1 Desalinated Water

Worldwide, most of seawater desalination plants are concentrated in the arid region countries where natural water resources are scare (Curto et al., 2021). The largest capacities of the seawater desalination plants are located in the GCC countries, where, in 2022, the volume of desalinated water reached about 7 billion m^3, however, the volume of rainfall of the region reached around 227 billion m^3 in 2019 (Statista, 2022). High number of water desalination plants are located in the middle east, north Africa, east China and south west Australia. The capacities ranged between 10,000 m^3/day to more than 250,000 m^3/day. In these plants, different desalination technologies are being used including, (i) Reverse Osmosis (RO) membrane, (ii) Multiple-Effect Distillation (MED), (iii) Multi-Stage Flash Desalination (MSF) and (iv) Electrodialysis Desalination (ED). Some countries (e.g. Kuwait and KSA)

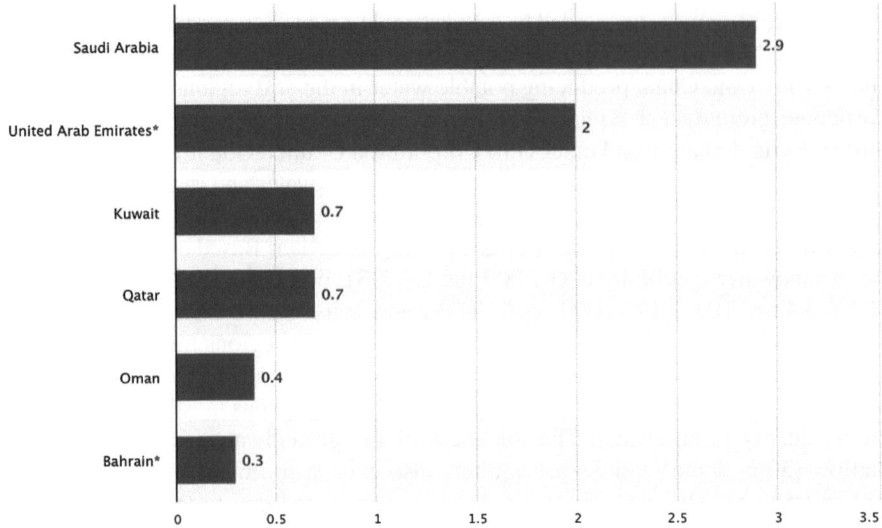

Fig. 3.1 Volume of desalinated water (billions cubic meters) produced in the GCC countries in 2022 (Statista, 2022)

tried to treat the brine water produced from seawater desalination plants through the process of zero liquid discharge technologies (Ahmed et al., 2023), which recrystallized all the salts. The desalination plants are also often used to desalinate brackish groundwater and high saline river water.

Seven billion of cubic meter water is desalinated in the GCC countries (Fig. 3.1), Kingdom of Saudi Arabia (KSA) is producing the largest volume (2.9 billion m^3/year), followed by United Arab Emirates (2 billion m^3/year) and the remaining countries produce less than 1 billion m^3/year (Statista, 2022). To produce potable water, the desalinated water in these countries is mixed with the groundwater and disinfected by chlorine before it is sent to potable water towers or subsurface portable storage tanks. From these tanks, the potable water is sent to the houses in the residential areas. Also, the potable water is sent to the agricultural farm areas and to the factories in the industrial and commercial sectors (Ahmed et al., 2023). Moreover, in some countries the desalinated water is injected into groundwater aquifers where it will be mixed with the groundwater to produce mineral potable groundwater (Curto et al., 2021). In general, the rich countries rely mainly on the high costly technology to produce desalinated water, while the poor countries usually try to use either groundwater or surface water to satisfy all water demands.

3.2.2.2 Treated Wastewater

Some arid countries (e.g. India, Pakistan, Iran, Jordan) have separate greywater and blackwater systems (Tchobanoglous et al., 2014). The blackwater (sewage water) is passed through sewage network through sewage pumping and lifting stations towards final sewage treatment plants, whereas the greywater is treated and reused in-situ (e.g. residential houses, government buildings, schools, mosques) for flushing toilet systems, landscape irrigation and other applications such as car washing etc. In these countries, the greywater network is separated from sewage network at source. Then, the greywater will be treated using technologies such as sand and carbon filtration and disinfection systems (Abusam et al., 2022).

The collection, production, treatment and reuse capacities of treated waste water (TWW) in different regions and countries varies (Jones et al., 2021) and ranging between 0.1 million m^3/year to more than 5 million m^3/year. The largest wastewater treatment plants are located in the arid regions where limited natural water resources exist such as GCC countries. The sewage water is passed from residential and industrial sectors connected to the pumping sewage stations.

In the arid regions, almost all consumed potable water is transferred to wastewater. For example, the potable water used in the factories to produce different products, and the output of the generated wastewater from these factories is considered as industrial wastewater. On the other hand, the potable water consumed in the residential areas for drinking, washing, cooking, swimming, are shifted to a domestic wastewater. Moreover, the potable water consumed for agricultural purposes in some cases led to increase the salinity of groundwater (Al-Qallaf et al., 2011).

3.2.2.3 Conventional Municipal Wastewater Treatment Method

The conventional wastewater treatment method for all plants is known as activated sludge processes, which involves several stages including, primary, secondary, and tertiary (The Ministry of Public Works, 2021). In the *primary* stage, the physical treatment processes are used, including metal screens and grit removal tanks. The screens filter out floating objects like cans and plastic materials from the municipal wastewater influent to prevent machinery fouling. The grit removal tanks allow heavy particles such as sand and gravel to settle at the base of the settling tank. The presence of these objects can damage treatment plant machinery and increase the load on liquid pumps, causing fouling and process failure. Additionally, oil and dissolved gases are removed during the preliminary stage. Primary settling tanks are only used when the raw wastewater is characterized by high concentrations of suspended solids.

The *secondary* stage is a biological treatment involving aeration tanks and clarifiers (secondary settling tanks) (The Ministry of Public Works, 2021). During aeration, air diffusers are usually placed at the bottom of the tank to activate the microbial community, which consumes mainly biological and nitrogenous pollutants, producing carbon dioxide and other byproducts. The microorganisms vary in mass density and require separation in the secondary clarifier (The Ministry of Public Works, 2021). This separation process removes floating or settled sludge from biologically treated wastewater.

The *tertiary* stage involves filtration (disc or sand filters) followed by disinfection to completely eliminate pathogenic microorganisms. However, disinfection of viruses is usually not addressed in the tertiary stage. The municipal wastewater treatment plants (WWTPs) commonly use chlorine or UV disinfection methods (The Ministry of Public Works, 2021). Tertiary TWW retains nutrients and minerals, making it suitable to irrigate urban landscape, and selected fodder crops. In the UAE, KSA, and Kuwait, the tertiary treatment stage is followed by membrane advanced filtration stage. The Sulaibiya wastewater treatment plant in Kuwait, the largest membrane plant in the world produces 460,000 m^3/day. The advance treatment stage at Sulaibiya plant includes ultra-filtration stage (UF) followed by reverse osmosis membrane (RO), which uses a porous membrane to purify water of particles larger than 0.1 nm. The quality of RO-treated water is comparable to drinking water in chemical composition but is currently used to irrigate crops in agricultural farms (Abdali and Al-Wafra areas).

The TWW is used for multiple applications, including irrigating urban landscapes, public parks, firefighting stations, oil exploration activities, cooling towers, groundwater recharge aquifers, shrimp/fish hatching, concrete block for buildings/houses, and irrigating fodder crops for animal use (The Ministry of Public Works, 2021). In the biological treatment stage, where there are anaerobic basins, sludge is treated anaerobically, consequently methane and hydrogen gases are generated and can be used as energy source for the treatment plants (The Ministry of Public Works, 2021).

3.2.2.4 Harvested Rain Water

Several countries in the arid regions are harvesting the rain water, store in reservoirs and use to irrigate crops. For example, in Australia, 25% of the water resources in the country are generated from rain water harvesting (Preeti et al., 2019). Recently, heavy rains in the Sultanate of Oman and UAE caused flooding and disasters (Aklan et al., 2024). In these countries, vast flooding is created in the residential areas, that has affected the sewage network and wastewater treatment plants in the form of over flow from wastewater tanks. Other countries such as Islamic Republic of Iran and KSA, created permanent dams to utilize the runoff during the summer drought periods for agricultural activities (Ahmad et al., 2024).

3.3 The Government Framework

There should be government framework for water governance to assure the water resources are used sustainably and rationally.

3.3.1 Legal Framework

Due to the limited natural water resources in the arid areas, the government of these countries started to setup laws and regulations to control the consumption of different water resources (desalinated water, groundwater, treated wastewater) for different applications such as agricultural, industrial, residential, construction and others. Also, the governments set a list of priorities for the available water resources in these arid countries. For example, most of the available groundwater resources in north Africa (Libya, Algeria, Tunisia) are limited for use in residential areas and not for vast agricultural activities (Ringler & Zhu, 2015).

3.3.2 Laws and Regulations

Kuwait, a hyperarid country sets up the regulations for the utilization of different water types. The Kuwait Environment Protection Authority (KEPA) has setup standards for the discharge of raw wastewater into the public sewage network (Table 3.1) and the reuse of TWW for irrigation purposes (Table 3.2) (Kuwait Environment Public Authority, 2021), assuming the waste water is safe for use from water salinity and sodicity hazards.

Further, the Ministry of Electricity and Water and Renewable Energy Kuwait (MEWRE, 2022) has setup the standards for the type, depth and quality of the groundwater aquifer consumption for agricultural activities.

3.3.3 Groundwater and Aquifer Recharge

Rainfall in the arid regions may lead to replenish the groundwater aquifers, raises the level and increase the water volume in the aquifer. In addition, new potable groundwater zones may be developed over the old brackish/saline groundwater. The groundwater is used for irrigation purposes (farms and trees) and to compensate water shortage during drought conditions (Al-Qallaf et al., 2011).

Studies have been conducted to recharge the groundwater using different sources of waters (treated water, desalinated water, tertiary treated wastewater and RO treated wastewater) (Mukhopadhyay et al., 2019). The aquifer recharge method often creates a new groundwater storage area in the aquifer and generate high quality mineral groundwater (Mukhopadhyay et al., 2019). In arid areas, however, aquifer recharge systems are used mainly during emergency periods, especially in summer seasons.

Table 3.1 Guidelines for water discharged into sewers (Kuwait Environment Public Authority, 2021)

Parameter	Symbol	Unit	Maximum value
pH	pH	–	6.5–8.0
Electrical conductivity	EC	µS/cm	2,000
BOD_5 (5 day, 20 °C)	BOD_5	mg/L	500
COD (dichromate)	COD	mg/L	700
Total suspended solids	TSS	mg/L	300
Ammonia	NH_3-N	mg/L	55
Phosphate	(PO_4-P)	mg/L	40
Total Kjeldahl nitrogen	TKN	mg/L	70
Phenol	C_6H_6O	mg/L	1
Fluoride	F	mg/L	15
Sulfide	S	mg/L	10
Arsenic	As	mg/L	0.1
Cadmium	Cd	mg/L	0.1
Cyanide	CN^-	mg/L	0.1
Chromium	Cr	mg/L	1
Nickel	Ni	mg/L	0.2
Mercury	Hg	mg/L	0.002
Copper	Cu	mg/L	0.5
Zinc	Zn	mg/L	2
Lead	Pb	mg/L	0.5
Silver	Ag	mg/L	4
Tar and tar oil	–	mg/L	Nil
Floating oil and grease	–	mg/L	5
Emulsified oil and grease	–	mg/L	5
Faecal coliform bacteria	F.C.	Cfu/100 mL	1000
Escherichia coli	E. coli	Cfu/100 mL	500
Faecal streptococci bacteria	F.S.	Cfu/100 mL	250
Egg parasites	–	Ova/L	Nil
Worm parasites	–	–	Nil

3.3.3.1 Brackish/Saline Water Groundwater

Not all the groundwater in the arid regions have potable water quality. In Kuwait there are four types of groundwaters (fresh, brackish, saline and brine) (Misak & Hussain, 2022). The brackish water is used for agricultural activities, whereas the saline water can be treated with low desalination technology costs for use in agricultural activities. The quality of the potable, brackish and saline waters are chemically characterized

Table 3.2 Guidelines to use water for irrigation (Kuwait Environment Public Authority, 2021)

Parameter	Symbol	Unit	Maximum values
pH	pH	–	6.5–8.5
BOD_5 (5 day, 20 °C)	BOD_5	mg/L	20.0
COD (dichromate)	COD	mg/L	100
Dissolved oxygen	DO	mg/L	>2.0
Residual chlorine	Cl_2	mg/L	0.5–1.0
Floatables	–	mg/L	Nil
Oil and grease	–	mg/L	5.0
Total suspended solids	TSS	mg/L	15.0
Total dissolved solids	TDS	mg/L	1500.0
Phosphate	$(PO_4\text{-}P)$	mg/L	30.0
Ammonia	$NH_3\text{-}N$	mg/L	15.0
Total Kjeldahl nitrogen	TKN	mg/L	30.0
Total nitrogen	TN	mg/L	65.0
Phenol	C_6H_6O	mg/L	1.0
Fluoride	F	mg/L	2.0
Sulfide	S	mg/L	0.1
Aluminum	Al	mg/L	5.0
Arsenic	As	mg/L	0.1
Barium	Ba	mg/L	2.0
Boron	B	mg/L	2.0
Cadmium	Cd	mg/L	0.01
Chromium	Cr	mg/L	0.15
Nickel	Ni	mg/L	0.2
Mercury	Hg	mg/L	0.002
Cobalt	Co	mg/L	0.2
Iron	Fe	mg/L	5.0
Antimony	Sb	mg/L	1.0
Copper	Cu	mg/L	0.2
Manganese	Mn	mg/L	0.2
Zinc	Zn	mg/L	2.0
Lead	Pb	mg/L	0.5
Total petroleum Hydrocarbons	TPH	mg/L	5.0
Faecal coliform bacteria	F.C.	Cfu/100 mL	100.0
Escherichia coli	E. coli	Cfu/100 mL	50.0
Faecal streptococci bacteria	F.S.	Cfu/100 mL	50.0
Egg parasites	–	–	Nil

as sodium/calcium bicarbonate water type, sodium/calcium sulphate water type and sodium/chloride water types, respectively (Misak & Hussain, 2022).

3.3.3.2 Treated Wastewater

Treated wastewater (secondary or tertiary treated) is mainly injected in the coastal groundwater aquifers to create hydrologic barrier to control seawater intrusion (Todd & Mays, 2005), e.g. this practice is common in some North African countries (Libya, Tunisia and Algeria). Some companies in the arid areas such as Kuwait Oil Company (KOC) inject the RO treated wastewater in the oil aquifers to stabilize the aquifers after gas and oil production (The Ministry of Public Works, 2021).

3.3.3.3 Water Production and Consumption

In Kuwait, the ministries supervising the activities of the desalination plants, drilling and production of groundwater as well as TWW usually produce annual statistics on the fresh water, groundwater (Tables 3.3 and 3.4) (MEWRE, 2022). The minimum daily gross consumption of fresh water during the years 2003 to 2022 has increased uniformly (Fig. 3.2).

3.4 Agricultural Production in Arid Areas

Agriculture in arid lands countries faces several problems including high temperatures and associated evapotranspiration, water scarcity, desertification, erosion and infertile soils (Shahid, 2023a, 2023b; Suleiman & Shahid, 2023a, 2023b), as well as the climate change impacts. Nonetheless, there are several potential ways to enhance the agriculture production in arid regions. In fact, arid areas can provide unique chances for agricultural plants adapted in dry environments. Soil can be exploited along with innovative irrigation technologies (Zaman et al., 2018a, 2018b), which needed for plant growth and production of large quantities of crops. New techniques such as utilization of aquaculture and vertical farming can be adopted under controlled protected agriculture system to improve the water use efficiency and to produce vegetables production throughout the year. Solar energy can be produced and used in extracting groundwater to irrigate crops. Moreover, scientific research along with the experiences of agricultural specialist should be used to improve the agricultural technologies in these areas and to increase productivity.

Table 3.3 Total production of fresh water (million imperial gallons) in Kuwait during 1993–2022 (MEWRE, 2022)

Year	Ministry of electricity and water and renewable energy	Kuwait national petroleum company	Total production	Percentage of annual increase/decrease
Million imperial gallons				
1993	49,743	1006	50,749	–
1994	57,203	984	58,187	14.7
1995	61,546	1151	62,697	7.8
1996	67,674	971	68,645	9.5
1997	73,306	1024	74,330	8.3
1998	78,367	1035.0	79,402	6.8
1999	84,214	1001.0	85,215	7.3
2000	88,475	957	89,432	4.9
2001	91,535	1151	92,686	3.6
2002	97,640	1036	98,676	6.5
2003	102,057	1065	103,122	4.5
2004	104,675	1033	105,708	2.5
2005	111,502	879	112,381	6.3
2006	114,600	914	115,514	2.8
2007	119,774	921	120,695	4.5
2008	128,061	720	128,781	6.7
2009	131,729	510	132,239	2.7
2010	134,128	365	134,493	1.7
2011	138,094	366	138,460	2.9
2012	140,423	330	140,753	1.7
2013	141,515	236	141,750	0.7
2014	144,897	138	145,035	2.3
2015	150,150	2	150,152	3.5
2016	158,113	0	158,113	5.3
2017	160,236	0	160,236	1.3
2018	159,828	0	159,828	−0.3
2019	160,430	0	160,430	0.4
2020	167,772	0	167,772	4.6
2021	164,663	0	164,663	−1.9
2022	164,265	0	164,265	−0.2

Table 3.4 Total production of brackish water (MIG) in Kuwait during 1993–2022 (MEWRE, 2022)

Year	Ministry of electricity and water and renewable	Kuwait oil company	Total production	Percentage of annual increase/decrease
Million imperial gallons				
1993	20,857	1113	21,970	–
1994	23,660	1319	24,979	13.7
1995	20,181	1497	21,678	−13.2
1996	22,010	1667	23,677	9.2
1997	23,978	1839	25,817	9.0
1998	26,076	1842	27,918	8.1
1999	26,121	1548	27,669	−0.9
2000	28,204	2283	30,487	10.2
2001	30,098	1882	31,980	4.9
2002	30,743	1978	32,721	2.3
2003	32,011	2066	34,077	4.1
2004	33,160	1791	34,951	2.6
2005	33,912	2134	36,046	3.1
2006	33,865	2349	36,214	0.5
2007	32,315	2088	34,403	−5.0
2008	29,284	2013	31,297	−9.0
2009	28,334	1946	30,280	−3.2
2010	22,628	1805	24,433	−19.3
2011	21,622	1617	23,239	−4.9
2012	21,231	1226	22,457	−3.4
2013	18,964	1528	20,492	−8.8
2014	17,996	1663	19,659	−4.1
2015	16,438	1826	18,264	−7.1
2016	17,008	1725	18,733	2.6
2017	15,006	1681	16,687	−10.9
2018	15,127	1578	16,705	0.1
2019	13,956	1775	15,731	−5.8
2020	14,669	1707	16,376	4.1
2021	14,327	1842	16,169	−1.3
2022	13,773	1760	15,533	−3.9

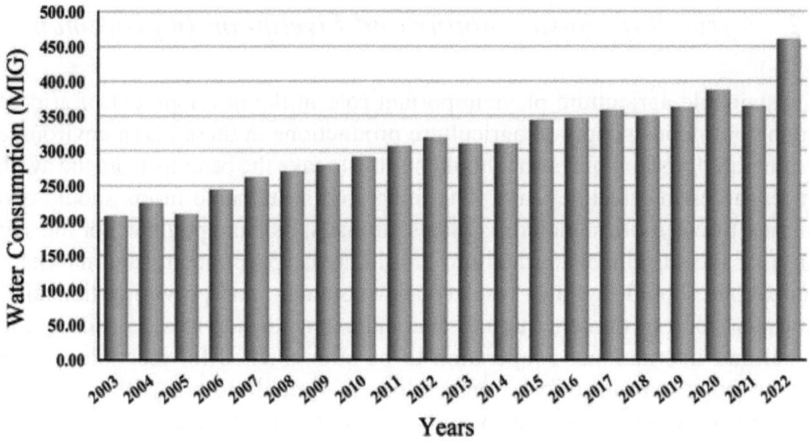

Fig. 3.2 Minimum daily gross consumption (MIG) of fresh water in Kuwait

3.4.1 Potential Agricultural Technologies for Arid Lands

Several technologies were developed and used in arid lands agriculture for efficient use of water resources to improve water productivities and crops intensification. These technologies include, use of modern irrigation systems (surface and subsurface drip, sprinkler etc.) using different brackish waters for crop production (Zaman et al., 2018a, 2018b), growing salt-tolerant and halophytes crops (Shahid & Alkandari, 2024), and managing salt-affected soils farms using integrated soil reclamation approach (Shahid & Rahman, 2011; Shahid et al., 2018b). The drip irrigation system is most useful for efficient irrigation and to control root zone soil salinity, as the system adds water in the root zone of the crop. Drip irrigation reduces the water losses, which are common in other irrigation systems (flood, sprinkler) through evaporation and wind action. The drip system adds water in the root zone based on the plants requirement and the water delivery is controlled through the installation of flow meters. In addition, it can be used effectively to provide nutrient fertilizers for plants, through fertigation (fertilizer added to irrigation water). Protected environment agriculture (greenhouses) technologies protect the environment and provide suitable temperature and humidity for plant growth. Following such methods, crops can be grown in arid areas, and protected from weather conditions such as high temperature and strong winds. Other technologies include, closed water cycle systems, which allows water to be recycled in efficient manner.

3.4.2 Agriculture Sustainability and Livelihood Improvement

The sustainable agriculture plays important role in the development of arid areas, through continuous supply of agriculture productions in these harsh environments. The integrated sustainable farms should be able to take the benefits from the available farm resources in effective ways. The integrated sustainable farms should diverse the agriculture activities, including plants, animals, fish and poultry. For the sake of supporting the sustainable agriculture in arid areas, governments and international organizations should invest in research, development, and providing financial and technical support to farmers through national extension services. In addition, local, regional and international cooperation to be strengthened for technology transfer and knowledge sharing to foster innovations in agriculture field. One of the direct benefits of agriculture farming in arid areas is to engage the work force among different age groups and women. This improves the level of income, especially in African countries. Therefore, sustainable agricultural production will improve the quality of farmers' lives.

3.4.3 Agriculture and Food Security

Agriculture plays important role in providing food security and other benefits to local farmers. It helps directly in protecting biological diversity, rationalization of water resources, utilization and reducing the magnitude of the desertification problem (Shahid, 2023a, 2023b; Shahid & Omar, 2001). Moreover, agriculture production provides food locally, reduces food import and saves national financial resources.

Water is essential for life and sustainable food production to meet the demand of escalating population. The processes connected with food security include food preparation (fish, shrimp and water organism's growth farms) and food production at commercial and domestic levels (Ericksen et al., 2010). As a result of the increase in world population, the need for water will increase in agriculture, industrial and household uses (drinking, shower, cooking). In addition, 80% of raw wastewater passed to the environment without treatment causing public health diseases (FAO-IWMI, 2018). Also, providing high quality water is very important for drinking, cooking, shower, wastewater engineering and personal care.

In some countries, the food crops (rice, wheat and sugarcane) in general are categorized under irrigated agriculture, whereas, other crops (wheat, corn and soya bean) as rainfed (Ringler & Zhu, 2015). Most of the soya bean is used for animal food (poultry, fish farms), while the remaining soya is used for vegetable oil production and as food for human consumption.

Water is also used for industrial activities and production processes. The main industrial connection between water security and food security is electricity generation, including water energy. Renewable energies (solar energy, wind energy, hydro-energy and wastewater energy) represent more than three-quarter of international renewable energy (Zeng et al., 2017). Renewable energy can be used for electricity for houses, pumping groundwater wells, operation of irrigation systems, pumping water tanks in the agriculture, fish and poultry farms. In general, food security for of a country in arid areas depends on several factors, including financial resources, water resources (seawater, surface water, groundwater, treated domestic and industrial wastewater), weather (temperature, rainfall), soil characteristics, types of crops planted, and food transportation facilities.

3.5 Strategies to Overcome Agriculture Challenges in Arid Areas

There is a set of general strategies to overcome the challenges of food production on global scale. The selection of objective oriented and tailor made strategies to address specific condition of region is essential to take maximum advantages from production systems and conservation of resources. These are briefly described below.

3.5.1 Water Stress and Water Conservation Strategy

Water stress is the share of freshwater withdrawal in available freshwater resources, after taking into account environmental water requirements. Nearly all countries in the Near East and North Africa (NENA) region have water stress levels close to or above 100%, with the higher shares in Kuwait, the United Arab Emirates and Saudi Arabia (FAO, 2023). These three countries are located in the Arabian Peninsula and are withdrawing each year 9 to almost 40 times their renewable freshwater resources available (FAO, 2023), this necessitates these countries to use non-renewable water resources which are diminishing rapidly. Share of water withdrawal by agriculture in total withdrawal of selected countries is presented in Table 3.5.

The water stress countries in the arid region need special attention for water conservation and uses. The water conservation strategies in the arid countries should be based on an integrated approach, that considers meeting future water demand, and projection of future water supplies. The strategy embraces all water supplies (natural resources, desalination and treated wastewater) and water used. To implement the strategy, objective oriented initiatives are to be developed for application over a period to achieve the water strategies goals. The strategy should improve the management of the nation's precious water resources and enhance their contribution to the economic growth of the country.

Table 3.5 Share of water withdrawal by agriculture in total withdrawal (%) in selected countries (FAO, 2023)

Country	% withdrawal in 2000	% withdrawal in 2020
Afghanistan	98.6	98.2
Australia	69.4	62.3
Bangladesh	87.8 (2010)	87.8
Bahrain	49.1	33.3
China	68.8	62.1
Egypt	86.4	79.2
Eritrea	70.1	94.5
India	91.5	90.4
Iran	93.8	92.2
Iraq	78.8	78.1
Jordan	66.2	51.6
Kuwait	53.6	62.3
Oman	90.4	80.8
Pakistan	94.3	94.0
Qatar	71.5	36.4
Saudi Arabia	87.7	81.6
Somalia	99.7	99.5
Sudan	96.2 (2015)	96.2
United Arab Emirates	78.51	46.5

3.5.2 *Water Withdrawal by Agriculture Compared to Total Water Withdrawal*

In 2020, the share of agricultural water withdrawals in total withdrawals was greater than 90% in 29 countries (FAO, 2023). The top water withdrawal countries in 2020 are, Somalia in Africa (over 99.7%), Afghanistan and Nepal in Asia (over 98.2% and 98.1% respectively) and Guyana in South America (94.4%). Comparison of percent water withdrawal by agriculture compare to total water withdrawal of selected water scarce countries is shown in Table 3.5.

3.5.3 *Improvement of Water Utilization*

Water scarcity is a general challenge in arid regions. Without sufficient water, agriculture can be at risk for cost-effective and optimum production. Therefore, it is important to improve the existing irrigation technologies in order to save water and utilize it efficiently. As stated above, all marginal quality water resources (brackish, saline, treated wastewater etc.) should be brought into production systems, along with the use of modern irrigation systems (Zaman et al., 2018b). It is to be noted that water scarcity is likely to increase due to climate change, only viable source of water which is likely to be increased over period of time is the wastewater, therefore, its use to be recognized and policy developed for sustainable uses without compromising the soil quality and risking human health. Use of modern irrigation system, and establishing mini-dams at farmers field should be encouraged, through the provision of financial support and subsidies.

3.5.4 Utilization of Suitable Crops

Climate smart and regenerative agriculture (Savills Research, 2021), region specific tailor made practices are to be adopted in harsh climatic conditions of arid regions for maximum crop production. In this regards, the responsible national extension services must provide training to farmers for technology transfer and use of new technologies. Drought and salt-tolerant crops under the theme of biosaline agriculture (Shahid & Rahman, 2011) to be demonstrated at farmer's field to gain farmers trust in the utilization of specific production system.

3.5.5 Utilization of High Yielding Crop Varieties

Technologies such as agriculture-modified genetics to improve crops ability to overcome harsh environments and consequently enhance productivity should be introduced. There is a huge responsibility of research and extension sectors to produce large quantities of seeds of high yielding crops and timely provision of seeds to farmers, in addition to other extension services including the use of seed rate, planting method, land preparation, type and time of fertilizer application, harvesting methods and access to the market.

3.5.6 Integrated Water Management

All water resources (including, seawater, surface water, rainwater, runoff, groundwater, treated domestic and industrial wastewater) should be identified and protected from contamination. They also should be utilized in effective ways, which allow their use for longer period. For example, rainwater and runoff should be collected during rainy seasons and stored in basin and dams for use during drought periods. Within Integrated Water Management (IWM), several solutions are suggested to overcome the problem of limited natural water resources in arid areas. They include the following:

- Mixing of desalinated water with brackish groundwater to produce potable water for drinking. Further, the desalinated water to be injected into groundwater aquifers to produce potable mineral water for firefighting, cooling towers, cleaning activities in industrial sectors, and for oil exploration activities.
- Utilization of both treated domestic and industrial wastewater in agricultural, industrial and residential activities. For example, the RO treated wastewater can be mixed with brackish groundwater to produce potable water to irrigate crops, vegetables for human consumption and grasses for animal food. In addition, the wastewater and sludge can be used for energy generation, fertilizer extraction (nitrogen and phosphorus) and soil enhancement products.
- Separation of treated greywater network from sewage network (e.g. in houses, governmental buildings and commercial malls) and recycling greywater for flushing toilets, garden irrigation and carwash. Such practices will reduce the cost and consumption of potable water in these buildings.
- More rainwater harvesting studies should be conducted in arid countries (Ali & Sang, 2023). Further, harvested rainwater should be used beneficially, e.g. drinking and or agricultural irrigation.
- Brackish and saline groundwater to be treated using small scale RO plants, and used in animal and poultry farms. The treated groundwater can be utilized for growth of fish and shrimp farms in desert areas as part of food security.

3.6 Conclusions and Recommendations

Treated wastewater is a future resource for arid land countries facing water scarcity for food production. This water should be monitored for gross pollutant parameters, and for use in agriculture its quality to be assessed for nutrients and microbiological characteristics. Sustainable agriculture production is challenge in drylands, where, there are harsh climatic conditions and in general water scarcity. However, with technological advancements to treat the marginal waters (brackish/saline and municipal wastewater), it is now possible to use these waters for crop production, and poultry and animal farms. Under harsh environmental conditions of arid regions, the extension services to work closely with farmers to introduce the adoption of climate smart and regenerative agriculture practices for long term and sustainable crop production and conservation of soil and water resources and environmental protection. Sustainable agriculture requires innovative thinking and utilization of new technologies to obtain sustainable agriculture production in arid areas. Mainly, it requires utilization of new irrigation systems, protection of water resources and using solar energy. In addition, the governmental and non-governmental organizations and farmers should cooperate in finding solutions and sustainable strategies to deal with limited resources, climatic changes, and soil enhancement in these regions. It is also important to create database for soil, lands, water, climate, and crops for arid countries.

Acknowledgements The authors would like to thank the management of the Kuwait Institute for Scientific Research for providing support and granting permission to publish the data related to Kuwait. A special thank goes to Mr. Naser Al-marri from National Scientific and Technical Information Center, KISR, and Mr. El-Sayed Ebrahim, from Water Research Center, KISR, for their assistance in information collection.

References

Abusam, A., Abdullah, H., Al-Haddad, A., Al-Ajeel, F., Dashti, F. M., & Khajah, M. (2022). *Investigations into the characteristics of household light greywater in Kuwait* (Final report No. 17590). Kuwait Institute for Scientific Research.

Adeel, Z., Safriel, U., Niemeijer, D., & White, R. (2005). Millennium ecosystem assessment. In *Ecosystem and human well-being*. Desertification synthesis. https://www.millenniumassessment.org/documents/document.355.aspx.pdf; https://interiorscience.tech/uk/map-of-the-rivers-in-the-world.html

Ahmad, B. A., Salar, S. G., & Shareef, A. J. (2024). An integrated new approach for optimizing rainwater harvesting system with dams site selection in the Dewana Watershed, Kurdistan Region, Iraq. *Heliyon, 10*(6), e27273. https://doi.org/10.1016/j.heliyon.2024.e27273

Ahmed, M. M., Alambi, R. K., Bhadrachari, G., Al-Muqahwi, S., & Thomas, J. P. (2023). Design and optimization of a unique pilot scale forward osmosis integrated membrane distillation system for seawater desalination. *Journal of Environmental Chemical Engineering, 11*(4), 109949. https://doi.org/10.1016/j.jece.2023.109949

Aklan, M., de Fraiture, C., & Hayde, L. G. (2024). Why we should revitalize indigenous water harvesting systems: Lessons learned. *International Soil and Water Conservation Research.* https://doi.org/10.1016/j.iswcr.2024.05.004.pp.12

Al-Haddad, A., Mukhopadhyay, A., Al-Qallaf, H., Rashid, T., Ahmed, M. E., & Mydlarczyk, A. (2019). *Groundwater monitoring and remediation in the oil refinery areas of Kuwait* (Final report KISR No. 15661). Kuwait Institute for Scientific Research.

Ali, S., & Sang, Y. F. (2023). Implementing rainwater harvesting systems as a novel approach for saving water and energy in flat urban areas. *Sustainable Cities and Society, 89*, 104304. https://doi.org/10.1016/j.scs.2022.104304

Al-Qallaf, H., Al-Haddad, A., Rashid, T., Bushehri, A., Bhandary, H., & Al-Salman, B. (2011). *The effect of using the RO treated wastewater on the soil and groundwater at Al-Abdally Farms* (Final report KISR No. 10781). Kuwait Institute for Scientific Research.

Curto, D., Franzitta, V., & Guercio, A. (2021). A review of the water desalination technologies. *Applied Sciences, 11*(2), 670. https://doi.org/10.3390/app11020670

Ericksen, P., Steward, B., Dixon, J., David, B., Loring, P. A., Anderson, M., & Ingram, J. S. (2010). The value of a food system approach. In J. Ingram, P. Ericksen, & D. Liverman (Eds.), *Food security, food systems and global environmental change* (pp. 25–45). Earthscan.

FAO. (2023). *AQUASTAT dissemination system.* AQUASTAT. https://data.apps.fao.org/aquastat/?lang=en. https://doi.org/10.4060/cc8166en-fig09

FAO-IWMI. (2018). More people, more food, worse water. Water pollution from agriculture. In J. Mateo-Sagasta, S. M. Zadeh, & H. Turral (Eds.), *A global review of water pollution from agriculture, Rome, Italy* (p. 224). ISBN: 978-92-5-130729-8 (FAO).

Gaur, M. K., & Squires, V. R. (2018). Geographic extent and characteristics of the world's arid zones and their peoples. In M. K. Gaur, & V. R. Squires (Eds.), *Climate variability impacts on land use and livelihoods in drylands* (Ch. 1, pp. 3–20). https://doi.org/10.1007/978-3-319-56681-8_1

Jones, E. R., van Vliet, M., Qadir, M., & Bierkens, M. F. P. (2021). Country-level and gridded estimates of wastewater production, collection, treatment and reuse. *Earth System Science Data, 13*(2), 237–254. https://doi.org/10.5194/essd-13-237-2021

Kuwait Environment Public Authority. (2021). https://epa.gov.kw/Portals/0/PDF/qarar12-2017.pdf

MEWRE (Ministry of Electricity and Water & Renewable Energy). (2022). https://www.mew.gov.kw/media/xsjdbhjx/water-annual-report-2022-for-qr.pdf

Misak, R., & Hussain, W. (2022). Groundwater in Kuwait. In *The geology of Kuwait* (pp. 199–214). Springer International Publishing. https://doi.org/10.1007/978-3-031-16727-0_9

Mukhopadhyay, A., Al-Khalid, A., Al-Senafy, M., Al-Haddad, A., Akber, A., Al-Fahad, K., Bhandary, H., Rashid, T., Al-Jumma, M., Al-Salman, B., & Shishter, A. (2019) *Field assessment of the use of RO-treated wastewater for artificial recharge of Dammam formation aquifer in Kuwait* (Final report KISR No. 15591). Kuwait Institute for Scientific Research.

Preeti, P., Jakhar, S., & Rahman, A. (2019). Rainwater harvesting in Australia using an Australia-wide model: A preliminary analysis. In *Proceeding of second international conference on water and environmental engineering (ICWEE 2019), 19–22 January, Dhaka, Bangladesh* (pp. 253–262).

Ringler, C., & Zhu, T. (2015). Water resources and food security. *Agronomy Journal, 107*(4), 1533–1538. https://doi.org/10.2134/agronj14.0256

Savills Research. (2021). *Regenerative agriculture.* UK-Rural February 2021. www.savills.co.uk/insight-and-opinion/

Shahid, S. A. (2023a). Potential threats to soil functions and mitigation options for sustainable uses. In M. K. Suleiman, & S. A. Shahid (Eds.), *Terrestrial environment and ecosystems of Kuwait* (Ch. 1, pp. 3–20). Springer. https://doi.org/10.1007/978-3-031-46262-7_1

Shahid, S. A. (2023b). Desertification: A central problem to restore ecosystems. In M. K. Suleiman, & S. A. Shahid (Eds.), *Terrestrial environment and ecosystems of Kuwait* (Ch. 5, pp. 119–137). Springer. https://doi.org/10.1007/978-3-031-46262-7_5

Shahid, S. A., & Alkandari, A. J. (2024). Halophytic crops as a solution for food security, land rehabilitation, and mitigating future water crises by utilizing marginal quality waters. In J. C. Dagar, S. R. Gupta, & A. Kumar (Eds.), *Halophytes vis-à-vis saline agriculture* (Ch. 17, pp. 441–478). Springer. https://doi.org/10.1007/978-981-97-3157-2_17

Shahid, S. A., & Omar, S. A. S. (2001). Causes and impacts of land degradation in the arid environment of Kuwait. In E. M. Bridges et al. (Eds.), *Response to land degradation* (pp. 76–77).

Shahid, S. A., & Rahman, K. (2011). Soil salinity development, classification, assessment and management in irrigated agriculture. In M. Passarakli (Ed.), *Handbook of plant and crop stress* (3rd ed., Ch. 2, pp. 23–39). CRC.

Shahid, S. A., Zaman, M., & Heng, L. (2018a). Introduction to soil salinity, sodicity, and diagnostics techniques. In Zaman et al. (Eds.), *Guidelines for salinity assessment, mitigation and adaptation using nuclear and related techniques* (Ch. 1, pp. 1-42). Springer. https://doi.org/10.1007/978-3-319-96190-3_1

Shahid, S. A., Zaman, M., & Heng, L. (2018b). Salinity and sodicity adaptation and mitigation options. In M. Zaman et al. (Eds.), *Guideline for salinity assessment, mitigation and adaptation using nuclear and related techniques* (Ch. 3, pp. 55–88). Springer. https://doi.org/10.1007/978-3-319-96190-3_3

Shahid, S. A., Burezq, H. A., & Baron, H. J. (2024). *Farm land salinity risk assessment to develop a management strategy for food security.* Project funded by Kuwait Foundation for the Advancement of Sciences (KISR Report No 18404). Kuwait Institute for Scientific Research, Kuwait.

Statista. (2022). *Production volume of desalinated water in the Gulf Cooperation Council region in 2022.* https://www.statista.com/statistics/800286/gcc-production-volume-desalinated-water-by-country/

Suleiman, M. K., & Shahid, S. A. (2023a). Prospective of agricultural farming in Kuwait and energy-food-water-climate nexus. In M. K. Suleiman, & S. A. Shahid (Eds.), *Terrestrial environment and ecosystems of Kuwait* (Ch. 15, pp. 363–391). Springer. https://doi.org/10.1007/978-3-031-46262_15

Suleiman, M. K., & Shahid, S. A. (2023b). Agricultural water footprint of major crops grown in Kuwait compared to the world average: A review. In M. K. Suleiman, & S. A. Shahid (Eds.), *Terrestrial environment and ecosystems of Kuwait* (Ch. 16, pp. 393–414). Springer. https://doi.org/10.1007/978-3-031-46262_16

Tchobanoglous, G., Stensel, H. D., Tsuchihashi, R., & Burton, F. (2014). *Wastewater engineering: Treatment and resource recovery* (5th ed., p. 155). Metcalf & Eddy, Inc. ISBN13: 9780073401188.

The Ministry of Public Works Portal. (2021). *Ministry projects*. Retrieved March 10, 2021, from http://www.mpw.gov.kw/sites/ar/Pages/InformationalServices/Projects.aspx?ProjectType=5

Todd, D. K., & Mays, L. W. (2005). *Groundwater hydrology* (3rd ed.). John Wiley & Sons.

Zaman, M., Shahid, S. A., & Heng, L. (2018a). *Guideline for salinity assessment, mitigation and adaptation using nuclear and related techniques* (p. 164). Springer. https://doi.org/10.1007/978-3-319-96190-3

Zaman, M., Shahid, S. A., & Heng, L. (2018b). Irrigation water quality. In M. Zaman et al. (Eds.), *Guideline for salinity assessment, mitigation and adaptation using nuclear and related techniques* (pp. 113–131). Springer. https://doi.org/10.1007/978-3-319-96190-3_5

Zeng, R., Cai, X., Ringler, C., & Zhu, T. (2017). Hydropower versus irrigation. An analysis of global pattern. *Environmental Research Letters, 12*(3), 034006.

[illegible]

Chapter 4
Impact of Climate Change on Grazing Lands and Response of Arid Lands Ecosystems to Livestock Industry

Mariam E. Al-Bahouh and Samira Mohammed

Abstract Globally the grazing lands cover about 6.12 billion hectares of the Earth surface and supporting about 360 million cattle and over 600 million sheep and goats. Globally, the grazing animals provide 10% beef and 30% sheep and goat meat. For an estimated 100 million people in arid areas, grazing livestock is the only potential source of livelihood. The grazing area coverage is approximately 1.5, 2.8, and 17 times more than the forests, cropland and urban settlements respectively. These grazing lands are located in different geographical regions, such as the countries of the humid climate (high rainfall, Europe) and low rainfall (arid regions). The arid region is plagued with low rain fall, high temperature, land degradation, prolonged drought and high vulnerability to climate change and this affects the grazing lands functionality to provide enough biomass to meet the feed requirements of the animals in the wild. Overall livestock sector contributes about 14.5% of global greenhouse gases (GHGs) emissions. On one side the grazing lands are diminishing by climate change impacts, and on the other side the human population is increasing unprecedentedly, and projected to be about 11.0 billion by 2100. This rapidly increasing population demands for livestock more than the existing requirements, as well as a general shift of human food towards protein-rich diets. Among the various sectors contributing to the GHGs emission, the livestock sector contributes to about 9.2% share of GHGs, a driving force of climate change. Thus, the adoption of policies to mitigate GHGs is essential to shift towards sustainable and environmentally friendly livestock production systems in arid regions. In this chapter a comprehensive review is completed to highlight the impact of climate change on grazing lands in the arid regions and how this will affect the livestock industry and ways are suggested to cope the climate change impact to sustain livestock industry.

M. E. Al-Bahouh (✉) · S. Mohammed
Food Security Program, Environment and Life Sciences Research Center, Kuwait Institute for Scientific Research, Safat, Kuwait
e-mail: mrbahouh@kisr.edu.kw

S. Mohammed
e-mail: smohammad@kisr.edu.kw

M. K. Suleiman and S. A. Shahid (eds.), *Fostering Arid Lands Agriculture in the Face of Climate Change*, Advances in Global Change Research 81,
https://doi.org/10.1007/978-3-031-98490-7_4

Keywords Livestock · Arid region · Feed · Fodder · Climate change · Greenhouse gas mitigation

4.1 Introduction

Arid lands are the most fragile and negatively affected by overexploitation and droughts. Arid lands represent about 30% of global lands (Fig. 4.1). Arid ecosystems lack water due to low precipitation and high evapotranspiration. The aridity index has been used to categorize arid lands into: hyperarid zone (about 8% of land surface), which is dry land with low vegetation and low rainfall (<100 mm); arid zone (about 16% of earth surface) that is often characterized by low organic matter and poor soil fertility (Jhariya et al., 2022).

Drought is a phenomenon of hyperarid, arid and semi-arid regions. In these regions, the systems of agricultural production are adapted to the variability and patterns of weather conditions (Diaz-Solis et al., 2016). Worldwide, about 30% of lands are used for livestock production that contribute to agricultural outputs of developed (40%) and developing countries (Steinfeld et al., 2006). Thus, it supports the livelihoods of about 1.3 billion people (Acosta, 2018). Pressure on agricultural landscapes and biodiversity has been increased due to population growth, higher incomes, changing diets, climate change and urbanization (Harrison et al., 2021; Wellesley et al., 2015). This makes livestock sector one of the largest producer of greenhouse gas (GHG) emissions (Harrison et al., 2021; Smith et al., 2014) due

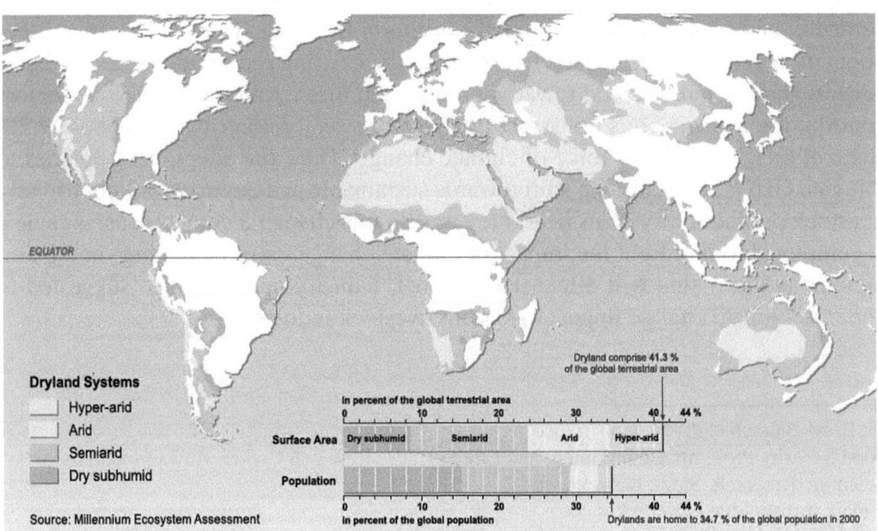

Fig. 4.1 Global map showing arid and semiarid areas (Wikimedia Commons, 2020)

to feed production, animal waste, enteric fermentation and land-use change (FAO, 2017). Global GHG emission (~14.5%) is linked to livestock sector (Gerber et al., 2013). The increase in GHG emission is also associated with the high rate of land degradation, pollution of water, air and biodiversity deterioration (Bellarby et al., 2013; Reynolds et al., 2010; Thornton, 2010). Consequently, as the global population and demand for livestock products increase, GHG emissions will continue to increase unless mitigation policies are developed and implemented. This could lead to the reduction of both livestock and crop yields due to the effect of climate change on natural resources (Kummu et al., 2021), and animal health (Niloofar et al., 2021).

4.2 Grazing Lands of Arid Regions

Grazing land refers to the designated areas where livestock are allowed to feed on grass and forages. It includes the terrain used for livestock grazing activities (Kuemmerla et al., 2013).

4.2.1 Asia

Grazing lands occupy about 1.72 billion hectares of Asian continent, representing to about 38% of total land area (Angerer et al., 2008). For countries like Sri Lanka, in southeast Asia, the rangeland proportion is relatively small. However, in northern and western Asia the rangelands represent a major category (about 40% of lands). These are the middle east lands which are located mainly in China, Mongolia, Russia, Turkey, Iran and Afghanistan (Angerer et al., 2008).

4.2.2 Africa

Grazing lands in Sub-Saharan Africa (SSA) have been reduced due to land degradation and conversion to urban and crop lands (Milne et al., 2016). It is reported to be one of the most degraded ecosystems in Africa due to mismanagement (Liniiger & Mekadaschi, 2019). In SSA, rangelands are dominated by grasses and/or bushes and trees, and are used as a natural ecosystem for livestock production (Allen et al., 2011). The rangelands used for livestock production in Africa (Fig. 4.2) are dependent on water availability. This is clear in the gradual transition from grassland to dense woodland at the wetter side of the spectrum (Liniiger & Mekadaschi, 2019).

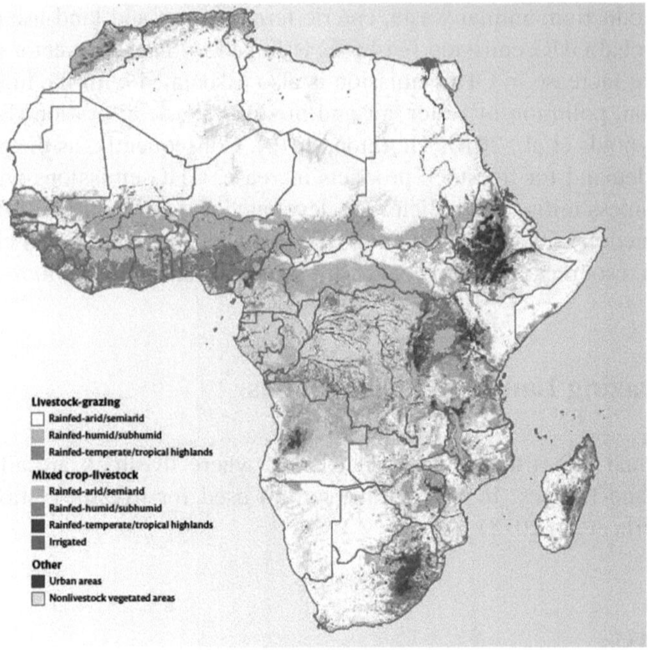

Fig. 4.2 Livestock production systems by climate zone. *Source* Sebastian (2014)

4.2.3 Arab Countries

The Arabian Peninsula, Jordan (eastern), Syria (south-eastern) and Iraq are characterized by their aridity, minimal water availability and sparse vegetation. Less extreme conditions are found in western Jordan, western and northern Syria, northern Iraq, Yemen and Oman (Bourn, 2003).

4.2.4 Other Arid Land Countries

Grazing lands are used for the production of fodder crops to support livestock. Grazing lands are found in many arid land regions including North and South America, northwest and central Australia. Arid and semi-arid lands represent about a third of western North America and about 30% of livestock grazing regions of Mexico, Canada and United States (Huntsinger & Starrs, 2006). In Australia, livestock grazing lands makes up to 54% (4.2 million km^2) of usable land (ACLUMP, 2016).

4.3 Climate Change Aspects of Arid Region

Climate change is a global phenomenon and no region is immune to climate change, however, different regions have variable vulnerability to climate change. Arid land countries are also facing the climate change impacts, affecting the crop production, farmer livelihood and grazing lands, especially dryland (hyperarid, arid, semiarid regions) countries including Africa.

4.3.1 GHG Emissions

Climate change is caused by the high emission rate of GHGs, which is linked to the high temperature of the atmosphere (Stocker, 2014). The impact of climate change could be extended to affect the production of livestock through competition for feed (quantity and quality) and natural resources. The GHGs will also result in heat stress, prevalence of livestock diseases, biodiversity loss as the demand for livestock might be doubled by the mid of the twenty-first century (Garnett, 2009).

4.3.2 Natural Habitats and Land Use

Since 1850s, natural habitats have been converted to croplands and pastures (Goldewijk & Battjes, 1997). Significant changes in natural landscape are the result of the high demand for livestock products. The conversion of land from forest to croplands and pastures resulted in land degradation due to the exhaustion of soil resources (Steinfeld et al., 2006). Natural carbon cycle could also be affected due to the change of land use. This in turn will increase the amount of carbon released into the atmosphere, which is directly related to the rate of GHGs emission. Natural habitats, compared to croplands and pastures, sequestered more carbon in vegetation cover and soil carbon sequestration. About 40% of global anthropogenic CO_2 emissions is sequestered by soil and terrestrial vegetation (Read et al., 2001).

Pasture-lands and croplands sequestered around 27 and 6% of global carbon, respectively (Watson, 2000). However, soil stores 1,100–1,600 billion-tons of carbon that is double than the amount sequestered by vegetation (Steinfeld et al., 2006; Sundquist, 1993). Thus, high amount of carbon is released into the atmosphere based on the conversion of forest to pasture and cropland (Steinfeld et al., 2006). Nevertheless, land conversion produces other emissions as it may reduce CH_4 oxidation by microorganism, which make the pasturelands a good source of CH_4 due to limitation of gas diffusion as a result of soil compaction from cattle hooves (Mosier et al., 2004). The complex relation between soil carbon sequestration and pasturelands results in a fact that pasturelands can either decrease or increase the emission of GHGs, based on the climate, ecosystem and grazing history (Rojas-Downing et al., 2017).

4.3.3 Feed Production

Livestock affects climate through feed production that emits carbon dioxide (CO_2), nitrous oxide (N_2O), and methane (CH_4). These gases are associated with climate change and make about 45% of global livestock GHG emission (Gerber et al., 2013). The use of synthetic fertilizers and manures for the production of forage and feed, transport and processing of feed products contribute to high emission of GHGs (Gerber et al., 2013; Thornton, 2010). The process of fertilizers manufacturing contributes annually to more than 40 million tons of CO_2 considering the amount used of fertilizers, and the process of packaging, transport and application (Steinfeld et al., 2006).

Ammonia volatilization from synthetic fertilizers is an indirect contributor to GHG emissions with an annual ammonia losses of 14% (Steinfeld et al., 2006). Fertilizers also increase the emissions of N_2O, with an average loss of 1% of N_2O-N. An estimated amount of 0.2 million tons of N_2O-N is contributed by livestock sector from mineral fertilizers, on annual basis (Bouwman, 1996; Steinfeld et al., 2006). Additional N_2O emission (0.5 million tons per year) is related to the leguminous crops that are produced for livestock including soybeans, alfalfa and clover. Hence, these two sources produce 0.7 million tons of N_2O per year (Steinfeld et al., 2006). It is expected that by 2030, the total N_2O emission from agriculture to increase by 35–60% (Bruinsma, 2017). On-farm fossil fuel used for operations and feed production (such as heating, cooling, pesticide production, irrigation) releases around 90 million tons of CO_2 annually, which is 50% more than the nitrogen fertilizers (Steinfeld et al., 2006).

4.4 Livestock and Climate Change

Livestock production is responsible for GHGs emission compared to other production systems. The GHG emissions are caused by the enteric fermentation, feed production, animal waste and the change in land use.

4.4.1 Livestock Production

Climate change has a significant effect on livestock production. As the increase in temperature, shifts in precipitation events cause more extreme events presented by the heat stress and limitation in water availability. These events are expected to adversely affect the performance and productivity of livestock both indirectly and directly.

4.4.2 *Impact of Climate Change on Livestock*

Impacts of climate change on the livestock are witnessed by its negative effects on feed, water demand, spread of diseases, animal physiological functions and performance, health, mortality rate, along with effects on animal breeding and biodiversity.

4.4.2.1 Feed Quantity and Quality

The increase in temperature and atmospheric level of CO_2 affect the quantity and quality of feed (Chapman et al., 2012). This depend on livestock system, species and location (Hatfield & Prueger, 2011; Thornton et al., 2009). The location and the extent of growing period determine the impacts of climate on forage quality and quantity (Polley et al., 2013). As the increased temperature (e.g. >2 °C) could result in different impacts (negative in semiarid region and positive in humid temperate areas) (Rojas-Downing et al., 2017).

Climate events like floods could affect roots (form and structure), leaf growth rate and yield (Baruch & Mérida, 1995). Variations in the growth rates also affect the composition of pastures (Thornton et al., 2009). However, the productivity could be improved with optimal environmental conditions such as temperature and precipitation (Grigorieva et al., 2023).

4.4.2.2 Water Demand

Water availability has an impact on the livestock industry since it is used for drinking by animals, growing fodder crops, and for the processing of final products (Thornton et al., 2009). Water consumption by animals may be increased 2 to 3 times more with the increase in atmospheric temperature (Nardone et al., 2010). Fresh water aquifers could be affected by sea level rise (SLR) and sea water intrusion (Ketabchi et al., 2016). Salinization is considered to be a source of contamination that could influence the animals' performance negatively (Nardone et al., 2010).

4.4.2.3 Diseases

Climate change and increased temperature could affect animal health directly or indirectly. The significance of the impact depends on the region, animal susceptibility and disease characteristics (Thornton et al., 2009). Increased temperature has a direct correlation with morbidity and mortality of animals. However, the microbial community could indirectly contribute to the incidence of diseases among livestock, which could have a detrimental effect on the performance and productivity (Thornton et al.,

2009; Tubiello et al., 2008). Climate change also induces the spread and outbreaks of new diseases (Thornton et al., 2009).

4.5 Effects of Heat Stress on Livestock

The increase in temperature affects the comfort zone of animal (Etim et al., 2013). The body temperature of animal is kept usually within a range of 0.5 °C (Henry et al., 2012) as the increase in temperature result in heat stress. During the heat stress, feed intake of animals is reduced, water intake and respiration rates are increased, which badly affects the production and reproduction efficiency of animals (Lacetera et al., 2003; Nardone et al., 2010). Heat reduces feed intake which is linearly associated with the rate of milk production, growth performance, feed conversion, health and mortality rate (Hahn, 1999).

4.5.1 Feed Utilization and Feed Intake

The feed provides nutrients to the animal which are required for their normal well-being and production (Thornton et al., 2009). Meeting the dietary needs is essential to prevent any metabolic and digestive issues (Mader, 2003). Heat stress decreases the feed intake, feed efficiency that lead to reduction in weight gain and performance due to negative energy balance (Henry et al., 2012; Lacetera et al., 2003; Mader & Davis, 2004).

4.5.2 Livestock Reproduction, Health Condition and Mortality Rate

Climate change and increase in temperature affect the reproduction efficiency negatively as it affects the oocyte growth, sperm concentration, impair the development of the embryo and pregnancy rate (Barati et al., 2008; Hansen, 2007; Kunavongkrit et al., 2005; Nardone et al., 2010). Harsh weather conditions also affect the animal health as it could affect the metabolic rate, oxidative status, liver function, and endocrine status (Bernabucci et al., 2006; Nardone et al., 2010; Ronchi et al., 1999; Webster, 1991) that could reflect negatively on energy, fitness and longevity of animals (King et al., 2006). High humid and warm conditions affect livestock mortality, an increase in temperature (1–5 °C) could directly induce the mortality rate (Howden et al., 2008).

4.5.3 Climate Change, Grazing Lands and Livestock Industry

Climate change and harsh conditions in different regions affect the biodiversity of animals and increase the risk of breed elimination. In developing regions (Third World countries, or non-industrialized countries) about 7–10% of mammalian species including livestock are in endangered risk category (Aguilar & Campuzano, 2009). Climate change reduces the productivity of livestock resulting in a significant financial loss (Nardone et al., 2010). St-Pierre et al. (2003) stated that up to 2.36 billion US dollar is the annual economic loss of livestock industry in the United States of America, which is 50% related to dairy cattle operations. The high producing cows take more heat stress resulting into reduced milk production (Berman, 2005; Kadzere et al., 2002).

Heat Index (temperature humidity index) affect the amount and composition of milk production of goat, ewe and buffalo (Olsson & Dahlborn, 1989; West, 2003). For a better yield and economic return, traditional animal production systems should be improved.

4.6 Response to Climate Change

Response to climate change on global level is addressed by adopting mitigation and adaptation strategies. Mitigation by reducing the GHGs emission to minimize the impacts of climate changes; and adaptation through actions to reduce the effects of climate change for sustainable environmental resources and production systems. Adoption of special protocols are essential for environmental health in order to have sustainable food systems that could contribute to food and nutrition security. These sustainable systems are essential to manage the natural resources for future generations.

4.6.1 Adaptation and Mitigation Strategies

Adaptation strategies are reflected by adopting different strategies related to livestock stocking rate, season of grazing, increase in reserve forages, selecting high-quality water resources, management of vegetation and nutritional value, herd size and composition with respect to grazing lands.

4.6.1.1 Conservative Stocking Rate

In arid and semi-arid regions, conservative stocking is the most flexible and successful tool that is adjusted based on the variability in droughts and precipitation levels

(Thomas et al., 2015). Using the conservative stocking rate and optimizing the size of herd are proactive tools that are used to face the negative effects of climate change (Diaz-Solis et al., 2009). It is reported that moderate stocking rates on native vegetation lands reduce the GHGs emission intensity (Liebig et al., 2010). Hence, it is detrimental to use a high herd size during the times of drought due to its negative effect on animal performance and revenue rate (Torell et al., 2010). As the heavier stocking rate could reduce the production of herbage due to its effect on the plant roots and water infiltration as a result of soil compaction (Adiku et al., 2010; Pulido et al., 2017; Rhoades & Locke, 1964).

4.6.1.2 Varied Season of Grazing

Grazing is known to increase land productivity and it is important for nutrient management. Varied season of grazing is one of the most reliable mitigation method for reducing GHGs emission from livestock (Boadi et al., 2002). Rotational grazing is a feasible strategy to reduce methane (CH_4) production per unit of milk or meat produced. As increasing animal productivity through nutrition and management will decrease CH_4 emission per unit of product (Johnson et al., 1996). This strategy is to be used in animal industry to produce the same amount of end product (e.g. meat, milk, etc.) based on the use of fewer animals.

4.6.1.3 Optimizing Herd Size and Composition

The composition and genetic potential of herd is important to optimize the production rate e.g. 10 goats could produce 25% more income compared to one buffalo (Gihad & El-Bedawy, 2000). Another mitigation strategy depends on the selection of animals that produce less CH_4 per unit of consumed feed. Ominski and Wittenberg (2005) reported that 27% of the variation in CH_4 emission was associated with the differences from animal-to-animal.

Adaptation and mitigation practices and evaluation of genetic capabilities are very crucial to ensure the sustainability of different breeds and species. These practices are essential for sustainable livestock production systems (Thornton et al., 2009).

4.6.1.4 Identifying Reserve Forage

Identifying reserve forage is a mitigation tool that help producers to adopt the negative impact of climate change. Climate change could affect the forage characteristics, distribution and production rate due to the imbalance in temperature and precipitation patterns (Baron & Bélanger, 2020). The availability of feed affects the efficiency of livestock production. To face the feed scarcity and un-favorable climatic conditions, nutritional interventions must be used, such as keeping forage for stressful periods and using feed alternatives along with antioxidants supplements (Tamboli et al.,

2023). Keeping reserve forage could have a positive effect on climate change. Moss et al. (2000) observed a reduction in CH_4 production due to ensiling, on the other hand, feeding silage (ryegrass and lotus) was found to increase CH_4 loss (Woodward et al., 2001).

4.6.1.5 Strategic Distribution of Water

Water quality and distribution along with its salinity level could create a significant problem for livestock production and productivity in dry regions. In dryland regions the availability and purity of fresh water resources are affected by climate change (Chedid et al., 2014; Heinke et al., 2020; Nicholson, 2017). Willms et al. (2002) concluded that water quality is critical to reduce CH_4 emission per unit of final product. Accordingly, the use of animals (e.g. goat, sheep and camels) with efficient grazing and drought resistant habits besides high feed conversion rates is highly recommended (Ciliberti et al., 2022). To address the issue of water scarcity, there is a need to shift toward using livestock systems that need less water to produce feed and raise animals (Nardone et al., 2010).

4.6.1.6 Proactive Vegetation Management in Grazing Lands

The amount of accumulated carbon in soils is affected by the timing and intensity of grazing and type of flora (Conant & Paustian, 2002; Freibauer et al., 2004). Carbon accumulation is greater on grazed lands compared to overgrazed or un-grazed lands (Liebig et al., 2005). However, the effect is inconsistent due to different factors including the type of plant species, climate and soil (Derner et al., 2006). Therefore, to increase carbon sequestration different tools could be used, including rotational grazing and excluding degraded pastures regions from livestock grazing areas (Keller et al., 2021).

4.6.1.7 Seasonal Distribution of Grazing Animals and Quantity of Forage

Based on the seasonal weather conditions and predictions of change in weather, stocking rate could be adjusted and forage supply increased to decrease the influence of drought on production rate of livestock (Kachergis et al., 2014). One of the rangeland management system that aims to improve vegetation, soil diversity and productivity is the high-density grazing. It has been known to be a key approach to mitigate climate change by increasing the amount of sequestered carbon in soils (Franke & Kotzé, 2022). The use of adaptive multi-paddock grazing system also proved to manage grazing lands and improve forage production. However both the paddock and the conventional grazing systems were approved to reduce GHGs emission considering the soil properties and environmental conditions (Ma et al., 2021).

4.6.1.8 Varied Forage

The livestock productivity is affected directly by climate change by varying the forage availability and the quality (Giridhar & Samireddypalle, 2015). Most feasible mitigation strategies to reduce CH_4 emission include the use of high quality forage and preservation of forage as a silage (Boadi et al., 2002). McCaughey et al. (1999) found that cows grazing the alfalfa-grass pastures had lower CH_4 emission compared to those grazed on grass pastures.

4.6.1.9 Forage Nutritional Value

The effects of climate change on forage nutritional value will affect nutrient capacity and livestock growth and development. The water availability, temperature and radiation led to the variations of forage nutritional value and quality along with production (Fu et al., 2022). Extreme drought and rainfall events are the effect of climate change which will affect the crop productivity and nutritional value (Easterling et al., 2000). Also, the increase in severity of over grazing and drought events could increase the conversion rate of forage from high to low quality (Yang et al., 2018).

4.7 Conclusions and Recommendations

Globally, livestock production is contributing to climate change through greenhouse gas emission. Livestock operation systems are connected to the problem of land degradation due to feed production and expansion of crop and pasture lands. Thus, about 9.2% of GHGs emissions are related to livestock productions systems. The intensification of livestock production is expected to increase its contribution to climate change, due to the need for more feed that is responsible for almost half of GHGs emission. Accordingly, the increase in livestock production along with feed production and enteric fermentation (largest GHG emission contributor) will increase the global emission of GHGs. Consequently, the climate change will affect the efficiency of livestock production, the nutritional contents of feed and the efforts to achieve food security especially in arid regions due to the prevalence of diseases and scarcity of water and feed resources. Climate change and livestock production are closely associated and there is a need to identify the ideal conditions and regions for livestock production to improve environmental health and achieve food security. Mitigation and adaptation practices with special regulations are critical to ensure the sustainability of livestock production systems. Special protocols and practices to face the climate change including the use of different fodder crop varieties, diversification of animal species are needed to improve the efficiency and control any outbreaks (e.g. diseases, pest). More studies in the field of livestock production and mitigation practices are needed for better understanding the response and the driving mechanisms of the earth to climate change and human activities.

Acknowledgements We would like to thank Wikimedia for allowing the free use of map shown in Fig. 4.1 under the Creative Commons CC0 License, and International Food Polity Research Institute (IFPRI) for granting permission to use map presented as Fig. 4.2.

References

ACLUMP. (2016). *Australian collaborative land use and management program-land use in Australia—At a glance*. www.abares.gov.au/landuse; http://www.agriculture.gov.au/abares/aclump/land-use/publications#land-use

Acosta, A. (2018). *World livestock: Transforming the livestock sector through the sustainable development goals* (p. 229). Food and Agriculture Organization of the United Nations.

Adiku, S. G. K., Dunn, G. H., Ahuja, L. R., Gunter, S., Bradford, J., Garcia, L., & Andales, A. A. (2010). Simulation of sandsage-bluestem forage growth under varying stocking rates. *Rangeland Ecology and Management, 63*(5), 546–552. https://doi.org/10.2111/REM-D-09-00145.1

Aguilar, A. G., & Campuzano, E. P. (2009). Informal sector. In R. Kitchin, & N. Thrift (Eds.), *International encyclopedia of human geography* (pp. 446–453). Elsevier. https://doi.org/10.1016/B978-008044910-4.00102-4

Allen, V. G., Batello, C., Berretta, E. J., Hodgson, J., Kothmann, M., Li, X., McIvor, J., Milne, J., Morris, C., Peeters, A., & Sanderson, M. (2011). The forage and grazing terminology committee. An international terminology for grazing lands and grazing animals. *Grass and Forage Science, 66*(1), 2–28. https://doi.org/10.1111/j.1365-2494.2010.00780.x

Angerer, J., Han, G., Fujisaki, I., & Havstad, K. (2008). Climate change and ecosystems of Asia with emphasis on Inner Mongolia and Mongolia. *Rangelands, 30*. https://doi.org/10.2111/1551-501X(2008)30[46:CCAEOA]2.0.CO;2

Barati, F., Agung, B., Wongsrikeao, P., Taniguchi, M., Nagai, T., & Otoi, T. (2008). Meiotic competence and DNA damage of porcine oocytes exposed to an elevated temperature. *Theriogenology, 69*(6), 767–772.

Baron, V. S., & Bélanger, G. (2020). Climate, climate-change and forage adaptation. *Forages: The Science of Grassland Agriculture, 2*, 151–186.

Baruch, Z., & Mérida, T. (1995). Effects of drought and flooding on root anatomy in four tropical forage grasses. *International Journal of Plant Sciences, 156*(4), 514–521.

Bellarby, J., Tirado, R., Leip, A., Weiss, F., Lesschen, J. P., & Smith, P. (2013). Livestock greenhouse gas emissions and mitigation potential in Europe. *Global Change Biology, 19*(1), 3–18.

Berman, A. (2005). Estimates of heat stress relief needs for Holstein dairy cows. *Journal of Animal Science, 83*(6), 1377–1384.

Bernabucci, U., Lacetera, N., Basiricò, L., Ronchi, B., Morera, P., Seren, E., & Nardone, A. (2006). Hot season and BCS affect leptin secretion of periparturient dairy cows. *Journal of Animal Science, 84*, 348. ISSN: 0021-8812.

Boadi, D., Wittenberg, K., & McCaughey, W. P. (2002). Effects of grain supplementation on methane production of grazing steers using the sulphur (SF_6) tracer gas technique. *Canadian Journal of Animal Science, 82*, 151–157. https://doi.org/10.4141/A01-038

Bourn, D. (2003). Livestock dynamics in the Arabian Peninsula. In *A regional review of national livestock resources and international livestock trade* (pp. 1–78). Environmental Research Group Oxford Limited.

Bouwman, A. F. (1996). Direct emission of nitrous oxide from agricultural soils. *Nutrient Cycling in Agroecosystems, 46*(1), 53–70. https://doi.org/10.1007/BF00210224

Bruinsma, J. (2017). *World agriculture: Towards 2015/2030: An FAO study* (p. 444). Routledge. https://doi.org/10.4324/9781315083858. eBook ISBN: 9781315083858.

Chapman, S. C., Chakraborty, S., Dreccer, M. F., & Howden, S. M. (2012). Plant adaptation to climate change—Opportunities and priorities in breeding. *Crop and Pasture Science, 63*(3), 251–268.

Chedid, M., Jaber, L. S., Giger-Reverdin, S., Duvaux-Ponter, C., & Hamadeh, S. K. (2014). Water stress in sheep raised under arid conditions. *Canadian Journal of Animal Science, 94*(2), 243–257.

Ciliberti, M. G., Caroprese, M., & Albenzio, M. (2022). Climate resilience in small ruminant and immune system: An old alliance in the new sustainability context. *Small Ruminant Research, 210*, 106662.

Conant, R., & Paustian, K. (2002). Potential soil carbon sequestration in overgrazed grassland ecosystems. *Global Biogeochem. Cycles, 16*. https://doi.org/10.1029/2001GB001661

Derner, J., Boutton, T., & Briske, D. (2006). Grazing and ecosystem carbon storage in the North American Great Plains. *Plant and Soil, 280*, 77–90. https://doi.org/10.1007/s11104-005-2554-3

Diaz-Solis, H., Grant, W. E., Kothmann, M. M., Teague, W. R., & Díaz-García, J. (2009). Adaptive management of stocking rates to reduce effects of drought on cow-calf production systems in semi-arid rangelands. *Agricultural Systems, 100*(1–3), 43–50. https://doi.org/10.1016/j.agsy.2008.12.007

Diaz-Solis, H., Grant, W., Kothmann, M. M., Teague, W. R., Paz Pellat, F., & Bolaños-González, M. (2016). Strategies to reducing GHG emissions in semi-arid rangelands of Mexico. *Terra Latinoamericana, 34*, 73–81.

Easterling, D., Meehl, G., Parmesan, C., Changnon, S., Karl, T. R., & Mearns, L. (2000). Climate extremes: Observations, modeling, and impacts. *Science (New York, N.Y.), 289*, 2068–2074. https://doi.org/10.1126/science.289.5487.2068

Etim, N. N., Williams, M. E., Evans, E. I., & Offiong, E. E. A. (2013). Physiological and behavioural responses of farm animals to stress: Implications to animal productivity. *American Journal of Advanced Agricultural Research (AJAAR), 1*(2), 53–61.

FAO. (2017). *Family farming knowledge platform*. Retrieved December 1, 2024, from https://www.fao.org/family-farming/home/en/

Franke, A. C., & Kotzé, E. (2022). High-density grazing in southern Africa: Inspiration by nature leads to conservation? *Outlook on Agriculture, 51*(1), 67–74. https://doi.org/10.1177/00307270221075060

Freibauer, A., Rounsevell, M. D. A., Smith, P., & Verhagen, J. (2004). Carbon sequestration in the agricultural soils of Europe. *Geoderma, 122*(1), 1–23. https://doi.org/10.1016/j.geoderma.2004.01.021

Fu, G., Wang, J., & Li, S. (2022). Response of forage nutritional quality to climate change and human activities in alpine grasslands. *Science of the Total Environment*, *845*(1), 157552. https://doi.org/10.1016/j.scitotenv.2022.157552

Garnett, T. (2009). Livestock-related greenhouse gas emissions: Impacts and options for policy makers. *Environmental Science and Policy, 12*(4), 491–503.

Gerber, P. J., Steinfeld, H., Henderson, B., Mottet, A., Opio, C., Dijkman, J., Falcucci, A., & Tempio, G. (2013). *Tackling climate change through livestock—A global assessment of emissions and mitigation opportunities* (p. 136). Food and Agriculture Organization of the United Nations (FAO). ISBN: 978-9251079201.

Gihad, E. A., & El-Bedawy, T. M. (2000). Contribution of goats to Egyptian small farmer food and income. In L. Gruner, & Y. Chabert (Eds.), *Proceedings of the seventh international conference on goats, France, 15–21 May 2000* (Vol. II, pp. 531–534). INRA Publication.

Giridhar, K., & Samireddypalle, A. (2015). Impact of climate change on forage availability for livestock. In *Climate change impact on livestock: Adaptation and mitigation* (pp. 97–112). Springer. https://doi.org/10.1007/978-81-322-2265-1_7

Goldewijk, C. G. M. K., & Battjes, J. J. (1997). *A hundred year (1890–1990) database for integrated environmental assessments (HYDE, version 1.1)* (Report No. 422514002). National Institute of Public Health and the Environment. Retrieved December 1, 2024, from https://www.pbl.nl/uploads/default/downloads/422514002.pdf

Grigorieva, E., Livenets, A., & Stelmakh, E. (2023). Adaptation of agriculture to climate change: A scoping review. *Climate, 11*(10), 202. https://doi.org/10.3390/cli11100202

Hahn, G. L. (1999). Dynamic responses of cattle to thermal heat loads. *Journal of Animal Science, 77*(Suppl. 2), 10–20.

Hansen, P. J. (2007). Exploitation of genetic and physiological determinants of embryonic resistance to elevated temperature to improve embryonic survival in dairy cattle during heat stress. *Theriogenology, 68*, S242–S249.

Harrison, M. T., Cullen, B. R., Mayberry, D. E., Cowie, A. L., Bilotto, F., Badgery, W. B., Liu, K., Davison, T., Christie, K. M., & Muleke, A. (2021). Carbon myopia: The urgent need for integrated social, economic and environmental action in the livestock sector. *Global Change Biology, 27*(22), 5726–5761.

Hatfield, J. L., & Prueger, J. H. (2011). Agroecology: Implications for plant response to climate change. In *Crop adaptation to climate change* (pp. 27–43).

Heinke, J., Lannerstad, M., Gerten, D., Havlík, P., Herrero, M., Notenbaert, A. M. O., Hoff, H., & Müller, C. (2020). Water use in global livestock production—Opportunities and constraints for increasing water productivity. *Water Resources Research, 56*(12), e2019WR026995.

Henry, B., Charmley, E., Eckard, R., Gaughan, J. B., & Hegarty, R. (2012). Livestock production in a changing climate: Adaptation and mitigation research in Australia. *Crop and Pasture Science, 63*(3), 191–202.

Howden, S. M., Crimp, S. J., & Stokes, C. J. (2008). Climate change and Australian livestock systems: Impacts, research and policy issues. *Australian Journal of Experimental Agriculture, 48*(7), 780–788.

Huntsinger, L., & Starrs, P. (2006). Grazing in arid North America: A biogeographical approach. *Sécheresse, 17*(1–2), 219–233. https://www.academia.edu/34386381/Grazing_in_arid_North_America_A_biogeographical_approach

Jhariya, M. K., Meena, R. S., & Meena, S. N. (2022). *Natural resources conservation and advances for sustainability*. Elsevier. ISBN: 978-0-12-822976-7. https://doi.org/10.1016/C2019-0-03763-6

Johnson, D. E., Ward, G. W., & Ramsey, J. J. (1996). In E. T. Korn (Ed.), *Livestock methane: Current emissions and mitigation potential*. Lewis Publishers.

Kachergis, E., Derner, J. D., Cutts, B. B., Roche, L. M., Eviner, V. T., Lubell, M. N., & Tate, K. W. (2014). Increasing flexibility in rangeland management during drought. *Ecosphere, 5*(6), art77. https://doi.org/10.1890/ES13-00402.1

Kadzere, C. T., Murphy, M. R., Silanikove, N., & Maltz, E. (2002). Heat stress in lactating dairy cows: A review. *Livestock Production Science, 77*(1), 59–91. https://doi.org/10.1016/S0301-6226(01)00330-X

Keller, K., Helgeson, C., & Srikrishnan, V. (2021). Climate risk management. *Annual Review of Earth and Planetary Sciences, 49*, 95–116. https://doi.org/10.1146/annurev-earth-080320-055847

Ketabchi, H., Mahmoodzadeh, D., Ataie-Ashtiani, B., & Simmons, C. T. (2016). Sea-level rise impacts on seawater intrusion in coastal aquifers: Review and integration. *Journal of Hydrology, 535*, 235–255. https://doi.org/10.1016/j.jhydrol.2016.01.083

King, J. M., Parsons, D. J., Turnpenny, J. R., Nyangaga, J., Bakari, P., & Wathes, C. M. (2006). Modelling energy metabolism of Friesians in Kenya smallholdings shows how heat stress and energy deficit constrain milk yield and cow replacement rate. *Animal Science, 82*(5), 705–716.

Kuemmerla, T., Erb, K., Meyfroidt, P., Müller, D., Verburg, P. H., Estel, S., Haberl, H., Hostert, P., Jepsen, M. R., Kastner, T., Levers, C., Lindner, M., Plutzar, C., Verkerk, P. J., van der Zanden, E. H., & Reenberg, A. (2013). Challenges and opportunities in mapping land use intensity globally. *Current Opinion in Environmental Sustainability, 5*(5), 484–493. https://doi.org/10.1016/j.cosust.2013.06.002

Kummu, M., Heino, M., Taka, M., Olli, V., & Viviroli, D. (2021). Climate change risks pushing one-third of global food production outside the safe climatic space. *One Earth, 4*. https://doi.org/10.1016/j.oneear.2021.04.017

Kunavongkrit, A., Suriyasomboon, A., Lundeheim, N., Heard, T. W., & Einarsson, S. (2005). Management and sperm production of boars under differing environmental conditions. *Theriogenology, 63*(2), 657–667.

Lacetera, N., Bernabucci, U., Ronchi, B., & Nardone, A. (2003). Physiological and productive consequences of heat stress. The case of dairy ruminants. In *Interaction between climate and animal production* (pp. 45–59).

Liebig, M. A., Johnson, H. A., Hanson, J. D., & Frank, A. B. (2005). Soil carbon under switchgrass stands and cultivated cropland. *Biomass and Bioenergy, 28*(4), 347–354. https://doi.org/10.1016/j.biombioe.2004.11.004

Liebig, M. A., Gross, J. R., Kronberg, S. L., & Phillips, R. L. (2010) Grazing management contributions to net global warming potential: A long-term evaluation in the Northern Great plains. *Journal of Environmental Quality, 39*(3), 799–809. https://doi.org/10.2134/jeq2009.0272

Liniiger, H., & Mekadaschi, S. R. (2019). *Sustainable rangeland management in Sub-Saharan Africa* (p. 412). A TerrAfrica Partnership Publication.

Ma, Z., Shrestha, B. M., Bork, E. W., Chang, S. X., Carlyle, C. N., Döbert, T. F., Sobrinho, L. S., & Boyce, M. S. (2021). Soil greenhouse gas emissions and grazing management in northern temperate grasslands. *Science of the Total Environment, 796*, 148975. https://doi.org/10.1016/j.scitotenv.2021.148975

Mader, T. L. (2003). Environmental stress in confined beef cattle. *Journal of Animal Science, 81*(14_suppl_2), E110–E119.

Mader, T. L., & Davis, M. S. (2004). Effect of management strategies on reducing heat stress of feedlot cattle: Feed and water intake. *Journal of Animal Science, 82*(10), 3077–3087.

McCaughey, W. P., Wittenberg, K., & Corrigan, D. (1999). Impact of pasture type on methane production by lactating beef cows. *Canadian Journal of Animal Science, 79*(2), 221–226. https://doi.org/10.4141/A98-107

Milne, E., Aynekulu, E., Bationo, A., Batjes, N. H., Boone, R., Conant, R., Davies, J., Hanan, N., Hoag, D., Herrick, J. E., Knausenberger, W., Neely, C., Njoka, J., Ngugi, M., Parton, B., Paustian, K., Reid, R., Said, M., Shepherd, K., & Nkonya, E. (2016). Grazing lands in Sub-Saharan Africa and their potential role in climate change mitigation: What we do and don't know. *Environmental Development, 19*, 70–74. https://doi.org/10.1016/j.envdev.2016.06.001

Mosier, A., Wassmann, R., Verchot, L., King, J., & Palm, C. (2004). Methane and nitrogen oxide fluxes in tropical agricultural soils: Sources, sinks and mechanisms. In *Tropical agriculture in transition—Opportunities for mitigating greenhouse gas emissions* (pp. 11–49).

Moss, A. R., Jouany, J.-P., & Newbold, J. (2000). Methane production by ruminants: Its contribution to global warming. *Annales de Zootechnie, 49*(3), 231–253. https://doi.org/10.1051/animres:2000119

Nardone, A., Ronchi, B., Lacetera, N., Ranieri, M. S., & Bernabucci, U. (2010). Effects of climate changes on animal production and sustainability of livestock systems. *Livestock Science, 130*(1–3), 57–69.

Nicholson, S. E. (2017). Climate and climatic variability of rainfall over eastern Africa. *Reviews of Geophysics, 55*(3), 590–635.

Niloofar, P., Francis, D. P., Lazarova-Molnar, S., Vulpe, A., Vochin, M.-C., Suciu, G., Balanescu, M., Anestis, V., & Bartzanas, T. (2021). Data-driven decision support in livestock farming for improved animal health, welfare and greenhouse gas emissions: Overview and challenges. *Computers and Electronics in Agriculture, 190*, 106406.

Olsson, K., & Dahlborn, K. (1989). Fluid balance during heat stress in lactating goats. *Quarterly Journal of Experimental Physiology: Translation and Integration, 74*(5), 645–659.

Ominski, K., & Wittenberg, K. (2005). Strategies for reducing enteric methane emissions in forage-based beef production systems. In *Climate change and managed ecosystems* (pp. 261–272). CRC Press. https://doi.org/10.1201/9780849330971.ch13

Polley, H. W., Briske, D. D., Morgan, J. A., Wolter, K., Bailey, D. W., & Brown, J. R. (2013). Climate change and North American rangelands: Trends, projections, and implications. *Rangeland Ecology & Management, 66*(5), 493–511.

Pulido, M., Schnabel, S., Contador, J. F. L., Lozano-Parra, J., Gómez-Gutiérrez, Á., Brevik, E. C., & Cerdà, A. (2017). Reduction of the frequency of herbaceous roots as an effect of soil compaction induced by heavy grazing in rangelands of SW Spain. *CATENA, 158*, 381–389.

Read, D., Beerling, D., Cannell, M., Cox, P., Curran, P., Grace, J., Ineson, P., Jarvis, P., Malhi, Y., & Powlson, D. S. (2001). *The role of land carbon sinks in mitigating global climate change* (Royal Society Policy Document 10/01). Royal Society.

Reynolds, C., Crompton, L., & Mills, J. (2010). Livestock and climate change impacts in the developing world. *Outlook on Agriculture, 39*(4), 245–248.

Rhoades, E. D., & Locke, E. H. (1964). Water intake on a sandy range as affected by 20 years of differential cattle stocking rates. *Rangeland Ecology & Management/Journal of Range Management Archives, 17*(4), 185–190.

Rojas-Downing, M. M., Nejadhashemi, A. P., Harrigan, T., & Woznicki, S. A. (2017). Climate change and livestock: Impacts, adaptation, and mitigation. *Climate Risk Management, 16*, 145–163. https://doi.org/10.1016/j.crm.2017.02.001

Ronchi, B., Lacetera, N., Bernabucci, U., Supplizi, A. V., & Nardone, A. (1999). Distinct and common effects of heat stress and restricted feeding on metabolic status of Holstein heifers. *Agriculture and Food Sciences*, Corpus ID: 87015667.

Sebastian, K. (Ed.). (2014). *Atlas of African agriculture research and development: Revealing agriculture's place in Africa.* International Food Policy Research Institute (IFPRI). https://doi.org/10.2499/9780896298460

Smith, P., Bustamante, M., Ahammad, H., Clark, H., Dong, H., Elsiddig, E. A., Haberl, H., Harper, R., House, J., & Jafari, M. (2014). Agriculture, forestry and other land use (AFOLU). In *Climate change 2014: Mitigation of climate change. Contribution of working group III to the fifth assessment report of the intergovernmental panel on climate change* (pp. 811–922). Cambridge University Press.

Steinfeld, H., Gerber, P. J., Wassenaar, T., Castel, V., Rosales, M., & De haan, C. (2006). *Livestock's long shadow: Environmental issues and options* (Vol. 24). Food and Agriculture Organization of the United Nations. ISBN: 978-92-5-105571-7.

Stocker, T. (2014). *Climate change 2013: The physical science basis: Working group I contribution to the fifth assessment report of the intergovernmental panel on climate change.* Cambridge University Press.

St-Pierre, N. R., Cobanov, B., & Schnitkey, G. (2003). Economic losses from heat stress by US livestock industries. *Journal of Dairy Science, 86*, E52–E77. https://doi.org/10.3168/jds.S0022-0302(03)74040-5

Sundquist, E. T. (1993). The global carbon dioxide budget. *Science, 259*(5097), 934–941.

Tamboli, P., Chaurasiya, A. K., Upadhyay, D., & Kumar, A. (2023). Climate change impact on forage characteristics: An appraisal for livestock production. In R. K. Singhal, S. Ahmed, S. Pandey, & S. Chand (Eds.), *Molecular interventions for developing climate-smart crops: A forage perspective* (pp. 183–196). Springer Nature Singapore. https://doi.org/10.1007/978-981-99-1858-4_10

Thomas, M. G., Mohamed, A. H., Sawalhah, M. N., Holechek, J. L., Bailey, D. W., Hawkes, J. M., Luna-Nevarez, P., Molinar, F., & Khumalo, G. (2015). Long-term forage and cow-calf performance and economic considerations of two stocking levels on Chihuahuan desert rangeland. *Rangeland Ecology & Management, 68*(2), 158–165.

Thornton, P. K. (2010). Livestock production: Recent trends, future prospects. *Philosophical Transactions of the Royal Society B: Biological Sciences, 365*(1554), 2853–2867. https://doi.org/10.1098/rstb.2010.0134

Thornton, P. K., van de Steeg, J., Notenbaert, A., & Herrero, M. (2009). The impacts of climate change on livestock and livestock systems in developing countries: A review of what we know and what we need to know. *Agricultural Systems, 101*(3), 113–127.

Torell, L. A., Murugan, S., & Ramirez, O. A. (2010). Economics of flexible versus conservative stocking strategies to manage climate variability risk. *Rangeland Ecology & Management, 63*(4), 415–425.

Tubiello, F., Schmidhuber, J., Howden, M., Neofotis, P. G., Park, S., Fernandes, E., & Thapa, D. (2008). Climate change response strategies for agriculture: Challenges and opportunities for the 21st century. In *Agriculture and rural development discussion paper* (Vol. 42).

Watson, R. (2000). Land use, land use change and forestry. In *A special report of the IPCC*. Cambridge University Press.

Webster, A. J. F. (1991). *Metabolic responses of farm animals to high temperature* (Vol. 55, pp. 15–22). EAAP Publication.

Wellesley, L., Happer, C., & Froggatt, A. (2015). Changing climate, changing diets: pathways to lower meat consumption. In *Chatham house report*. Retrieved December 2, 2024, from https://www.chathamhouse.org/2015/11/changing-climate-changing-diets-pathways-lower-meat-consumption

West, J. W. (2003). Effects of heat-stress on production in dairy cattle. *Journal of Dairy Science, 86*(6), 2131–2144. https://doi.org/10.3168/jds.S0022-0302(03)73803-X

Wikimedia Commons. (2020). *Map of dry lands geomorphology*. https://commons.wikimedia.org/wiki/File:Drylandmap.jpg

Willms, W., Kenzie, O., Mcallister, T., Colwell, D., Viera, D., Wilmshurst, J., Entz, T., & Olson, M. (2002). Effects of water quality on cattle performance. *Journal of Range Management, 55*, 452–460. https://doi.org/10.2307/4003222

Woodward, S. L., Waghorn, G. C., Ulyatt, M. J., & Lassey, K. R. (2001). Early indications that feeding lotus will reduce methane emissions from ruminants. *Proceedings of the New Zealand Society of Animal Production, 61*, 23–26.

Yang, C., Gao, P., Hou, F., Yan, T., Chang, S., Chen, X., & Wang, Z. (2018). Relationship between chemical composition of native forage and nutrient digestibility by Tibetan sheep on the Qinghai-Tibetan Plateau. *Journal of Animal Science, 96*(3), 1140–1149. https://doi.org/10.1093/jas/sky002

Part II
Agriculture and Food Security

Chapter 5
Sustaining Food Security in Arid Lands Countries-Augmenting Domestic Agricultural Production Through Corporate Farming and Food Imports: A Trade-Off

Majda Khalil Suleiman and Shabbir Ahmad Shahid

Abstract Globally farming is accomplished on various sized farms including small, medium and large farms. Among the total farms, over 475 million farms occupy area of less than 20,000 m^2 per farm and operate at about 12% of the world's agricultural lands. Whereas family farms operate about 75% of the world's agricultural lands. The number of small farms looks significant, but, these farms operate only on small share of global farmland, but produce 1/3rd of world food. The small farms are generally occurring in developing arid land countries and belong to poor farmers, these are mainly subsistence farms, providing food to the families with small share to sell in the market. According to FAO estimates farm sizes are getting smaller and smaller per family member in most regions (especially Africa and China) except Europe. Many farmers sell their land, migrate to cities work in factories to earn for their livings. The countries which are located in hyperarid environment with scarcity of arable land and water, but are capital rich due to oil resources, these countries depend on imported food as main source, in addition to limited local food production. Therefore, these countries explore trade with potential countries in two ways, food import and opportunity for corporate farming by leasing arable land for specific period of time. Such trade developments provide opportunity to the citizens of the host country for employment and infrastructure development. In general, corporate farming leads to a more efficient industry and more social benefits in both countries, the host and the foreign investor, although if not managed properly, corporate farming may have negative effects too in the form of affecting rural economies. In some developing countries with economic deficit and to earn foreign exchange corporate farming

M. K. Suleiman (✉) · S. A. Shahid
Food Security Program, Environment and Life Sciences Research Center, Kuwait Institute for Scientific Research, Safat, Kuwait
e-mail: mkhalil@kisr.edu.kw

S. A. Shahid
e-mail: sshahid@kisr.edu.kw

M. K. Suleiman and S. A. Shahid (eds.), *Fostering Arid Lands Agriculture in the Face of Climate Change*, Advances in Global Change Research 81,
https://doi.org/10.1007/978-3-031-98490-7_5

is becoming more prevalent, which in one way has concerns of this shift, on the other hand there are opportunities for the host countries, however, both the countries are in *win–win* situation and then such understandings can go for a long way. In authors opinion establishment of new cities in the unused, remotely located marginal lands and simultaneously practicing corporate farming with the technical support of national soil science societies to address marginality issue is viable option to increase food production and release pressure of urbanization on existing overcrowded cities.

Keywords Food security · Marginal lands · Soil science societies · Virtual water · Water footprint · Food trade · Food policies · Foreign investor · Economic development

5.1 Introduction

The crops, animals, water, energy and food security are interrelated and go along with each other in a chain format with positive synergies (Suleiman & Shahid, 2023a). The loose link in this chain affects each other and have significant consequences in achieving food demand to feed current and ever increasing population especially in Asian and African countries (IPCC, 2023), and to achieve the ambitious United Nations Sustainable Development Goals (SDGs) especially SDG-2 (end hunger by 2030). Agriculture is backbone of many countries, some have surplus of food commodities, others have food deficit, this leads to trade food commodities among countries of different regions. Globally there are over 608 million farms of different sizes (Lowder et al., 2021), 84% of the total farms are smaller than 2 ha, in contrast the large farms (>50 ha) are 1%. Analyses of agriculture land between 2000 and 2021 has revealed a general decrease of agriculture land globally and in all regions to various extents and reasons. This suggests to either intensify crop production from existing agriculture lands or bring the marginal lands into agriculture production after removing the marginality, a costly process but doable with investment and using modern technologies.

In poor countries poverty and malnutrition is high, the rural population is depending on agriculture for substantial parts of their income and food security, directly or indirectly. These rural poor's need sustainable policies for economic growth to be permanently out of poverty. But, their continuous use of farm lands is marginalized and not producing to their full capacities. Improving the quality of marginal lands of rural poor's is essential. In addition, bringing the unused marginal lands into production will not affect the smallholder farmers, as they will continue growing crops on their lands.

To meet the food demands of growing population, the food deficit countries need to trade with potential countries with surplus food commodities, especially, the cereals (staple food of many poor populations in Asia and Africa). Most people living in marginal areas of Asia and Sub-Saharan Africa have the lowest per capita land holdings and agricultural GDP (World Bank, 2008). In the light of above, and to increase food production, we have to have a paradigm shift in our understanding and use marginal lands for corporate farming by mobilizing the financial resources locally and involving international investors. The national soil science societies can provide scientific support to their governments to bring the marginal lands into production and establish corporate farming to increase local production, minimize the import and increase the export. The *pros and cons* of establishing corporate farms must be fully investigated and deliberated prior to final decision, and policy development.

The 2021 figure of world agriculture emission is 7.8 Gt CO_2eq., of which 53% (4.13 Gt CO_2eq.) emission is from livestock related activities, and 37% (2.9 Gt CO_2eq.) from enteric fermentation in ruminant livestock (FAO, 2023a).

5.2 Global Farm Sizes and Areas Under Agricultural Lands

Globally, farming is accomplished over 570 million farms in 167 countries (FAO, 2014; Lowder et al., 2016), where 84% farms are less than 2 ha (20,000 m^2) size, and control 12% of the land. Farms (less than 1 ha) account for 70% of all farms, but operate only 7% of all agricultural land, whereas farms between one and two hectares account for 14% of all farms and control 4% of the land (Lowder et al., 2021). Resultantly, it can be summarized that there are 510 million farms of less than 2 hectares size and account for 84% of all farms and occupy 12% of global agricultural land, and nearly 40% of land is found on farms larger than 1000 ha (Lowder et al., 2021). Only 1% farms are bigger than 50 ha but these farm control 65% of agricultural land. Australia has by far the largest average holdings' area since the 1990 census round (between 3,000 and 4,500 ha) (FAO, 2023b). Currently Lowder et al. (2021) using updated estimate that is based on methodology comparable to that used for the estimate presented in FAO (2014) reported more than 608 million farms (179 countries). However, whether these are small or large farms, these are mainly categorized as family-owned and operated farms. The small holder subsistence farming community is growing crops for their own use and buy other food products from the market. Other farmers to some extent sell their food products in the market with variable prices based on the season and market demand. The business as usual has led the world to food crises especially in the poor African countries vulnerable to climate change impacts. Recently, Lowder et al. (2021) has provided latest census stating that globally small farms account for 84% of all farms, occupying about 12% of all agricultural land, and produce roughly 35% of the world's food. Whereas, the largest farms (> 50 ha) are 1% of global farms but occupy more than 70% of the world's farmland. It is also established that average farm sizes have increased in the developed world and decreased in the developing world (Dawe, 2015; Vliet et al., 2015). Exceptions being in some land abundant countries in Africa (Jayne et al., 2014), and parts of Asia (Masters et al., 2013) where farm size increased, as well as increase in farm size from 1960 to 2,000 in some upper-middle-income countries (Lowder et al., 2016). Latest data on average farm size (FAO, 2014; cf. Lowder et al., 2021) shows the smallest average farm size in South Asia (1.4 ha), East Asia and

Pacific (1.5 ha), Sub-Saharn Africa excluding South Africa (1.6 ha) and Middle East and North Africa (3.6 ha).

There could be many reasons of decrease in farm size, such as, increase in family size and inheritance, population density, soil salinization and farms abandonment, and socio-economic factors etc. It is also visualized that smallholder farmers are mainly subsistence farmers, and their contribution to food production is lower than the large farms. Therefore, increase in large farms (farm land) may raise the concerns that smallholder farmers are not being appreciated and their livelihood may be weakened. Agriculture development with smallholder farmers may provide opportunities to poor farmers to improve their livelihood and wellbeing.

5.2.1 Food Production by Small and Large Farms

Small farms have been defined in two ways (i) farms where labor is provided by the family (Lowder et al., 2016), and (ii) where size of the farms is less than 2 ha (Lowder et al., 2021). A number of attempts have been made to determine the share of the world's food produced by smallholders (FAO, 2014; Herrero et al., 2017; Ricciardi et al., 2018) and by family farms (FAO, 2014; Graeub et al., 2016). The smallholder farmers (less than 2 ha) are estimated to produce 28–31% total food production and 30–34% of food supply on 24% of gross agricultural area (Shroff, 2022). Therefore, empowering the smallholder farmers with best agriculture production innovative technologies is the key to feed the current and escalating population. Another report stated small farms (less than 2 ha) produce 35% of the world food (Lowder et al., 2021).

The regional assessment of food production has shown small farms (≤20 ha) are producing more than 75% of most food commodities in Sub-Saharan Africa, Southeast Asia, South Asia, and China, whereas, very small farms (≤2 ha) owned by millions of smallholder farmers are important and have local significance in Sub-Saharan Africa, Southeast Asia, and South Asia, and are contributing about 30% of most food commodities (Herrero et al., 2017). Recently, Lowder et al. (2021) has presented the latest census on the number of farms based on sizes, and food production. They stated that globally over 608 million farms are reported, of which over 90% are family farms, occupying 70–80% of farmland and producing about 80% of the world's food in value terms. Our analyses show that Lowder et al. (2021) has extended the previous work (Lowder et al., 2016) and presented census about farms differently, using new datasets, the first analysis was reported based on the 2010 census (FAO, 2014). The increase of farms from 570 to 608 million is likely be due to economic growth and agricultural development, as well as supporting United Nations Sustainable Develop Goal 2 (end hunger by 2030).

5.3 Food Import Phenomenon in Capital Rich Water Scarce Countries

Some countries in the world are capital rich due to abundant natural resources (oil and gas) but have water and arable land scarcity for crop production to meet food demand, thus, depending on food import from countries who have sufficient food production to meet their local demand and to trade with other countries for financial gains. These capital rich countries face no food insecurity for their population. This suggests that import–export of food is a global demand to feed the current and rapidly increasing population. In the recent past, the trend of corporate farming is gaining importance to increase crop production, a way forward to break the inertia in small scale farming to intensify crop production.

5.3.1 Water Foot Print, Food Import and Virtual Water

Water footprint is the sum of the volume of freshwater (blue, green, and gray) used to produce the product along its supply chain (Hoekstra et al., 2011). The WF concept was developed by Hoekstra in 2002 (Hoekstra, 2003). ***Green*** WF is the rainwater and soil moisture consumed by plants and crops in their cultivation; ***Blue*** WF is the surface and groundwater used for irrigation (i.e., evaporated or incorporated into a crop) during cultivation and ***Gray*** WF *is* freshwater required to dilute pollution and bring the water resource up to safe water quality standards.

Globally, the food is floating *to and fro* among various countries based on the countries need to feed their population. The country's national statistics determines which food product and quantity is to be imported and exported and, what are the potential countries having abundant food resources of specific product like cereals, vegetables and other crops. Globally there have been trade of export worth 1,663,365 million US$, including fruits and vegetable 308,406 million US$ and cereals 243,605 million US$ (FAO, 2022a).

Virtual water is the volume of water used to grow crops and produce food products. During the import–export trade, food importing countries also bring virtual water to the country, thus the exporting countries along with food products also export their national water resources. The crops internal and external water footprints (total national consumption: Eq. (5.1) are recently described by Suleiman and Shahid (2023b), expressing total water used to grow crops internally (internal, WF), and the water imported (virtual) by importing food products from other countries (external WF).

As an example, a country is importing 1 ton of tomato from food exporting country. The global water footprint to produce 1 kg of tomato is 214 L (Mekonnen & Hoekstra, 2011) therefore, importing country is also importing 214 cubic meter of water. In case the country is growing tomato internally, the country would have used 214 cubic meter of water within national boundary.

$$\sum \text{WF consumption national} = \sum \text{WF consumption internal} + \sum \text{WF consumption external} \tag{5.1}$$

5.4 Constraints for Food Production in Capital Rich Arid Region Countries

The food production in capital rich arid region countries (oil and gas producing) is constrained due to multiple factors, harsh climatic conditions, high temperature, low rainfall, poor soil qualities and water scarcity. Therefore, these countries have practiced open field agriculture (selected crops) in winter months when temperature is low, and produce cash crops (vegetables) in protected environment agriculture throughout the year. However, this practice requires high energy to cool the greenhouses. These countries produce food locally to small extent and major portion of food is imported.

5.4.1 Water Stress Constrained Internal Crop Production

Any country facing water stress internally poses significant limitations to grow crops within their national boundaries to offset the national food requirements. Water stress is recently defined (FAO, 2023c) as the share of freshwater withdrawal in available freshwater resources, after taking into account environmental water requirements. The countries most affected by water stress are predominantly located in the Western and Central Asia as well as Northern Africa. The most acute water stressed countries reported in 2000 are Kuwait, United Arab Emirates and Saud Arabia withdrawing annually 9 to almost 40 times their renewable freshwater resources available (FAO, 2023a). Such a high level of water withdrawal may lead to fast depletion of renewable water resources.

The water stressed countries in general grow crops to a limited extent internally, but mainly import food from potential countries where food resources are abundant. The acute water stressed countries grow specific crops in open field agriculture especially forages throughout the year and selected cash crops during the winter season. In addition, cash crops (vegetables) are grown throughout the year under protected environment agriculture (PEA) system (FAO, 2021).

Another factor which constrained local crop production in acute water stressed countries is the quality of the groundwater which is becoming salty over the years and this reduces choices of crops to grow. At extreme salinity conditions the farm land may be abandoned. Globally, during 2020 and 2021, about 86 million hectares agriculture land has reduced (FAO, 2023d). In contrast the global harvested area of primary crops went up 24% between 2000 and 2021 to 1.5 billion ha (FAO, 2022b).

The cultivation of cereals crops represents over half of the world's harvested area during the period, even though their share declined to 50% in 2021 (FAO, 2023a). The significant decline in the cereal harvested area is a great concern to future food security as in many countries cereals are staple food. This significant decline could be either due the impact of climate change or the shifting trend of crops by the farmers towards high income cash crops. Recently, it has been observed that in some Asian countries the cost of production for cereal (wheat) production was much higher than the price declared by the government to buy from the farmers, and the government imported the wheat at lower price to meet their food demand, ignoring the local purchase from the farmers.

5.4.2 Crop Land Area per Capita per Region (FAO, 2023d)

Global population is projected to reach 10.35 billion by the end of 2100 (IPCC, 2023), against the one planet Earth, which has limited global arable land, and decreasing due to soil degradation and urbanization (FAO-ITPS, 2015). The rapid population growth is in Asia and Africa regions. The combination of population growth, biotic and abiotic factors have affected the croplands, resulting into per capita cropland decreased to 0.2 ha per capita in almost all regions between 2000 and 2021, culminating into world average decline by 18% (FAO, 2023d). The highest decline is in Africa (25% per capita to 0.21 ha per capita). Three countries (Kazakhstan, Australia and Canada) have the highest per capita cropland. Recent data (FAO, 2023d) has revealed a decrease of 2% (0.09 billion ha) global agriculture land in 2021 (4.70 billion hectares) compared with 2000. The report further stated that about 1/3rd of the total agricultural land was cropland (1.58 billion ha in 2021). The FAO (2023a) report after analyzing large database shared key facts (i) the production of primary crops was 9.5 billion tonnes in 2021, a 54% more than in 2000, (ii) four crops (sugar cane, maize, wheat and rice) account for about half of global primary crop production, (iii) cereals are the most traded commodity by quantity in 2021 (the Americas and Europe are the largest exporters and Asia is the largest importer) (FAO, 2022a).

It is apparent (Table 5.1) that in the last 2 decades, there is general decrease of agriculture land area per capita. This can be explained by two statements, (i) decrease in area may be due to increase in population from 2000 to 2021, where land holdings become small due to inheritence distribution, (ii) agriculture land may be degraded due to soil salinization, waterlogging, pollution or soil erosion. Such a reduction in agriculture land, if continues in the comig years, will have significant consequence on global food security, especially the countries located in drylands (hyperarid, arid, semiarid, dry subhumid).

Table 5.1 Change in cropland area per capita between 2000 and 2021 (regional perspectives). Cf. FAO (2023d)

World/regions	Cropland area per capita (ha) 2000	Cropland area per capita (ha) (2021)	Percent decrease in cropland area per capita (2000–2021)
World	0.24	0.20	−16.7
Africa	0.28	0.21	−25.0
Americana	0.44	0.37	−15.9
Asia	0.15	0.13	−13.33
Europe	0.42	0.39	−7.14
Oceania	0.82	0.77	−6.10

5.5 Global Food Trade (Export and Import)

The import and export through regional and international trade is the method to maintain the food reserves to meet the food demand among countries. During the 2000 and 2021 the monetary value of food export increased by 4.4 times, that is from US$ 380 billion in 2000 to US$ 1.66 Trillian (FAO, 2022a). The high increase being in oil and fats prices. The world and regional cereals commodities (wheat, maize and rice) export and import is shown in Table 5.2, where Table 5.3, depicts whether the region is net importer or exporter in certain commodities.

Table 5.2 illustrates that there is regional difference in import and export of cereals. Therefore the cereals commodities float between the regions. Here, when the export is more than the import, the region is called net-exporter and when the import is higher that export, called net-importer (Table 5.3). The Table 5.2 shows a difference of 1.18% between import and export of commoditites in the world, which, in reality should match exactly. The differennce is due to rounding the values in one decimal. The values are in thousands, it is assumed that the values less then 50 tons may be rounded to zero.

Table 5.2 World and regional cereals imports and exports by main commodities (thousand tonnes) (FAO, 2022a)

World/ region	Imports (000 tonnes)				Exports (000 tonnes)			
	Wheat	Maize	Rice	Other crops	Wheat	Maize	Rice	Other crops
World	201,010.1	199,321.1	50,919.1	63,185.9	198,139.3	196,075.4	50,654.8	63,495.6
Africa	44,919.7	17,554.7	17,628.7	4,043.7	2,89.3	4,071.4	1,038.7	244.2
Americana	23,048.7	44,653.2	5,496.1	3,725.0	57,402.5	131,210.7	5,800.6	17,371.1
Asia	96,385.2	101,256.1	22,565.9	40,666.1	10,463.4	6,576.1	41,448.3	2,012.2
Europe	35,707.5	35,740.5	4,510.7	14,649.0	104,419.1	54,127.7	2,276.4	33,127.8
Oceania	949.0	116.6	717.8	102.0	25,565.1	89.5	90.8	10,740.2

Table 5.3 Net cereals and other crops exporting and importing regions (FAO, 2022a)

Regions	Cereal crops			Other crops
	Wheat	Maize	Rice	
Africa	Importer	Importer	Exporter	Importer
Americana	Exporter	Importer	Exporter	Exporter
Asia	Importer	Importer	Importer	Importer
Europe	Importer	Exporter	Importer	Exporter
Oceania	Importer	Importer	Importer	Exporter

Table 5.4 Percent agriculture land area equipped for irrigation (World and regional perspective) (FAO, 2023d)

World/region	Agriculture land by use (000 hectares) Year 2021	Area equipped for irrigation (000 hectares) Croplands year 2021	% area equipped for irrigation Year 2021
World	1,579,878.1	352,497.0	22.3
Africa	293,253.0	17,013.0	5.8
Americana	377,345.1	56,719.0	15.0
Asia	587,038.9	248,240.0	42.3
Europe	288,071.5	27,212.0	9.4
Oceania	34,169.7	3,312.0	9.7

Table 5.4 shows variable agriculture land area coverage by irrigation in different regions, the lowest being in Africa, where most of the food poduction comes from rainfed agriculture, and the highest in Asia. In Asian countries, such as India and pakistan, there are natural rivers due to snow melting at Himalays which makes tributaries and convert to rivers of various capacities. There is an excellent example of Pakistan, where World largest irrigation system exists due to water flowing from five rivers (Sindh, Satlej Chenab, Jhelum and Ravi). In summer, there is enough water in these rivers, due to monsoon season, when high floods are very common. In winter due to low rainfall, and snow at mountains, less water is available in the rivers. In some years, country faces prolonged droughts affecting wheat crop significantly. Under such conditions, it is recommended to build different sized dams, and harvest water for later use when water crises come due to drought condition.

5.5.1 Import and Export of Food Commodities in Arid Land Countries

The arid land countries are plagued with many constraints (natural or anthropic) hindering meeting the food requirement of the population by growing crops and rearing livestock within national boundaries. The main constraints are, water and arable land scarcity, high temperature, low rainfall (aridity index), desertification and soil salinization (Shahid, 2013, 2023; Shahid et al., 1999, 2018; Zaman et al., 2018). These factors have led the arid land countries to trade food commodities with other countries where water and land resources are abundant and have surplus food after meeting their local food demand. Our analyses of latest data available (FAO, 2023a, 2023b, 2023c) has revealed that no country is self-sufficient in all food commodities (cereals, vegetables, oil crops, pulses and other crops etc.). In order to present latest picture of import–export by main arid land (Middle east countries) countries, latest data is retrieved from FAO (2023a, 2023b, 2023c) which is cited from FAO (2022a), and presented in Table 5.5.

Table 5.5 Export and import of food commodities by Middle East Countries (FAO, 2022a)

Countries↓	Import food commodities (000) tonnes				Export food commodities (000) tonnes				Remarks
Food commodities→	Wheat	Maize	Rice	Other	Wheat	Maize	Rice	Other	
Middle East countries									
Bahrain	91.9	19.0	63.9	6.1	–	0.0	0.0	0.0	Net importer
Iran	7075.2	9778.8	883.2	3342.4	–	81.9	2.6	6.0	Net importer
Iraq	1261.9	1063.3	1255.4	392.6	0.6	–	0.1	0.0	Net importer
Jordan	853.1	778.2	177.9	745.0	42.5	57.3	1.3	0.0	Net importer
Kuwait	484.9	118.6	199.9	527.8	–	0.0	0.0	0.0	Net importer
Lebanon	754.2	439.8	60.8	52.3	8.4	14.3	32.3	4.6	Net importer
Oman	2803.6	335.3	215.2	89.5	2.2	72.3	12.1	0.5	Net importer
Qatar	188.9	72.2	160.9	182.9	–	0.1	0.0	–	Net importer
Saudi Arabia	2277.2	2935.1	1172.5	5395.7	0.0	0.1	3.4	0.0	Net importer
Syria	31.7	124.3	180.9	103.2	38.3	0.4	0.1	3.6	Net importer
United Arab Emirates	1410.2	656.8	561.5	540.5	87.3	35.7	58.8	61.9	Net importer
Yemen	3065.1	660.8	640.3	5.9	0.8	10.6	0.0	1.4	Net importer

The import/export analyses of Middle East Countries has shown that all countries are net importer of cereals and other crops. Table 5.6 clearly shows that except India, all other countries are net importer of cereals and other crops.

It is clear from Table 5.7 that Australia, Canada, France, Russian Federation, Ukraine and United States of America are the main countries who have the capacity for high export of wheat, whereas, Argentina, Brazil, Ukraine and USA are high exporter of maize and India for rice export. Overall, the listed countries import lesser quantities than exporting commodities, and can be listed as net exporter, exceptions are for some commodities for certain countries. Except Brazil and Paraguay, all countries are net exporter of other commodities (export > Import).

5.6 Are Large or Corporate Farms Solution to Food Security?—Pros and Cons

The analyses of published literature on number and sizes of farms and their food production capacities, has revealed that large farms (>50 ha) are 1% of 608 million farms occupying 65% of farm land and producing 70% food. Here is the point, should the government develop policies to promote smallholder famers or large farm holders. In authors opinion, both farms holders to be encouraged to produce food to meet their own needs and sell small quantities to the market (subsistence farmers), and large farms to produce agriculture commodities at large scale to feed the local population and to export extra food to food importing countries. The Tables 5.5 and 5.6 clearly show, the countries in need to large quantities of food, such as the capital rich GCC countries.

The question rises, if the corporate farming is a solution to feed the current and escalating population. Our analyses (Table 5.1) of crop land per capita (2000–2021) clearly show that there is regional decline of land per capita, such as a decrease of 6.1, 13.33 and 25.0% is found in Oceania, Africa and Asia respectively. This can be interpreted, as if the decline is due to number of reasons including natural (climate change), biotic (diseases/pests) and abiotic (drought, salinity) and urbanization of existing croplands etc. It can also be hypothesized that, the available land in different countries has already been used for agriculture and there is no land left to put in agriculture.

Having said this, there are rich countries, who wish to make investment in agriculture production and want to opt large farms, some called land grabbing, others called *win–win* situation by both the investors and the host country, where the host country will have financial gains, infrastructure development and employments to poor population. If, however, both parties develop consensus that, the host country will allow some portion of food to take out of the country, after meeting the local requirements, especially at the time of pandemics, there is no harm in developing

Table 5.6 Import and export of food commodities by Arid region and other countries (FAO, 2022a)

Countries↓	Import food commodities (000) tonnes				Export food commodities (000) tonnes				Remarks
Food commodities→	Wheat	Maize	Rice	Other	Wheat	Maize	Rice	Other	
Bangladesh	6982.4	1774.9	2579.1	14.7	0.0	20.2	12.5	0.1	Net importer
China	10,987.2	32,705.8	5339.8	22,350.0	4.4	6.8	2615.7	28.1	Net importer
Egypt	5773.7	6982.0	153.7	8.8	0.0	4.8	31.5	5.0	Net importer
India	0.1	24.4	3.6	62.2	6090.6	3615.8	21,034.7	170.9	Net Exporter
Malaysia	1318.1	3555.9	1142.0	48.3	0.5	11.1	96.7	0.3	Net importer
Morocco	4668.5	2171.9	50.5	322.2	0.0	0.0	3.5	0.0	Net importer
Nepal	265.3	584.5	1098.3	22.8	0.0	0.0	0.0	0.1	Net importer
Pakistan[a]	2486.2	29.0	7.7	18.0	0.0	556.7	3933.0	3.0	Net importer
Sri Lanka	1306.6	8.2	147.1	17.7	8.6	0.0	6.8	0.2	Net importer

[a] Net exporter in maize and rice

Table 5.7 Main net cereals exporting countries (FAO, 2022a)

Countries↓	Import food commodities (000) tonnes				Export food commodities (000) tonnes				Remarks
Food commodities→	Wheat	Maize	Rice	Other	Wheat	Maize	Rice	Other	
Argentina	0.1	5.5	2.2	14.4	9485.2	36,912.0	274.6	4441.8	Net exporter
Australia	1.1	2.5	203.7	4.4	25,563.0	86.9	73.7	10,739.5	Net exporter
Brazil	6225.1	3205.6	686.8	523.6	1129.3	20,429.6	770.9	4.2	Net exporter
Bulgaria	49.7	88.1	51.6	8.2	5119.2	1759.3	73.9	448.4	Net exporter
Canada	217.8	2844.3	410.8	354.3	21,546.5	1688.2	17.5	5479.3	Net exporter
France	263.8	568.2	572.8	153.8	16,090.5	4303.0	45.3	6961.2	Net exporter
India	0.1	24.4	3.6	62.2	6090.6	3615.8	21,034.7	170.9	Net exporter
Kazakhstan	1048.2	8.2	184.7	99.7	3814.5	57.5	122.1	1486.9	Net exporter
Paraguay	0.1	71.8	0.4	0.5	337.5	1902.9	643.7	0.8	Net exporter
Poland	594.9	222.0	212.8	290.4	3591.5	2286.0	38.4	2587.0	Net exporter
Romania	953.1	961.4	63.5	414.2	6097.7	6904.4	18.8	21,140.0	Net exporter
Russian Federation	123.2	35.0	190.7	13.1	27,366.4	2936.4	117.4	4436.2	Net exporter
Serbia	1.4	4.9	12.4	7.2	1070.9	2300.6	0.8	129.2	Net exporter
Ukraine	17.7	18.6	76.2	64.9	19,394.9	24,539.5	8.9	5696.7	Net exporter
United States of America	1450.8	651.5	953.9	1942.7	24,014.2	70,041.4	2837.6	7083.7	Net exporter

such understanding. Such deals will give chance to host countries for economic development, provision of jobs, and supporting other industries (Coker, 2008; Shahid & Behnassi, 2014).

5.6.1 Potential Countries for Corporate Faming to Bridge Gap for Food Security

The literature survey on agriculture lands, import and export trade among different countries, and natural resources has revealed that corporate farming cannot be accomplished in all countries. Such a high and strategic investment comes from local and foreign investors, based on multiple factors including, legal aspects of the host country, climatic conditions, local and foreign market access, availability of local manpower, suitability of soils for agriculture production, available water resources, and knowledge to rectify marginality of lands etc. These are the countries where corporate farming has the potential, amidst some constraints (https://www.quora.com/).

USA—Large agricultural area, variable climate, modern technologies. In general, foreign investment is possible, in some states may be restricted.

Canada—Good soil quality and strong agricultural sector, foreign investment permitted to own specific number of hectares.

Australia—Large agriculture area, good environmental conditions and healthy agriculture export market. Legal deal can be made with the government to own land, without compromising the local farmers.

New Zealand—Conducive climate for agriculture, and favorable export market. Foreigners can own land; some conditions apply to buy large area.

Brazil—Vast arable land, multiple opportunities for agriculture, growing demand for food export. Investors have to negotiate regulations with government.

Argentina—Strong agriculture sector and good quality fertile land. Investors have to study political and economic viability.

Spain—Favorable climate for diversified crop production, strong agriculture sector and accessibility to European Union market. Local legalities to be completed.

Ukraine—A niche for foreign investment, noteworthy agriculture potential and sector liberalized. Investor has to study geopolitical situation.

Foreigners can also lease lands of various sizes for long time in Asian and African countries (net food importer) after negotiating the legal deals, with the condition to protect the local farmers and the market. However, there are some important points, which any foreign investors must consider while making the land deals for corporate farming in any country, including but not necessarily limited to: legal implications, restriction for foreign ownership of land, existing infrastructure and availability of essential resources (land, water, labor), local and foreign market, risks assessment regarding political and economic issues.

5.6.2 Corporate Farming

It has been realized that large farms are producing significantly more food than small farms. This suggests adoption of corporate farming to boost crop production through increasing cultivated land on marginal (never used before) and land currently in use for cultivation but either not producing economically or abandoned due to increase in marginality, lack of knowledge and financial resources to address the marginality. For an easy understanding the concept of corporate farming, we briefly defined corporate farming without affecting the smallholder livelihoods "*large scale cultivation of marginal unused, and abandoned agricultural lands, by local and foreign investors using climate smart and regenerative agriculture practices in a win–win scenario among the host country and the investors*". If this definition is well adopted, then the roles of both parties to be established, policies are developed for in-country use or export of agriculture commodities, infrastructure development and employments to local communities. The definitions of corporate farming are greatly debated, some views, it affects the smallholder farming, other looks from an angle that, in some countries land was given to small farmers, who were not able to cultivate due to lack of financial support and remote areas. Under such conditions, if the government takes this land back and convert to corporate farms, provided, replacement of good quality land is given to small farmers whose land has been retrieved by the government, this should be a good solution to bring the unused lands under corporate farming. Therefore, the corporate farming should be seen in a context.

5.6.3 Developing Corporate Farms

Developing corporate large size farms is a viable option to increase crop production, however, there are different opinions of different groups on such development. One school of thought is of the opinion that small farms are to be transformed to corporate farms, second school of thought suggests bringing potential marginal lands into crop production. The authors are pro-second school of thought. This way the smallholder farmers (subsistence farmers) continue living on their farms, meeting their food demands and contributing to food security. The second school of thought is more appealing, due to increased cropland area, which will ultimately increase per capita share of agriculture land, which has declined between 2000 and 2021 (FAO, 2023d). Bringing marginal lands to crop production is a challenge, and the cost of rectifying marginality increases based on the level and extent of marginality.

This model has recently been suggested in Pakistan. Critics say the corporate farming legislations that invite investment and regulate farm operations are heavily focused on big investors, squeezing small farmers. With this corporate model it is assumed that Pakistan can potentially bring around 22 million acres (8.96 million hectares) under cultivation through corporate farming, and currently, Pakistan's import of agri-products stands at more than US$10 billion per annum (https://www.dawn.com/news/1809409) which is huge pressure on foreign-exchange bills. Increase in local food production is essential in Pakistan to feed the current and growing population. Pakistan total export and import of food products during 2021 valued 4709 and 9138 million US$ respectively, a difference of 4429 million US$ (FAO, 2022a), suggesting Pakistan is net importer of food products.

5.6.4 Understanding Marginality

There are multiple definitions of marginal lands.

- Where cost-effective production is not remunerated (Peterson & Galbraith, 1932).
- An area with poor agronomic characteristics (OECD, 2001).
- Limitations of land for sustained application of a given use (CGIAR/TAC, 2000).
- Poor combination of physical and chemical characteristics of the soils for the productivity potential (USDA-NRCS, 2010).
- Untapped potential of areas (less favored) that are farther from markets (e.g. lack of transport infrastructure), although with crop production potential (World Bank, 2003).
- Lands supporting a yield of up to 40% of its crop potential as marginal (FAO & UNEP, 2010).
- Marginal lands as areas where "cost-effective production is not possible" (Dauber et al. (2012).

Considering the above definitions, it can be speculated that the population living in marginal areas may be facing extreme poverty (Gatzweiler et al., 2011) due to uneconomical use of marginal areas for crop production (high cost of production) and other uses. Even the land under cultivation can be of marginal quality for agriculture if not properly used, e.g., marginality due to salinization, and vulnerability to soil erosion. It is therefore, essential to determine the level and cause of marginality before taking a decision to transform barren land to agriculture, pros and cons need to be full understood and means to mitigate marginality identified and implemented using modern technologies to gain agriculture benefits, conserve the soils and protect the environment.

The marginal lands could be (i) land never been used due to reasons (remote area without road infrastructure and access to market, (ii) lack of water for irrigation, (iii) uneven landscape which is uneconomical to level etc. Such lands can be termed as Less Favored Agricultural Areas (LFAAs) (Barbier & Hochard, 2014), where about 1.75 billion people worldwide (about 38% of the rural populations) lived on remote LFAAs, and are facing poverty. Nearly 1.6 out of 1.75 billion people are living in marginal areas of developing countries (Barbier & Hochard, 2014). Another estimate shows marginal lands are about 21% of the total land resources (Alexandratos & Bruinsma, 2012). Other estimates (Kang et al., 2013; Wood et al., 2000) have shown marginal lands are about 36% of global agriculture land (1.3 billion ha) and supporting about one-third of global population. Campbell et al. (2008) reported that global area of abandoned agriculture is 385–472 million ha. Besides the marginal lands currently not in any use, of 1.6 billion hectares cultivated land, 0.3 billion (19%) are marginal lands, 0.44 billion ha (28%) are prime lands, and 0.82 billion ha (52%) are good lands (FAO, 2011). It is apparent that besides the marginal lands which have never been used, there are marginal lands within currently used agriculture land. We should make efforts to reverse the marginality of abandoned and marginal agricultural lands through scientific diagnostics of causes of abandonment and implementation of integrated soil management program.

In order to fully utilize the marginal lands, it is strongly suggested to consult national soil science societies for technical assistance who can better translate the mode of marginality and suggest cost-effective and environment friendly solutions to use the marginal lands for agriculture.

5.7 Food Security, Its Component and National Food Security Policies

Any country on the globe wishes to have sufficient availability of nutritious food within national boundaries, as the food security is basic right of the human well-being to sustain healthy life, economic development and political stability. This is only possible when the country has sufficient natural resources (soil, water), conducive environment for diversified crop production, access to advance production technologies, and trained human to produce food. However, this is not the case in many poorly developing arid countries of Africa and Asia, and these countries depend on other countries food resources. To have sufficient food available all the time in the country, easily accessible and at affordable price, the governments have to design food security policies after expert's consultation, to assure timely availability of food even during political instabilities and pandemics. This requires strong commitments from the governments, to allocate budget in case food is to be imported, develop guidelines and share with farmers to grow strategic crops to meet the local requirements and financial earning through food export.

There are four components of food security:

(i) Food availability at all times.
(ii) Accessibility (easy access to food at near markets, with transportation facilities).
(iii) Affordability (food available at affordable prices for all income groups to meet basic food requirements).

(iv) Nutritious (nutritious, balanced and healthy diets to keep the society heath to perform).

In the lights of above, it is pertinent that population and food production grow simultaneously. To achieve all above, it is essential that the government to develop robust food security policies based on the following essential points.

- Consult national soil science societies, and identify unused marginal remote areas (not easily accessible), never been used for agriculture or other purposes.
- Develop new cities in marginal lands, and establish infrastructures and basic facilities (roads network, basic facilities, hospitals, schools, IT services, shopping areas etc.).
- Allocate the marginal lands around the new cities to local and foreign investors for corporate farming, and generate employments to the residents of new cities.
- Engage the investors with soil science societies for technical guidance to reduce the marginality of lands and grow crops successfully.
- The food produced from these new corporate farms is additional food grown from existing farming areas.

This is the ideal framework to improve food production at country and global scale, knowing that agriculture land per capita has been reduced in all the regions (developed and developing). It is hoped that this ambitious approach will mitigate the impact of climate change and anthropogenic impacts hindering meeting the food demand as envisaged in the United Nations Sustainable Development Goal No. 2 (End hunger by 2030).

5.8 Conclusions and Recommendations

The analyses of published literature on food production has revealed that both the small and large scale farms are contributing significantly to food production and have pivotal role in food and nutritional securities, and are key to conserve crops diversity. The production of these farms may be affected due to natural (climate change) and human effects. Therefore, the farmers are to be equipped with innovative technologies (plants, irrigation, harvesting and processing) to maintain and intensify the pace of food production (crop per drop) to meet the food demand of escalating population. In order to be sustainable in local crop production, policies are to be articulated to grow crops based on the national needs and to export surplus food to other counties. The analyses further determined that there is regional variation of import–export of cereals products and, thus cereals float from region to region for financial gains to increase national capital to strengthen financial base, and to do business with international market. The idea floated in this chapter to build new cities in remote area and providing basic facilities to attract the local population which is surging in the already crowded cities, and attract local and foreign investors to take the initiative to grow crops in these marginal areas through corporate farming to increase food production and financial gains. Biophysically agricultural marginal lands if used for cultivation without addressing the marginality will not produce optimally and consistently, therefore, these local and foreign investors are to be engaged with national soil science societies for technical support to address the marginality issues holistically and leading to high production, and increasing the food production above the normal food production capacities of the existing small, medium and large farms. Thus, these scenarios will not only produce more food, but also release pressure on big cities to over populate, which have significant environmental consequences.

Acknowledgements We would like to thank the FAO management to allow, copying, redistribution and adaptation for non-commercial purposes, provided that the work is appropriately cited. FAO further stated that work is made available under the Creative Commons Attribution-NonCommercial-ShareAlike 3.0 IGO licence (CC BY-NC-SA 3.0 IGO; https://creativecommons.org/licenses/by-nc-sa/3.0/igo/legalcode). In this chapter proper citation is given to FAO publications.

References

Alexandratos, N., & Bruinsma, J. (2012). *World agriculture towards 2030/2050, the 2012 revision* (ESA Working Paper No. 12-03). Food and Agriculture Organization of the United Nations. https://www.fao.org/3/ap106e/ap106e.pdf

Barbier, E. B., & Hochard, J. P. (2014). Land degradation, less favored lands and the rural poor: A spatial and economic analysis. In *A report for the economics of land degradation initiative.* http://www.edwardbbarbier.com/Projects/ELD/Economics_of_Land_Degradation_Initiative_Barbier_and_Hochard_abridged.pdf

Campbell, J. E., Lobell, D. B., Genova, R. C., & Field, C. B. (2008). The global potential of bioenergy on abandoned agriculture lands. *Environmental Science & Technology, 42*(15), 5791–5794.

CGAIR/TAC. (2000). *CGIAR research priorities for marginal lands.* CGIAR (Consultative Group on International Agricultural Research)/Technical Advisory Committee (TAC). https://cgspace.cgiar.org/handle/10947/332

Coker, M. (2008, September 10). UN chief warns on buying farms. *The Wall Street Journal.*

Dauber, J., Brown, C., Fernando, A. L., Finnan, J., Krasuska, E., Ponitka, J., Styles, D., Thrän, D., Groenigen, K. J. V., Weih, M., & Zah, R. (2012). Bioenergy from "surplus" land: Environmental and socio-economic implications. *BioRisk, 7*, 5–50https://doi.org/10.3897/biorisk.7.3036

Dawe, D. (2015). *Agricultural transformation of middle-income Asian economies: Diversification, farm size and mechanization* (Vol. 15, pp. 1–56). FOA.

FAO. (2011). *The state of the world's land and water resources for food and agriculture (SOLAW)—Managing systems at risk.* Food and Agriculture Organization of the United Nations and Earthscan. https://doi.org/10.4324/9780203142837

FAO. (2014). The state of food and agriculture. In *Innovation in family farming.* FAO.

FAO. (2021). *Unlocking the potential of protected agriculture in the countries of Gulf Cooperation Council—Saving water and improving nutrition, Cairo* (p. 193). https://doi.org/10.4060/cb4070en

FAO. (2022a). *Trade: Crops and livestock products.* FAOSTAT. [Cited October 2023]. https://www.fao.org/faostat/en/#data/TCL; FAO. (2023). (1976–2021). FAO [Cited October 2023]. https://www.fao.org/fishery/statistics-query/en/trade/trade_value

FAO. (2022b). *Production: Crops and livestock products.* FAOSTAT. [Cited October 2023]. https://www.fao.org/faostat/en/#data/QCL. https://doi.org/10.4060/cc8166en-fig61

FAO. (2023a). *World food and agriculture—Statistical yearbook 2023, Rome.* https://doi.org/10.4060/cc8166en

FAO. (2023b). *Structural data from agricultural censuses.* FAOSTAT [Cited October 2023]. https://www.fao.org/faostat/en/#data/WCAD. https://doi.org/10.4060/cc8166en-figb4

FAO. (2023c). *AQUASTAT dissemination system.* AQUASTAT [Cited October 2023]. https://data.apps.fao.org/aquastat/?lang=en

FAO. (2023d). *Land use.* FAOSTAT [Cited October 2023]. https://www.fao.org/faostat/en/#data/RL. https://doi.org/10.4060/cc8166en-fig59

FAO and ITPS. (2015). *Status of the world's soil resources (SWSR)—Technical summary* (p. 77). Food and Agriculture Organization of the United Nations and Intergovernmental Technical Panel on Soils.

FAO and UNEP. (2010). *A decision support tool for sustainable bioenergy: An overview* (p. 50). United Nations Energy Publication. ISBN: 978-92-5-106638-6.

Gatzweiler, F. W., Baumüller, H., Husmann, C., & von Braun, J. (2011). *Marginality: Addressing the root causes of extreme poverty* (ZEF Working Paper Series 77). Center for Development Research, University of Bonn.

Graeub, B. E., Chappell, M. J., Wittman, H., Ledermann, S., Kerr, R. B., & Gemmill-Herren, B. B. (2016). The state of family farms in the world. *World Development, 87*, 1–15.

Herrero, M., Thornton, P. K., Power, B., Bogard, J. R., Remans, R., Fritz, S., Gerber, J. S., Nelson, G., See, L., Waha, K., Watson, R. A., West, P. C., Samberg, L. H., van de Steeg, J., Stephenson, E., van Wijk, M., & Havlík, P. (2017). Farming and the geography of nutrient production for human use: A transdisciplinary analysis. *The Lancet Planetary Health, 1*(1), e33–e42. https://doi.org/10.1016/S2542-5196(17)30007-4

Hoekstra, A. Y. (2003). Virtual water trade. In *Proceedings of the international expert meeting on virtual water trade, Delft, The Netherlands, 12–13 December 2002* (Value of water research report series no. 12). UNESCO-IHE.

Hoekstra, A. Y., Chapagain, A. K., Maite, M., Aldaya, M. M., & Mekonnen, M. M. (2011). *The water footprint assessment manual—Setting the global standard.* Earthscan Ltd.

IPCC. (2023). Summary for policymakers. In Core Writing Team, H. Lee, & J. Romero (Eds.), *Climate change 2023: Synthesis report. Contribution of working groups I, II and III to the sixth assessment report of the intergovernmental panel on climate change* (pp. 1–34). IPCC. https://doi.org/10.59327/IPCC/AR6-9789291691647.001

Jayne, T. S., Chamberlin, J., & Headey, D. D. (2014). Land pressure, the evolution of farming systems, and development strategies in Africa: A synthesis. *Food Policy, 48*, 1–17.

Kang, S., Post, W. M., Nichols, J. A., Wang, D., West, T. O., Bandaru, V., & Izaurralde, R. C. (2013). Marginal lands: Concept, assessment and management. *Journal of Agricultural Science, 5*(5), Article 129.

Lowder, S. K., Skoet, J., & Raney, T. (2016). The number, size, and distribution of farms, smallholder farms, and family farms worldwide. *World Development, 87*, 16–29. https://doi.org/10.1016/j.worlddev.2015.10.041

Lowder, S. K., Sánchez, M. V., & Bertini, R. (2021). Which farms feed the world and has farmland become more concentrated? *World Development, 142*(2), 105455. https://doi.org/10.1016/j.worlddev.2021.105455

Masters, W. A., Djurfeldt, A. A., De Haan, C., Hazell, P., Jayne, T., Jirström, M., & Reardon, T. (2013). Urbanization and farm size in Asia and Africa: Implications for food security and agricultural research. *Global Food Security, 2*(3), 156–165. https://doi.org/10.1016/j.gfs.2013.07.002

Mekonnen, M. M., & Hoekstra, A. Y. (2011). The green, blue and grey water footprint of crops and derived crop products. *Hydrology and Earth System Sciences, 15*(5), 1577–1600. https://doi.org/10.5194/hess-15-1577-2011

OECD. (2001). *The organisation for economic co-operation and development—Marginal land definition.* https://stats.oecd.org/glossary/detail.asp?ID=1591

Peterson, G. M., & Galbraith, J. K. (1932). The concept of marginal land. *Journal of Farm Economics, 14*(2), 295.

Ricciardi, V., Ramankutty, N., Mehrabi, Z., Jarvis, L., Chookolingo, B. (2018). How much of the world's food do smallholders produce? *Global Food Security, 17*, 64–72. https://doi.org/10.1016/j.gfs.2018.05.002

Sallustio, L., Pettenella, D., Merlini, P., Romano, R., Salvati, L., Marchetti, M., & Corona, P. (2018). Assessing the economic marginality of agricultural lands in Italy to support land use planning. *Land Use Policy, 76*, 526–534. https://doi.org/10.1016/j.landusepol.2018.02.033

Shahid, S. A. (2013). Developments in salinity assessment, modeling, mapping, and monitoring from regional to submicroscopic scales. In S. A. Shahid, M. A. Abdelfattah, & F. K. Taha (Eds.), *Developments in soil salinity assessment and reclamation-innovative thinking and use of marginal soil and water resources in irrigated agriculture* (Ch. 1, pp. 3–43). Springer Dordrecht Heidelberg. https://doi.org/10.1007/978-94-007-5684-7_1

Shahid, S. A. (2023). Desertification: A central problem to restoring ecosystems. In M. K. Suleiman, & S. A. Shahid (Eds.), *Terrestrial environment and ecosystems of Kuwait* (Ch. 5, pp. 119–137). Springer. https://doi.org/10.1007/978-3-031-46262-7_5

Shahid, S. A., & Behnassi, M. (2014). Climate change impact in the Arab region: Review of adaptation and mitigation potential and practices. In M. Behnassi, M. S. Mteng'e, G. Ramachandra, & K. Shelat (Eds.), *Vulnerability of agriculture, water and fisheries to climate change—Toward sustainable adaptation strategies* (Ch. 2, pp. 15–38). Springer. https://doi.org/10.1007/978-94-017-8962-2_2

Shahid, S. A., Omar, S. A. S., & Al-Ghawas, S. (1999). Indicators of desertification in Kuwait and their possible management. *Desertification Control Bulletin, 34*, 61–66. UNEP, ISSN: 0379-2455.

Shahid, S. A., Zaman, M., & Heng, L. (2018). Salinity and sodicity adaptation and mitigation options. In M. Zaman et al. (Eds.), *Guideline for salinity assessment, mitigation and adaptation using nuclear and related techniques* (Ch. 3, pp. 55–88). Springer. https://doi.org/10.1007/978-3-319-96190-3_3

Shroff, J. (2022). Why smallholder farmers are central to new food security interventions. *World Economic Forum.* Retrieved December 16, 2024, from https://www.weforum.org/stories/2022/09/smallholder-farmers-key-achieving-food-security/

Suleiman, M. K., & Shahid, S. A. (2023a). Prospective of agricultural farming in Kuwait and energy-food-water-climate nexus. In M. K. Suleiman, & S. A. Shahid (Eds.), *Terrestrial environment and ecosystems of Kuwait* (Ch. 15, pp. 363–391). Springer. https://doi.org/10.1007/978-3-031-46262_15

Suleiman, M. K., & Shahid, S. A. (2023b). Agricultural water footprint of major crops grown in Kuwait compared to the world average: A review. In M. K. Suleiman, & S. A. Shahid (Eds.), *Terrestrial environment and ecosystems of Kuwait* (Ch. 16, pp. 393–414). Springer. https://doi.org/10.1007/978-3-031-46262_16

USDA-NRCS. (2010). United States Department of Agriculture-Natural Resources Conservation Services (USDA-NRCS). National Soil Survey Handbook 430-VI.

Vliet, J. A. V., Schut, A. G. T., Reidsma, P., Descheemaeker, K., Slingerland, M., Van de Ven, G. W. J., & Giller, K. E. (2015). De-mystifying family farming: Features, diversity and trends across the globe. *Global Food Security, 5*, 11–18. https://doi.org/10.1016/j.gfs.2015.03.001

Wood, S., Sebastian, K., & Scherr, S. J. (2000). *Pilot analysis of global ecosystems: Agroecosystems.* International Food Policy Research Institute. https://www.wri.org/research/pilot-analysis-global-ecosystems-agroecosystems

World Bank. (2003). *World development report 2003* (p. 250). Oxford University Press. https://openknowledge.worldbank.org/handle/10986/5985

World Bank. (2008). *Word development report 2008: Agriculture for development*. The World Bank. https://openknowledge.worldbank.org/handle/10986/5990

Zaman, M., Shahid, S. A., & Heng, L. (2018). *Guideline in salinity assessment, mitigation and adaptation using nuclear and related techniques* (p. 164). Springer. https://doi.org/10.1007/978-3-319-96190-3

Chapter 6
Potential Technological Innovations to Advance Arid Lands Agriculture in the Context of Climate Change

Shabbir Ahmad Shahid and Majda Khalil Suleiman

Abstract Crop production in the arid lands of developing countries has been significantly constrained due to the impact of climate change. The increase in temperatures and greenhouse gases emissions induced by climate change have a direct impact on agriculture activities, due to water scarcity, increased soil salinization and subsequent reduction in crops choices, as well as increased droughts and dry spells. Other factors affecting agricultural performance in arid lands are multi-faceted, including but not necessarily limited to, lack of improved climate-resilient crop varieties, lack of timely availability of inputs and farm machinery for crop harvesting, and storage facilities, etc. The persistent climate change impact without mitigation measures will surely hamper global agriculture in general and arid land agriculture in particular, where it affects both irrigated and rainfed agriculture. Innovations in agricultural technologies and their adoption by farmers at large have always been important and will be even more vital in the context of climate change. Potential innovative biotechnological research ideas (introduction of biological nitrogen fixation character in non-legume plants, water absorption by stomata from atmosphere, enhancing photosynthesis rate by doubling sunlight absorption by plants leaves, opening and closing of stomata at high and low temperatures to control CO_2 absorption and reduction in transpiration, reducing postharvest food losses and waste) are proposed for the green revolution to feed the current and escalating population. These technologies once developed, tested and proven successful must be transferred to farmers for wider adoption in over 571 million farms, especially in arid regions. In order to develop innovative technologies, effective policy options need to be designed, and the government should support and encourage farmers to boost crop production for food security and agricultural development. Finally, it should be noted that the development of new innovative technologies needs the interest of governments and international organizations to

S. A. Shahid (✉) · M. K. Suleiman
Food Security Program, Environment and Life Sciences Research Center, Kuwait Institute for Scientific Research, Safat, Kuwait
e-mail: sshahid@kisr.edu.kw

M. K. Suleiman
e-mail: mkhalil@kisr.edu.kw

M. K. Suleiman and S. A. Shahid (eds.), *Fostering Arid Lands Agriculture in the Face of Climate Change*, Advances in Global Change Research 81,
https://doi.org/10.1007/978-3-031-98490-7_6

invest in agricultural research to boost crop production to feed the growing population. Only such an approach can enable vulnerable communities of arid lands to use the natural resources in a sustainable manner to ensure food security in the future.

Keywords Agriculture · Innovations · Climate change · Business as usual · Policy development · National agriculture research system · Paradigm shift

6.1 Introduction

Agriculture is the backbone of many countries and contributes to the national gross domestic product (GDP) at various levels. Agricultural production in the past many years has been constrained by multiple factors, including the impact of climate change and anthropogenic effects. Agriculture has a long history and since then, improvements have been made in agricultural technologies, related to plants, soil, water, fertilizers, diseases, and insect pests. Even though the farmers in poor arid land countries, mainly in Asia and Africa, are not able to achieve the potential yields and face yield gaps, there exists a considerable gap between progressive farmers and subsistence farmers. In addition, soil degradation due to erosion and salinization is also at the peak, where 20% (62 million ha) of all irrigated land (340 million ha) is salinized (Qadir et al., 2014). The global salt-affected soil map (FAO, 2021, 2022) has revealed that 424 million hectares (ha) of land are affected by surface salinity (0–30 cm) and 833 million ha by subsurface salinity (30–100 cm). It looks we are at the state of *Insanity (doing the same thing over and over again and expecting different results).* Thus, there is a need for a paradigm shift *(seeing something in a new and different way that creates a huge change in thinking and behavior)* in our understanding by adopting high-level technologies including but not limited to the use of remote sensing (RS) and geographic information system (GIS), drone assessment of crop performance, precision agriculture, and climate smart agriculture (CSA) combined with regenerative agriculture (RA) practices. The approach should be on a science-based resource assessment, leading to informed decisions to select crops that are climate smart and resilient. There are three big challenges related to climate change (IPCC, 2023), (i) adapting and building resilience to a more challenging climatic environment, (ii) achieving a substantial reduction in atmospheric greenhouse gases (GHGs) concentrations, and (iii) intensifying crop production.

Without the adoption of innovative technologies, the UN Sustainable Goals (SDGs) seem very ambitious to achieve by 2030. Especially SDG-2 (end hunger) which is intended to be achieved by 2030. Target 2.3 states that by 2030, double the agricultural productivity and incomes of small-scale food producers, and Target 2.4 by 2030 ensures sustainable food production systems and implements climate-resilient agricultural practices that progressively improve land and soil quality. These targets can be achieved in collaboration with farming communities, national agricultural research systems (NARS) and with the support and commitment of international organizations, by adopting regenerative and climate-resilient agricultural approaches.

Climate change has a strong connection with agricultural productivity. Climate change is likely to reduce agricultural production and stability in countries where food insecurity already exists. Therefore, the adoption of CSA is crucial to achieve food security and achieve climate change goals. Poor developing countries need significant changes to achieve food security in response to climate change. In addition, escalating population growth, food consumption and dietary patterns require an increase in agricultural production by at least 70% by 2050 to meet food demand.

The CSA has three important pillars, (i) sustainability and increasing agricultural production, (ii) adapting and building resilience to climate change and, (iii) reducing GHG emissions. The CSA includes various management components (crops, livestock, soil and water, agroforestry and integrated food energy systems). These management systems are described in the following sections, as well as the innovative technological ideas proposed, which the authors believe are the future hopes to bring a revolution to the agriculture sector, to mitigate climate change and crop intensification for food security.

Policy recommendations and developments in technology transfer to farming communities are essential to facilitate climate change adaptation and mitigation as effective means of producing climate-resilient plants communities in arid areas and promoting agricultural production for food security.

6.2 The Aridity Index, Climate and Climate Change

The ratio of total precipitation per year (P) and potential evapotranspiration (PET), both expressed in terms of mm/annum (P/PET) is called the climatic aridity index (Zomer et al., 2022) and determines various levels of indices for different regions, such as hyper-arid (<0.05), arid (0.05–0.20), and semi-arid (0.2–0.5). The hyper-arid zone covers 4.2%, the arid zone 14.6%, and the semi-arid zone 12.2% of the global land area, which is about 1/3 of the world's total land area. In the hyper-arid region, rainfall is less than 100 mm and rarely exceeds 100 mm (Shahid & Alkandari, 2024). Climate is a short term (day or weak) average weather condition over time and space. Whereas climate change is a long-term shift in average weather conditions in terms of change in temperature, rainfall, and wind.

6.3 Constraints to Agriculture Due to Changing Climate

Climate change impacts have been seen as the main cause of reduced crop production through the following ways:

- Prolonged warming-induced droughts
- Reduction in water resources
- Desertification

- Soil salinization.

To address the impacts of climate change, we need to address these issues holistically, leading to reduction in GHGs emissions (mitigation) and science-based adaptations to address the impact of climate change in order to sustainably intensify crops for food security. Crops intensification is essential as available arable land in most countries is either already under agriculture farming or cannot be brought to crop production due to lack of sufficient water resources to offset the water requirements of crops.

6.4 Potential Innovative Biotechnological Research Ideas

Researchers have been working since decades to breed high yielding, heat and disease resistant, varieties, most likely through conventional methods, such as crossing different lines of breeds to evolve hybrid varieties, etc. This is the time to look beyond conventional research and bring a revolution through innovative research, which has never been practiced in the past, thinking that it may not be possible to innovate some of the traits to be introduced in the new crops varieties or lines.

In the last few decades, there have been significant advancements in technologies, and researchers have used modern tools to conduct research. Given below are the innovative ideas (not currently practiced), which once accomplished, may lead to another revolution in crop intensification for food security to feed the current and rapidly growing population.

6.4.1 Exploitation of Atmospheric Nitrogen

Nitrogen is an essential and critical limiting element for plant growth and production. It is a major component of chlorophyll, the most important pigment needed for photosynthesis, as well as amino acids, the key building blocks of proteins (Wagner, 2011).

Earth's atmosphere is composed of about 78% nitrogen, 21% oxygen, 0.9% argon, and 0.1% other gases, with trace amounts of carbon dioxide, methane, water vapor, and neon being some of the other gases that make up the remaining 0.1% (Joseph, 2023). The composition of the air atmosphere above the ground at different altitude varies. For example, in the case of plant photosynthesis activity, the respiration by terrestrial living organisms takes place within the range of 12 km above the ground surface (Lee & Loiacono, 2024; NASA, 2024).

The atmosphere above the Earth has a total mass of about 5.15×10^{18} kg (David, 1996). Atmospheric air contains different gases in different proportions in terms of moles, such as nitrogen (78.08%), oxygen (20.95%), argon (0.93%), carbon dioxide (0.04%) and other trace gases. There is a huge source of nitrogen in the atmosphere, which is materialized through industrial conversion to nitrogen fertilizers. It is noteworthy that there is no raw material of nitrogen to manufacture nitrogenous fertilizers other than the atmospheric nitrogen.

Atmospheric nitrogen is extracted and reacted with the hydrogen from the natural gas (CH_4) at high temperature and pressure to form ammonia (NH_3), which then reacts with carbon dioxide exothermically to form carbamate salt, which then through heating transformed to urea fertilizer [($CO(NH_2)_2$].

6.4.1.1 Proposed Biotechnological Innovation—I: Innovating Biological Nitrogen Fixation in Non-Legume Crops

The history of biological nitrogen fixation (BNF) dates back to 1901 when Beijerinck discovered it and explained that it is carried out by a specialized group of microorganisms called prokaryotes (Beijerinck, 1901). He further explained that these organisms, after utilizing nitrogenase (enzyme), catalyze the conversion of atmospheric nitrogen (N_2) to ammonia (NH_3). There is a diversity of microorganisms [(Azotobacter, Azospirillum, Rhizobium, Actinomycetes and Cyanobacteria (Anabaena)] that have the ability to fix atmospheric nitrogen (N_2) in the roots of legume crops (Havlin et al., 2014).

6.4.1.2 Mechanism of Biological Nitrogen Fixation in the Roots of Legume Crops

Legumes are special type of plants, which present ball like structure on the roots called a *nodule*. There is diversity of bacteria in different quantities depending on the soil properties and the history of plants grown. The seeds of legume plants germinate, the roots start to grow, and the bacteria rupture the roots and enter into the roots and multiply (first step) there. The roots respond by forming a tumor like structure called nodules on the surface of roots (second step). Rhizobia (bacteria) absorb nitrogen (N_2) from the atmosphere and convert to NH_4 in nodules which in turn can be used by plants (Jones et al., 2007). The bacteria require 16 mol of adenosine triphosphate (ATP) to reduce each mole of nitrogen (Hubbell & Kidder, 2009) to form ammonia, which is taken from the host plants. Both the plants and rhizobia are mutually benefitted (symbiotic relationship), because rhizobia use energy (sugars, carbohydrates, ATP) from the host plant to fix N_2 and the rhizobia provides NH_4 to plants. It is to be noted that there are numerous nitrogen fixing bacteria (rhizobia) in soil which require specific host plant. Each rhizobium and plant has a different ability to fix atmospheric nitrogen.

6.4.1.3 Rational Behind Biological Nitrogen Fixation in Non-Legume Plants

Nitrogen is an essential plant nutrient and is incorporated into soil based on the plant's requirement. To offset the nitrogen requirement of the crop grown globally, over 110 million tons of nitrogen fertilizers are used (Zhao et al., 2022). However, if this industrial nitrogen is compensated through biological nitrogen fixation, there must be a significant cut on nitrogen fertilizer production and reduction in GHG emission. However, scientists have dreamed of enabling nitrogen fixation by rhizobia in non-leguminous crops to mitigate the increasing demand for nitrogen fertilizer (Zhao et al., 2022). It is noteworthy that during the industrial process of fertilizer production, huge quantities of GHGs are emitted, which causes climate change and global warming.

6.4.1.4 Innovating Biological Nitrogen Fixation by Rhizobia in the Roots of Non-Legume Crops

From the above, it is clear that only leguminous plants have the ability to host rhizobia bacteria for atmospheric nitrogen in a symbiotic relationship. The paradigm shift from this natural process is that, if BNF character is introduced in non-leguminous plants, and if this materializes, there will be an agricultural revolution in terms of crops intensification, mitigation of GHGs emission, significant reduction in nitrogen fertilizer production, etc. The authors are of the opinion that at this time of advancement in technological developments, the breeders should focus more on innovation rather than doing business as usual (BAU), which is not leading us to achieve UN-SDG-2 (end hunger) by the end of 2030. With this innovative idea, a literature review was conducted to understand whether such an approach has been investigated and/or adopted previously.

A study has presented some results on the impact of Rhizobium strains that colonize tomato and pepper roots and promote plant growth, yield, and quality of seedlings and fruits (García-Fraile et al., 2012). The study further revealed that, to date, there are no commercial biofertilizers for non-legumes based on rhizobia, although their safety for humans has been proven after decades of legume inoculation (Bashan, 1998; Bhattacharjee et al., 2009) and they have potential as non-legume plant growth promoters (Bhattacharjee et al., 2009; Lugtenberg & Kamilova, 2009). The literature has revealed that some scientists have visualized of enabling rhizobial nitrogen fixation in non-leguminous crops, however, despite decades of research, rhizobial nitrogen fixation in non-host plants has not been demonstrated (Zhao et al., 2022). A preliminary study has shown that a combination of N_2-fixing rhizobium and a clubroot pathogen (*Plasmodiophora brassicae*) exhibited a synergistic effect on nitrogen fixation in cruciferous plants (Zhao et al., 2022). Bacteria from the family of *Rhizobiaceae* and *Bradyrhizobiaceae* were detected by 16S rDNA high-throughput sequencing analysis in root cortical tissues of *P. brassicae*-infected rapeseed plants (Zhao et al.,2017). However, -N_2-fixing rhizobia nodulate legume plants almost exclusively, and four *Parasponia* species are the only known exceptions of rhizobium-induced nodule symbiosis with a non-leguminous species (Op den Camp et al.,2012). *Earlier work has also shown that Parasponia* is the only non-legume host plant known to be nodulated by rhizobia (Akkermans et al., 1978; Trinick, 1973).

The above description has clearly shown that there is a research gap for the inoculation of rhizobium to non-legume crops. However, preliminary results are promising for further research in this direction to expand non-legume plants for BNF, other than *Parasponia* species.

6.5 Efficient Plants Nutrients Use and Management

Apart from the induction of BNF character in non-legume plants, for efficient use of nutrients, the United Nations Food and Agricultural Organization recommends using 4R nutrients stewardship. This idea emerged after realizing that there is high quantity of fertilizers that are lost during the crop production process through various process (volatilization, leaching, soil fixation etc.). The 4R nutrient stewardship is recognized worldwide as the best nutrient management, however, there are constraints in poor arid land countries, inhibiting its full utilization. These are, (i) proper education for the farmers, (iii) timely availability of fertilizers, and (iii) lack of financial support to the farmers. Following are the key points of 4R nutrients stewardship.

6.5.1 Principles of Nutrients Stewardship

6.5.1.1 R1: Right Source

The right source of fertilizers matching the specific crop requirements and soil conditions is essential for efficient use of fertilizers. For example, fertilizer containing NO_3–N applied to sandy soils has high chances of nitrate leaching and pollution of groundwater. However, if NO_3–N polluted water is drained to lakes or drainage canals, eutrophication will occur and cause environmental problems.

Therefore, the following principles are recommended for right source of fertilizer application (SPRPN, 2021).

- Suit soil physical and chemical properties
- Recognize synergisms among nutrient elements and sources
- Recognize blend compatibility of materials
- Recognize benefits and sensitivities to associated elements
- Control effects of non-nutritive elements.

6.5.1.2 R2: Right Rate

Every crop requires certain quantities of nutrients at different plant growth stages. Therefore, science-based and location-specific recommendations given by the local extension services should be followed strictly for timely and proper application of fertilizers. It should be noted that each crop field contains residual nutrients, which were not utilized by the previous crop, these quantities should be taken into account while making fertilizer application plans. There are soil and water testing labs in almost all dryland countries, which provide soil and water testing services to farmers at a nominal cost, and farmer must avail these facilities ahead of time for timely and science-based recommendations on fertilizer use in the agricultural field.

The following are the recommended principles for the right rate of fertilizer application (SPRPN, 2021).

- Assess plant nutrient demand
- Assess soil nutrient supply
- Predict fertilizer use efficiency
- Consider soil resource impacts
- Consider economics and the law of diminishing returns.

6.5.1.3 R3: Right Time

Each plant nutrient has specific functions in plant growth, shoot and root development, seed and fruit formation and energy transfer. Therefore, fertilizer nutrients are added to the crop field based on the nutrient requirements of the crop at a specific growth stage. Generally, nitrogen application is divided into three doses, while phosphorus and potassium are mainly applied at the time of land preparation (basal dose).

The following are the recommended principles for the right time of fertilizer application (SPRPN, 2021).

- Assess timing of plant uptake
- Assess dynamics of soil nutrient supply
- Recognize dynamics of soil nutrient loss
- Evaluate logistics of field operations.

6.5.1.4 R4: Right Place

In order to enhance nutrient use efficiency, it is appropriate to place fertilizer in a place where plant can easily access. In some crops, plant to plant distance is wider (cotton crop 1–2 m), there is an empty area between the plants. Under such cropping pattern, if the farmer is using conventional broadcasting method for fertilizer application, it is highly likely that the fertilizer falling in the empty area will not be accessible to plants, reducing the fertilizer use efficiency. The current modern technique of efficient fertilizer use is called fertigation (fertilizer application to plants after mixing the fertilizer with irrigation water). Fertigation using drip irrigation system is recommended for arid lands. Foliar spray of fertilizer mixed with irrigation water may cause foliar injury due to hot weather in arid area. Phosphorus in soil can be largely fixed in the soils of arid land countries due to the presence of different quantities of calcium carbonates equivalents, thus placement of the right quantity of phosphate fertilizer nearby seeds (band placement) is highly recommended.

Recommended principles for the right place of fertilizer application (SPRPN, 2021) are given as under.

- Consider where plant roots are growing
- Consider soil chemical reactions
- Suit the goals of the tillage system
- Manage spatial variability.

6.6 Proposed Biotechnological Innovation-II: Photosynthesis Enhancement

The green plants (green chlorophyll) absorb red and blue sunlight (photosynthetic active radiation-PAR) to synthesize nutrients, and convert water, carbon dioxide, and minerals into oxygen and energy-rich organic compounds (carbohydrates), the core process of photosynthesis (Reaction – 1).

$$\begin{array}{l} \textit{Sunlight \& Chlorophyll} \\ \qquad\quad\downarrow \\ \qquad\qquad\qquad 6CO_2 + 6H_2O \rightarrow C_6H_{12}O_6 + 6O_2 + 6H_2O \end{array} \tag{6.1}$$

It should be clear that a small portion of sunlight is used to complete photosynthesis process, and not the whole sunlight, however, the best visible light wavelengths for photosynthesis are blue (425–450 nm) and red (600–700 nm). Blue (shortest wavelengths) carries the most energy, the red (longest wavelengths) carries the least energy.

6.6.1 Sunlight Absorbed by Plants

The plant absorbs sunlight based on many factors, including plant stage, canopy, chlorophyll contents, exposure to sunlight, and weather conditions, etc. However, if we look at the fraction of sunlight absorbed by plants to make photosynthesis process, it is very minute and about 0.1%. Researchers and innovators are exploring different and new ways and methods to enhance sunlight absorption by plants to increase photosynthesis intensity.

6.6.1.1 Proposed Innovation to Double Crop Production?

Here is the point which emerges to innovate photosynthesis process, through innovative biotechnological research. The proposed innovative research idea, thus, emerges from the discussion that if the biotechnologists/breeders can induce character in the newly developed crop varieties that enhances sunlight use from 0.1 to 0.2%, crop production may be doubled, provided other plant growth factors are fully understood and CSA practices are adopted globally through international awareness campaigns and technology transfer.

6.7 Proposed Biotechnological Innovation-III

Water is an essential part of human and plant life. Crop production in many arid countries is accomplished either by irrigating the crops by groundwater, surface water (rivers and canals) or by rainwater (rainfed agriculture), or a combination of both where appropriate. It has been recognized globally that there will be physical water scarcity in many arid land countries, and these countries will be the most suffered from climate change impacts. Conventional research has been progressing since decades on the efficient use of irrigation water using modern irrigation systems, including surface and subsurface drip, sprinkler, bubbler, and central pivot. Each system has its *pros and cons*, which must be fully understood and examined while making a decision on the final selection of right system. Since, in this section, we are proposing an innovative way of water use by plants, it is appropriate to briefly describe the different types of modern irrigation systems. The reader is referred to Zaman et al. (2018) for more details about the irrigation systems and zones of salinity development in the root zone and surrounding areas in the irrigated field.

6.7.1 Conventional Irrigation Systems

Following are the conventional irrigation systems.

6.7.1.1 Flood Irrigation System

Agriculture fields are irrigated through small water channels (lined or unlined) linked to canal-river system (Fig. 6.1a–e) or a small scale water source (water reservoirs). The size of the plot to be irrigated depends on the type of crops grown and soil texture. The area in the plot where the water enters the field receives more water compared to the other end of the plot. This system is suitable for fine textured soils where infiltration is low, and water moves faster in the field than in sandy soil. Water is not distributed uniformly.

(a) Flood irrigation (b) Border irrigation (c) Bubbler irrigation

d) Lines water channel (aflaj) (e) Unlined water channel (Aflaj) (f) Drip irrigation system

Fig. 6.1 Conventional and modern irrigation systems

6.7.1.2 Border Irrigation System

In this system crop field is divided into land parcels of various sizes and the irrigation is accomplished in the same way as flood irrigation through water channels, or through piped irrigation. The border system is more efficient than the flood irrigation in the large plot.

6.7.1.3 Surge Irrigation System

In this type of irrigation system, water is stopped by creating a barrier at the entry point of the channel, once the water has accumulated to a certain depth, the barrier is then removed and the water moves rapidly in the plot and cover the entire plot. Such a system is adopted in coarse textured soil (sandy soil) in small plots.

6.7.1.4 Furrow Irrigation System

Here, the soil for seeding or planting (seed beds) is prepared in the form of beds of different shapes (flat or sloppy) and sizes. The furrows associated with these soil beds are used to irrigate the crops, either by flowing water in both furrows or flowing water in alternate furrows. In each case, the soil salinity will occur in different soil zone (Shahid, 2013; Zaman et al., 2018). The farmer should be aware of the area of high salt accumulation before placing the seeds or transplanting, to avoid yield loss due to irrigation induced salinization.

6.7.1.5 Basin Irrigation System

This irrigation system is used for date palm/trees irrigation. A round basin s prepared around the tree to about 1-m radius (Fig. 6.1c). The sides are raised with soil to protect water overflow. A bubbler is attached to the water reservoir and trees are irrigated and the flow rate is controlled using flow meter.

6.7.2 Modern Irrigation Systems

Following are the modern irrigation systems used around the globe for efficient use of water resources to irrigate crops.

6.7.2.1 Drip Irrigation System

Drip irrigation system is the most efficient among other irrigation systems. It provides water to the plants at the root zone through the pipes consisting of small holes and each hole has emitter installed to release the water to irrigate the plants (Fig. 6.1f). The flow of water can be controlled for optimum irrigation to the selected crop. This leads to high water use efficiency and productivity. The entire system relates to water source (water tank) and through flow meters the delivery of the water is controlled. Drip irrigation works well when fresh water (low salinity) is used for irrigation, however, the holes in the pipes can be blocked with salts when brackish water is used to irrigate the crops (Zaman et al., 2018), thus it needs periodic flushing with fresh water to clean the pipes from salts. The pipes can be used at surface to irrigate the crops (surface drip) and can be buried at certain depth (subsurface drip) for efficient use of water by avoiding high surface evaporation (subsurface drip irrigation). Drip irrigation system is preferred over sprinkler system to avoid leaf burn due to salt water spray on leaves and to avoid the strong wind action which can push the water to the area where the crop does not exist.

6.7.2.2 Sprinkler Irrigation System

In the sprinkler irrigation system, water is sprayed with pressure through the nozzles installed at the top of sprinkler pipe head. The water is sprayed through the air to the field. The use of brackish water may cause leave burn due to toxic effects of ions (Na, Cl) in the irrigation water (toxicity effect). The sprinkler irrigation system is beneficial in terms of leaching the surface salts from the root zone to lower depths, when irrigation is accomplished more than the water requirement of the plants (evapotranspiration).

6.7.2.3 Central Pivot System

The central pivot (CP) irrigation system is a low to medium pressure fully mechanized automated of permanent assemble. It is different to that of sprinkler system, as the irrigation pipes hanging from the central pipe of the pivot are irrigating the plants near to the plants (also called water-wheel and circle irrigation) and not through spraying at sprinkler head. The central pivot system is designed in a way that irrigates the entire plot in a circle while rotating at certain time interval. The plots are prepared in a circle.

6.8 Proposed Biotechnological Innovation—III to Optimize Water Uptake by Plants to Mitigate Drought

In this proposed innovation, it is proposed that biotechnologists are to focus on research to develop a special character in cultivars to enhance absorption of water from the atmosphere and small quantities through root zone. It may sound strange, but in the era of technological advancements, innovations are expected and cannot be over-ruled, otherwise the impact of climate change in terms of global warming and associated prolonged drought will have a significant impact on crop production.

6.8.1 Moisture Level of Air Atmospheric

In addition to gaseous composition of atmosphere, the moisture in the atmosphere is present in different forms (water vapors, clouds). The quantity of water vapors and clouds vary from place to place and at variable distance from the ground surface. This depends on the humidity capacity of air determined by temperature. As an example, at 30 °C, a volume of air can contain up to 4.0% water vapor. At −40 °C, however, it can hold no more than 0.2% (https://www.britannica.com/science/humidity). Humidity is expressed in two ways, (i) absolute humidity and (ii) relative humidity. Absolute humidity is the mass of water vapors present in a given volume of air, and expressed in terms of milligrams/liter (mg/l) or grams per cubic meter (g/m^3), and the quantity of water vapors in atmosphere is determined by atmospheric temperature. With the increase of temperature water vapors are increasing and vice versa. For example, fully saturated air at 20 °C contains about 17 g m^{-3}, whereas at 37 °C it contains 44 g m^{-3} when saturated. It can be expressed numerically as, absolute humidity = mv/V where mv is mass of vapors (or vapor concentration) and V is the volume of air.

6.8.2 Physiological Mechanisms of Water Uptake by Plants

Water is considered an essential element of life, and there will be no life without water. Water is considered the most important and key abiotic factor for plants survival, growth and productivity. Water is a principal determinant of vegetation distributions worldwide (McElrone et al., 2013). If it is essential for plants to grow and produce, then why would plants waste so much of it through transpiration? Here is brief description, how the plants consume water for growth. The plants extract water from soil by its roots system and can retain upto 5% for cell expansion and plant growth. Essentially all of the water used by land plants is absorbed from the soil by roots (McCully, 1999). The water absorbed by the roots from soil travels to stems through the xylem, and water enters leaves via petiole (Sack & Holbrook, 2006; Sack & Tyree, 2005), and lost to atmosphere through stomatal opening, called transpiration.

To complete the photosynthesis process (described earlier), plants absorb CO_2 from the atmosphere by opening stomata (small pores in leaves), at this stage prolific amount of water is lost to atmosphere (transpiration) relative to the absorption of CO_2. For example, across plant species an average of 400 water molecules are lost for each CO_2 molecule gained (McElrone et al., 2013). Therefore, in nature, the plants keep balance between CO_2 absorption for photosynthesis to make sugars and subsequent water loss through stomata opening risking plants dehydration. The transpiration takes place through negative pressure commonly referred to as Cohesion-Tension (C-T) mechanism (Wheeler & Stroock, 2008).

6.8.3 Theory of Change for Innovation (TOC)

The TOC here can be described as, the biotechnologists to focus on innovative research to introduce following characters through breeding and/or genetic engineering.

(i) Introduce character in leaves to enhance water absorption from the atmosphere air and less demand on water absorption from soil by roots.
(ii) Develop character, changing the opening and closing the stomata, at high temperature reduce the stomata opening and at low temperature increase the stomata opening.
(iii) Enhance CO_2 absorption by stomata per unit time and reducing the time of stomatal opening to loose water during CO_2 absorption.

These are the innovative research ideas to be transformed into action to address the impacts of climate change in terms of plants adaptation to high temperature and mitigating the drought impacts and managing water scarcity in arid lands countries.

6.9 Managing Soils of the Saline Environment Through Biotechnological Innovative Research

Water scarcity in the arid land countries necessitates the use of marginal quality waters (saline/brackish) to irrigate crops. From an agronomical perspective, soil salinity restricts the area of agricultural land (Sardo & Hamdy, 2005) through economic losses either by farmland abandonment, or by limiting the productivity of agricultural crops, and impacts livelihood choices and land use strategies (Anik et al., 2018), further threatening food security (Rewald et al., 2013). In addition to decreasing crop yields, and from ecological context it decreases microbe population (Folli-Pereira et al., 2013) in soil, and destroys ecosystems diversity (Chaves et al., 2009), accelerates soil and environmental degradation processes (Estrada et al., 2013) and, consequently affects the entire food chain. Soil salinity has caused severe agronomical and ecological problems, particularly in arid and semiarid regions (Estrada et al., 2013; Himabindu et al., 2016). The damage is mainly due to the accumulation of water-soluble salts in the root zone causing osmotic variations, reducing the ability of plant root cells to absorb water from soil (Munns & Tester, 2008; Kamran et al., 2021; Stavi et al., 2021).

The use of marginal quality water (brackish) has led to salinization on 20% (62 million hectares) of the 310 million ha of irrigated lands (Qadir et al., 2014). Furthermore, it has recently been reported (IWMI, 2014) that we are losing 2,000 ha of land due to soil salinity in over 75 countries. The global salt-affected soils map (FAO, 2021) has mapped 424 million ha affected at surface (0–30 cm) and 833 million ha at subsurface (30–100 cm). This shows that soil salinization is not a recent phenomenon but has a long history (Shahid et al., 2018a). The civilizations of Mesopotamia (Jacobson & Adam, 1958) and Viru valley in Peru (Wiley, 1953) have failed due to irrigated soil salinization (Shahid et al., 2018a). Soil salinity has now degraded areas in the Aral Sea Basin in Central Asia, the Yellow River Basin in China, the Murray-Darling Basin in Australia, and the San Joaquin Valley in California at an alarming pace in the last 50 years (Chang & Silva, 2014). The rapid spread of salinization in irrigated agriculture fields requires holistic approach through innovative research to tackle this ailment to crop production.

In the light of the proposed biotechnological innovations described in the earlier section, it is believed that soil salinization problem can be controlled significantly, while the plants leaves start absorbing water from atmosphere and less water uptake (salty water) from the root zone. In addition, higher rates of photosynthesis through doubling the CO_2 absorption by plants from atmosphere is likely to double the crop production. Control on the opening of leaves stomata is likely to control transpiration. Thus, this is the challenge to focus research on the proposed innovative ideas to make a difference and break the *inertia* to current conventional agricultural practices (business as usual) to lead for another green revolution to feed the escalating population and to mitigate the impact of climate change, through increased soil carbon sequestration.

Currently to manage soil salinization in irrigated agriculture, scientific diagnostics of the problem at the farm level need to be made to implement integrated soil reclamation program to address the soil salinization and sodicity hazards to crop production (Shahid et al., 2008; Shahid & Rahman, 2011; Shahid, 2018a, b).

6.9.1 Proposed Innovation—IV: Develop Salt-Tolerance Character in Glycophytes

Plants based on their salt tolerance capacity are classified into two main classes, (i) glycophytes, and (ii) halophytes. The glycophytes are plants which grow well on soil conditions which are free or low in salt contents and have limited salt-tolerance mechanisms. Halophytes are plants which can tolerate salinity in soil and grow well in saline environment (Shahid & Alkandari, 2024) and avoid salt damage. However, both halophytes and glycophytes can undergo damage due to high soil salinity (Pan et al., 2020) when the root zone soil salinity is higher than the salinity threshold value of a specific crop (Ayers & Wescot, 1985; Shahid, 2013; Zaman et al., 2018). To offset the effects of salts, plants have evolved various mechanisms, including modulating ion homeostasis, ion compartmentalization and export, and the biosynthesis of osmo-protectants (Balasubramaniam et al., 2023). Salt-tolerant plants survive salt stress by vacuolization of toxic Na^+ and Cl^- in leaves, secrete excess salts by glands (Shahid, 2012), and accumulate osmolytes (proline and glycine betaine) as reported by Roy & Chakraborty (2014). The role of both Glycine betaine and proline in a variety of plant species have long been recognized to cope with environmental stresses, such as drought, salinity, extreme temperatures, UV radiation and heavy metals (Ashraf & Foolad, 2007). The challenge is how to increase the capacity of glycophytes to accumulate these two organic osmolytes, here is the role of biotechnologists to establish this through innovative research.

6.9.1.1 Theory of Change (TOC)

The theory of change in the proposed innovative research is to innovate a specific character of salt-tolerance existing in halophytes to glycophytes. With technological advances and recent developments in genomic and proteomic technologies, it may be possible to identify genes and proteins involved in plant salt-tolerance mechanisms (Balasubramaniam et al., 2023), and transferred to glycophytes. Such innovations will improve glycophytes salt tolerance level, and allow to cope the salinity impacts, enhance crop production and quality of major crops cultivated under saline conditions in arid and semiarid regions of the world.

6.10 Post-Harvest Losses and Waste in the Context of Food Supply Chain

Another area which requires immediate attention is the reduction of food losses and wastage. In poor African countries, farmers are not equipped with modern facilities to store the harvest for a few days before transporting it to the market. About 1/3rd of the food produced globally for human consumption (corresponding to about 1.3 billion tons) is lost or wasted each year along the food supply chain (FAO, 2011; Nicastro & Carillo, 2021).

The Director General of FAO Mr. Qu Dongyu at the 5th International Day of Awareness of Food Loss and Waste global event (27 September 2024) highlighted the critical need for financing and collaboration across the agrifood value chain. He further emphasized that reducing food loss and waste is crucial for improving food security and nutrition (FAO, 2024).

From the above discussion, it is clear that saving food from losses and waste can feed a significant number of the global population, especially African countries where a significant number of people go hungry on daily basis. There is an urgent need to reduce food losses and waste to feed the current population and the estimated 10 billion by 2050, which requires an increase in food production of at least 70% (FAO, 2009, 2019). Thus, there is a need to educate and train farmers with latest technologies to reduce food losses and waste from farm to fork.

6.11 Policy Development for Innovative or Biotechnology Research

The above innovative research ideas can only be developed and implemented on a large scale, provided the scientists and researchers lobby the policy developers and decision makers to appreciate the need for such a high-level biotechnological research. The decision makers must then raise their voices at the national and international levels to generate green funds and develop projects for interested countries where crop intensification is a high priority. The FAO, IFAD and CGIAR among others, are the potential international bodies to take the lead to generate funds and further innovative research to increase crop production.

6.12 Enhancing the Research-Extension-Farmers Link

One of the reasons for low crop production and high yield gaps is the lack of education and training to smallholder farmers (less than 2 ha farm land), representing 84% of the total 570 million farms globally (FAO, 2014; Lowder et al., 2016). These farmers need immediate attention for education and transfer of the latest technologies for crops intensification, which will improve livelihood and food security. Therefore, strengthening this link is vital to increase crop production and ensure food security.

6.13 Conclusions and Recommendations

There is an incessant trend of rapid population growth in Asian and African countries where drylands (hyperarid, arid and semiarid) predominate. These marginal lands are highly vulnerable to biotic and abiotic stresses and farmers are unable to achieve potential crop yields, resulting in a gap between current yield and potential yield. This has affected the food security of drylands substantially. One of the abiotic stresses is increasing soil salinization, which has affected 62 million hectares of irrigated land. The past historical trend and the current projection by IPCC have clearly shown a decline rather than an increase in crop yields during the twenty-first century. The business as usual is not leading us to feed the current and ever-growing population, so we require a paradigm shift in our understanding and a transition from conventional to innovative research. This will make a difference to utilize marginal land and water resources sustainably to intensify crops production vertically, rather horizontally due to lack of arable lands in most of the arid land countries, either due to water scarcity for irrigation or arable lands have already been used. The innovative researchable ideas proposed in this manuscript can be achieved provided the international organizations such as FAO, IFAD and CGIAR make a strong commitment to generate extraordinary funds and work closely with the National Agricultural Research Systems (NRAS) of arid countries. Along with international organizations, NARS may also lobby their governments to fund innovative research projects aimed at ensuring national food security to feed the current and ever-increasing population.

References

Akkermans, A.D.L., Abdulkadir, S., & Trinick, M. J. (1978). N2-fixing root nodules in Ulmaceae: Parasponia or (and) Trema spp. *Plant and Soil, 49*, 711–715.

Anik, A. R., Ranjan, R., & Ranganathan, T. (2018). Estimating the impact of salinity stress on livelihood choices and incomes in rural Bangladesh. *Journal of International Development, 30*(4), 1414–1438. https://doi.org/10.1002/jid.3364

Ashraf, M., & Foolad, M. R. (2007). Roles of glycine betaine and proline in improving plant abiotic stress resistance. *Environmental and Experimental Botany, 59*(2), 206–216. https://doi.org/10.1016/j.envexpbot.2005.12.006

Ayers, R. S., Westcot, D. W. (1985). Water quality for agriculture. *FAO Irrigation and Drainage Paper 29*, Rev. 1. Rome, FAO. www.fao.org/3/T0234E/T0234E07.htm

Balasubramaniam, T., Shen, G., Esmaeili, N., & Zhang, H. (2023). Plants' response mechanisms to salinity stress. *Plants, 12*, 2253. https://doi.org/10.3390/plants12122253

Bashan, Y. (1998). Inoculants of plant growth-promoting bacteria for use in agriculture. *Biotechnology Advances, 16*, 729–770. https://doi.org/10.1016/S0734-9750(98)00003-2

Beijerinck, M. W. (1901). *Über Oligonitrophile Mikroben. Zbl. Backt., 7*, 561–582.

Bhattacharjee, R. B., Singh, A., & Mukhopadhyay, S. N. (2009). Use of nitrogen-fixing bacteria as biofertiliser for non-legumes: Prospects and challenges. *Applied Microbiology and Biotechnology, 80*(2), 199–209. https://doi.org/10.1007/s00253-008-1567-2

Chang, A. C., & Silva, B. D. (2014). *Salinity and drainage in San Joaquin Valley, California: Science, technology, and policy*. Springer. https://doi.org/10.1007/978-94-007-6851-2

Chaves, M. M., Flexas, J., & Pinheiro, C. (2009). Photosynthesis under drought and salt stress: Regulation mechanisms from whole plant to cell. *Annals of Botany, 103*(4), 551–560. https://doi.org/10.1093/aob/mcn125

David, L. R. (1996). Handbook of chemistry and physics. *Boca Raton, FL: CRC, 1996*, 14–17.

Estrada, B., Aroca, R., Azcón-Aguilar, C., Barea, J. M., & Ruiz-Lozano, J. M. (2013). Importance of native arbuscular mycorrhizal inoculation in the halophyte Asteriscus maritimus for successful establishment and growth under saline conditions. *Plant and Soil, 370*, 175–185. https://doi.org/10.1007/s11104-013-1635-y

FAO. (2009). Global agriculture towards 2050. Retrieved November 17, 2024, from http://www.fao.org/fileadmin/templates/wsfs/docs/Issues_papers/HLEF2050_Global_Agriculture.pdf

FAO. (2011). Global food losses and food waste. Extent, causes and prevention; Food and Agriculture Organization of the United Nations: Rome, Italy, 2011. Retrieved November 17, 2024, from http://www.fao.org/3/a-i2697e.pdf

FAO. (2014). The state of food and agriculture: Innovation in family farming. FAO Rome.

FAO. (2019). The state of food and agriculture—Moving forward on food loss and waste reduction; Food and Agriculture Organization of the United Nations: Rome, Italy, 2019. Retrieved November 17, 2024, from http://www.fao.org/3/ca6030en/ca6030en.pdf

FAO. (2021). Global map of salt affected soils Version 1.0. Retrieved December, 2021, from https://www.fao.org/soils-portal/data-hub/soil-maps-and-databases/global-map-of-salt-affected-soils/en/

FAO. (2022). Halt soil salinization, boost soil productivity—Proceedings of the global symposium on salt-affected soils, 20–22 Oct 2021, Rome.

FAO. (2024). Tackling food loss and waste from the farm to the table and beyond. Retrieved November 17, 2024, from https://www.fao.org/newsroom/detail/tackling-food-loss-and-waste-from-the-farm-to-the-table-and-beyond/en#:~:text=Currently%2C%20over%2013%20percent%20of,percent%2C%20according%20to%20UNEP%20statistics

Folli-Pereira, M. D. S., Meira-Haddad, L. S. A., Houghton, C. M. N. S. D. V. D. C., & Kasuya, M. C. M. (2013). Plant–microorganism interactions: Effects on the tolerance of plants to biotic and abiotic stresses. In K. R. Hakeem, P. Ahmad, & M. Ozturk (Eds.), *Crop improvement* (pp. 209–238). Springer.

García-Fraile, P., Carro, L., Robledo, M., Ramírez-Bahena, M. H., Flores-Félix, J. D., Fernández, M. T., Mateos, P. F., Rivas, R., Igual, J. M., Martínez-Molina, E., Peix, Á., & Velázquez, E. (2012). Rhizobium promotes non-legumes growth and quality in several production steps: Towards a biofertilization of edible raw vegetables healthy for humans. *PLoS ONE, 7*(5), Article e38122. https://doi.org/10.1371/journal.pone.0038122

Havlin, J. L., Tisdale, S. L., Nelson, W. L., & Beaton, J. D. (Eds.). (2014). *Nitrogen-soil fertility and fertilizers: An introduction of nutrient management. Published by Ghosh AK* (p. 122). PHI Learning Private Limited.

Himabindu, Y., Chakradhar, T., Reddy, M. C., Kanygin, A., Redding, K. E., & Chandrasekhar, T. (2016). Salt-tolerant genes from halophytes are potential key players of salt tolerance in glycophytes. *Environment and Experimental Botany, 124*, 39–63. https://doi.org/10.1016/j.envexpbot.2015.11.010

Hubbell, D. H., Kidder, G. (2009). Biological nitrogen fixation. University of Florida IFAS Extension Publication SL16, pp. 1–4.

IPCC. (2023). Summary for policymakers. In Core Writing Team, H. Lee, & J. Romero (Eds.), *Climate change 2023: Synthesis report. Contribution of working groups I, II and III to the sixth assessment report of the intergovernmental panel on climate change*. IPCC, Geneva, Switzerland, pp. 1–34, https://doi.org/10.59327/IPCC/AR6-9789291691647.001

IWMI. (2014). Press Release: World losing 2,000 hectares of irrigated farmland to salt-induced degradation daily. Retrieved November 4, 2024, from https://www.iwmi.cgiar.org/news/press-release-world-losing-2000-hectares-irrigated-farmland-salt-induced-degradation-daily/

Jacobson, T., & Adams, R. M. (1958). Salt and silt in ancient Mesopotamian agriculture. *Science,* (New Series) *128*(3334), 1251–1258.

Jones, K. M., Kobayashi, H., Davies, B. W., Taga, M. E., & Walker, G. C. (2007). How rhizobial symbionts invade plants?: The *Sinorhizobium-Medicago* model. *Nature Reviews Microbiology, 5*(8), 619–633. https://doi.org/10.1038/nrmicro1705

Joseph, A. (2023). Surface environment evolution for Venus, Earth, and Mars—the planets which began with the same inventory of elements. In A. Joseph, (Ed.), *Water Worlds in the Solar System.* Elsevier, pp. 359–398. ISBN 9780323957175. https://doi.org/10.1016/B978-0-323-95717-5.00014-1

Kamran, M., Parveen, A., Ahmar, S., Malik, Z., Hussain, S., Chattha, M. S., Saleem, M. H., Adil, M., Heidari, P., & Chen, J. T. (2021). An overview of hazardous impacts of soil salinity in crops, tolerance mechanisms, and amelioration through selenium supplementation. *International Journal of Molecular Sciences, 21*(1)148. https://doi.org/10.3390/ijms21010148

Lee, K., & Loiacono, M. (2024). What is Earth's atmosphere? Retrieved November 17, 2024, from https://www.nasa.gov/general/what-is-earths-atmosphere/

Lowder, S. K., Skoet, J., & Raney, T. (2016). The number, size, and distribution of farms, smallholder farms, and family farms worldwide. *World Development, 87*, 16–29. https://doi.org/10.1016/j.worlddev.2015.10.041

Lugtenberg, B., & Kamilova, F. (2009). Plant-growth-promoting rhizobacteria. *Annual Review of Microbiology, 63*(1), 541–556. https://doi.org/10.1146/annurev.micro.62.081307.162918

McCully, M. E. (1999). Roots in soil: Unearthing the complexities of roots and their rhizospheres. In R. L. Jones (Ed.), *Annual Review of Plant Physiology and Plant Molecular Biology*, 50. Annual Reviews Inc., Palo Alto, pp. 695–718.

McElrone, A. J., Choat, B., Gambetta, G. A., & Brodersen, C. R. (2013). Water uptake and transport in vascular plants. *Nature Education Knowledge, 4*(5), 6.

Munns, R., & Tester, M. (2008). Mechanisms of salinity tolerance. *Annual Reviews of Plant Biology, 59*, 651–681. https://doi.org/10.1146/annurev.arplant.59.032607.092911

NASA. (2024). Earth's atmosphere. Retrieved November 11, 2024, from https://www.nasa.gov/general/what-is-earth-atmosphere/

Nicastro, R., & Carillo, P. (2021). Food loss and waste prevention strategies from farm to fork. *Sustainability, 3*, 5443.https://doi.org/10.3390/su13105443

Op den Camp, R. H., Polone, E., Fedorova, E., Roelofsen, W., Squartini, A., Op den Camp, H. J., Bisseling, T., & Geurts, R. (2012). Nonlegume *Parasponia andersonii* deploys a broad rhizobium host range strategy resulting in largely variable symbiotic effectiveness. *Molecular Plant-Microbe Interactions, 25*(7), 954–963. https://doi.org/10.1094/MPMI-11-11-0304

Pan, J., Peng, F., Tedeschi, A., Xue, X., Wang, T., Liao, J., Zhang, W., & Huang, C. (2020). Do halophytes and glycophytes differ in their interactions with arbuscular mycorrhizal fungi under salt stress? A meta-analysis. *Botanical Studies, 61*, 13. https://doi.org/10.1186/s40529-020-00290-6

Qadir, M., Quillérou, E., Nangia, V., Murtaza, G., Murari Singh, M., Thomas, R. J., Drechsel, P., & Noble, A. D. (2014). Economics of salt-induced land degradation and restoration. *Natural Resources Forum, 38*(4), 282–295.

Rewald, B., Shelef, O., Ephrath, J. E., & Rachmilevitch, S. (2013). Adaptive plasticity of salt-stressed root systems. In P. Ahmad, M. M. Azooz, M. M. V. Prasad (Eds.), *In book: Ecophysiology and responses of plants under salt stress.* Springer, pp. 169–201. https://doi.org/10.1007/978-1-4614-4747-4_6

Roy, S., & Chakraborty, U. (2014). Salt tolerance mechanisms in salt tolerant grasses (STGs) and their prospects in cereal crop improvement. *Botanical Studies, 55*, 3. https://doi.org/10.1186/1999-3110-55-31

Sack, L., & Holbrook, N. M. (2006). Leaf hydraulics. *Annual Review of Plant Biology, 57*(1), 361–381. https://doi.org/10.1146/annurev.arplant.56.032604.144141

Sack, L., & Tyree, M. T. (2005). Leaf hydraulics and its implications in plant structure and function. In N. M. Holbrook, & M. A. Zwieniecki (Eds.), *Vascular transport in plants*. Elsevier Academic Press, pp. 93–114.

Sardo, V., & Hamdy, A. (2005). Halophytes—a precious resource. In A. Hamdy, F. Gamal, N. Lamaddalena, C. Bogliotti, R. Guelloubi (Eds.), *Non-conventional water use: WASAMED project*. CIHEAM/EU DG Research, Bari, pp. 119–128.

Shahid, S. A. (2012). Mangroves—adaptation and occurrence, climate change and coastline protection. *Farming Outlook, 11*(3), 26–31.

Shahid, S. A. (2013). Irrigation-induced soil salinity under different irrigation systems–assessment and management, short technical note. *Climate Change Outlook and Adaptation, 1*(1), 19–24.

Shahid, S. A., & Alkandari, A. J. (2024). Halophytic crops as a solution for food security, land rehabilitation, and mitigating future water crises by utilizing marginal quality waters. In J. C. Dagar, S. R. Gupta, & A. Kumar (Eds.), Halophytes vis-à-vis saline agriculture (Ch. 17, pp. 441–478). Springer International Publishing. https://doi.org/10.1007/978-981-97-3157-2_17

Shahid, S. A., & Rahman, K. (2011). Soil salinity development, classification, assessment and management in irrigated agriculture. In M. Pessarakli (Ed.), *Handbook of Plant and Crop Stress* (3rd ed., Ch. 2, pp. 23–39). CRC Press.

Shahid, S. A., Dakheel, A. H., Mufti, K. A., & Shabbir, G. (2008). Automated in-situ salinity logging in irrigated agriculture. *European Journal of Scientific Research, 26*(2), 288–297.

Shahid, S. A., Zaman, M., & Heng, L. (2018a). Introduction to soil salinity, sodicity and diagnostics techniques. In M. Zaman, S. A. Shahid, & L. Heng (Eds.), *Guideline for salinity assessment, mitigation and adaptation using nuclear and related techniques* (Ch. 1, pp. 1–42). Springer International Publishing. https://doi.org/10.1007/978-3-319-96190-3_1

Shahid, S. A., Zaman, M., & Heng, L. (2018b). Soil salinity: Historical perspectives and a world overview of the problem. In M. Zaman, S. A. Shahid, L. Heng (Eds.), *Guideline for salinity assessment, mitigation and adaptation using nuclear and related techniques* (Ch. 2, pp. 43–53). Springer International Publishing. https://doi.org/10.1007/978-3-319-96190-3_2

SPRPN. (Scientific Panel on Responsible Plant Nutrition). (2021). Furthering 4R nutrient stewardship. Issue Brief 03. Retrieved November 11, 2024, from www.sprpn.org

Stavi, I., Thevs, N., & Priori, S. (2021). Soil salinity and sodicity in drylands: A review of causes, effects, monitoring, and restoration measures. *Frontiers in Environmental Sciences, 9*, Article 712831. https://doi.org/10.3389/fenvs.2021.712831

Trinick, M. J. (1973). Symbiosis between Rhizobium and the non-legume, Trema aspera. *Nature, 244*, 459–460. https://doi.org/10.1038/244459a0

Wagner, S. C. (2011). Biological nitrogen fixation. *Nature Education Knowledge, 3*(10), 15.

Wheeler, T. D., & Stroock, A. D. (2008). The transpiration of water at negative pressures in a synthetic tree. *Nature, 455*(7210), 208–212. https://doi.org/10.1038/nature07226

Wiley, G. R. (1953). Prehistoric settlement patterns in the Viru Valley, Peru. *Bureau of American Ethnology Bulletin, 155*, 1–453.

Zaman, M., Shahid, S. A., Lee, H. (2018). Irrigation systems and zones of salinity development. In M. Zaman., S. A. Shahid, & L. Heng (Eds.), *Guideline for salinity assessment, mitigation and adaptation using nuclear and related techniques* (Ch. 4, pp. 91–111). Springer International Publishing. https://doi.org/10.1007/978-3-319-96190-3_4

Zhao, Y., Gao, Z., Tian, B., Bi, K., Chen, T., Liu, H., Xie, J., Cheng, J., Fu, Y., & Jiang, D. (2017). Endosphere microbiome comparison between symptomatic and asymptomatic roots of *Brassica napus* infected with *Plasmodiophora brassicae. PLoS ONE, 12*, Article e0185907. https://doi.org/10.1371/journal.pone.0185907

Zhao, Y., Gao, L., Gao, Z., Tian, B., Chen, T., Xie, J., Cheng, J., Fu, Y., Li, Y., Xiao, S., Bisseling, T., & Jiang, D. (2022). Exploring a rhizobium to fix nitrogen in non-leguminous plants by using a tumor-formation root pathogen. *Phytopathology Research 4*, 48, https://doi.org/10.1186/s42483-022-00154-w

Zomer, R. J., Xu, J., & Trabucco, A. (2022). Version 3 of the global aridity index and potential evapotranspiration database. *Science Data, 9*, 409. https://doi.org/10.1038/s41597-022-01493-1

Chapter 7
Soil Salinity Dynamics in Arid Lands and Impact on Agriculture for Food Security

Shabbir Ahmad Shahid, Amal Jaafer Alkandari, and Fahad Meshari Alnajdi

Abstract Soil salinity in the general landscape or in the farm lands is not static but dynamic based on the rainfall conditions, underground water movement, soil type and the change in management practices. In humid regions where the rainfall is over 1000 mm per annum, and the rainfall pattern is uniformly distributed along with good drainage conditions, there should not be issue of soil salinity due to continuous leaching of salts. However, in arid lands (hyperarid, arid and semi-arid) soil salinity is developed in the landscape and in the root zones of plants grown, due to low rainfall (rainfed agriculture) and irrigated agriculture where the irrigation is accomplished through marginal quality waters (saline, saline-sodic) and poor salinity management practices. It is visualized that the climate change can severely increase salinization due to prolonged droughts and overexploitation of aquifers and making them more saline due to water recycling and low aquifer recharge. The soil salinity in irrigated agriculture fields is not a new phenomenon, it can be traced back to Mesopotamia where the early civilization first flourished and then failed due to human-induced salinization. Another example is Viru valley in Peru where society based on irrigated agriculture failed due to soil salinization. Globally, significant number of national and international institutes are struggling to offset the effects of soil salinity on crop production and food security. Simultaneous to these efforts, globally on a daily basis 2000 ha farm lands is lost due to salinity only in over 75 countries, costing 27 billion dollars due to decreased crop yield, and with this pace by 2100, 1/3rd of total irrigated lands will be out of farmers hands, and this will lead to food crises to meet the food demand of escalating global population, especially the developing world located in arid regions. Considering the impacts of soil salinization on food security, an international network on salt-affected soils (INSAS) has been created by the UN-FAO sponsored Global Soil Partnership, and now promoting the declaration of 2028 as year of saline agriculture, with the aim to promote climate resilient agriculture

S. A. Shahid (✉) · A. J. Alkandari · F. M. Alnajdi
Food Security Program, Environment and Life Sciences Research Center, Kuwait Institute for Scientific Research, Safat, Kuwait
e-mail: sshahid@kisr.edu.kw

M. K. Suleiman and S. A. Shahid (eds.), *Fostering Arid Lands Agriculture in the Face of Climate Change*, Advances in Global Change Research 81,
https://doi.org/10.1007/978-3-031-98490-7_7

for sustainable production systems and healthy ecosystems in salt-affected areas. In this chapter recent developments in soil assessment and dynamics monitoring, economic costs due to decreased crop yield as well as solutions to mitigate soil salinity development and adaptation strategies are presented to help farmers in the arid lands to use their farm lands for maximum production capacity.

Keywords Arid lands · Salinity dynamics · Climate resilient · Saline agriculture · Food security · Climate change · Resilient

7.1 Introduction

Soil salinity is a complex environmental problem that poses a serious threat to global agricultural production, especially in arid and semi-arid regions that depend on irrigation (Ivushkin et al., 2019). The scarcity of water and arable lands in arid region countries, and salinity problem raises special concerns for national and global food security systems. Understanding soil and groundwater salinization process and dynamics through science based diagnostics can lead to sustainable management and revitalization of salt-affected soils (reversing salinity affects) for food production and developing national salinity management strategies globally, and especially in the drylands countries. It is to be noted that where annual rainfall is over 1,000 mm, the soil salinity does not exist, due to regular washing of salts from the root zone and rain water is sufficient to grow crops.

Historical perspectives of soil salinization have been the focus of recent publication (Shahid et al., 2018a). In the drylands countries, soil and groundwater salinization is a major risk and challenge to crop production. The recent estimates by FAO (2021) revealed that globally 424 million hectares of topsoil (0–30 cm) and 833 million hectares of subsoil (30–100 cm) are salt-affected to various degrees. This estimate is based on 73% of land mapped for soil salinization. Another estimate shows 20% (62 million hectares) of all irrigated land is affected by soil salinization to various levels (Qadir et al., 2014). This shows that these salt induced improvised irrigated farms are either out of crop production (farms abandonment) or crop production is significantly reduced. Recently, it has been reported that on daily basis 2,000 hectares (20 km^2) land has been abandoned due to soil salinity in over 75 countries (IWMI, 2014). At the current rate of farmland degradation and abandonment (2,000 ha daily) by the year 2100, 1/4 of the total irrigated land (310 million ha) will be out of production (Shahid & Al-Kandari, 2024). In the Middle East and North Africa (MENA) region, losses due to salinity are close to one billion USD per annum (UNEP, 2020). Salinity impact is not only confined to agricultural farms; it has also affected plant ecology and biogeography (Bui, 2013). Therefore, it is desirable to enhance the salinity management program systematically both at national (Shahid et al., 2023) and global scales (FAO, 2021). To address the salinization at national and international levels, a well-organized International Network of Salt-Affected Soils (INSAS) has been launched, which encourages the National Agricultural Research Systems (NARS) to work on soil salinity for informed decisions on crop selection to achieve higher production from the improvised farms (Choukr-Allah et al., 2023). The review of literature revealed that among the Gulf Cooperation Council (GCC) countries, Oman (MAF, 2012) and Abu Dhabi emirate (EAD, 2018) have assessed their farm land salinity dynamics. Recently, a proposal on farm land salinity is approved by Kuwait Foundation for the Advancement of Sciences-KFAS (Shahid et al., 2023) to develop salinity management strategy for the farm land of Kuwait.

With the global human population estimated at 7.6 billion in 2017 and projected to grow up to 11.12 billion in 2100 (UN, 2017), responsible consumption and production of food are considered one of the most pressing issues included in the Sustainable Development Goals. These figures alert us that irrigated agriculture resources are being depleted at a rate that will certainly not allow future generations to meet their own food demands, unless new, innovative and regenerative approaches are adopted to manage these resources.

Salinization is likely to have a negative impact on the economy, living conditions and ecosystem functions of affected regions (Ruto et al., 2018) especially the water scarce region (Drylands). Agriculture is highly dependent on land and water conditions, which are increasingly affected by climate change and climate change induced soil salinization mainly due to global warming (high evapotranspiration). However, properly managed salt-affected soils allow for considerably more agricultural production (de Vos et al., 2016; Hirich et al., 2021; Rao et al., 2017; Shahid et al., 2008). With regards to rainfed agriculture system, such as Middle East and North Africa (MENA) and African countries soil degradation is caused by wind and water erosion, while in irrigated systems the farming practices themselves are responsible for *soil salinity and sodicity* (OECD-FAO, 2018). Thus, the soil is a major concern for the sustainability of irrigated agriculture and water management. Detailed information about the soil/water salinity status of agricultural farms in many arid lands countries does not exist, however, Oman (MAF, 2012), Abu Dhabi Emirate (EAD, 2018) and Kuwait (Shahid et al., 2023) took the initiative to assess soil and groundwater salinization in the agricultural farms for food security.

7.2 Salts Effects on Plants and Adjustments

Shahid and Alkandari (2024) has briefly described the basics for salt absorption and tolerance by plants in saline environments. There are two effects of salts on plants, (i) osmotic effect and (ii) specific ion effect. The *osmotic effect* is evident when the salts are high in soil water, the water in the plants cells is pulled out by the saline soil environment resulting into decreased turgor pressure in the cell leading to wilting of plants and plant death. At higher salt levels (specific ion effect) plants metabolism process is decreased.

There are three ways by which the plants adjust salts during plant growth, (i) *Salts excretion from leaves*—salts are secreted by salt glands (e.g. mangrove) in the leaves, (ii) *salts exclusion at the roots level*—the salts are filtered at the roots level, thus, accumulation of salts occurs in the very near vicinity of the roots and (iii) *accumulation of salts in cell vacuoles*—the salts move from the cytoplasm to the large vacuoles.

7.3 Causes and Threats of Soil Salinity

There are multiple causes of soil salinization including, natural, (i) weathering of soil parent material and release of salts into solution, and crystallization through evaporation, (ii) sea water intrusion into coastal areas and subsequent evaporation concentrate salts in soil surface and subsurface, and (ii) anthropogenic (human impacts), whereby the man induced soil salinization is due to mismanagement of soil and water resources without care of soil health (overexploitation of groundwater resource) to manage soil salinization in agricultural fields. Land use change is another cause of soil salinization, for example, in Australia agriculture land was established by clearing the forests, the latter have high water absorption capacity than the formers. Over long period of time the groundwater level was increased and the forest land which was once nonsaline, converted to saline land, and currently, Australia hosts over 1 million hectares of saline land.

The soil salinization is also significant issue in Central Asian countries, and the area around Aral Sea is highly salinized (Mirzabaev, 2016; Shahid & Alkandari, 2024), and earlier World Bank executed a project in Aral Sea basin for water management (World Bank, 1998). These highly salinized areas are without vegetation cover and has no pastoral value (Gintzburger et al., 2003), in places non-grazable halophytes are present. In the irrigated areas in Central Asia (CA), secondary salinization is the major cause of land degradation (Gupta et al., 2009; Mirzabaev et al., 2015). Qadir et al. (2009) has reported that 40–60% of irrigated lands in the region were affected by secondary salinization. A hypothetical salinity development model has been established by Shahid et al. (2010), where a detailed salinity development, assessment, and management methods in irrigated agriculture fields are described by Shahid (2013) and Shahid and Rahman (2011).

The soil salinity threats sustainable uses of soil resources, be it for agriculture, horticulture crops, construction or grazing. The saline lands create nuisance value to the environment. The impacts can be seen in soil degradation and farm land salinization, where, once soil salinity is increased to a maximum level, even the farm land is abandoned. Such a development on farm land affects crop production and the food security.

7.4 Historical Perspectives of Soil Salinity

The soil salinization is not a recent development; it has been with mankind since centuries. There are two historical examples where the civilization was failed due to soil salinization. Mesopotamia is currently called Iraq, where the soil salinization was developed. An earlier work of Jacobson and Adams (1958), presented 'Salt and silt in ancient Mesopotamian agriculture' highlighted the history of salinization in Mesopotamia. It is very well reported in earlier publications that there is good evidence for Mesopotamia that early civilizations flourished and then failed due to human-induced salinization. The subject of soil salinization in the context of crops failure in Mesopotamia is very well described by Shahid et al. (2018a, b, c).

In the Viru valley of Peru, the earlier literature shows (Wiley, 1953) that irrigated agriculture began between 800 BC and 30 AD, and by 800 AD, the population was at a peak. Then from 1200 AD, the population declined appreciably and the residents of the once densely populated Viru valley bottom relocated to more narrow upper reaches of the valley. The historians partly attribute this relocation to increased salinity and a rising water- table, together with inadequate soil drainage (Armillas, 1961). Tanji (1990) draws a historical perspective of irrigation-induced salinity in several regions. In a similar perspective, general remarks by Wiley (1953) clearly reveal that 'many societies based on irrigated agriculture have failed', e.g. Mesopotamia and the Viru valley of Peru.

7.5 Global, Regional and National Level Extent and Distribution of Salt-Affected Soils

Old estimate based on soil map of the worlds (FAO/UNESCO 1974) reported 1 billion ha area as salt-affected (Abrol et al., 1988). Later, Oldeman et al. (1991) reported 45 million ha of irrigated land and 32 million ha world's rainfed croplands affected by salinity or sodicity. Recently launched report on global status of salt-affected soils (FAO, 2024) shows the total area of salt-affected soils of the world amounts to 1,381 million ha (13.81 million km^2), or 10.7 percent of the total land area (Table 7.1). The largest areas are observed in Australia (357 million ha) and Argentina (153 million ha) (Table 7.1).

The salt-affected area in Asian countries accounts for Bangladesh 1.056 million ha, China 36.0 million ha, India 6.74 million hectares, Pakistan 6.67 million hectares, and Thailand 0.806 million ha.

Current estimates (FAO, 2021) shows 424 million hectares of topsoil (0–30 cm) and 833 million hectares of subsoil (30–100 cm) are salt-affected to various degrees. Considering the high level of land degradation (33%) as reported by FAO-ITPS (2015) and 20% irrigated agriculture area is salt-affected, future plans of feeding the escalating population and accomplishing SDG 2 "*End hunger*" would be an

Table 7.1 Regional areas of salt-affected soils (FAO 2024)

Region	Area of salt-affected soils (km^2)
Pacific	3,570,000
Europe and Eurasia	2,378,209
Latin America and Carribbean	2,352,857
Near East and North Africa	2,303,461
Asia	1,543,269
Sub-Saharan Africa	883,795
North America	779,912
Total	**13,811,503**

ambitious dream. The extent of salt-affected soils has been reported by Qadir et al. (2014), Shahid et al. (2018a, b, c) and Shahid and Alkandari (2024).

7.6 Socioeconomic and Environmental Impacts of Soil Salinity

The impacts of salinity could be huge and in diversified directions, such as loss in crop production, disturbance of socio-economic aspects, environmental degradation, and loss farm land resources etc. These impacts are listed below and are built up on those published earlier (Shahid, 2013).

Loss of crop yield and impact on food security—the salts in soil inhibit seeds germination. The crop produced over salty soils do not grow well due to salinity stress, inhibiting water absorption from soil, thus ends up to either low crop yield or at extreme salinity levels a complete crop failure, thus affecting food security and food supply chain.

Socio-economic impacts—the crops grown on saline fields require high inputs (water, fertilizers, seeds rate etc.), which in many cases are not affordable by farmers, leading to high cost of production, resulting into low farmer income, inhibiting to access medical facilities and children education.

Environmental Impact—the soil salinization can also cause significant environmental impacts, in terms of low vegetation cover, low soil carbon sequestration capacity, desert environment, poor grazing capacity and wildlife habitats.

7.7 Classification of Salt-Affected Soils

There are a number of classifications with respect to salt-affected soils or specifically addressing soil salinity and sodicity. The FAO-UNESCO (1974) define saline soils as *Solonchaks* and sodic soils as *Solonetz*, where both names are Russian. The most widely used classification of salt-affected soils was established in 1954 (US Salinity Lab. Staff, 1954), which sets the limit of electrical conductivity of extract from saturated soil paste ≥4 mS/cm to call as saline. Later Soil Science Division Staff (2017) revised the classification and sets the limit at ECe 2 mS/cm, this makes sense that at ECe between 1–4 mS/cm numerous sensitive crops yield is reduced.

In the author's opinion, even setting the limit of ECe 2 mS/cm is higher, because the yield of certain sensitive crops (celery, cabbage, potato, pepper, lettuce, radish, onion, carrot, beans, and turnips) is reduced between ECe 1 and 2 mS/cm (Ayers & Westcot, 1976, 1985). Therefore, to be on safe side and to provide accurate information to farmers to choose crops based on farm salinity levels, we strongly propose the limit of ECe > 1 mS/cm, this lower limit is decreased due to the fact that growth of sensitive plants is reduced to varying extents at ECe between 1 and 2 mS/cm, such is the case of vegetable crops (celery, cabbage, potato, pepper, lettuce, radish, onion, carrot, beans, and turnips), fruit trees (orange, peach, apricot, grape, almond, plum, blackberry, and strawberry), all of which present a threshold salinity value below 2 mS/cm (ECe) (Ayers & Westcot, 1976, 1985). Following classification is proposed for use in farm land salinity management and crops selection.

In the authors point of view, the classifications of saline soil as proposed by US Salinity Lab Staff (1954) with ECe 4 mS/cm, and Soil Science Division Staff (2017) with ECe 2 mS/cm should be used for soil salinity mapping to publish national and regional soil salinity maps from land use point of view. However, at farm level, the proposed salinity classification (Tables 7.2 and 7.3) is recommended, which is crop specific and applicable at farm levels where sensitive crops are being practiced (grown), even by irrigating with fresh water.

Based on our experience in dealing with saline soils for crop production, the US Salinity Lab Staff (1954) and Soil Science Division Staff (2017) salinity classes are modified and following are proposed (Table 7.2). It is to be noted that global soil science community agreed on the soil sodicity level of 15 ESP to consider a sodic soil. It is to be noted that soil pH is a complex phenomenon and controlled by soil constituents with the capacity of developing buffering capacity (resist in change of soil pH) of soils, and slightly changed by salt contents and types of salts in soil. Therefore, pHs is not considered as one of the criterion to classify either saline or sodic soils.

Table 7.2 Objective oriented proposed soil salinity classification

Salinity class	ECe (mS/cm)	Remarks
Non-saline (C_0)	<1.00	Suitable for all crops sensitive to tolerant categories
Very slightly saline (C_1)	1 to <2	Will show impact on sensitive crops, whose salinity threshold (T) is between ECe 1 and 2 mS/cm (e.g., celery, cabbage, potato, pepper, lettuce, radish, onion, carrot, beans, turnips ets), fruit trees (orange, peach, apricot, grape, almond, plum, blackberry, strawberry etc.). To manage salinity regular leaching is recommended to keep the ECe < 1 mS/cm
Slightly saline (C_2)	2 to <4	Suitable for crops with salinity threshold (T) between ECe 2 and 4 mS/cm. Leaching is required to maintain root zone salinity at or below threshold salinity level of specific crop. Cyclic use of water is recommended (alternate irrigation with fresh and salty water)
Moderately saline (C_3)	4 to <8	Suitable for crops with salinity threshold between ECe 4 to 8 mS/cm. Leaching is required to maintain root zone salinity at or below threshold salinity level of specific crop for optimum yield. Cyclic use of water is recommended (alternate irrigation with fresh and salty water)
Strongly saline (C_4)	8–16	Suitable for highly salt tolerant crops. Regular leaching is recommended
Very strongly saline (C_5)	>16	Recommended for halophytes crops usually non-edible, unless the reduction in yield is acceptable to the farmer (Shahid & Alkandari, 2024)

7.8 Latest Developments in Soil Salinity Assessment and Monitoring

Since the publication of USDA Handbook no 60 in the year 1954 by the US Salinity Lab Staff (1954) there have been significant technological developments and advancements in assessing and monitoring soil salinity from farm level to global scale (Shahid, 2002). Following methodologies are in use in most of the countries.

7.8.1 Conventional and Modern Methods of Soil Salinity Assessment and Monitoring

- Soil salinity assessment by measuring EC of soil suspension (1:1, 1:2.5, 1:5 soil:water, weight/volume).
- Soil salinity assessment by measuring EC of extract from saturated soil paste (ECe).
- In-situ EC measurement by salinity probe (Fig. 7.1a).

Table 7.3 Soil salinity (root zone) tolerance levels of different crops and yield losses (10, 25 and 50%) at different average root zone salinity levels (Maas & Hoffman, 1977; Ayers & Westcot, 1985)

	Yield potential↓				Maximum[3] ECe (mS/cm)
	100%	90%	75%	50%	
Field crops	ECe[1,2] mS/cm	ECe mS/cm	ECe mS/cm	ECe mS/cm	
Barley (*Hordeum vulgare)*[4]	8.0	10.0	13.0	18.0	28.0
Sugar beet *(Beta vulgaris)*[5]	7.0	8.7	11.0	15.0	24.0
Sorghum *(Sorghum bicolor)*	6.8	7.4	8.4	9.9	13.0
Wheat *(Triticum aestivum)*[4,6]	6.0	7.4	9.5	13.0	20.0
Wheat, durum *(Triticum turgidum)*	5.7	7.6	10.0	15.0	24.0
Soybean *(Glycine max)*	5.0	5.5	6.3	7.5	10.0
Cowpea *(Vigna unguiculata)*	4.9	5.7	7.0	9.1	13.0
Groundnut *(Arachis hypogaea)*	3.2	3.5	4.1	4.9	6.6
Rice (paddy) *(Oriza sativa)*	3.0	3.8	5.1	7.2	11.0
Sugarcane *(Saccharum officinarum)*	1.7	3.4	5.9	10.0	19.0
Corn (maize) *(Zea mays)*	1.7	2.5	3.8	5.9	10.0
Broadbean *(Vicia faba)*	1.5	2.6	4.2	6.8	12.0
Bean *(Phaseolus vulgaris)*	1.0	1.5	2.3	3.6	6.3
Vegetable crops					
Beet, red *(Beta vulgaris)*[5]	4.0	5.1	6.8	9.6	15.0
Broccoli *(Brassica olerace botrytis)*	2.8	3.9	5.5	8.2	14.0
Tomato *(Lycopersicon esculentum)*	2.5	3.5	5.0	7.6	13.0
Cucumber *(Cucumis sativus)*	2.5	3.3	4.4	6.3	10.0
Spinach *(Spinacia oleracea)*	2.0	3.3	5.3	8.6	15.0
Celery *(Apium graveolena)*	1.8	3.4	5.8	9.9	18.0
Cabbage *(Brassica oleracea capitate)*	1.8	2.8	4.4	7.0	12.0
Potato *(Solanum tuberosum)*	1.7	2.5	3.8	5.9	10.0
Corn, sweet (maize) (Zea mays)	1.7	2.5	3.8	5.9	10.0
Sweet potato (Impomoea batatas)	1.5	2.4	3.8	6.0	11.0
Pepper *(Capsicum annum)*	1.5	2.2	3.3	5.1	8.6
Lettuce *(Lactuca sativa)*	1.3	2.1	3.2	5.1	9.0
Radish *(Raphanus sativus)*	1.2	2.0	3.1	5.0	8.9
Onion *(Allium sativus)*	1.2	1.8	2.8	4.3	7.4
Carrot (Daucus carota)	1.0	1.7	2.8	4.6	8.1
Bean *(Phaseolus vulgaris)*	1.0	1.5	2.3	3.6	6.3
Turnip *(Brassica rapa)*	0.9	2.0	3.7	6.5	12.0
Forage crops					
Wheatgrass, tall *(Agropyron elongatum)*	7.5	9.9	13.0	19.0	31.0

(continued)

Table 7.3 (continued)

Field crops	Yield potential↓ 100% ECe[1,2] mS/cm	90% ECe mS/cm	75% ECe mS/cm	50% ECe mS/cm	Maximum[3] ECe (mS/cm)
Wheatgrass ((*Agropyron cristatum)*	7.5	9.0	11.0	15.0	22.0
Bermuda grass *(Cynodon dactylon)*[7]	6.9	8.5	11.0	15.0	23.0
Barley (forage) *(Hordeum vulgare)*[4]	6.0	7.4	9.5	13.0	20.0
Ryegrass, perennial *(Lolium perenne)*	5.6	6.9	8.9	12.0	19.0
Trefoil *(Lotus comiculatus tenuifolium)*[8]	5.0	6.0	7.5	10.0	15.0
Harding grass *(Phalaris tuberosa)*	4.6	5.9	7.9	11.0	18.0
Fescue, tall *(Festuca elatior)*	3.9	5.5	7.8	12.0	20.0
Wheatgrass *(Agropyron sibiricum)*	3.5	6.0	9.8	16.0	28.0
Vetch *(Vicia angustifolia)*	3.0	3.9	5.3	7.6	12.0
Sudan grass *(Sorghum Sudanese)*	2.8	5.1	8.6	14.0	26.0
Cowpea (forage) *(Vigna unguiculata)*	2.5	3.4	4.8	7.1	12.0
Sesbania *(Sesbania exaltata)*	2.3	3.7	5.9	9.4	17.0
Alfalfa *(Medicago sativa)*	2.0	3.4	5.4	8.8	16.0
Lovegrass *(Eragrostis sp.)*[9]	2.0	3.2	5.0	8.0	14.0
Corn (forage)(maize) *(Zea mays)*	1.8	3.2	5.2	8.6	15.0
Clover, berseem *(Trifolium alexandrinum)*	1.5	3.2	5.9	10.0	19.0
Orchard grass *(Dactylis glomerata)*	1.5	3.1	5.5	9.6	18.0
Clover, red *(Trifolium pratense)*	1.5	2.3	3.6	5.7	9.8
Clover, alsike *(Trifolium hybridum)*	1.5	2.3	3.6	5.7	9.8
Clover, ladino *(Trifolium repens)*	1.5	2.3	3.6	5.7	9.8
Clover, strawberry *(Trifolium fragiferum)*	1.5	2.3	3.6	5.7	9.8
Fruit crops[10]					
Date palm *(Phoenix dactylifera)*	4.0	6.8	11.0	18.0	32.0
Grapefruit *(Citrus paradisi)*[11]	1.8	2.4	3.4	4.9	8.0
Orange *(Citrus sinensis)*	1.7	2.3	3.3	4.8	8.0
Peach *(Prunus persica)*	1.7	2.2	2.9	4.1	6.5
Apricot *(Prunus ameniaca)* [11]	1.6	2.0	2.6	3.7	5.8
Grape *(Vitus sp.)* [11]	1.5	2.5	4.1	6.7	12.0
Almond *(Prunus dulcis)*[11]	1.5	2.0	2.8	4.1	6.8
Plum, prune *(Prunus domestica)* [11]	1.5	2.1	2.9	4.3	7.1
Blackberry *(Rubus sp.)*	1.5	2.0	2.6	3.8	6.0

(continued)

Table 7.3 (continued)

	Yield potential↓				Maximum[3] ECe (mS/cm)
	100%	90%	75%	50%	
Field crops	ECe[1,2] mS/cm	ECe mS/cm	ECe mS/cm	ECe mS/cm	
Boysenberry *(Rubus ursinus)*	1.5	2.0	2.6	3.8	6.0
Strawberry *(Fragaria sp.)*	1.0	1.3	1.8	2.5	4.0

[1] Adapted from Maas and Hoffman (1977). These data should only serve as a guide to relative tolerances among crops. Absolute tolerances vary depending upon climate, soil conditions and cultural practices. In gypsiferous soils, plants will tolerate about 2 mS/cm higher soil salinity (ECe)
[2] ECe means average root zone salinity as measured by electrical conductivity of the saturation extract of the soil, reported in milli Siemens per centimetre (mS/cm) at 25 °C
[3] The zero yield potential or maximum ECe indicates the theoretical soil salinity (ECe) at which crop growth ceases
[4] Barley and wheat are less tolerant during germination and seedling stage; ECe should not exceed 4–5 mS/cm in the upper soil during this period
[5] Beets are more sensitive during germination; ECe should not exceed 3 mS/cm in the seeding area for garden beets and sugar beets
[6] Semi-dwarf, short cultivars may be less tolerant
[7] Tolerance given is an average of several varieties; Suwannee and Coastal Bermuda grass are about 20 percent more tolerant, while Common and Greenfield Bermuda grass are about 20 percent less tolerant
[8] Trefoil seems less tolerant than Narrowleaf Birdsfoot Trefoil
[9] Tolerance given is an average for Boer, Wilman, Sand and Weeping Love grass; Lehman Lovegrass seems about 50 percent more tolerant
[10] These data are applicable when rootstocks are used that do not accumulate Na^+ and Cl^- rapidly or when these ions do not predominate in the soil
[11] Tolerance evaluation is based on tree growth and not on yield

(a) **(b)**

Fig. 7.1 In-situ measurement of root zone salinity using salinity probe (**a**) and Electromagnetic-38 (**b**)

7.8.2 *Assessing Soil Salinity Dynamics Using Remote Sensing and Geographical Information System*

Soil salinity at a broader landscape for a general understanding can be assessed by integrating the remote sensing (RS) with geographic information system (GIS). GIS is a computer application that involves the storage, analysis, retrieval, and display of data that are described in terms of their geographic location. It is to be noted that each technology has drawbacks, and no one technology is complete, therefore, integration of results from more than one technology are used for a better understanding. Remote Sensing imagery is well suited to map the surface expression of salinity (Spies & Woodgate, 2004), for example, poor vegetation cover could be an indication of salinity, however, in some areas there is good vegetation cover in saline areas due to the growth of salt-tolerant crops or halophytes (Dagar et al., 2024), such information may misguide the scientists. Therefore, in such specific landscapes, ground truthing is recommended for the correlation analysis with different indices image bands values (Gad et al., 2021) to assure the nature of vegetation in the area (Shahid et al., 2010), as well as collection of soil samples for salinity assessment in the salinity testing laboratory (Shahid, 2013; Zaman et al., 2018a, b).

7.8.3 *Real Time Dynamic Salinity Logging System (RDSLS)—Sensor Technology*

In addition to conventional methods of salinity assessment (Shahid et al., 2018b; US Salinity Lab. Staff, 1954), recently there have been advancements in salinity assessment through sensor installments at location of measurement. Similar real time salinity assessment and dynamics system was tested by Shahid et al. (2008). The very first system of its own kind was installed at forage production field at Dubai based International Center for Biosaline Agriculture, UAE (Shahid et al., 2008). The system measures apparent soil salinity in the plants root zones on a timely basis as programmed, thus, root zone salinity status can be viewed any time for informed decisions to manage salinity to improve crop production. Reader is referred to Shahid et al. (2008) for more details on RDSLS and its operation.

7.8.4 *Morphological and Micro-Chemical Characterization of Soil Salinity*

The measurement of ECe is sufficient to have information about salinity status of a location and to determine leaching requirement or selection of suitable crops to grow in specific locations. The morphological assessment coupled with micro-chemical analyses are modern techniques for further detailed exploration about the nature of salts and their composition. Such assessments are of academic interest for students studying for higher studies (Shahid, 2013). These are briefly described below. The reader is referred to Shahid (2002, 2013) for more details on these techniques.

Macromorpgology—naked eye visual assessment of salinity in the field.

Mesomorphology—assessment using a conventional hand lens in the field or using binocular in the laboratory.

Micromorphology—studying salts in soil thin sections using a polarizing microscope (Shahid, 1988).

Sub-microscopic—studying salts at submicroscopic levels using scanning electron microscope at high resolution (Bisdom, 1980).

Micro-chemical analyses- studying selected features of salts using Energy Dispersive X-Rays Analysis (EDXRA), or Electron Probe Micro Analyses-EPMA, to determine chemical composition of salts, line scanning for elements distribution and elemental mapping. The above techniques are very well described in various publication (Shahid, 1988, 1992; Shahid & Jenkins, 1994; Shahid et al., 1990).

7.9 Innovative Agricultural Practices to Cope Salinity Effects on Crop Production

Soil salinity and sodicity are two salts related ailments affecting plants growth, crop yield and soil properties. The soil salinity can be reduced using leaching requirement/ leaching fraction concepts. A hypothetical model of salinity development is presented by Shahid (Shahid et al., 2010). Another problem in the plant root zone or soil profile may be development of soil sodicity. This also affects soil properties by sealing the conducting pores with the accumulation of dispersed clay (Na-clay) and reduction in soil hydraulic conductivity (Shahid & Jenkins, 1991a, b). It appears from above that soil salinity and sodicity are complex phenomenon, and should be dealt with a combination of technologies or methods holistically called "integrated approach".

The integrated approach of soil reclamation includes, physical, chemical, hydrological, and biological methods (Shahid & Rahman, 2011) for use based on the intensity and types of problems in the area where salinity and sodicity are diagnosed and causing significant effects on crop production and the value of lands. Here these methods are briefly described.

7.9.1 Physical Methods

The physical methods include land leveling for uniform distribution of water, salt-scraping to remove surface salt crust, deep ploughing or subsoiling to break root restricting layer hindering plants roots penetration and water movement, seed bed preparation to direct salts to an area where salts have less effects on plants roots,

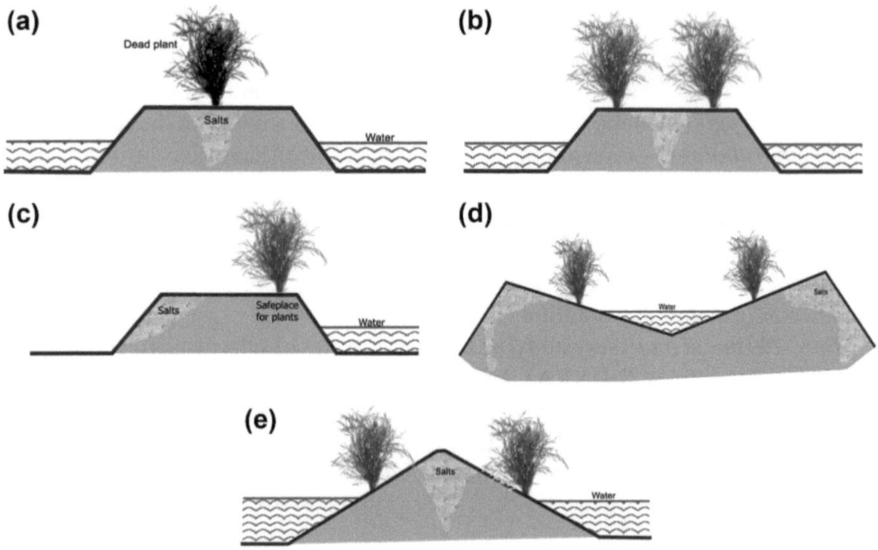

Fig. 7.2 Salinity management, **a** dead plant in high salinity zone, **b** flat top bed, both sides furrow irrigation, **c** one side furrow irrigation, **d** sloppy bed, middle furrow irrigation, **e** sloping bed, safe zone for plants, both furrows irrigated (author's own creation)

placement of seeds or transplanting in less salty-zone (Fig. 7.2), mixing sand with soil which have high clay concentration.

7.9.2 Chemical Methods

There is no chemical method to reclaim soil salinity, however, such methods exist to reduce soil sodicity, by adding gypsum ($CaSO_4.2H_2O$) based on the gypsum requirement of sodic soil. The addition of gypsum adds calcium which replaces sodium from the ion exchange complex of soil and reduce sodium adsorption and increase calcium adsorption, thus reducing exchangeable sodium percentage (ESP) to a safe limit (<15). Elemental sulfur is another chemical amendment which is used only in calcareous sodic soils, where through biochemical oxidation reaction, S is oxidized to SO_2, SO_3 and finally sulfuric acid (H_2SO_4) is formed, which reacts with $CaCO_3$ in soil and mobilize calcium to replace sodium from soil exchange complex and reduces soil ESP.

7.9.3 Hydrological Methods

The hydrological methods include selection of modern irrigation systems (drip-surface and subsurface, sprinkler, bubbler and central pivot system etc.). It also includes the use of drainage system (horizontal, vertical) to reduce water logging in the area. Comparative salinity accumulation zones under different irrigation systems are shown in Fig. 7.2.

7.9.3.1 Comparative Salinity Zones with Different Irrigation Systems

Each irrigation system transmits irrigation water in soil in different directions and space based on the volume of water added and the wetting zone on different texture soils, where salinity is built-up. A combination of physical and hydrological methods to manage soil salinity when irrigation water is brackish or saline is demonstrated in Fig. 7.2. Safe zone of seeding or planting away from salt-build-up is evident. Farmers are required to follow such guidelines when using furrow irrigation system and brackish water.

7.9.4 Biological Methods

Biological methods include, (i) growing salt-tolerant crops in salinized field or agriculture field irrigated with brackish water, (ii) using organic fertilizers to improve physical soil health, and (iii) serial biological concentration, briefly described below.

7.9.4.1 Growing Salt-Tolerant Crops (Saline and Biosaline Agriculture)

Each crop has specific salt-tolerance level (threshold salinity-T) below that crops perform optimally for optimum crop yield (Roy & Chakraborty, 2014). Once the salinity in the root zone exceeds threshold limit, it causes specific % yield decline compared to yield from non-saline soils (called slope). There is a linear percent reduction of yield by each ECe unit increase over the threshold salinity of crop grown. The saline agriculture is growing salt tolerant crops in saline environments (Qureshi & Barrett-Lennard, 1998), whereas biosaline agriculture is a set of technologies used to have maximum production from a specific crop under saline conditions. These technologies include, land preparation, selection of salt-tolerant crops, selection of irrigation system, leaching and drainage, recycling drainage water, use of gypsum to address soil sodicity hazard, precision agriculture (inputs used where needed), insect-pests management etc.

The concept of serial biological concentration (SBC) (Heuperman, 1995) is an alternate salinity management option, especially where there is shallow water table (fresh to slightly saline water) and where pumping the water to lower water table to a safe limit and safe disposal is main issue. Under such circumstances, the plants are grown in series (sensitive, moderately tolerant and high salt-tolerant). The prerequisite for SBC is that each crop plot is underlain by a tile drain to collect drained water to use for next stage. In this sequence, volume of water at the final crop or evaporation pond is significantly reduced, whereas salt level is increased. Based on the salt level of final water, it can be used to grow halophytes or accumulated in water pond for fish farming, which ultimately leads to evaporation pond for salt-harvesting. Table 7.3 presents list of crops and their yield potential at different root zone (ECe) salinity levels.

7.10 Prediction of Crops Yield Loss in Irrigated Agriculture Fields

Farmers are not aware of scientific knowledge published in books and journals regarding the salt-tolerance level of crops and level of yield decline with rising secondary salinity in irrigated fields. It is only through national extension services the technologies are transferred to the farmers for better adoption of crops for optimum yield. Following equation (Maas, 1990; Maas & Grattan, 1999) may be used as a guide to determine level of yield loss at certain root zone salinity level.

$$\mathrm{Yr} = 100 - \mathrm{s}(\mathrm{ECe} - \mathrm{ECt}) \tag{7.1}$$

where:

Yr = yield under saline soil environment relative to non-saline environment.
s = slope (% linear yield decline with 1 ECe (mS/cm) unit increase above ECt.
ECe = average root zone soil salinity during crop season.
ECt = threshold soil salinity of crop (at or below ECt crop yield do not reduce).

Example: Crop Maize (Zea mays).

Average root zone salinity (ECe 5 mS/cm).
Threshold salinity (ECt 1.7 mS/cm).
Slope (12%).
Yr = 100–12(5.0–1.7) = 60.4%

Based on the above calculation, it is apparent that at average root zone salinity of ECe 5 mS/cm in maize crop field, there will be 39.6% crop yield decline. Therefore, the farmer must reduce root zone salinity by leaching salts using fresh water to have optimum crop production, or get himself ready to bear this loss. Similarly, yield loss can be determined for other crops, provided predicted root zone salinity, slope and threshold ECt is known for the crop.

7.11 Salinity Effects on Food Security and Economic Losses of Crop Production

The economic losses to crops yield due to secondary and dryland salinization are huge. The global annual cost of salt-induced land degradation in irrigated areas could be US$ 27.3 billion because of lost crop production (Qadir et al., 2006, 2009, 2014; Quillérou & Thomas, 2012). Earlier estimate by Ghassemi et al. (1995) has shown 45.4 million hectares irrigated area as salt-affected (which is 20% of total 227 million hectare irrigated area). Based on this affected area in 1990, the annual cost of salt induced degradation was estimated as 264 US$ ha^{-1}. In 2002 total irrigated area increased to 277 million hectares, with estimated economic cost of 14.6 billion US$ compared to 12 billion US$ in 1990. In 2013 total irrigated area increased to 310 million hectares (FAO-AQUASTAT, 2013), assuming 20% irrigated area as salt-affected (62 million hectares), inflation adjusted cost of salt-induced degradation is 441 US$ ha^{-1}, making a total economic losses of 27.3 billion US$ (Qadir et al., 2014). In general, the social and economic dimensions of salt-induced land degradation have, however, received little attention in comparison to its biophysical aspects (Zekri et al., 2010).

Secondary salinization in Uzbekistan results in an annual loss of about 1 billion US$ (Sutton et al., 2007), and in Kazakhstan such losses amount to 6.2 billion US$ (Saigal, 2003). Mirzabaev (2016) estimated the losses due to the abandonment of cropland in Central Asia about 0.1 billion US$. Significant soil salinization problem has also been reported in a number of counters including Australia (Northcote & Skene, 1972), India, Iraq, Kazakhstan, Pakistan, United States, Uzbekistan, and Spain etc. Reader is referred to Qadir et al. (2014) for more details of economic losses due to salt-induced degradation in a number of countries. Provided these losses in irrigated agriculture fields are addressed holistically, the productivity of these salty fields can be increased. However, if the salinity issue is not tackled timely, significant decline in crop yield will affect food security in irrigated agriculture field of many countries, thus reducing the national capacity to meet the local food demand and export capacity. A comprehensive review of the economic costs of salt-induced land degradation in different parts of the world is given by Qadir et al. (2014).

7.12 Viewing Sustainable Development Goal-2 in the Face of Increasing Salinity in Irrigated Fields and Food Security

Among the 17 Sustainable Development Goals, SDG 2 is directly related to agriculture. All countries and States members of the UN are working together to achieve SDGs by 2030 (Pedersen, 2018; UN, 2016). Soils are not specifically connected to SDG-2, but it is connected to SDG 2, 3, 6, 13 and 15, that is End hunger (SDG 2), human health through nutrition (SDG 3), clean water (SDG 6), climate change (SDG

13) and life on land (SDG 15). However, SDG-2 is directly linked to food production and soil (*To end hunger, achieve food security and improved nutrition and promote sustainable agriculture).* SDG-2 has five targets, the most important to food security and soil quality are Target 2.3 and 2.4.

SDG 2—End hunger, achieve food security and improved nutrition and promote sustainable agriculture.

Target 2.3: by 2030 double the agricultural productivity and the income of small-scale food producers.

Target 2.4: by 2030 ensure sustainable food production systems and implement resilient agricultural practices that progressively improve land and soil quality.

The SDG 2 and the associated targets are highly relevant to arid lands countries, where agriculture is constrained due to water scarcity, high temperature, low rainfall, drought and soil salinization. To achieve the above-cited SDG and the activities by 2030 in arid regions, it is essential to adopt Sustainable Soil Management (SSM) through investing in soils, especially enhancing understanding of spatial salinity distribution (salinity mapping and regular monitoring) to create salinity-based crops zones "*that is what we must attempt to achieve in the context of arid lands agriculture countries*" (Shahid et al., 2024a, b). In the absence of such adoption in the arid lands countries, there would be a danger that farm productivity will decline and that may increase food imports costing billions of dollars. Therefore, it is essential to measure soil salinity to manage on scientific grounds.

7.13 Linking Soil Salinity to Climate Change

IPCC (2023) has highlighted that adverse impacts from human-caused (anthropogenic) climate change will continue to intensify in future. This has already reduced food security and affected water security and hindering efforts to achieve SDG-2 (end hunger by 2023). It is noteworthy, that overall global agricultural productivity has increased, however, the climate change has slowed this growth over the past 50 years globally, with related negative impacts mainly in mid-and low latitude regions but positive impacts in some high latitude regions (IPCC, 2023). As stated earlier that salinity has profound impact on irrigated agriculture, improvising farm land quality for crop production and even abandonment of farms. Due to water scarcity and projected water demand for agriculture, it is likely that groundwater salinity will increase through recycling groundwater for irrigation. The increased soil salinity will affect plant growth due to reduced water absorption in the root zone. In addition, the increased evapotranspiration due to global warming will increase water demand, and climate change induced drought may also affect crop production, unless, drought resistant crop varieties are evolved and introduced to the farming community at large scale. Adaptation to climate change through improved varieties (drought resistant and salt-tolerant) can generate multiple benefits in improving agriculture productivity and food security of salt-affected environments (Qadir et al., 2008). It is to be noted that salinity is not a new phenomenon generated from changing climate, it has long history from Mesopotamia (Shahid et al., 2018a). Therefore, salinity ailment to irrigated agriculture field has to be tackled through a holistic approach, and the countries vulnerable to climate change impact to develop national salinity management strategies to be implemented on top priority basis. Considering the importance of such strategies to improve crop production for food security, The Kuwait Foundation for the Advancement of Sciences (KFAS) has provided funds to develop a science based strategy by March 2026 (Shahid et al. 2023) for informed decision to improve farm productivities in Kuwait. It is visualized that the implementation of this strategy will increase farm productivities for local crop production, and lead to bridge gap between local production and food import, especially in the form of fodder production. Effective adaptation options to tackle salinity problems include, cultivar improvements, on-farm water management and storage, soil moisture conservation, irrigation, community-based adaptation, farm and landscape level diversification in agriculture, sustainable land management approaches, use of agro ecological principles and practices and other approaches that work with natural processes (IPCC, 2023).

7.14 International Network on Salt-Affected Soils (INSAS)

Considering the importance of salinity diagnostics and management in irrigated agriculture fields for optimum production in specific and ecosystems functions and services in general, INSAS was launched in 2019 at Dubai based International Center for Biosaline Agriculture UAE during the First Global Form on Innovations for Marginal Environments (GFIME). The INSAS is technical network of UN-sponsored Global Soil Partnership. The mission of INSAS is *"to support and facilitate joint efforts towards the sustainable management of salt-affected soils for food security, agricultural sustainability and climate change mitigation."* The INSAS is working under the UN-GSP with eight objectives (https://www.fao.org/global-soil-partnership/insas/en).

7.15 Conclusions and Recommendations

Soil salinity after water scarcity in arid land countries is recognized as the main constraint to crop production for food security. Salinity is very simple to assess in the field and analyze in the soil testing labs, but very difficult to manage under field conditions due to number of reasons including scarcity of fresh water sources, necessitating to use marginal quality water which is brackish or saline, low aridity index, high temperature and poor economic resources of farmers. The historical perspectives have shown that salinity spread is a continuous phenomenon and likely to increase in future if significant mitigation efforts are not made through using integrated form of technological packages along with farmers education and provision of funds. Adoption of salt-tolerant crops into production systems can help conserve freshwater resources, which can then be allocated for more productive purposes to sustain food security. The economic losses due to salinity induced decrease crop yield are huge, and farmers are abandoning their farms due to high cost to perform agriculture in a sustainable manner. It is recommended that salinity problem to be assessed from region, national and farmers field scales for a better understanding of the problem and to develop an integrated and holistic approach to mitigate the problem. There is an urgent need to strengthen the link between research-extension and the farmers for rapid transfer of new technologies to the potential users. Adaptation to climate change through improved varieties (drought resistant and salt-tolerant) using biosaline agriculture approach can generate multiple benefits in improving agriculture productivity and food security in arid lands countries.

References

Abrol, I. P., Yadav, J. S. P., & Massoud, F. I. (1988). Salt-affected soils and their management. FAO Soils Bulletin 39, Food and Agriculture Organization of the United Nations, Rome, Italy, p. 131.

Armillas, P. (1961). Land use in pre-Columbian America. In L. D. Stamp (Ed.), *A History of Land Use in Arid Regions* (Vol. 17, pp. 255–276). UNESCO Arid Zone Research.

Ayers, R. S., & Westcot, D. W. (1976). *Water quality for agriculture.* FAO Irrigation and Drainage Paper 29, FAO, Rome.

Ayers, R. S., Westcot, D. W. (1985). *Water quality for agriculture.* FAO Irrigation and Drainage Paper 29, Rev. 1. Rome, FAO. www.fao.org/3/T0234E/T0234E07.htm

Bisdom, E. B. A. (1980). A review of the application of submicroscopy techniques in soil micromorphology. 1: Transmission Electron Microscope (TEM) and Scanning Electron Microscopy (SEM). In E. B. A. Bisdom (Ed.), *Submicroscopy of Soils and Weathered Rocks* (pp. 67–116). Pudoc.

Bui, E. N. (2013). Soil salinity: A neglected factor in plant ecology and biogeography. *Journal of Arid Environments, 92*, 14–25.

Choukr-Allah, R., El Mouridi, Z., Benbessis, Y., & Shahid, S. A. (2023). Salt-affected soils and their management in the Middle East and North Africa (MENA) region: A holistic approach. In R. Choukr-Allah & R. Ragab (Eds.), *Biosaline agriculture as a climate change adaptation for food security* (pp. 13–45). https://doi.org/10.1007/978-3-031-24279-3_2

Dagar, J. C., Gupta, S. R., & Kumar, A. (Eds). (2024). *Halophytes vis-a-vis saline agriculture: Perspectives and opportunities for food security* (p. 541). Springer. https://doi.org/10.1007/978-981-97-3157-2

de Vos, A., Bruning, B., van Straten, G., Oosterbaan, R., Rozema, J., & van Bodegom. (2016). Crop salt tolerance under controlled field conditions in The Netherlands, based on trials conducted at Salt Farm Texel. Salt Farm Texel. https://edepot.wur.nl/409817

EAD. (2018). Farm soil salinity. In *Groundwater atlas of Abu Dhabi Emirate*. Environment Agency Abu Dhabi, United Arab Emirates, pp. 104–113.

FAO. (2000). Extent and causes of salt-affected soils in participating countries. Global network on integrated soil management for sustainable use of salt-affected soils. FAO-AGL website.

FAO. (2021). *Global map of salt affected soils version 1.0*. Retrieved December, 2021, from https://www.fao.org/soils-portal/data-hub/soil-maps-and-databases/global-map-of-salt-affected-soils/en/

FAO. (2022). *Halt soil salinization, boost soil productivity—Proceedings of the global symposium on salt-affected soils*, 20–22 Oct 2021, Rome.

FAO. (2024). Global status of salt-affected soils-Main report. *Rome*. https://doi.org/10.4060/cd3044en

FAO-AQUASTAT. (2013). *Area equipped for irrigation and percentage of cultivated land*. Retrieved September 16, 2013, from http://www.fao.org/nr/water/aquastat/globalmaps/index.stm

FAO-UNESCO. (1974). FAO-Unesco soil map of the world. 1:5 000 000, UNESCO Paris, 10 volumes.

FAO and ITPS. (2015). *Status of the World's Soil Resources (SWSR)–Main Report*. Food and Agriculture Organization of the United Nations and Intergovernmental Technical Panel on Soils. Rome.

Gad, M. M. E., Mohamed, M. H. A., & Mohamed, M. R. (2021). Soil salinity mapping using remote sensing and GIS. *Geomatica, 75*, 295–309. https://doi.org/10.1139/geomat-2021-0015

Ghassemi, F., Jakeman, A. J., & Nix, H. A. (1995). *Salinisation of land and water resources: Human causes, extent, management and case studies*. CABI Publishing.

Gintzburger, G., Toderich, K. N., Mardonov, B. K., & Mahmudov, M. M. (2003). *Rangelands of the arid and semi-arid zones in Uzbekistan* (p. 426). V. Howe and CIRAD Editing and Publications Service.

Gupta, R., Kienzler, K., Martius, C., Mirzabaev, A., Oweis, T., de Pauw, E., Qadir, M., Shideed, K., Sommer, R., Thomas, R., Sayre, K., Carli, C., Saparov, A., Bekenov, M., Sanginove, S., Nepesov, M., & Ikramov, R. (2009). Research prospectus: A vision of sustainable land management research in central asia. ICARDA Central Asia and Caucasus Program. Sustainable Agriculture in Central Asia and the Caucasus Series No. 1. CGIAR-PFU, Tashkent, Uzbekistan.

Heuperman, A. F. (1995). *Salt and water dynamics beneath a tree plantation growing on a shallow water table*. Report of the Department of Agriculture, Energy and Minerals Victoria, Institute for Sustainable Irrigated Agriculture, Tatura Center, Australia, p. 61.

Hirich, A., Choukr-Allah, R., Ezzaiar, R., Shahid, S. A., & Lyamani, A. (2021). Introduction of alternative crops as a solution to groundwater and soil salinization in the Laayoune area, South Morocco. *Euro-Mediterranean Journal for Environmental Integration 6, 52*. https://doi.org/10.1007/s41207-021-00262-7

IPCC. (2023). *Intergovernmental panel on climate change report*. Sixth Assessment Report.

Ivushkin, K., Bartholomeus, H., Bregt, A. K., Pulatov, A., Kempen, B., & de Sousa, L. (2019). Global mapping of soil salinity change. *Remote Sensing of Environment, 231*, 111260, https://doi.org/10.1016/j.rse.2019.111260.

IWMI. (2014). *World losing 2,000 hectares of irrigated farmland to salt-induced degradation Daily*. International Water Management Institute Press Release, October 28, 2014. Retrieved January 4, 2025, from https://www.iwmi.org/news/press-release-world-losing-2000-hectares-irrigated-farmland-salt-induced-degradation-daily/

Jacobson, T., & Adams, R. M. (1958). Salt and silt in ancient Mesopotamian agriculture. *Science (New Series), 128*(3334), 1251–1258.

Maas, E. V. (1990). Crop salt tolerance. In *Agricultural salinity assessment and management*. New York: American Society of Civil Engineers.

Maas, E. V., & Grattan, S. R. (1999). Crop yields as affected by salinity. In R. W. Skaggs, & J. van Schilfgaarde (Eds.), *Agricultural drainage agronomy* (pp. 55–108). Monograph No. 38. ASA.

Maas, E. V., & Hoffman, G. J. (1977). Crop salt tolerance—current assessment. *Journal of Irrigation Drainage Division ASCE, 103*(IR2), 115–134.

MAF. (2012). *Oman salinity strategy*. Volume 1: Main report. Ministry of Agriculture and Fisheries, Sultanate of Oman.

Mirzabaev, A. (2016). Land degradation and sustainable land management innovation in Central Asia. In F. Getzweiler, & J. von Braun (Eds.), *Technological and institutional innovations for marginalized smallholders in agricultural development*. Springer.

Mirzabaev, A., Goedecke, J., Dubovyk, O., Djanibekov, U., Le, Q., & Aw-Hassan, A. (2015). Economics of land degradation in Central Asia. *ZEF Policy Briefs,* 1–6.

Northcote, K. H., & Skene, J. K. M. (1972). *Australian soils with saline and sodic properties*. CSIRO Australia, Soil publication no. 27. CSIRO, Melbourne.

OECD-FAO Agricultural Outlook 2018–2027. OECD Publishing, Paris/Food and Agriculture Organization of the United Nations, Rome. https://doi.org/10.1787/agr_outlook-2018-en

Oldeman, L. R., Hakkeling, R. T. A., & Sombroek, W. G. (1991). *World map of the status of human-induced soil degradation. An explanatory note* (p. 31). Second revised edition. International Soil Reference and Information Center (ISRIC).

Pedersen, C. S. (2018). The UN sustainable development goals (SDGs) are a great gift to business! *Procedia CIRP, 69*, 21–24. https://doi.org/10.1016/j.procir.2018.01.003

Qadir, M., Noble, A. D., Schubert, S., Thomas, R. J., & Arslan, A. (2006). Sodicity-induced land degradation and its sustainable management: Problems and prospects. *Land Degradation and Development, 17*(6), 661–676. https://doi.org/10.1002/ldr.751

Qadir, M., Tubeileh, A., Akhtar, J., Larbi, A., Minhas, P. S., & Khan, M. A. (2008). Productivity enhancement of salt-affected environments through crop production. *Land Degradation and Development, 19*, 429–453.

Qadir, M., Noble, A. D., Qureshi, A. S., Gupta, R. K., Yuldashev, T., & Karimov, A. (2009). Salt-induced land and water degradation in the Aral Sea Basin: A challenge to sustainable agriculture in Central Asia. *Natural Resources Forum, 33*(2), 134–149. https://doi.org/10.1111/j.1477-8947.2009.01217.x

Qadir, M., Quillérou, E., Nangia, V., Murtaza, G., Murari Singh, M., Thomas, R. J., Drechsel, P., & Noble, A. D. (2014). Economics of salt-induced land degradation and restoration. *Natural Resources Forum, 38*(4), 282–295.ff10.1111/1477–8947.12054ff.ffhal-01954413f.

Quillérou, E., & Thomas, R. J. (2012). Costs of land degradation and benefits of land restoration: A review of valuation methods and suggested framework for inclusion into policy making. CAB reviews: Perspectives in agriculture, veterinary science. *Nutrition and Natural Resources, 7*(060), 1–12. https://doi.org/10.1079/PAVSNNR20127060

Qureshi, R. H., & Barrett-Lennard, E. G. (1998). *Saline agriculture for irrigated land in Pakistan: A Handbook*. ACIAR.

Rao, N. K., McCann, I., Shahid, S. A., Butt, K. R., Al Araj, B., & Ismail, S. (2017). Sustainable use of salt-degraded and abandoned forage production using halophyte grasses. *Crop and Pastures Science, 68*, 483–492. https://doi.org/10.1071/CP16197

Roy, S., & Chakraborty, U. (2014). Salt tolerance mechanisms in salt tolerant grasses (STGs) and their prospects in cereal crop improvement. *Botanical Studies, 55*, 3. https://doi.org/10.1186/1999-3110-55-31

Ruto, E., Gould, I., Bosworth, G., & Wright. (2018). *The state, causes and impact of soil salinization: A global overview*. SALFAR Work Package 3 Baseline January 2018 Report. Lincoln.

Saigal, S. (2003). Kazakhstan: Issues and approaches to combat desertification. TA-5941-REG. The Asian Development Bank and the Global Mechanism. Retrieved March 10, 2019, from www.droughtmanagement.info/literature/ADB_kazakhstan-combat-desertification-2003.pdf

Shahid, S. A. (1988). *Studies on the micromorphology of salt-affected soils in Pakistan*. PhD thesis, University of Wales, Bangor UK. 2 Volumes.

Shahid, S. A. (1992). Use of submicroscopic technique (SEM/EDXRA) for *in situ* micromorphological and microchemical analysis of salt-efflorescence from a *typic salorthid* in Pakistan. *Pakistan Journal of Agriculture, 1*(2), 82–87.

Shahid, S. A. (2002). Recent technological advances in characterization and reclamation of salt-affected soils in Arid zones. In N. M. Al-Awadhi & F. K. Taha (Eds.), *New Technologies for Soil Reclamation and Desert Greenery* (pp. 307–329). Publishers.

Shahid, S. A. (2013). Developments in salinity assessment, modeling, mapping, and monitoring from regional to submicroscopic scales. In S. A. Shahid, M. A. Abdelfattah, & F. K. Taha (Eds.), *Developments in soil salinity assessment and reclamation-innovative thinking and use of marginal soil and water resources in irrigated agriculture*. Springer Dordrecht Heidelberg, pp. 3–43. https://doi.org/10.1007/978-94-007-5684-7_1

Shahid, S. A., & Alkandari, A. J. (2024). Halophytic crops as a solution for food security, land rehabilitation, and mitigating future water crises by utilizing marginal quality waters. In J. C. Dagar, S. R. Gupta, & A. Kumar (Eds.), *Halophytes vis-à-vis saline agriculture*. Springer, pp. 441–478. https://doi.org/10.1007/978-981-97-3157-2_17

Shahid, S. A., & Jenkins, D. A. (1991a). Mechanisms of hydraulic conductivity reduction in the Khurrianwala soil series. *Pakistan Journal of Agricultural Sciences, 28*(4), 369–373.

Shahid, S. A., & Jenkins, D. A. (1991b). Effect of successive waters of different quality on hydraulic conductivity of a soil. *Journal of Drainage and Reclamation, 3*(2), 9–13.

Shahid, S. A., & Jenkins, D. A. (1994). Mineralogy and micromorphology of salt crusts from the Punjab, Pakistan. *Development in soil science 22* (pp. 799–810). Elsvier.

Shahid, S. A., & Rahman, K. (2011). Soil salinity development, classification, assessment and management in irrigated agriculture. In M. Pessarakli (Ed.), *Handbook of plant and crop stress* (3rd ed., pp. 23–39). CRC Press.

Shahid, S. A., Qureshi, R. H., & Jenkins, D. A. (1990). Salt-minerals in the saline-sodic soils of Pakistan. In Proc. Indo-Pak Workshop on Soil Salinity and Water Management, Feb. 10–14, 1990 Islamabad. *Pakistan, 1*, 175–191.

Shahid, S. A., Dakheel, A. H., Mufti, K. A., & Shabbir, G. (2008). Automated in-situ salinity logging in irrigated agriculture. *European Journal of Scientific Research, 26*(2), 288–297.

Shahid, S. A., Abdefattah, M. A., Omar, S. A. S., Harahsheh, H., Othman, Y., & Mahmoudi, H. (2010). Mapping and monitoring of soil salinization-remote sensing, GIS, modeling, electromagnetic induction and conventional methods–case studies. In M. Ahmad, & S. A. Al-Rawahy (Eds.), *Proceedings of the international conference on soils and groundwater salinization in arid countries*. Oman, Sultan Qaboos University, 1, pp. 59–97.

Shahid, S. A., Zaman, M., & Heng, L. (2018a). Soil salinity: Historical perspectives and a world overview of the problem. In M. Zaman, S. A. Shahid, & Heng, L. (Eds.), *Guideline for salinity assessment, mitigation and adaptation using nuclear and related techniques*. Springer, (Ch. 2, pp. 43–53). https://doi.org/10.1007/978-3-319-96190-3_2

Shahid, S. A., Zaman, M., & Heng, L. (2018b). Introduction to soil salinity, sodicity, and diagnostics techniques. In M. Zaman, S. A. Shahid, & L. Heng (Eds.), *Guideline for salinity assessment, mitigation and adaptation using nuclear and related techniques*. Springer, (Ch. 1, pp. 1–42). https://doi.org/10.1007/978-3-319-96190-3_1

Shahid, S. A., Zaman, M., Heng, L. (2018c). Salinity and sodicity adaptation and mitigation options. In M. Zaman, S. A. Shahid, & L. Heng (Eds.), *Guideline for salinity assessment, mitigation and adaptation using nuclear and related techniques* (Ch. 3, pp. 55–88). Springer. https://doi.org/10.1007/978-3-319-96190-3_3

Shahid, S. A., Burezq, H. A., & Baron, H. J. (2023). Farm land salinity risk assessment to develop a management strategy for food security. Project funded by Kuwait Foundation for the Advancement of Sciences. Kuwait Institute for Scientific Research, Kuwait (KISR # 18404).

Shahid, S. A., Alkandari, A., Baron, J., Othman, A., Burezq, H., Al-Najdi, F., Abdulraheem, A., Al-Jeri, M., Thomas, M., & Quoreshi, A. (2024). Farm land salinity risk assessment to develop a management strategy for food security. Annual Progress Report No 1, Kuwait Institute for Scientific Research, Kuwait (KISR # 18906).

Soil Science Division Staff. (2017). *Soil survey manual*. USDA Handbook no 18. United States Department of Agriculture, U.S. Government Printing Office, Washington, DC 20402, p. 437.

Spies, B., & Woodgate, P. (2004). *Technical report: salinity mapping methods in the Australian context*, prepared for the Programs Committee of the Natural Resources Management, Ministerial Council through Land and Water Australia and the National Dryland Salinity Program.

Sutton, W., Whitford, P., Stephens, E. M., Galinato, S. P., Nevel, B., Plonka, B., & Karamete, E. (2007). *Integrating environment into agriculture and forestry. Progress and prospects in Eastern Europe and Central Asia*. World Bank.

Tanji, K. K. (1990). Nature and extent of agricultural salinity. In K. K. Tanji (Ed.), *Agricultural salinity assessment and management*. ASCE manuals and reports on engineering practice no 71. ASCE New York, USA, pp. 1–17.

UN. (2016). *Sustainable consumption and production* [WWW document]. 2016. UN Sustainable Development Knowledge Platform. https://sustainabledevelopment.un.org/sdg12.

UN. (2017). *World population projected to reach 9.8 billion in 2050, and 11.12 billion in 2100* [WWW Document]. UN Department of Economic and Social Affairs. Retrieved January 6, 2025, from https://www.un.org/development/desa/en/news/population/world-population-prospects-2017.html.

UNEP. (2020). *United nations environment program mediterranean action plan*. Report of the inception meeting of the Mediterranean Sea Program (MedProgram) Enhancing Environmental Security (GEF ID 9607), Video conference, 20–22 July 2020.

US Salinity Lab Staff. (1954). *Diagnosis and improvement of saline and alkali soils* (p. 160). Handbook 60. United States Department of Agriculture.

Wiley, G. R. (1953). *Prehistoric settlement patterns in the Viru Valley*. Peru. Smithsonian Institute, Bureau of American Ethnology, Bulletin 155 Washington DC, USA XXII+454 pp. (+60 plates).

World Bank. (1998). Aral Sea Basin Program (Kazakhstan, Kyrgyz Republic, Tajikistan, Turkmenistan, and Uzbekistan)–Water and Environment Management Project. Project Document. Washington DC, USA, Global Environment Division, World Bank, p. 55.

Zaman, M., Shahid, S. A., & Heng, L. (2018a). *Guideline in salinity assessment, mitigation and adaptation using nuclear and related techniques*. Springer, Cham, p. 164. https://doi.org/10.1007/978-3-319-96190-3

Zaman, M., Shahid, S. A., & Heng, L. (2018b). Irrigation water quality. In M. Zaman et al. (Eds.), *Guideline for salinity assessment, mitigation and adaptation using nuclear and related techniques*. Springer, (Ch. 5, pp. 113–131). https://doi.org/10.1007/978-3-319-96190-3_5

Zekri, S., Al-Rawahy, S. A., & Naifer, A. (2010). Socio-economic considerations of salinity: Descriptive statistics of the Batinah sampled farms. In A. Mushtaque, S. A. Al-Rawahi, & N. Hussain (Eds.), *Monograph on management of salt-affected soils and water for sustainable agriculture*. Sultan Qaboos University, pp. 99–113.

Chapter 8
Open-Field and Controlled Environment Agriculture Systems for Food and Environmental Security in Arid Regions: Integration of Emerging Technologies

Binson M. Thomas and Narayana R. Bhat

Abstract Ensuring desirable level of food and nutrition securities for the growing population amidst undeniable climate change impacts and shrinking natural resources base is an overwhelming challenge in arid land countries. There are ample indications that the global food production might not keep up the pace with escalating population growth and, several countries in the arid region, Gulf Cooperation Council countries in particular, will continue to rely on significant food import. Furthermore, any disruption in the food supply chain will have significant, sudden negative consequences on food and nutritional securities. Water scarcity, land degradation, and soil salinization are rendering more agricultural lands unproductive every year. A reliable data suggested that nearly 2000 ha irrigated land is abandoned on a daily basis in 75 countries due to soil salinity alone. The current systems of food production in the arid region are not aligned with available resources or prevailing climatic conditions. The development and adoption of innovative, climate-resilient agri-food production systems are crucial to produce food sustainably without endangering the environment or the natural resource base of the arid countries. In this context, farmers are now shifting to alternative production systems like controlled environment agriculture, where they can grow crops with limited quantities of potable water and irrespective of prevailing climatic conditions. This chapter presents a comprehensive review of current agricultural practices and several steps already initiated or required to be implemented to advance towards regenerative, low-emission agri-food production systems contributing to food security and a sustainable future for everyone who relies on agriculture. The use of newer communication, data gathering,

B. M. Thomas (✉)
Food Security Program, Environment and Life Sciences Research Center, Kuwait Institute for Scientific Research, Safat, Kuwait
e-mail: bmthomas@kisr.edu.kw

N. R. Bhat
Ex-Principal Research Scientist, Food Security Program, Environment and Life Sciences Research Center, Kuwait Institute for Scientific Research, Safat, Kuwait

M. K. Suleiman and S. A. Shahid (eds.), *Fostering Arid Lands Agriculture in the Face of Climate Change*, Advances in Global Change Research 81,
https://doi.org/10.1007/978-3-031-98490-7_8

data handling, and artificial intelligence technologies can assist farmers to make right decisions to ensure resources conservation and enhance efficiency and sustainability of food production systems.

Keywords Climate-smart agriculture · Climate change · Global warming · Precision agriculture · Soilless agriculture · Sustainable farming systems · Food security

8.1 Introduction

Drylands are a vital resource occupying nearly 41% of total earth's surface and providing livelihood to over 2.0 billion people (Millennium Ecosystem Assessment, 2005). Additionally, drylands support 50% of the world's livestock population and account for 44% of global food production (Millennium Ecosystem Assessment, 2005; Gaur and Squires, 2018; Kimura & Moriyama, 2024). Drylands have significant economic potential and environment consequences. The low and erratic precipitation and frequent prolonged droughts make it impossible to grow any crop without irrigation, and further complicates the situation. As a result, the agricultural sector in these countries consumes huge volumes of the available water resources (Wehrey et al., 2023).

Arid lands are associated with aridity. Aridity reflects the relationship between average precipitation and evapotranspiration (Guar & Squires, 2018). The aridlands are characterized by the severe lack of available water, to the extent of hindering or preventing normal growth and development of plants and animals (Guar & Squires, 2018). There are ample evidences to indicate that humanity co-existed with aridity where a number of ancient civilizations in the arid and semiarid region flourished and expanded. Though they represent poorly managed and overused habitats, they are the home for a large number of precious biodiversity (Guar & Squires, 2018).

Globally, arid regions cover around 28 million km^2 while semi-arid regions cover around 46 million km^2 (FAO, 2009). These areas are located between 15° and 30° on both sides of the equator and distributed in Northern and Southern America, Northern Africa, the Sahelian region, Africa South of the Equator, the Near East and the Asia and the Pacific regions. The Gulf Cooperation Council (GCC) countries (Bahrain, Kuwait, Oman, Qatar, Saudi Arabia and United Arab Emirates) are stretched over vast arid lands. In contrast, areas affected by desertification in this region are expanding at an alarming rate (Gamo et al., 2013; Shahid, 2023; Shahid & Omar, 2001; Shahid et al., 1999; Wehrey et al., 2023).

Common features of arid lands include aridity index, high average temperatures and solar radiation, hot dry winds during summer, short winter with moderate temperatures, erratic and low precipitation, low atmospheric humidity, frequent drought events, sandy soils with low organic matter and water holding capacity, deep saline groundwater, and sparse xerophytic vegetation with inherently low biomass production (Gamo et al., 2013; Omar et al., 2001). The virgin soils are non-saline, however, these soils have the tendency to be saline when these soils are irrigated with brackish quality water to irrigate crops. Recent estimates (FAO, 2021) revealed nearly 424 million ha of top (0–30 cm) and 833 million ha of subsoil (30–100 cm) is affected by salt globally with 10% additional agricultural area being added every year, and approximately 50% of all arable land is expected to be affected by salts (FAO, 2021; Negacz et al., 2022; Shahid et al., 2018a; Tarolli et al., 2024).

A considerable proportion of salt-affected soils occur in the arid regions of Middle East and Africa. Therefore, soil salinization is considered as one the most critical threats to agriculture and food security in this region and hampering the achievement of UN-Sustainable Development Goal # 2 (end hunger and ensure sufficient food for all people) (Tarolli et al., 2024). Some of the salt affected soils can be reclaimed and salinity can be reversed for crop production (Shahid et al., 2018b, 2018c; Shahid & Rehman, 2011; Zaman et al., 2018a, 2018b; Rao et al., 2009, 2017; Hirich et al., 2021; Shahid & Alkandari, 2024). Other options to use salt-affected soils are (i) adoption of salt-tolerant crops (saline agriculture), (ii) nature-based solutions (placing natural barriers, maintaining buffer zones along canals and ditches, integrating with aquaculture), and (iii) bioengineering strategies (use of organic fertilizers, crop rotation, microbe-based solutions and precision or conservation agriculture) (Negacz et al., 2022; Shahid & Alshankiti, 2013; Tarolli et al., 2024).

The effects of climate change (extreme temperatures, droughts, erratic rain, and flashfloods) are prevalent to several countries, and are caused by nature and human-caused environmental devastation and ravenous resource extraction (Wehrey et al., 2023). Most of the arid land countries are exposed to climate change impacts (soaring heat waves, mobile/unstable surface layer, declining precipitation, extended drought periods, more intense sandstorms and floods, and rising sea levels). The consequences of these events are experienced unevenly across the region, because countries that lack adequate resources will not be able to implement mitigation measures properly and timely to reduce the negative impacts. Changing global climate has direct impact on the production and productivity of crops (Varshney et al., 2018).

8.2 Global Food Insecurity in Developing Countries

A recent report highlighted the global food insecurity of some counties, stating globally, over 2.33 billion people are facing moderate/severe food insecurity, which is 45% increase from 1.6 billion in 2015 (https://www.visualcapitalist.com/sp/mapped-food-insecurity-by-country/). Moderate food insecurity occurs when there is uncertainty to obtain the food, and unwillingly compromising the quality and quantity of food consumed, and when the individuals cannot access food for atleast one day (severe food insecurity). During 2021 and 2023, in Nigeria over 161 million people, and 106 million people in Pakistan suffered moderate and severe food insecurity. Eighty percent people in Afghanistan, South Sudan and Sierra Leone were food insecure (2021–2023).

8.3 Growing Challenges for Agriculture in the Arid Region

Food security is ensured when all people at all times have access to enough, safe and nutritious food that meets their dietary needs and food preferences for an active and healthy life (World Food Summit, 1996). Food security is often under constant threat due to unpredictable climatic conditions and geopolitical environments. Food security of the existing and escalating growing population is undeniably one of the most compelling global challenges of the twenty-first century. This is due to the fact that the dietary preference is continuously changing, natural resources base is progressively shrinking and the severity of climate change impact is increasing. Therefore, there are ample indications that global food production might not keep the pace with population growth. It is predicted that 50% more food will be needed to feed a population expected to be 9 billion in 2050 (Godfray et al., 2010). Nevertheless, increasing local agricultural production is key to ensuring long-term food security for communities (Kaiser, 2011) and as such, any disruptions in the domestic agriculture sector will have far reaching impact on food security. Consequently, a radical shift in food production strategy, like adoption of climate-resilient and precision agriculture is urgently needed to produce more food from available water and land resources (Chacha et al., 2023).

There are ample challenges in both open field (OF) and protected environment agriculture (PEA) in water scarce arid region countries. Reasons being, anthropic and natural impacts, including water scarcity, high temperatures, low precipitation, lack of access to inputs (fertilizers, pesticides) etc. Thus the current production systems are at stake and unsustainable to meet the food demand of increasing population.

8.3.1 Factors Influencing Plant Growth and Productivity in the Arid Region

Biotic and abiotic factors have significant effects on crop production. Suboptimal environmental conditions impose stress conditions and eventually, impact both the growth and geographical distribution of crops. Xerophytes grow in places where water shortage is of frequent occurrence. Similarly, prolonged drought can severely limit growth of most mesophytic plants. Plants are exposed to several stress factors (stressors), e.g., drought causes irreversible changes in plants (small, thick leaves, thick cuticle, fewer branches, flower/fruit drop, embryo abortion). Other stresses (high salinity and nutrients deficiency) inhibit growth and weaken plants' ability to fight back disease or pest attack. The plants might recover partially or fully when the stress factor is removed. Biotic and abiotic (environmental) stress factors that influence plant growth and development include light, temperature, water, humidity, pollution, salinity and nutrition. Depending on the nature and severity of environmental factors, plants are able to adapt themselves to these stresses by effecting internal changes. Therefore, it may be possible to modify/regulate plants' response to these stresses to our advantage. This requires proper understanding of the mechanisms of the plants response to various stresses, and then develop such characters through breeding programs.

Plants are also exposed to biotic stress factors (viruses, fungi, insects, nematodes and vectors) that cause direct damage to plants and negatively impact plant vigor and crop productivity (Lal et al., 2023). Both abiotic and biotic stresses affect different physiological processes (seed germination, growth, flowering, fruiting, leaf/flower senescence, fruit drop) and eventual death or decline of horticultural plants (Lal et al., 2023). Concurrent exposure of plants to various biotic and abiotic stresses causes severe damage to above ground (aerial) parts, nutrient assimilation, root growth, reproductive growth and reduction in biomass production.

There have been moves to investigate the effects of various abiotic and biotic stress conditions on physiological, biochemical and molecular processes in horticultural plants. This is essential to understand the nature, genetic control, and severity of negative impacts and, to develop suitable tools/practices/techniques to mitigate their negative impacts on crop growth and productivity (Lal et al., 2023). The mitigation measures include external application of biostimulants, growth regulators, secondary metabolites, antimicrobial peptides and novel chemicals like nano nutrient solutions and nanoencapsulations/nanobiosensors (Altaf et al., 2022; Singh et al., 2021).

8.4 Smart Technologies for Mitigating the Impacts of Climate Change and Aridity in Plant and Animal Production

The global food production is not matched to prevailing population growth, which is expected to exceed 9 billion by 2050 (United Nations, 2019). Nearly, 12% of world's total farmlands are <2.0 ha, but they contribute approximately to 35% of the total food production. Nearly, 10.8% of the total world population is malnourished, whereas approximately, 2 billion people lack essential micronutrients such as vitamin A and iron, and 2 billion are obese (United Nations, 2019, 2020). Hence, considering this scenario, it becomes imperative to secure the livelihood of small-scale farmers by providing them with technical instructions in diverse farming approaches (FAO and UNICEF, 2020; Lowder et al., 2021; Chauhan et al., 2021; Vishnoi & Goel, 2024). The traditional food production systems relying on area expansion and overuse of resources like land, mineral fertilizers, irrigation water, synthetic and harmful pesticides, are unable to meet the food/nutrition demand of population. Other factors that have rendered the present food system unsustainable include shrinking arable land area, land use change, deforestation to expand the area for crop production, deterioration in soil health and its productive capacity, and depletion in both quantity and quality of natural water resources.

Some crops in the present food production system (especially rice grown in ponding water) emits methane and contributes to warming, however, if rice is cultivated as direct seeding similar to wheat cultivation, methane emission can be mitigated (Liu et al., 2022). Despite constraints in the production systems, several countries will require crops intensification from current farming lands, as the expansion is not viable option due to the dearth of arable land and water resources. In the Middle East and North Africa (MENA) region, agriculture sector accounts for 67% of the total renewable water resources (Zaki et al., 2022). In many areas, the abstraction of groundwater far exceeds annual recharge, thus depleting the strategic water reserves.

In addition to population increases, growing affluence in GCC and other countries has led to increased demand for animal-based food items like meat and dairy products (high water footprint products) (Suleiman & Shahid, 2023a, 2023b) under arid environments, and has far greater negative environmental impacts (Pelletier et al., 2011). Therefore, there is a need to develop innovative strategies/practices to produce food without endangering environment or natural resource base of arid countries. This calls for adoption of climate-responsive, precision agricultural practices as well as facilitating their adoption by majority of local farmers through demonstrating technical and environmental feasibility as well as training of local farmers. This can be accomplished by adapting food production systems that meet goals of feeding the world, evolving climate change resilient processes/practices and drastically reducing/eliminating greenhouse gas emissions. Several countries have adopted climate-smart agriculture (CSA) by employing farming methods that respond to the effects of climate change on agriculture (and build resilience to climate change), increase agricultural productivity and ensure food and nutritional security for the growing population (Chauhan et al., 2021; De Pinto et al., 2020). Application of climate-smart technologies in OFs and PEA is discussed in the following sections.

8.4.1 Open Field Agriculture (OFA)

Agronomic crops are predominantly cultivated in the OFs, whereas horticultural crops (vegetables) are produced both in the field (generally winter season), and under protected environment throughout the year. Economic viability of crops produced in the OF depends on several factors including prevailing climatic conditions, soil health and water quality and availability (Bhat, 2011). To overcome the constraints, farmers are using high yielding varieties (hybrids, genetically modified crops). In addition, harsh environmental conditions require regular watering to offset plants water requirement, as well as use of heavy inputs (synthetic fertilizers, pesticides, growth stimulants etc.). These practices have severely degraded the environment and natural resources base (soils productive capacity, soil erosion due to water and wind, ground water pollution, salinization of arable lands), emergence of new pests/pathogens with greater pathogenicity, and increased greenhouse gases (GHGs) emission. These have endangered the health of agricultural workers and negatively impacted safety of the produce for human consumption. These practices have also rendered OF agriculture untenable and unsustainable. Therefore, it is high time that the producers protect and save the natural resources by reducing their exploitation and avoiding the use of harmful inputs. The best agricultural management practices (BAMPs) designed to minimize harmful effects and maintain economic and environmental sustainability are desirable. These practices ensure product safety for consumers and farm workers. These practices include soil conservation, sustainable water management and irrigation, integrated pest management, pesticide use, nutritional management, organic and inorganic waste management, and energy use and conservation. There are no standard BAMP's employed in all farms or situations, nevertheless, individual farmers can choose the combinations that are most appropriate for their farm/situation. Some examples of BAMP's in OF agriculture would include, (i) selection of appropriate crop/variety (ies) and cropping systems, (ii) land preparation that would conserve or enhance soil properties, (iii) appropriate methods of planting, (iv) efficient irrigation systems and precision irrigation, (v) efficient nutrient management using 4R concept, (vi) eradication of weeds to reduce competition with crops, (vii) integrated pest management (IPM) to maintain the diseases and pest at a manageable level, (viii) proper harvesting and postharvest handling practices, and (ix) recycling/reuse of organic farm wastes into compost and organic fertilizers. In essence, crop production and protection practices need to be integrated to achieve optimum and sustainable productivity and product quality.

Productivity of crop plants depends on land resources (e.g., soil type, quality and health), climatic variables (temperature, light, precipitation, humidity, wind, pollution, etc.) and agricultural production practices. It is necessary to estimate dynamic climatic variables and their resultant impact, and, understand the impact of natural resources (land and water) and agricultural practices to predict plant growth, yield and quality as precisely as possible. Because these factors vary both in time and space, thus, it is necessary to monitor these variables on a real time basis, which can be easily done by remote sensing (Panda et al., 2023).

8.4.1.1 Remote Sensing and Climate Prediction

Remote sensing (RS) is being used to forecast crop progress and production, assessment of crop health, crop identification and estimation, crop damage assessment, crop yield modelling and estimation, identification of pests and diseases, estimation of soil moisture, soil mapping, drought monitoring, water resources mapping, vegetation assessment, and evapotranspiration at field and landscape (community) levels (Sinha et al., 2018; Sishodia et al., 2020). Precision agriculture (PA) has been one of the advanced agricultural systems using Global Positioning System (GPS) to deliver the inputs in right quantities, right place and at right time (Stafford, 2000). Unlike in traditional farming, a fewer number of people can manage a large field more efficiently with PA system.

Recent advances in satellites and drones have greatly improved applicability of RS in PA (Inoue & Yokoyama, 2019; Inoue, 2020). High resolution cameras mounted on Unmanned Aerial Vehicle (UAV) take thousands of images to generate

3D surface models to collect valuable data (Verhoeven, 2011; Inoue & Yokoyama, 2019). Infrared thermal cameras are used to measure canopy temperature, brightness and emissivity of the vegetation cover and open surface from a distance. The data generated are used to estimate canopy evapotranspiration and evaporation from open surface (Bhat, 2022). Synthetic aperture radars (SAR) are used for monsoon forecasting and other agricultural applications, whereas satellite backscattering signatures are used to determine canopy traits (Bouvet & Le Toan, 2011; Santi et al., 2012). Microwave signatures are used to estimate soil water content (Martínez-Fernández et al., 2016).

8.4.1.2 Recarbonization of Agricultural Land and Decarbonization of Agri-Food System

Carbon dioxide (CO_2) is a GHG accounting for more than three-fourth of all GHGs emissions and contributing to global warming. The carbon dioxide reduction (CDR) is the main focus to mitigate climate change. Healthy soils can play key role in soil carbon sequestration (SCS) and increasing carbon sink. According to the recently concluded scientific forum (November 6–7, 2024), experts from International Maize and Wheat Improvement Centre (CIMMYT), Inter-American Institute for Cooperation on Agriculture (IICA) and the Trust Funds for Agriculture (FRA) recommended the adoption of regenerative/resilient-centred practices. These practices include, cover crops, organic and biofertilizers, and agroforestry to recarbonize agricultural soils to improve soil health, agri-food systems and value chains (Anon, 2024).

8.4.1.3 Crop Modelling

Crop modeling predicts plant growth and development based on environmental conditions and production inputs. Certain ecophysiological measurements are used to simulate a crop's response (growth or yield) to interaction among various environment, management, water, weather, and soil factors (Adarsh et al., 2023). Modeling is used to manage planting date, planting density, crop variety, fertilization, and irrigation, soil aspects (drainage class, pH, organic matter content, and texture i.e., sand/silt/clay content), and crop physiology (leaf area index, total above ground biomass, plant height, canopy, and vegetation cover). The model is calibrated using the most important parameter(s). Similarly, an independent set of data is used to validate the model and fine-tune inputs. Climatic attributes (atmospheric temperature, photoperiod, light intensity, precipitation, wind speed) are being used as input parameters for predicting plant growth and development.

8.4.1.4 Precision Agriculture System

Precision agriculture (PA) is advanced agricultural production system where the production inputs (water, nutrients and other agrochemicals) are optimized to enhance productivity, product quality, sustain soil health and the environment (Gebbers & Adamchuk, 2010). The Precision irrigation (PI) and efficient nutrient management (site-specific nutrient application) are the two main components of PA and involve optimal management of spatial and temporal variations and combining it with the optimal performance of the application systems/strategies (Bhat, 2022). In contrast, the traditional practices consider a 'whole-field' approach (Balafoutis et al., 2017; Smith et al., 2010). Studies on variable rate application have shown that substantial water and fertilizer savings ranging from 0 to 50%, and 8 to 20%, respectively can be realized in individual years and over several year periods (Bhat, 2022; Smith et al., 2010). Although tools and technologies (Variable Rate Irrigation, Regulated Deficit Irrigation, Partial Root-zone Drying, vertical irrigation, fertigation, etc.) required for the site-specific (SS) systems are currently available, substantial research and development efforts to integrate and optimize them in one package are required for realization of maximum impacts (Bhat, 2022; Speer et al., 2007). The precision agriculture involves the following steps:

Crop and Varietal Selection—Crops vary widely in their climatic and cultural requirements. Similarly, most of the modern-day varieties have been evolved for a specific situation or use. Therefore, selection of crops/varieties largely determines the success of any farming operations, especially under OF conditions. It is imperative that the crops/varieties selected are high yielding and must match the prevailing climatic conditions and available resources (land and water). Furthermore, skills/capabilities of farmers and marketability of the produce must be taken into consideration while selecting the crop/variety.

Precision irrigation—Soil receives water from rainfall and irrigation and lose to the atmosphere via evapotranspiration (ET). Plants absorb the required volumes of water from the soil to meet its transpiration needs. The ET is influenced by canopy and air temperature, solar radiation, wind speed, prevailing relative humidity, and soil water availability. So, the crop water need is the amount of water lost through evapotranspiration. An efficient irrigation management requires appropriate technologies to forecast crop water needs, an efficient system to deliver irrigation water in required quantities and at the right place and time. Furthermore, there is need for proper management strategy to maintain and operate the irrigation system both efficiently and optimally.

Irrigation scheduling using estimates of ET_0 (potential ET) requires knowledge of water contents at saturation, field capacity, and temporary and permanent wilting points (Bhat et al., 2011; Binson et al., 2022) and these characteristics depend greatly on the soil texture. Readily available water (RAW) to plants is the volume of water held between field capacity and a lower level at which the crop starts to experience stress (wilting point).

Precision Nutrient Management—Irrigation and nutrition are two most important management practices influencing yield, fruit quality and water- and nutrient-use efficiencies. In arid regions, where rainfall is low and infrequent, crop water requirement is met mostly by irrigation from different sources (groundwater, treated waste water etc.). The fertilizer nitrogen (N) use efficiency represents the percentage of applied N that is taken up by the plant, and is influenced by plant demand for N, the rate and timing of application, growing conditions (soil and environmental conditions), and irrigation water management. With the adoption of micro-irrigation techniques (drip irrigation) by growers, fertigation (applying soluble fertilizers through irrigation) plays a major role in nutrient management. In this context, it is important to develop an integrated, precision irrigation and nutrient management practices (e.g., date palm) to secure optimal yield, fruit quality and at the same times, minimize NO_3 leaching into groundwater, which will ultimately improve the nitrogen use efficiency (Bhat, 2022). Application of phosphorus as band placement and micronutrients through foliar application can improve nutrient use efficiencies.

Integrated Pest Management (IPM)—A pest, disease or weed can be described as any unwanted organism (insects, insect-like organism, weeds, mollusc, vertebrates, nematodes and microorganisms) that injures humans, domestic animals, useful plants and competes in any manner with them. They also spread disease to humans, animals or plants. The IPM is a long-term strategy or combination of different practices to maintain pest level to the minimum with rational use of pesticides. It is a location specific or even field specific combination of all suitable techniques. It is considered as an integral part of crop production involving all agronomic practices and promotes safe handling and application of pesticides for the protection of health and the environment. Following are the benefits of using IPM.

- Cost of pesticide is greatly reduced and use of unnecessary pesticides are avoided
- Input use in crop production is optimized and productivity is sustained
- Development of pest resistance is eliminated
- Biodiversity is protected
- Product and human safety are ensured
- Environmental quality is protected

The IPM includes mechanical, cultural, hygienic, biological and chemical methods of pest management. The use of individual method depends on crop, pest situation, available resources and infrastructure under which the crop is produced. It is also important to remember that any crop is a dynamic system—it is constantly changing. The IPM has to be dynamic to take into consideration the various developmental stages of plant growth that changes with time and changing environment. The pest level varies with season and year as well as with production practices followed by farmers.

8.4.1.5 Integrated Farming System (IFS)

In IFS crop production is combined with livestock, poultry, aquaculture, and other complementary operations/activities. The symbiotic relationships that exist among various components of farming are utilized to make the best use of available resources to optimize economic and environmental viability, sustain biodiversity and provide safety and year-round employment to people who rely on farming. The major benefits of integrated farming system include optimization of farm productivity, maintenance of soil health and fertility, avoidance of environmental pollution, and the wastes/by-products from one component serving as a valuable resource for another component of the integrated system. The key elements of integrated farming system are biogas, farm ponds, biological control, solar energy, recycling of organic farm waste (composting), rain water harvesting and watershed management, as well as crops and animals.

In addition, zero-tillage or no-tillage, integrated nutrient management (organic and mineral fertilizers), site-specific nutrient and irrigation management, crop rotation, use of biofertilizers, solar energy, and maintenance of farm inventory/data management system to track and monitor various farm activities are used in the IFS. The adoption of IFS depends on soil and climatic conditions, availability of resources, current level of resources use, skills available at farm level, and economic viability. Various IFS systems in use are (i) crop-livestock farming, (ii) greenhouse vegetable—aquaculture, (iii) fish—livestock farming, (iv) fish-duck production, (v) fish—rabbit farming, (vi) crop production—bee farming, (vii) crop production-livestock production and vermicomposting, (viii) range management-livestock production, (ix) livestock-biogas production, (x) fish farming-paddy production, and (xi) fruit and vegetable production-food processing. There are some limitations for successful use of IFS, including, (i) lack of awareness about sustainable farming systems, (ii) lack of resources, (iii) unavailability of marketing facilities for various farm products, (iv) lack of storage facilities, and (v) lack of farm credit.

8.5 Protected Environment Agriculture (PEA)

The PEA is the cultivation of crops under protected-controlled environments, wherein the indoor climatic conditions (temperature, relative humidity, light intensity and quality, light duration, CO_2 concentration) and production inputs (growing substrates, water, nutrients, agro-chemicals, labor) are optimized to improve production efficiency, maximize productivity and product quality and extend the growing season. Decades of research on innovative growing structures, growing systems, substrates, choice of varieties, energy saving, information technology and scientific data have created considerable potential for PEA, especially in the arid regions. Now, PEA has become a key element of modern-day agriculture in several countries representing the most intensified system of food production in terms of resources used, energy consumed, yield per unit area, GHG emission and cost. At the same time, it should be recognized that while PEA offered greater opportunities, it created more challenges in the utilization of resources, especially energy and water.

8.5.1 Global and Regional Expansion of PEA

Globally greenhouse crop production covered 405,000 ha (FAO, 2013), whereas, protected agriculture is being carried out in an area of 5,630,000 ha worldwide (Karaca, 2020), and the area for vegetables production was estimated to be 497,815 ha globally (Hickman, 2018).

Majority of the PEA occurred under plastic and temporary structures. However, using the artificial intelligence, Tong et al. (2024) presented the global assessment of greenhouse coverage and mentioned that greenhouse structures covered 1.3 million ha in 2019. This included 61% large (>1500 ha) and 39% small (<1500 ha) greenhouses clusters, with the majority of a recent upsurge in greenhouse cultivation occurring in the Global South, including a dramatic increase in China, accounting for 60% of the global coverage. A typical commercial greenhouse is a large, *gutter* connected (multi-ridge) structure known as a greenhouse range. The stand-alone *Quonset or gothic* shaped greenhouses with an approximate floor area of 300 m^2 account for a small segment of the total greenhouse acreage under cultivation (Cook & Calvin, 2005). Greenhouse ranges in the west tend to be high-tech, integrating the latest technologies/systems automation and agronomy, to control everything from light intensity, spectral quality, photoperiod, carbon dioxide content, watering, fertigation/feeding, IPM, harvesting, grading, packing and storage of produce. However, greenhouses in the arid regions require a different strategy/approach than in the tropical or temperate regions to ensure optimal plant development, productivity and product quality. The greenhouse design (geometry and width:length ratio) is influenced by crop type, crop variety (indeterminate vs. determinate), the cooling system employed and the prevailing climatic condition, especially wind speed and direction, and sand movement (Tong et al., 2024). As the width:length ratio affects the temperature and humidity gradients across the greenhouses, a different strategy needs to be employed for different growing conditions, especially in arid climates to achieve desired cooling efficiency, reduce water consumption and uniformity of temperatures inside the enclosed structures (Ghani et al., 2019).

8.5.2 Prospects of PEA in Arid Countries

Given the harsh climatic conditions, including high temperatures, strong winds coupled with sand storms, most farmers in the arid region have shifted from OF agriculture to protected agriculture. Protected agriculture and hydroponics are an intensive form of agriculture that uses, (i) less than half of the water used by conventional agriculture practices, uses less fertilizers, permits extended growing season, enables intensive production in small spaces, and prevents soil degradation. In addition, it achieves better quality and a higher market value of the produce because of the controlled environment coupled with regular monitoring of nutrient compositions, and elimination or reduced use of herbicides, pesticides and soil-borne diseases.

Protected agriculture allows growers to produce more of high-value vegetable crops such as tomato, pepper, cucumber, strawberry and watermelon, with higher plant densities and on a year-round basis. Pilot trials in the Arabian Peninsula showed a significant increase in yields when comparing greenhouse horticulture with OF cultivation. For instance, tomato attains maximum yields of up to 12.5 kg/m^2 when using hydroponics under greenhouse conditions, nearly four times higher than the average yield obtained in the OF (ICARDA/APRP, 2010).

8.5.3 Water Saving in PEA Compared to OFA

Crop production under PEA conserves water and enhance agriculture productivity. The greenhouse water savings come from reduced ET, reduction in water consumption compared to open-air farming for four primary reasons (i) plants lose less water through ET, due to the decreased wind, increased humidity, and regulated temperature, (ii) drip irrigation systems improve the efficiency of water usage, (iii) crops are spaced more closely, thus reducing open/vacant surface area from where water is wasted, and (iv) the crop cycle in a greenhouse is typically shorter than in the OF resulting in less water used over the entire growing season. Nevertheless, enormous amount of water is used for cooling of greenhouses, especially under arid climates. The PEA produces far greater yields per drop of water used (high water productivity).

8.5.4 Temporal Dynamics of Greenhouses

Developing novel PE food production techniques have drawn the attention of scientists globally. The ability of greenhouse to provide optimum growing environment under the varying ambient weather conditions depends on the structure (structural materials, shape, orientation, width: length ratio, height, size, covering material, etc.), the efficiency of the cooling system employed and energy and water sources used (Ghani et al., 2019). An ideal greenhouse design for countries in the arid region must offer protection against environmental factors, such as sand, dust and strong winds, intense solar radiation, extended day length and achieve high cooling efficiency, uniform temperature regimes across the length and breadth of the greenhouse, low operating and maintenance cost, and low water consumption, all of which are associated with enhanced economic viability. Some reviews on the high-tech use in PEA have highlighted the challenges in achieving desired indoor growing conditions under harsh arid climatic conditions (Al-Bassam et al., 2007; Ghani et al., 2019; Kassem, 1994).

Greenhouse covering materials serve two functions,(i) allowing maximum visible light from the sun to enter the greenhouse and, (ii) minimizing the entry of infrared radiation to protect the crop from excess heat (Fadel et al., 2016; Hemming et al., 2016). This property of the covering material not only keeps the heat gain due to the radiation to its minimum level and ensures that the optimum growth and development of the plants is achieved.

Nevertheless, the sunlight entering the greenhouse is trapped in the greenhouse and as a result, the inside temperature becomes 15 °C or more, higher than the dry bulb temperature (DBT) on a normal sunny day. Thus, for most of the time, enclosed structure will not be suitable for growing crops unless excess heat trapped in the greenhouse is continuously removed. Use of conventional air conditioning like in residential, commercial and office buildings is the best choice, but the capital and operating costs of such a system prohibits their applicability in a greenhouse (Mastalrez, 1997). Also, the relative humidity of air cooled by conventional refrigeration system will be low for normal plant growth. Alternatively, lower wet-bulb temperature (WBT) and large wet bulb depression (Δtdw), the difference between ambient DBT and WBT, particularly in the interior of arid region offer good opportunity for the cooling system working on humidification process an optimum choice. More importantly, for most of the summer season, during the peak hours, the Δtdw exceeds 20 °C. This period coincides with solar radiation leading to the maximum heat gain in the PE structure. Likely for this reason, most of the farmhouses in Kuwait are in the interior region of the country (Al-Hadban & Maheswari, 2006; Al-Bassam et al., 2007).

In areas where good quality water is not available for cooling, closed greenhouses are considered a viable option for growing food crops in controlled environments (Zaragoza & Buchholz, 2008). This is the area where most of the current developments in greenhouses are taking place. These structures allow CO_2 enrichment, thereby serving as carbon sink along with increased productivity and minimized pesticide use. However, the main challenge is to minimize the amount of energy needed for cooling. Concerted efforts are being made to enhance the self-reliance and sustainability of PE food production systems by utilizing high-grade energy produced from solar radiation for greenhouse cooling and desalination of saline water (El-Awady et al., 2014; Hassanien et al., 2016; Nayak & Tiwari, 2009; Sonneveld et al., 2009). Incorporation of energy efficient features is also being investigated to minimize heat gain thereby reducing energy and water demands for the cooling system.

Besides the indoor environment, the selection of crops and growing systems are also important in sustainable greenhouse food production, especially under arid conditions. In PEA, crops are grown in the soil (soil-based production system), soilless substrates (gravel, sand, peat moss, coir peat, vermiculite, perlite) or in nutrient solutions (hydroponic culture).

8.5.5 Soil-Based Plant Production

Soil based plant production system is the conventional and most popular across the globe. However, conventional cultivation systems based on soil are not efficient or sustainable in the long-term, mainly because of leaching of water or nutrients out of the root zone by deep percolation, erosion or evaporation and progressive salinization of soil, among other factors (Bhat et al., 2007). In addition, rising water table, chemical contamination of ground water and water logging have been wide-spread in soil-based production systems with irreversible negative consequences on productivity and product quality. Colla et al. (2005) found the traditional soil technique produced much lower WUEs than the soilless techniques in Zucchini. Similarly, Rouphael et al. (2005) showed higher yields in summer squash (*Cucurbita pepo* L.) grown in soilless culture. Elevated water-use in soil-based cultivation methods is due to the water losses through infiltration below the root zone. Similar studies on sweet peppers, *Capsicum annuum* (Albaho et al., 2009), summer squash, *Cucurbita pepo* L. (Rouphael et al., 2005), cucumber (Jovicich et al., 2007), colored peppers (Jovicich et al., 2005) and tomatoes showed higher yields in soilless than in soil culture (Bhat et al., 2007, 2009).

Diseases that are often associated with the use of growth media for seed germination, and stock propagation range from seedling damping-off, leaf spots and stem cankers, fire blight, crown gall, root rots and flower blights. Damping-off and related diseases are most frequently caused by *Rhizoctonia solani*, and also by *Pythium* and *Phytophthora* spp. Besides these fungal and bacterial diseases, there are a number of soils inhabiting nematodes that can cause a variety of damages to grown plants.

8.5.6 Soilless Production Systems

Broadly speaking, soilless culture is the cultivation of crops in growing substrates that are devoid of soil (Albaho et al., 2015; Bhat et al., 2007, 2009). Lack of arable land and good-quality irrigation water together with the difficulty and cost of managing soil-borne diseases and soil salinity have led to the development of soilless substrates (Bhat et al., 2007; El-Hady & El-Dardiry, 2006). Considerable progress has been achieved in the recent past in developing an economically feasible soilless production system viz, the hydroponic or solution culture (nutrient film culture, deep water culture, aeroponic system, closed insulated plant system, recirculating aquaponic system, artificial light plant factory) and soilless substrate culture (gravel, coco peat, sand, gravel, peat moss, rockwool, trough system, growbox, growbag, containerized production). Hydroponic system in one or the other form is widely used commercially accepted system in several countries across the world. In most of the countries, soilless production systems are mainly employed for the production of vegetables and cut-flowers. In hydroponic systems, physical support is provided by other inert materials, such as rock wool, perlite, gravel, etc., while nutrients are supplied through solutions containing mineral elements. Hydroponic systems not only save on water and fertilizer, but also reduce pollution problems associated with leaching of nutrients, pesticides, and other materials used in protected agriculture.

Nutrient Film Technique (NFT)—Offers several advantages, (i) easy to install, (ii) readily adaptable to any structure or crop, (iii) requires a short time for changeover between crops, (iv) conserves water and nutrients, (v) eliminates the need for steam or chemical sterilization, and (vi) has greater potential for complete automation (Burrage, 1993). In comparison with conventional ground-bed cultivation, NFT resulted in earlier flowering and higher flower yields per plant without loss of quality and with considerable savings in the amount of water used in carnation production. Benoit and Ceustermans (1991, 1993, 1995) suggested considerable reduction in crop duration and improvements in finished product quality in Lamb's lettuce, watercress and chervil when produced using the NFT system. However, not all crops adapt well to NFT; for example, the fruit quality (texture or firmness) in strawberry produced by NFT was poor compared with those produced in substrate-based systems.

Aeroponics—In this system, plant roots, free from supporting material, are misted with nutrient solution at regular intervals. This technology enables growers to realize higher production levels, both in terms of quantity and quality, reduce manpower consumption and extend the growing season considerably. Mateus-Rodrgues et al. (2014) compared potato mini-tubers production in aeroponics with other systems (semi-hydroponic, fiber-cement tiles and soil systems) and reported several benefits of aeroponics including high multiplication rates (up to 1:45) and high production capacity per area (>900 tubers/m^{-2}).

Aquaponics—Another promising food production system for the arid region, which is a variant of hydroponics, a recirculating aquaponics system (RAS), which integrates hydroponic vegetable production with fish production (Goddek et al., 2015; Love et al., 2015). In this system, fish is grown in indoor ponds to produce nutrient-rich wastes that acts as a food source for the plants grown in hydroponic system. The plants, on their part, purify and filter the wastewater that gets recycled directly to the fishponds. This closed-cycle system has great potential for optimization of all aspects of fish and crop production and adoption in PEA in the arid region.

Artificial Light Plant Factory (ALPF)—Is one of the most advanced completely closed vertical growing system that holds good potential for environmentally constrained countries in the arid region. It saves up to 70% of water and the space used for food production. It uses eco-friendly methods making farming a reality anywhere, even under difficult environments (Kozai et al., 2015; Vyas, 2018). It uses a hydroponic method of cultivation and combines several innovative technologies to produce quality food more sustainably. The ALPF enables rapid turnover of crops of desirable quality under predetermined artificial environment with precise and automated environmental control regardless of the natural environmental conditions. Plant factories have drawn attention in many countries due to its capability of environmental control not only to augment yield and quality, but also for increased functional components and bioactive compounds in the produce. High-performance production systems unconstrained by weather conditions have recently been developed to produce high-quality transplants under artificial light (Kozai et al., 2015). Closed systems for plant production with artificial light have several potential benefits, such as higher quality of transplants, shorter, precise production period, and lower use of resources as compared with conventional systems.

Closed Insulated Pallet System (CIPS)—Is an environmentally friendly growing system patented by Oregon State University, US (Green & Schneckenburger, 1992). It reduces resource inputs (mainly water, fertilizer and other chemicals) and labor needs for growing plants in containers (Green et al., 1992). A basic characteristic of CIPS is the plant-driven movement of water from a reservoir below the plant container through a capillary wick and then through the growth medium to the roots. The efficiency and simplicity of the system, and recyclability of many of its components, in addition to the fact that it requires minimal labor throughout the entire growing period, attracts more end-users who want to pursue new technologies for use on their farms. The reflective nature of the CIPS's surface prevents many insects from approaching plants grown using CIPS, thus, preventing many diseases carried by the notorious vectors. This system also allows retention of fertilizer reserves within the root zone to increase fertilizer availability to plants by ion diffusion and direct root contact (Green et al., 1992). Since there is no open or exposed space to support weed growth, there is no need for herbicide application.

Open Container Soilless System—Is the simplest and comprises of soilless growing media filled in high-density or low-density polyethylene (HDPE or LDPE) containers. Flexible polyethylene containers of desirable (5–25 L) capacity made from reusable HDPE are used for this purpose (Bhat et al., 2007, 2009).

Growbag Culture—Technique is one of the simplest and most user-friendly soilless cultivation systems with potential to be adopted in the GCC region (Albaho et al., 2009; Bhat et al., 2007, 2009). Cultivation by bag culture using soilless substrates was introduced in the mid 1970s in Belgium. Despite the initial failure due to unbalanced nutrient management and high salinity, the system gained momentum and became the main growing method in Central Europe during the 1990s. The growbag system has enabled growers to get better fruits, fewer disease problems and a crop that matures 7–10 days earlier than conventional methods and no longer have to worry about nematodes, *Verticillium* wilt or *Fusarium* crown rot.

Growbox Production Technique (GPT)—Is an economically feasible, soilless, eco-friendly, and water-use efficient production technique with a growing medium that can support two to three crops, and a HDPE container, which can withstand harsh environmental conditions and stressed handling during repeated solarization and cropping. The GPT is essentially a modification of the internationally accepted growbag production technique integrated with various soilless and organic growing substrates (Albaho et al., 2015). The GPT is highly suitable to the delicate agro-ecosystem of Kuwait, and the GCC region.

8.6 Functional Food Production

Functional food (nutraceuticals), offers certain health benefits (nutritional supplements, additional nutrients, disease resistance designed to improve health) in addition to their nutritional value (Ghazanfar et al., 2023; Granato et al., 2020). Some examples of functional food are vitamin fortifications, mineral supplements, probiotics, fiber, antioxidants, etc. The hydroponic systems can be used with great efficiency to produce functional foods.

8.7 Conclusions and Recommendations

Globally, agriculture sector is facing the daunting task of feeding the mankind with changing dietary and nutritional preferences to effectively address widespread malnutrition and hunger in many parts of the world. Some food insecure countries have to overcome frequent disruptions in the food supply-chain. The global food production is not keeping pace with population expansions due to climate change impacts and shrinking natural resources base is an undeniable fact. Hence, several countries, especially in the arid region, have historically relied on food importation, which is vulnerable to frequent political unrest, unexpected sudden compelling circumstances in the food exporting countries and consequently, cannot be sustained too long. The present food production systems in most countries in the arid region are also not sustainable due to water scarcity, harsh climate and poorly developed soils. This calls for a radical change in the way the food is currently produced by developing and adopting innovative, climate-smart farming practices or approaches. This is crucial to enhance sustainability, profitability and environmental safety of crop production operations. In the arid region crops are grown either in the open when the climatic conditions are moderate (as is the case with winter months) or in protected environments with various levels of sophistications. The modern-day field crop production in agriculturally developed countries utilize several innovative practices and tools to predict changing weather patterns and their impact on plant growth. Based on the data gathered, farmers are now able to make science-based decisions on crops/variety selection and employ integrated production and protection approaches to deliver inputs/resources precisely to a right place, in right quantities and at right time. The resultant productivity and product quality is highly predictable and sustainable, while minimizing environmental and resource degradation.

The cultivation of crops under protected (controlled) environments allows for optimization of indoor climatic conditions (temperature, relative humidity, light intensity and quality, light duration, CO_2 concentration) and production inputs (growing substrates, water, nutrients, agro-chemicals, labor) to secure greater productivity and product quality while at the same time improve production efficiency, and extend the growing season. The choice of an appropriate growing structure and growing system, substrates, choice of varieties and growing techniques as well as adoption of innovative energy saving techniques, application of information technology, AI, machine learning, and data science have increased the utility and potential of PEA in in the production of fresh commodities, especially in harsh climates (arid regions). Now, PEA has become a key component in modern agriculture in several countries. As it is the most intensified system of food production in terms of resources used, energy consumed, yield per unit area, greenhouse gas emission and cost, more and more farmers are moving to controlled environment agriculture, whereby they can grow crops over the year using less volume of fresh water. Adequate supply of food commodities in required quantities and at the highest possible quality necessitates the utilization of better food production technologies and sustainable management of natural resources. Protected agriculture industry must assess food production challenges in the given agroclimatic and geopolitical setting and, develop strategic management practices to evolve new possibilities and opportunities. The use of efficient plant production technologies together with innovative communication and digital techniques/tools, internet of things and artificial intelligence technologies will ensure resources conservation and greater efficiency in food production systems than ever before.

References

Adarsh, S., Biju, M. S. K., Nayana, N. L., & Deb, A. (2023). Crop modelling: Applications in climate change adaptation and food security. *Advances in Agronomy, 2*, 13–34.

Albaho, M. S., Bhat, N. R., Abo-Rezq, H., & Thomas, B. M. (2009). Effect of three different substrates on growth and yield of two cultivars of *Capsicum annuum*. *European Journal of Scientific Research, 28*(2), 227–233.

Albaho, M. S., Thomas, B. M., Naseeb, A., Ghloum, D., & Jacob, J. (2015). *Development and evaluation of growbox technique for sustainable vegetable production in protected environment*. Kuwait Institute for Scientific Research, Final Report, KISR Report No.13154, Kuwait.

Al-Bassam, E., Maheshwari, G. P., & Alasseri, R. R. (2007). *Optimization of water and electricity consumption in cooling towers-Phase I*. Kuwait Institute for Scientific Research, Final Report, KISR Report No 8767, Kuwait

Al-Hadban, Y., & Maheshwari, G. P. (2006). *Development of effective and water-efficient condenser pre-cooling system*. Kuwait Institute for Scientific Research, Proposal No 7333, Kuwait.

Altaf, M. A., Shahid, R., Ren, M., Naz, S., Altaf, M. M., Khan, L. U., Lal, M. K., Tiwari, R. K., & Shakoor, A. (2022). Melatonin mitigates cadmium toxicity by promoting root architecture and mineral homeostasis of tomato genotypes. *Journal of Soil Science and Plant Nutrition, 22*(1), 1112–1128. https://doi.org/10.1007/s42729-021-00720-9

Anon. (2024). Recarbonization of agricultural soils and decarbonization of agri-food systems: towards a sustainable future. Press Release, Climate adaptation and mitigation, First Scientific Forum of International Maize and Wheat Improvement Centre (CIMMYT), Inter-American Institute for Cooperation on Agriculture (IICA) and Trust Fund for Agriculture (FIRA), November 6–7, 2024, Texcoco, State of Mexico.

Balafoutis, A., Beck, B., Fountas, S., Vangeyte, J., Van der Wal, T., Soto, I., Gómez-Barbero, M., Barnes, A., & Eory, V. (2017). Precision agriculture technologies positively contributing to GHG emissions mitigation, farm productivity and economics. *Sustainability, 9*, 1339. https://doi.org/10.3390/su9081339

Benoit, F., & Ceustermans, N. (1991). Growing early cucumbers on a few ecologically sound substrates. *Acta Horticulturae, 287*, 299–308. https://doi.org/10.17660/ActaHortic.1991.287.34

Benoit, F., & Ceustermans, N. (1993). *Low pressure UV disinfection also effective for NFT-lettuce.* European Vegetable Research and Development Center, 2860 Sint-Katelinjne-Waver, Belgium, 9 pp, CABI Record Number: 19930322077.

Benoit, F., & Ceustermans, N. (1995). Horticultural aspects of ecological soilless growing methods. *Acta Horticulturae, 396*, 11–24. https://doi.org/10.17660/ActaHortic.1995.396.1

Bhat, N. R. (2022). Guidelines for best irrigation management practices for date palm for ARASIA member state parties (RAS5089). Report submitted to International Atomic Energy Agency, Vienna, Austria. Kuwait Institute for Scientific Research, KISR Final Report No. 17789, 44–86, Kuwait.

Bhat, N. R., Albaho, M., Suleiman, M. K., Al-Mulla, L., Christopher, A., Thomas, B., Ali, S. A., George, P., Lekha, V. S., Jacob, S., Al-Zalzaleh, M., & Bellen, R. (2007). Standardization of growing substrates and fertilizer application for organic greenhouse vegetable production. Kuwait Institute for Scientific Research, KISR Report No. 8912, Kuwait.

Bhat, N. R., Albaho, M., Suleiman, M. K., Al-Dosery, S., Thomas, B., Ali, S. I., George, P., Bellen, R., & Al-Zalzaleh, M. (2009). Optimization and pilot scale organic greenhouse vegetable production. Kuwait Institute for Scientific Research, KISR Report No. 9724, Kuwait.

Bhat, N. R., Suleiman, M. K., Al-Menaie, H., Al-Mulla, L., Thomas, B., Lekha, V. S., George, P., Ali, S. I., & Al-Zalzaleh, M. (2011). Irrigation studies in Date Palm (*Phoenix dactylifera L.*). Kuwait Institute for Scientific Research, KISR Report No 10663, Kuwait.

Binson, M. T., Bhat, N. R., Al-Mansour, H., Khalil, M., Almulla, L., Mazahrih, N., Hani, N. B., Alwan, A., Dalgamony, S., Amarat, O., Eid, L., & Ahmed, H. S. (2022). Enhancing the sustainability of date palm production in states parties through climate-smart irrigation, nutrient and best management practices (ARASIA) (RAS5089). Kuwait Institute for Scientific Research, KISR Final Report No. 17789, Kuwait.

Bouvet, A., & Le Toan, T. (2011). Use of ENVISAT/ASAR wide-swath data for timely rice fields mapping in the Mekong River Delta. *Remote Sensing of Environments, 115*, 1090–1101. https://doi.org/10.1016/j.rse.2010.12.014

Burrage, S. W. (1993). Nutrient film technique in protected cultivation. *Acta Horticulturae, 323*. https://doi.org/10.17660/ActaHortic.1993.323.1

Chacha, J. M., Thirumalai, M., Idawa, O. K., Chilwea, J. P., Kilamba, C. J., Hussy, B. D., Ishaq, H., & Rajendran, K. (2023). Greenhouse and open-field farming: A comparison through yield and growth parameters investigated in Dar es Salaam, Tanzania. *Innovations in Agriculture*, ISSN: 2617-1295. https://doi.org/10.25081/ia.2023-02

Chauhan, A. S., Rani, A., Kumari, A., Singh, R. N., & Kumari, K. (2021). Climate-smart agriculture: an integrated approach for attaining agricultural sustainability. *Climate Change and Resilient Food Systems*, Chapter 5, 141–189. https://doi.org/10.1007/978-981-33-4538-6_5

Colla, G., Fanasca, S., Cardarelli, M., Rouphael, Y., Saccardo, F., Graifenberg, A., & Curadi, M. (2005). Evaluation of salt tolerance in rootstocks of cucurbitaceae. *Acta Horticulturae, 697*, 469–474.

Cook, R., & Calvin, L. (2005). Greenhouse tomatoes change the dynamics of the North American fresh tomato industry. Economic Research Report No. 2, Economic Research Service, USDA Washington, DC, pp. 81.

De Pinto, A., Cenacchi, N., Kwon, H., Koo, J., & Dunston, S. (2020). Climate smart agriculture and global food-crop production. *PLoS One, 15*, Article e0231764. https://doi.org/10.1371/journal.pone.0231764

Degerli, B., Nazir, S., Sorgüven, E., Hitzmann, B., & Özilgen, M. (2015). Assessment of the energy and exergy efficiencies of farm to fork grain cultivation and bread making processes in Turkey and Germany. *Energy, 93*(P1), 421–434. https://doi.org/10.1016/j.energy.2015.08.095

El-Awady, M. H., El-Ghetany, H. H., & Latif, P. M. A. (2014). Experimental investigation of an integrated solar green house for water desalination, plantation and wastewater treatment in remote arid Egyptian communities. *Energy Procedia, 50*, 520–527. https://doi.org/10.1016/j.egypro.2014.06.063

El-Hady, A. M., & El-Dardiry, E. I. (2006). Improving hydrophysical properties quality of compost. *Journal of Applied Sciences Research, 2*(12), 1137–1141.

Fadel, M. A., Hammad, B., Al-Hosany, F., & Iwaimer, O. (2016). Greenhouse covering materials: A comparative study. *Agricultural Engineering International, CIGR Journal, 18*, 48–57.

FAO. (2009). Global map of aridity. Spatial resolution of 10 arc minutes and temporal resolution of 1961–1990; Food and Agriculture Organization of the United Nations (FAO), Rome, Italy.

FAO. (2021). *Global map of salt affected soils.* Version 1.0. Retrieved November 8, 2024, from http://www.fao.org/salt-portal/data-hub/soil-map-and-databases/global-map-of-salt-affected-soils/en/

FAO-UNICEF. (2020). *The state of food security and nutrition in the world 2019—Safeguarding against economic slowdowns and downturns.* FAO, Rome. ISBN 978-92-5-31570-5, https://openknowledge.fao.org/handle/20.500.14283/ca5162en

Food and Agriculture Organization (FAO). (2013). *Good agricultural practices for greenhouse vegetable production: Principles for Mediterranean areas.* Food and Agriculture Organization, Plant Production and Protection Paper No. 217. Rome.

Gamo, M., Shinoda, M., & Maeda, T. (2013). Classification of arid lands, including soil degradation and irrigated areas, based on vegetation and aridity indices. *International Journal of Remote Sensing, 34*, 6701–6722.

Gaur, M. K., & Squires, V. R. (Eds.). (2018). *Climate variability impacts on land use and livelihoods in drylands* (348 pp.). Springer.

Gebbers, R., & Adamchuk, V. I. (2010). Precision agriculture and food security. *Science, 327*(5967), 828–831. https://doi.org/10.1126/science.1183899

Ghani, S., Bakochristou, F., El Bialy, E., Rashwan, M., Abdel-halim, A., & Ismail, S. (2019). Design challenges of agriculture greenhouses in hot and arid environments—A review. *Engineering in Agriculture, Environment and Food, 12*(1), 48–70. https://doi.org/10.1016/j.eaef.2018.09.004

Ghazanfar, S., Ali, G. M., Abid, R., Farid, A., Akhtar, N., Batool, N. A., Khalid, S., Okla, M. K., Al-Amri, S. S., Alwasel, Y. A., & Hameed, Y. (2023). An overview of functional food. *Current Topics in Functional Foods.* https://doi.org/10.5772/intechopen.103978

Goddek, S., Delaide, B., Mankasingh, U., Ragnarsdottir, K. V., Jijakli, H., & Thorarinsdottir, R. (2015). Challenges of sustainable and commercial aquaponics. *Sustainability, 7*, 4199–4224.

Godfray, H. C. J., Beddington, J. R., Crute, I. R., Haddad, L., Lawrence, D., Muir, J. F., Pretty, J., Robinson, S., Thomas, S. M., & Toulmin, C. (2010). Food security: The challenge of feeding 9 billion people. *Science, 327*, 812–818.

Granato, D., Barba, F. J., Kova, D. B., Lorenzo, J. M., Cruz, A. G., & Putnik, P. (2020). Functional foods: Product development, technological trends, efficacy testing, and safety. *The Annual Review of Food Science and Technology, 11*, 95–118.

Green, J. L., & Schneckenburger. (1992). *Pallet system for container-grown plants.* U.S. Patent 5, 170, 581. Assigned to the State of Oregon, acting by and through the State Board of Higher Education on behalf of Oregon State University. Issued 2 June.

Green, J. L., Robbins, J. A., & Briggs, B. A. (1992). A closed, insulated pallet production system (Abstract). *HortScience, 27*, 569. ISSN: 2327-9834, https://doi.org/10.21273/HORTSCI.27.6.569f

Hassanien, R. E., Li, M., & Lin, W. D. (2016). Advanced applications of solar energy in agricultural greenhouses. *Renewable and Sustainable Energy Reviews, 54*, 989–1001.

Hemming, S., Swinkels, G. L. A. M., Breugel, V. A. J., & Ghani, V. M. (2016). Evaluation of diffusing properties of greenhouse covering materials. *Acta Horticulturae, 1134*, 306–309.

Hickman, G. W. (2018). International Greenhouse Vegetable Production—Statistics. Available online: http://www.cuestaroble.com/listghproducers.html

Hirich, A., Choukr-Allah, R., Ezzaiar, R., Shahid, S. A. & Lyamani, A. (2021). Introduction of alternative crops as a solution to groundwater and soil salinization in the Laayoune area, South

Morocco. *Euro-Mediterranean Journal for Environmental Integration, 6*(52). https://doi.org/10.1007/s41207-021-00262-7

ICARDA/APRP. (2010). Annual Report 2010. International Center for Agricultural Research for Dry Areas (ICARDA), Aleppo, Syria. http://hdl.handle.net/20.500.11766/7505

Inoue, Y. (2020). Satellite- and drone-based remote sensing of crops and soils for smart farming—A review. *Soil Science and Plant Nutrition, 66*(6), 798–810.

Inoue, Y., & Yokoyama, M. (2019). Drone-based optical, thermal, and 3D sensing for diagnostic information in smart farming-systems and algorithms. In *International Geoscience and Remote Sensing Symposium* (pp. 7266–7269), (IGARSS), 28 July–02 August 2019, Yokohama, Japan. IEEE. https://doi.org/10.1109/IGARSS.2019.8898736

Jovicich, E. J., VanSickl, J. J., & Cantliffe, D. J. (2005). Greenhouse-grown colored peppers: A profitable alternative for vegetable production in Florida. *Horticulture Technology, 15*(2), 355–369.

Jovicich, E. J., Cantliffe, D. J., & Simone, E. H. (2007). Comparative water and fertilizer use efficiencies of two production systems for cucumbers. *Acta Horticulturae, 731*, 235–249.

Kaiser, M. L. (2011). Food security: An ecological-social analysis to promote social development. *Journal of Community Practice, 19*, 62–79.

Karaca, C. (2020). Current situation, problems and solution of protected cultivation in antalya. In A. Atılgan, B. Saltuk (Eds.), *Current researches in agriculture, forestry and aquaculture science.* Duvar Publishing. ISBN: 978-625-7680-04-2

Kassem, A. S. (1994). Energy and water management in evaporitive cooling systems in Saudi Arabia. *Resources Conservation and Recycling, 12*, 135–146. https://doi.org/10.1016/0921-3449(94)90002-7

Kimura, R., & Moriyama, M. (2024). Recent global distribution of aridity index and land use in arid regions. *Scientific Online Letters on the Atmosphere (SOLA).* https://doi.org/10.2151/sola.2024-011

Kozai, T., Niu, G., & Takagaki, M. (eds.). (2015). *Plant factory: An indoor vertical farming system for efficient quality food production.* Academic Press. https://books.google.com/books?id=z-C7DwAAQBAJ

Lal, M. K., Tiwari, P. K., Altaf, M. A., Kumar, A., & Kumar, R. (2023). Editorial: Abiotic and biotic stress in horticultural crops: Insight into recent advances in the underlying tolerance mechanism. *Frontiers in Plant Science, 14*, 1212982. https://doi.org/10.3389/fpls.2023.1212982

Liu, Y., Liu, W., Geng, X., Liu, B., Fu, X., Guo, L., Bai, J., Zhang, Q., Geng, Y., & Shao, X. (2022). Direct-seeded rice reduces methane emissions by improving root physiological characteristics through affecting the water status of paddy fields. *Rhizosphere, 24*, Article 100628. https://doi.org/10.1016/j.rhisph.2022.100628

Love, D. C., Jillian, P., Fry, P., Li, X., Hill, E. S., Genello, L., Semmens, K., & Thompson, R. E. (2015). Commercial aquaponics production and profitability: Findings from an international survey. *Aquaculture, 435*, 67–74.

Lowder, S. K., Sanchez, S. M. V., & Bertini, R. (2021). Which farms feed the world and has farmland become more concentrated? *World Development, 142*, Article 105455.

Martínez-Fernández, G., Denman, S. E., Yang, C., Cheung, J., Mitsumori, M., & McSweeney, S. (2016). Methane inhibition alters the microbial community, hydrogen flow, and fermentation response in the rumen of cattle. *Frontiers in Microbiology, 7*, 1122.

Mastalrez, J. W. (1997). *The greenhouse environment* (p. 629). Wiley and Sons.

Mateus-Rodrgues, J., de Haan, S., Andrade-Piedra, J. L., Maldonaldo, L., Hareau, G., Barker, I., Chuquillanqui, C., Otazu, V., Frisancho, R., Bastos, C., Pereira, A. S., Medeiros, C. A., Montesdeoca, F., & Benitez, J. (2014). Technical and economic analysis of aeroponics and other systems for potato mini-tuber production in Latin America. *American Journal of Potato Research, 90*(4). https://doi.org/10.1007/s12230-013-9312.5

Millennium Ecosystem Assessment. (2005). *Ecosystems and human well-being.* World Resources Institute. 25 pp.

Nayak, S., & Tiwari, G. N. (2009). Theoretical performance assessment of an integrated photovoltaic and earth air heat exchanger greenhouse using energy and exergy analysis methods. *Energy and Building, 41*, 888–896.

Negacz, K., Malek, Z., de Vos, A., & Vellinga, P. (2022). Saline soils worldwide: Identifying the most promising areas for saline agriculture. *Journal of Arid Environments, 203*, Article 104775.

Omar, S. A. S., Misak, R., King, P., Shahid, S. A., Abo-Rezq, H., Grealish, G., & Roy, W. (2001). Mapping the vegetation of Kuwait through reconnaissance soil survey. *Journal of Arid Environments, 48*(3), 341–355.

Panda, S., Maram, B. R., & Shirisha, K. (2023). Remote sensing in agriculture: applications in crop monitoring, yield prediction and precision farming. In S. R. Kadam, L. Yosung, H. M. Chishi, T. Malemngambi, & H. A. Sushma (Eds.), Chapter 8. *In Advances in Agronomy* (Vol. 2, pp. 114–132). Bigfoot06 Publications (OPC) Pvt. Ltd. ISBN Print Book 978-81-19201-41-9.

Pelletier, N., Audsley, E., Brodt, S., & Garnet, T. (2011). Energy intensity of agriculture and food systems. *Annual Review of Environment and Resources, 36*. https://doi.org/10.1146/annurev-environ-081710-161014

Rao, N. K., Shahid, M., & Shahid, S. A. (2009). Alternative crops for diversifying production systems in the Arabian Peninsula. *Arabian Gulf Journal of Scientific Research, 27*(4), 195–203.

Rao, N. K., McCann, I., Shahid, S. A., Butt, K. R., Al Araj, B., & Ismail, S. (2017). Sustainable use of salt-degraded and abandoned forage production using halophyte grasses. *Crop and Pastures Science, 68*, 483–492. https://doi.org/10.1071/CP16197

Rouphael, Y., Colla, G., Fanasca, M., Salerno, A., Rivera, C. M., Rea, A., & Karam, F. (2005). Water-use efficiency of greenhouse summer squash in relation to the method of culture: Soil vs. soilless. *Acta Horticulturae, 697*, 81–86.

Santi, E., Fontanelli, G., Montomoli, F., Brogioni, M., Macelloni, G., Paloscia, S., Pettinato, S., & Pampaloni, P. (2012). The retrieval and monitoring of vegetation parameters from COSMO-SkyMed images. In *International geoscience and remote sensing symposium*, 22–27 July, 2012, Munich, Germany (pp. 7031–7034) (IGARSS). IEEE. https://doi.org/10.1109/IGARSS.2012.6351952

Shahid, S. A. (2023). Desertification: A central problem to restoring ecosystems. In M. K. Suleiman, & S. A. Shahid (Eds.), *Terrestrial environment and ecosystems of Kuwait* (Ch. 5, pp 119–137). Springer. https://doi.org/10.1007/978-3-031-46262-7_5

Shahid, S. A., & Alkandari, A. J. (2024). Halophytic crops as a solution for food security, land rehabilitation, and mitigating future water crises by utilizing marginal quality waters. In J. C. Dagar, S. R. Gupta, & A. Kumar (Eds.), *Halophytes vis-à-vis saline agriculture* (Ch. 17, pp. 441–478). Springer. https://doi.org/10.1007/978-981-97-3157-2_17

Shahid, S. A., & Alshankiti, A. (2013). Sustainable food production in marginal lands—Case study GDLA member countries. *International Soil and Water Conservation Research, 1*(1), 24–38.

Shahid, S. A., & Omar, S. A. S. (2001). Causes and impacts of land degradation in the arid environment of Kuwait. In E. M. Bridges et al. (Eds.), *Response to land degradation* (pp. 76–77).

Shahid, S. A., & Rahman, K. (2011). Soil salinity development, classification, assessment and management in irrigated agriculture. In M. Pessarakli (Ed.), *Handbook of plant and crop stress* (3rd ed., Ch. 2, pp. 23–39). CRC Press.

Shahid, S. A., Omar, S. A. S., & Al-Ghawas, S. (1999). Indicators of desertification in Kuwait and their possible management. *Desertification Control Bulletin, 34*, 61–66.

Shahid, S. A., Zaman, M., & Heng, L. (2018a). Soil salinity: Historical perspectives and a world overview of the problem. In M. Zaman, S. A. Shahid, L. Heng (Eds.), *Guideline for salinity assessment, mitigation and adaptation using nuclear and related techniques* (Ch. 2, pp. 43–53). Springer. https://doi.org/10.1007/978-3-319-96190-3_2

Shahid, S. A., Zaman, M., & Heng, L. (2018b). Introduction to soil salinity, sodicity, and diagnostics techniques. In M. Zaman, S. A. Shahid, & L. Heng (Eds.), *Guideline for salinity assessment, mitigation and adaptation using nuclear and related techniques* (Ch. 1, pp. 1–42). Springer. https://doi.org/10.1007/978-3-319-96190-3_1

Shahid, S. A., Zaman, M., & Heng, L. (2018c). Salinity and sodicity adaptation and mitigation options. In M. Zaman, S. A. Shahid, & L. Heng (Eds.), *Guidelines for salinity assessment, mitigation and adaptation using nuclear and related techniques* (Ch. 3, pp. 55–88). Springer. https://doi.org/10.1007/978-3-319-96190-3_3

Singh, A., Rajput, P., Chopra, R., Singh, A. S., Singh, A., & Singh, P. (2021). Applications of nanomaterials in agriculture and environment management. In S. S. Shende, V. D. Rajput, A. V. Gorovtsov, T. M. Minkina, S. N. Sushkova (Eds.), *Microbial Synthesis of Nanomaterials* (pp. 163–178). Nova Science Publishers Inc.

Sinha, N. K., Mohanty, M., Somasundaram, J., Shinogi, K. C., Hati, K. M., & Chaudhary, R. S. (2018). *Application of remote sensing in agriculture.* Harit Dhara, 1, 15–16, ICAR-Indian Institute of Soil Science, Bhopal, Madhya Pradesh

Sishodia, R. P., Ray, R. L., & Singh, S. K. (2020). Applications of remote sensing in precision agriculture: A review. *Remote Sensing, 12*, 3136. https://doi.org/10.3390/rs12193136

Smith, R. J., Baillie, J. N., McCarthy, A. C., Raine, S. R., & Baillie, C. P. (2010). *Review of precision irrigation technologies and their application.* Sustainable Irrigation. National Center for Engineering in Agriculture, Publication 1003017/1, University of Southern Queensland, Toowoomba, Australia.

Sonneveld, P. J., Swinkels, G. L. A. M., Bot, G. P. A., & Flamand, G. (2009). Feasibility study for combining cooling and high-grade energy production in a solar greenhouse. *Biosystems Engineering, 105*(1), 51–58. https://doi.org/10.1016/j.biosystemseng.2009.09.012

Spreer, W., Nagle, M., Neidhart, S., Carle, R., Ongprasert, S., & Muller, J. (2007). Effect of regulated deficit irrigation and partial root zone drying on the quality of mango fruits (*Mangifera indica* L., cv. 'Chok Anan'). *Agricultural Water Management, 88*, 173–180. https://doi.org/10.1016/j.agwat.2006.10.012

Stafford, J. V. (2000). Implementing precision agriculture in the 21st century. *Journal of Agricultural Engineering Research, 76*, 267–275.

Suleiman, M. K., & Shahid, S. A. (2023a). Prospective of agricultural farming in Kuwait and energy-food-water-climate nexus. In M. K. Suleiman, & S. A. Shahid (Eds.), *Terrestrial Environment and Ecosystems of Kuwait* (Ch. 15, pp. 363–391). Springer. https://doi.org/10.1007/978-3-031-46262_15

Suleiman, M. K., & Shahid, S. A. (2023b). Agricultural water footprint of major crops grown in Kuwait compared to the world average: A review. In M. K. Suleiman, & S. A. Shahid (Eds.), *Terrestrial Environment and Ecosystems of Kuwait* (Ch. 16, pp. 393–414). Springer. https://doi.org/10.1007/978-3-031-46262_16

Tarolli, P., Luo, J., Park, E., Barcaccia, G., & Masin, R. (2024). Soil salinization in agriculture: Mitigation and adaptation strategies combining nature-based solutions and bioengineering, *iScience, 27*, 108830. https://doi.org/10.1016/j.isci.108830

Tong, X., Zhang, X., Fensholt, R., Jensen, P. R. D., Li, S., Larsen, M. N., Reiner, F., Tian, F., & Brandt, M. (2024). Global area boom for greenhouse cultivation revealed by satellite mapping. *Nature Food, 5*(6), 1–11. https://doi.org/10.1038/s43016-024-00985-0

United Nations. (2019). *The sustainable development goals report.* United Nations. https://unstats.un.org/sdgs/report/2019/

United Nations. (2020). *The sustainable development goals statistical annex report.* United Nations Statistics Division; Development Data and Outreach Branch New York, NY, USA. https://unstats.un.org/sdgs/report/2020/

Varshney, R. K., Singh, V. K., Kumar, A., Powell, W., & Sorrells, M. E. (2018). Can genomics deliver climate-change ready crops? *Current Opinion in Plant Biology, 45*, 205–211. https://doi.org/10.1016/j.pbi.2018.03.007

Verhoeven, H. (2011). Climate change, conflict and development in Sudan: Global neo-malthusian narratives and local power struggles. *Development and Change, 42*, 679–707. ISSN 0012-155X, https://doi.org/10.1111/j.1467-7660.2011.01707.x

Vishnoi, S., & Goel, R. K. (2024). Climate smart agriculture for sustainable productivity and healthy landscapes. *Environmental Science & Policy, 151*, Article 103600.

Vyas, K. (2018). *13 Vertical farming innovations that could revolutionize agriculture*. Retrieved October 12, 2021, from https://interestingengineering.com/13-vertical-farming-innovations-that-could-revolutionize-agriculture

Wehrey, F., Dargin, J., Mehdi, Z., Muasher, M., Yahya, M., Kayssi, I., Hassan, Z., Andrews, M., Madain, M., Al-Mailam, M., Hamzawy, A., Yerkes, S., Clasen, H., & Yabi, G. (2023). *Climate change and vulnerability in the Middle East—Carnegie Endowment for International Peace*. United States of America. Retrieved December 16, 2024, from https://coilink.org/20.500.12592/7vg9ts. COI: 20.500.12592/7vg9ts

World Food Summit. (1996). Rome, Italy. November 13–17, 1996.

Zaki, N. A., Klöve, B., & Haghighi, A. T. (2022). Expanding the irrigated areas in the MENA and central Asia: Challenges or opportunities. *Water, 14*(16), 2560. https://doi.org/10.3390/w14162560

Zaman, M., Shahid, S. A., & Heng, L. (2018a). *Guideline in salinity assessment, mitigation and adaptation using nuclear and related techniques* (p. 164). Springer. https://doi.org/10.1007/978-3-319-96190-3

Zaman, M., Shahid, S. A., & Heng, L. (2018b). Irrigation water quality. In M. Zaman, S. A. Shahid, & L. Heng (Eds.), *Guidelines for salinity assessment, mitigation and adaptation using nuclear and related techniques* (Ch. 5, pp. 113–131). Springer. https://doi.org/10.1007/978-3-319-96190-3_4

Zaragoza, G., & Buchholz, M. (2008). Closed greenhouses for semi-arid climates: Critical discussion following the results of the watergy prototype. *Acta Horticultureae, 797*(797), 37–42. https://doi.org/10.17660/ActaHortic.2008.797.2

Chapter 9
The Challenge of Feeding Escalating Population in Arid Land Countries

Abul Kalam Samsul Huda and Shabbir Ahmad Shahid

Abstract The chapter surveys selected literature related to the challenge of feeding escalating population in arid land countries. Strengths, weaknesses, opportunities, and constraints of physical and socio-economic environments are analysed. Environmental features include low rainfall, highly variable timing of rainfall events with great variation in the amount of rainfall per event, dry environment, extreme temperatures, prolonged drought, increased evaporation, and soil infertility. Lessons learnt from recent extreme climate events and/or other national/global supply chain disruptions in feeding local population in arid land countries are recognized. Strategies used in producing sufficient food locally versus importing to feed their population are considered. Adopting capital-intensive high-tech production methods, such as greenhouse and hydroponic systems, are essential for increasing vegetable production in some situations. Many countries are already utilizing these systems while also importing vegetables to supplement local production to reach at a target self-sufficiency level. Challenges and opportunities posed by climate variability, climate change and future climate projection scenarios need deeper investigation in developing future strategies and action plan towards minimising risks and maximizing opportunities. Climate-smart and regenerative agricultural and land management practices (e.g., choosing appropriate crops and their planting window, timing, method and amount of water, nutrient application) lead to improved soil health, enhanced production of food with better quality and reduced environmental pollution. Favourable government policies including soil and water resources governance, human capacity building and knowledge sharing program, provision of adequate infrastructure, effective extension services, easy accessibility to financial support, and collaborative public/private partnership engagement would go a long way in

A. K. S. Huda
School of Science, National Vegetable Protected Cropping Centre, Western Sydney University, Sydney, Australia
e-mail: s.huda@westernsydney.edu.au

S. A. Shahid (✉)
Food Security Program, Environment and Life Sciences Research Center, Kuwait Institute for Scientific Research, Safat, Kuwait
e-mail: sshahid@kisr.edu.kw

M. K. Suleiman and S. A. Shahid (eds.), *Fostering Arid Lands Agriculture in the Face of Climate Change*, Advances in Global Change Research 81,
https://doi.org/10.1007/978-3-031-98490-7_9

developing successful strategies and programs to feeding escalating population with quality food.

Keywords Arid lands · Food security · Escalating population · Climate constraints · Mitigation options · Capacity building · Extension services · Climate-smart agriculture · Regenerative practices

9.1 Introduction

Arid lands are characterised by prolonged dryness, low and erratic rainfall, high wind and water erosion, extreme heat, frequent drought, and degraded soil with salinity issues in some parts (Mares, 1999; Middleton & Thomas, 1992). These areas occupy 41% of the earth's land surface and are home to roughly 2.5 billion people whose livelihoods depend on arid land ecosystem services (Gaur & Squires, 2018). Most of these countries are sparsely populated except Indian arid zone. Some nations in the arid lands (e.g., Gulf Cooperation Council Countries) are capital-rich and have many different options in feeding their people even if their natural resources are not suitable to make them self-sufficient for food security. Strategies (investment in research, infrastructure, capacity building) used by some countries in producing sufficient selected (mainly fresh produce and perishable products) food locally versus importing to feed their population are considered. There are many other countries which may have suitable natural resources but must depend on support from international sources (including aid) for much of their food supply for many other pressing reasons. Burundi, Central African, Democratic Republic of the Congo, Lesotho, Madagascar, Niger, Somalia, South Sudan, and Yemen have alarming Global Hunger Index (2023) scores in 2023. The lower the intensity of the crises, the less resources must be used to cope with the consequences. Climate-smart agricultural practices when adopted by these countries will result in improved soil health, and enhanced crop production possibly relieving the hunger to a great extent.

9.2 Climate Variability and Climate Change (Rainfall Variability, Rise in Temperature)

Current information regarding climate variability and climate change strongly suggests mean and extreme temperatures are rising. Evaporation is expected to increase with increased temperature and thus affecting water availability. Impact of these changes on ecological systems and crop production needs to be assessed.

Recent example of extreme climate events includes record-breaking heavy rains in April 2024 in the United Arab Emirates resulting into severe impacts on supply chain disruption and infrastructure damage. The highest rainfall (255 mm in a day)

was received in the Khatm Al Shakla area in Al Ain, as per the National Centre of Meteorology, resulting widespread flooding in all seven emirates (Findlay, 2024).

Different scenarios for the climate projection may be considered and evaluated. Long-term daily climate data (rainfall, maximum and minimum air temperature, solar radiation, evaporation, relative humidity, wind) may be collected and analysed at respective research institutes, and if possible, at selected locations in the region towards developing Climate Adaptation program.

The climate data analysis will be helpful, and this can be achieved by using existing data to be carrying out in-depth research in relation to vegetation, land, water, crop, and animal sectors. The basic data on fauna and flora, soil (depth, physical, chemical properties, mapping), water (quality, nutrient contents), and climate need to be collected regularly. Collection of climate data can be achieved by installing weather station. The weather station should be equipped with built-in sensors. These include wind transducers, precipitation sensors, pressure sensor, humidity, and temperature sensors. Additional equipment (Class-A pan to compute reference ET) can be installed along with the weather station. Care should be taken in selecting appropriate location for establishing weather station. For example, the location should be free from any potential turbulence caused by nearby objects, such as trees and buildings. It is generally accepted that wind measurement will not be remarkably disturbed by any object if the minimum distance of 10 times of the object's height is maintained. An open area of at least a 150 m should be maintained in all directions from the mast.

These data should be centralized, digitally maintained and accessible to potential users. The crop production techniques/practices developed at research stations be packaged for technology transfer to potential stakeholders for optimum resource use towards enhancing production, quality, and economic values of crops.

9.3 Major Crop Production Constraints in Arid Regions

Low crop productivity is commonly observed in arid land countries because of many factors including lack of irrigation facilities and research infrastructure (Kayan et al., 2017; Ryan, 1999). Challenges to feed escalating populations include enhancing agricultural production, productivity, and profitability. Improved and suitable agricultural systems are available for these societies to adopt in such marginal environments; for that capacity building in using those technologies will go a long way. Low and erratic rainfall is common feature (Frost et al., 2007). Crops suffer from water deficits with unpredictable and uneven rainfall distribution particularly when it coincides with sensitive growth stages (flowering and grain-filling) affecting crop development and yield. In addition to these natural/physical constraints, the other shocks include some disturbances that may occur with neighbouring countries disrupting supply chain. The dispute that Qatar had with its neighbouring countries (beginning in 2017) leading to the blockade of food supply is one such recent example of threat to improving local food security. Lessons learnt from Qatar as to what measures, policies, and decisions they made including investments in research, infrastructure, and capacity building towards reaching their self-sufficiency targets for fresh and perishable products will be useful to many other nations in the event of facing similar situations.

9.4 Identifying Suitable Crops/Animals Adapted to Arid Land Farming

The countries that have access to the soil mapping results can develop potential land use plan using data on soil profile characteristics, soil properties including depth, structure, texture, water-holding capacity, nutrient availability/fertility, soil pH, porosity, salinity, slopes, and topography. The countries who do not have those data may collect them to prepare land suitability maps for demarcating good agricultural farming area.

Nutrient use efficiency (kg of economic yield or biomass per unit of nutrients applied) may be calculated and used as a guide to determine the schedule of nutrient application. For example, the quantity of fertilizer that needs to be applied may be calculated based on target yield of specific crop, quality of produce, initial soil conditions, irrigation water quality, and potential losses. Foliar spray of micronutrient (Fe, Mn, Cu, Zn) may be considered to increase the availability to plants based on crop requirements, because the addition of micronutrients to soil may lead to fixation due to high soil pH and calcareous nature of soils.

Similarly, water productivity (kg of economic yield or biomass per unit of water applied) may be given due consideration in calculating efficiency of irrigation systems. Fertilizer and Fertigation Use Efficiency across different crops and production systems may be documented. However, it should be taken into consideration that the fertilizer induced salinity of fertigated water is not increased above the threshold salinity level of the test crops. Physiological traits should be measured to mechanistically understand the impact of fertigation on plant growth, yield, and quality. Rates of plant development, photosynthesis, respiration, and leaf-level water loss may be considered where appropriate. Developing relationships between plant growth (development) and physiological traits are essential for modelling the impact of water and fertilizer uses on crop yield.

The 4R (Right-Time, Source, Rate, Place) nutrient stewardship should be taken into consideration for precision nutrients management. Integrated nutrient management practices are recommended to improve soil fertility and physical properties of soils. Treated Sewage Effluent (TSE), where used, nutrients present in them may be compensated while developing fertilizer treatments for different crops. Salinity of the water used in fertigation be analysed to avoid salinity effects on crops. Real- time Data-driven Precision irrigation strategy, where feasible, may be developed to apply irrigation based on plant growth stage using crop coefficient of specific crops.

9.5 Potential for Economically Viable and Sustainable Expansion of Agriculture, Food Systems and Mechanisms

It is useful to develop medium and long-term climate adaptation plan towards enhancing resilience to climate challenges. Existing information may be used to develop some suitable indices for sector-wise vulnerability. However, with reference to agriculture sector, the climate adaptation plan should aim to bring together best practices derived from several past and on-going initiatives.

Researchers are optimizing plant growth and shielding the crop from adverse weather conditions and temperature fluctuations in some countries through the precise application of a controlled environment (protected cultivation). The diverse configurations of protective cultivation structures will provide farmers and growers with the capability to finely regulate variables such as temperature, humidity, and light. This manipulation will allow the creation of optimal conditions conducive to crop production. The systematic utilization of protected cultivation plays a pivotal role in mitigating risks associated with unpredictable weather patterns throughout crop season. The success of this approach underscores its significance in ensuring stable and resilient agricultural practices, offering a promising avenue for sustainable crop production in the face of climatic uncertainties.

The cultivation of crops varieties, particularly those endowed with drought tolerance and adaptability to extreme environmental conditions, has significantly diminished the likelihood of crop failure. The careful selection of crop varieties recognized for their resilience in the face of challenging climatic conditions proves instrumental in augmenting overall yield. This practice serves to enhance farmers' productivity and also functions as a viable strategy to mitigate potential vulnerabilities associated with crop cultivation. Encouraging crop diversification, particularly with resilient varieties, emerges as a prudent approach for farmers to fortify their agricultural systems, fostering sustainability and strengthening resilience against the uncertainties posed by fluctuating environmental conditions.

The optimization of irrigation practices achieved through a thoughtful design (drip irrigation) has proven to be both effective and efficient. It delivers water precisely to the root zone of the plant, thereby realizing dual benefits: a) minimizing evaporation and b) reducing salt accumulation in the root system. The integration of precision irrigation technologies and soil moisture sensors is anticipated to yield even more favourable outcomes. These technological advancements serve as valuable tools for farmers. By incorporating such technologies, farmers are empowered to enhance water efficiency and, consequently, elevate overall agricultural productivity.

The integration of trees with crops (agroforestry) and livestock holds the potential to deliver many benefits including environmental. It is expected to yield multi-faceted advantages, including erosion reduction, improved soil health, and enhanced moisture-retaining capacity. This holistic approach aligns with sustainable agricultural practices, fostering ecological resilience and contributing to the overall health and vitality of the agroecosystem. Through diligent land management strategies that interlink various elements of agroforestry and conservation agriculture, environmental sustainability may be promoted while concurrently optimizing agricultural productivity.

The meticulous implementation of precision nutrient management, rooted in an initial soil analysis to ascertain site-specific nutrient requirements prior to fertilizer application, has yielded significant benefits. This approach has optimized nutrient utilization and enhanced nutrient use efficiency while concurrently mitigating the risk of environmental pollution arising from nitrate leaching. The adoption of the 4R principles has been pivotal in achieving these outcomes. Additionally, the incorporation of alternative fertilization methods, such as soil amendment through the incorporation of organic fertilizers derived from manure and compost, will further contribute to minimizing adverse environmental impacts. This comprehensive strategy reflects a commitment to sustainable agricultural practices, ensuring agricultural productivity while safeguarding environmental integrity.

The augmentation of overall productivity can be achieved through the adaptation of climate-smart agriculture (CSA) practices by considering judicious use of the genetic traits inherent in the selected seedling varieties exhibiting robust resistance to abiotic and biotic stress. This success underscores the strategic adoption of plant breeding techniques along with efficient natural resource management practices. By leveraging these advancements, farmers can significantly enhance their resilience to climatic variations, thereby fortifying agricultural sustainability.

9.6 Integrating Practices for Climate Adaptation Program: Role of Modelling

The pivotal role of modelling in the integration of practices for climate adaptation programs is underscored by its capacity to provide a systematic and analytical framework. Modelling serves as a fundamental tool in the effort to harmonize agricultural practices with strategies designed to adapt to changing climatic conditions. This often involves the sound understanding of crop physiology, agronomy, soils, water, environment, economics, mathematical, and computational skills. It allows to simulate a diverse range of scenarios, predict outcomes, and assess the impacts of various practices.

Modelling offers an invaluable approach to understanding and addressing the climate challenges in the realm of agriculture and food production. Various players particularly the researchers leverage these models to analyze the intricate interactions between climatic variables and other factors including water scarcity, soil erosion, salinization issues, and agricultural practices to understand the overall dynamics of food production. Thus, by incorporating real-world data as applicable to the area and scientific principles, modelling will allow exploration of potential scenarios, enabling scientists, policymakers, and related stakeholders to make informed decisions. These may include year-round local food production with specific yield targets and generate scenarios for achieving these targets through informed management decisions.

The application of models in agriculture extends beyond predicting climate impacts; it encompasses optimizing resource allocation, enhancing crop management strategies, and developing resilient agricultural systems. The ability to simulate and analyze different agricultural practices under varying climatic conditions aids in identifying adaptive measures that can mitigate risks, provide insights into potential return of resource uses and associated management planning decisions. These include water requirements for different planting scenarios and improve overall productivity.

Modelling contributes significantly to the formulation and implementation of adaptive strategies in devising effective policies and guiding sustainable practices towards ensuring food security. In overall, modelling serves as a comprehensive decision support system, contributing to sustainable and efficient agricultural practices by aligning farming practices with water, nutrients and energy requirements. Interpretation of results involves considering the interplay of various factors and tailoring management strategies accordingly to foster a more robust and climate-resilient food production landscape.

9.7 Ability of Adapted Crops/Animals to Further Adjust Under Climate Change Scenarios

There are many tools available to assess the ability of adapted crops/animals to further adjust climate change. Soil-weather-crop-management data driven model (e.g., FAO AquaCrop, Steduto et al., 2009) provides a framework to integrate data collected from the field experiments in simulating crop performance including yields which may help making improved decisions in terms of water and fertilizer applications.

In arid regions using water efficiently for agriculture is the primary goal due to water scarcity. AquaCrop is a user-friendly water productivity model used for:

• Simulating yields of selected vegetable crops (cucumber, eggplant, capsicum, and tomato etc.)

- Simulating yields for other crops including potato, wheat, barley, maize etc.
- Generating irrigation schedules
- Identifying research gaps
- Evaluating the alternative management options to bridge the yield gap.

The AquaCrop model requires a climate data file, crop file, soil file, and management practices as input to simulate the yield and biomass based on water productivity. Some crop files are calibrated in the AquaCrop model. These crop files can be used as an input to generate results. New Crop files can be created and calibrated with the input information required by the model. The input information of the crop files can be obtained from research experiments and from other sources. The AquaCrop model can help users generate irrigation schedules, improve water productivity, and minimize water consumption.

The daily values of the weather conditions such as temperature, wind speed, rainfall, relative humidity, evaporation, and solar radiation are required to create a climate file. Default files for soil and crop are available and can be used if the local field research is not available. The soil and management data files can be selected from the default parameters provided in the AquaCrop model.

Date of planting, plant density and crop duration are required inputs. The AquaCrop model simulates crop growth and production using initial canopy cover, crop growth coefficient, maximum canopy cover, and crop decline coefficient. Alternatively, one can target the yield required, the model can be used to generate scenarios for management decision including water and fertilizer nutrients to be applied. Thus the AquaCrop model is a valuable tool in agricultural planning and management, providing insights into water use efficiency, crop yields, and optimal irrigation practices and allowing the assessment of how effectively water is utilized in crop production.

Similarly, the model simulated yields can be compared with reported yields to identify and quantify yield gaps, indicating areas for improvement. Interestingly, the model aids in optimizing inputs such as water, nutrients, and soil management to achieve desired yields and assessing the impact on crop production by creating different scenarios. This will inform policy decisions on year-round local food production, considering water availability and climate conditions. Consequently, the model will enable planners to set specific yield targets and generate scenarios for achieving these targets through informed management decisions. Finally, the results from AquaCrop provide insights into potential crop yields and associated water requirements for different planting scenarios.

In overall, AquaCrop serves as a comprehensive decision support system, contributing to sustainable and efficient agricultural practices by aligning water use with crop requirements. Interpretation of results involves considering the interplay of various factors and tailoring management strategies accordingly.

9.8 Policy Implications and Challenges in Feeding Population with Quality Food

Promising technologies developed in research stations including soilless and protected agriculture in many countries can be used to mitigate soil, water, and climate related constraints in cost-effective way.

9.8.1 Water

Due to low rainfall and aquifer recharge, the underground water resources are depleting very fast. This water scarcity may be compensated by using the TSE (for restricted crops), which is considered a viable source, which will increase over time with increasing population of Arid countries. To use the water efficiently, the farm workers should be educated regarding best irrigation management practices, this is likely to decrease water footprint of crops significantly. Blending of brackish water with TSE and RO (Reverse Osmosis) water may be adopted to mitigate the salinity effects on seed germination and crop performance. The advanced irrigation technique may be used based on the crop type to increase water use efficiency and salinity management.

9.8.2 Crops

Technologies developed and evaluated at respective research stations in Open Field crops (Date palm, Onion, Potato); Fodder (Napier grass); Protected Cropping (Cucumber, Tomato, Capsicum, Lettuce) in some arid land countries (e.g., Kuwait, Qatar) have shown promising preliminary results and have the potential to go long way in addressing the challenges and providing opportunities towards enhanced food security. Evaluation of feasibility of alternate management practices (e.g., planting dates, plant density, water and nutrient amount and timing) for optimum resource use and enhanced food security may be considered. Use of validated crop models (e.g., AquaCrop) may be considered in this activity to help guide taking improved and informed decisions.

Many alternate crops including Alfalfa and Napier grass, Rhodes grass may be considered provided irrigation efficiency is increased under sandy soil condition by using organic soil amendments. This perennial grass is relatively drought tolerant. The well managed Rhodes grass crop has the potential to be harvested about 8 times a year.

9.9 Role of Collaborative Public/Private Partnership Engagement

Collaborative public/private partnership engagement would go a long way in developing successful strategies and programs to feed escalating population with quality food through investment in developing infrastructures for protected agriculture, greenhouse technologies and Plant Factory with Artificial Lighting (PFAL).

The *pros* for PFAL include its potential ability to supply throughout the year with required quantity and nutrition quality of the produce. Considering the system's water use efficiency, higher crop water productivity may be achieved by acquiring higher amount of produce from using limited amount of water. Pesticide free products can be made available. The systems can be established on marginal lands (e.g., hard pans) not suitable for crops, and can be enhanced by automatic operation (fertilizer, irrigation). The *cons* for PFAL include, high initial capital investment, high uninterrupted energy supply and maintenance cost, lack of highly skilled technicians, and infrastructure to manage post-harvested products.

The *pros* of the Cooled Greenhouse include crop-specific micro-climate management; regular supply of agricultural products throughout the year; diversified crops compared to open field; low energy requirements compared to plant factory; and desired quality of the products can be achieved. The *cons* of the Cooled Greenhouse are potential plant lodging due to the physical effect of the wind on plants from blowing fans; non-uniform microclimate (positioning of the fan); heterogeneity in plant evapotranspiration; labour intensive efforts; energy and water intensive during summer.

Sensitivity analysis needs to be done to assess the economic feasibility and viability of the technology by considering the risk of instability in output and input prices prior to commercialization. Cost of inputs, capital, operational costs, and marketing and revenues generated from selling of produce and other bio-products need to be used to calculate profits and cost-effectiveness. Modelling crop production using crop, environment, management, commodity, and market information may be considered as a potential tool for open field and protected agriculture.

The crop production techniques/practices developed and optimized for open field and/or protected agriculture in some arid land countries, as well as farms waste management may be packaged for technology transfer to potential stakeholders.

9.9.1 New Generation Greenhouse (NGGH)

The NGGH is a fully enclosed system dependent on solar energy as the main source of light and power, may be a prudent choice blessed with ubiquitous availability of sunlight round the year in arid land countries. The NGGH will serve as a carbon sink, improving the carbon credit rating—a global commitment to the future for many nations.

9.9.2 Greenhouse

Research on different greenhouse prototypes (structure, cladding, cooling) may be conducted prior to selecting the appropriate one to mitigate the harsh environmental conditions of arid countries and to address the influences of climate change impact. Advanced controls and sensor-equipped systems in some advanced facilities enable users to tailor fertilizer applications to the unique requirements of their crops as well as to their stages of growth and the surrounding environment. Energy-saving, and different lighting sources can be tested and optimized using the modular agricultural production system (MAPS).

9.9.3 *Protected Agriculture*

Huge potential of using solar energy for protected agriculture exists under hot arid conditions, however it needs to be tested to assure its feasibility during short days with less radiation (winter season). Research area may be on greenhouse cover material to block excess solar radiation. This should ensure that enough sunlight is available for photosynthesis. These are to assure that sufficient light is passed to the greenhouse to promote optimum plant growth during both vegetative and reproductive stages. Using solar energy to reduce the energy demand can also be an option. Emerging glass technologies which have the potential to re-emit solar radiation (blocking of excess solar radiation) may be considered to keep the environment cooler. Arid climates are blessed with ample solar radiation offering significant prospects for agricultural activities in enclosed greenhouses.

9.9.4 *Crop and Animal-Based Residue Management*

Developing guidelines for rapid crop and animal-based residue management under harsh environmental conditions is an important forward-looking action. To manage the waste sustainably it is essential to establish a system to collect, segregate, and store the waste appropriately to avoid moisture losses and mitigate environmental effects. The segregated waste can be used as animal feed where appropriate; the high lignin segregated materiel (wood) may be recycled to biochar through pyrolysis process. The fresh biomass may be composted in isolation or through co-composting with biochar to improve the durability in the field.

It is recommended to test different composting methods under arid conditions prior to upscaling, especially with regards to nutritional quality and duration of composting process.

9.9.4.1 Low Level—Individual Farm Based

Individual farmers can produce compost on their own from organic waste at farm level based on the agriculture residue availability. The appropriate pile dimensions (height, width, length) can be selected. During composting process moisture, temperature, aeration may be maintained to promote aerobic decomposition.

9.9.4.2 Medium Level—Farming Community Based

The ideal pile height, and width are about 1.2–2.5 m and 3–6 m with a suitable length of the pile based on the available quantity of agriculture waste. During composting process moisture, temperature, aeration may be maintained to promote aerobic decomposition. In case of farming community-based composting, the piles can be set up for fermentation based on consensual agreement with the farming community.

9.9.4.3 High Level—National Level

Windrow composting is a commonly used as a large-scale composting method. The windrow system consists of organic waste materials like fronds, green grasses, and other potential feedstock with periodical overturning. Prior to setting up composting system it is essential to get approval from local authorities to abide the rules and regulations and ensuring biosecurity.

9.10 Conclusions and Recommendations

Forward looking government policies including cooperative public/private partnership engagement would go a long way in developing successful strategies and programs to feed escalating population. Following are the key recommendations, which may be considered on case-by-case basis where appropriate by respective authorities in arid countries. Current crop production procedures may be improved by following crop and location specific climate smart precision agriculture, in terms of using 4 R nutrient stewardship, improved soil moisture sensors and using soil amendments to sequester soil carbon for soil health improvement. Create soil and climate database at respective research stations and provide free access to potential users as a reference through portal and technical flyers, and brochures. Manage multiple types of feedstocks available to recycle into organic fertilizers using appropriate composting technologies. Improve ongoing soil, crops, and fertilizer management practices by using locally produced organic fertilizers through recycling different types of agricultural waste. Irrigation, soil, and nutrient management practices tested and proved successful, may be packaged, and shared with the agricultural community in the respective regions. Diagnostic-based reclamation procedure depending on salinity and water logging condition may be considered using integrated approach incorporating physical, chemical, biological, and hydrological methods where appropriate. AquaCrop model (FAO) may be considered provided sufficient input data is available from different sources. Crop, climate, soil, water, and management data may be used for this purpose. Techdocs/booklets on greenhouse technologies and Plant Factory with Artificial Lighting (PFAL) may be prepared for knowledge translation into industry practice for large-scale adoption leading to enhanced food security in selected Arid countries. Economic feasibility analysis may be conducted for the promising technologies prior to scaling up. Demonstration of the selected outstanding research results to farmers/agriprenuers/stakeholders in various ways including but not necessarily limited to Open Day/Farmer's Day would help faster adaptation of technologies by the industry practitioners. Strengthening collaboration with public and private institutions, engaging with early career researchers/University Students, and building capacity of potential stakeholders involved in agribusiness sectors would help feeding escalating population.

Acknowledgements Thanks to Kuwait Institute for Scientific Research (KISR) for the invitation and support in preparing this book chapter.

References

Findlay, M. (2024, April 17). UAE confirms the end of worst ever storm. Timeout UAE. Archived from the original on 19 April 2024. Retrieved 19 April 2024. UAE witnesses record-breaking rains, highest in 75 years. Khaleej Times. Archived from the original on 17 April 2024. Retrieved 17 April 2024.

Frost, P., Campbell, B., Mutamba, M., Mandondo, A., & Kozanayi, W. (2007). In search of improved rural livelihoods in semi-arid regions through local management of natural resources: Lessons from case studies in Zimbabwe. *World Development, 35*, 1961–1974. https://bit.ly/3xVuOWu

Gaur, M. K., & Squires, V. R. (2018). Geographic extent and characteristics of the world's arid zones and their peoples. In M. Gaur, & V. Squires (Eds.), *Climate variability impacts on land use and livelihoods in drylands.* Springer. https://doi.org/10.1007/978-3-319-56681-8_1. Source: https://link.springer.com/chapter/10.1007/978-3-319-56681-8_1

Global Hunger Index (GHI). (2023). https://en.wikipedia.org/wiki/Global_Hunger_Index, https://www.welthungerhilfe.org/fileadmin/pictures/publications/en/studies_analysis/2023-ghi-global-hunger-index-synopsis_EN.pdf

Kayan, N., Kutlu, I., Ayter, N. G., & Adak, M. S. (2017). Effects of different tillage systems and soil residual nitrogen on chickpea yield and yield components in rotation with wheat under dry farming areas. *International Journal of Agriculture and Biology, 19*, 517–522. https://bit.ly/2UrzoOM

Mares, M. A. (1999). *Encyclopaedia of deserts.* University of Oklahoma Press. https://bit.ly/3g8SsJo

Middleton, N. J., & Thomas, D. S. (1992). *World atlas of desertification.* United Nations Environment Program (UNEP), ISBN: 0340555122, London.

Ryan, J. (1999). Desert and dryland development: Challenges and potential in the new millennium. In *Proceedings of the 6th International Conference on the Development of Drylands*, 22–27 August 1999, Cairo, Egypt. International Centre for Agricultural Research in the Dry Areas (ICARDA), Aleppo, Syria. Xiv+655 pp. ISBN: 92-9127-130-0.

Steduto, P., Hsiao, T. C., Raes, D., & Fereres, E. (2009). AquaCrop—The FAO crop model to simulate yield response to water: I. *Concepts and Underlying Principles. Agronomy Journal, 101*(3), 4–437.

Chapter 10
Revitalizing Indigenous and Native Plants for Food Security of Arid Lands Countries in Changing Climate Conditions

Tareq A. Madouh and Merlin K. Davidson

Abstract Climate change heavily impacts global food production, posing threats of insecurity in arid regions. A notable limitation of sustainable food production in arid land regions is the inadequacy of arable land and water resources, making food availability and accessibility harder for the escalating population. The alternative preference for communities living in rural areas is to abide by local food resources viz., indigenous and native traditional wild plants, as substitutes to mitigate hunger and malnutrition through exploiting traditional food systems. There are evidences that wild herbs, vegetables, fruits and nuts, are nutritionally rich with a high profile of secondary metabolites serving as phytochemicals that are imperative for endorsing healthy and balanced diet that may solve malnutrition especially in young children. The use of native plants in the food management system requires specialized historical knowledge about indigenous edible plants and their safety for direct consumption as herbs by human or through feeding the animals as pastures. This requires a broad analysis of the ethnic indigenous and native vegetation in the arid regions and to highlight those having the potential to be added into the food management systems for food and nutritional securities. There is a huge prospect to filter potential plants, which are underutilized for food purposes, especially the indigenous inhabitants of arid region. Such a comprehensive review on the list of potential indigenous and local plants, with their nutritional facts, geographical distribution and the areas where these are already part of the food production scheme are the focus in this chapter.

Keywords Food security · Adaptation · Mitigation · Nutrition insecurity · Desert plants · Arid · Semiarid · Nutritional facts

---

T. A. Madouh (✉) · M. K. Davidson
Food Security Program, Environment and Life Sciences Research Center, Kuwait Institute for Scientific Research, Kuwait City, Kuwait
e-mail: tmadouh@kisr.edu.kw

M. K. Davidson
e-mail: mdavidson@kisr.edu.kw

M. K. Suleiman and S. A. Shahid (eds.), *Fostering Arid Lands Agriculture in the Face of Climate Change*, Advances in Global Change Research 81,
https://doi.org/10.1007/978-3-031-98490-7_10

10.1 Introduction

Globally, there is a captivating sign that food systems are adversely affected by climate change (Nalau et al., 2018), and are prominently impacted in the arid regions. Arid and semiarid regions, including the desert biome are featured by unfavorable atmospheric settings that comprise low and inconsistent precipitation, intense heat radiation, high wind speed and greater quantity of water loss by evapotranspiration throughout the year (Madouh & Quoreshi, 2023a; Modarres, 2007). The arid and semiarid regions are expected to have amplified threats and obligations with regard to change in climate (Quoreshi and Madouh, 2023). Agriculture and food production in these regions is facing manifold and complex encounters leading to decline in crop production for food security. Variations in temperature and rainfall regimes have amplified the decline rate by shrinking the crop and rangeland productivity (Almutawa & Alfraih, 2023; Aniah et al., 2019; Madouh, 2023). The consequent impacts are anticipated to intensify the warm and dry arid climate. The ensuing variations in agro-ecological conditions are in fright, besides, they generate promising circumstances for novel pests and diseases infesting agricultural crops and livestock. Noteworthy progress is made in dealing the threats of secured food supply; however, the existing paths are inadequate to meet the climatic demands (Makondo, 2018). Consequently, the progressions that endorse concerted thoughts and intercessions between researchers, to fortify coping strategies and adaptive dimensions are urgently required. In this outlook, the acquaintance and proficiencies of ethnic societies in food security system deserve specific attention. The rising tendency to adapt the traditions of ethnic folks can be ascribed to elevate dominances of climate change in food security for the community and acknowledging the key role of ethnic knowledge in the food systems (Shayanowako et al., 2021).

10.2 Global Distribution of Arid Lands

On the global scale, approximately 41% land area is covered by arid and semi-arid regions and hosting 2.5 billion people. Deserts are arid land with sparse vegetation, a desolate area comprising large groups of drylands characterized by meagre crop productivity, inadequate water resources, with feeble agricultural practices (Madouh and Quoreshi, 2023b). The largest arid desert of the world is reported to be the Sahara Desert, in the African continent, and the most affected by the global food crisis (Shen et al., 2020). Figure 10.1 illustrates arid zone distribution and the severity of aridity around the globe (Hanan et al., 2021). Arid ecosystems are characterized by a high evaporation than precipitation (Madouh, 2022). The climate in arid lands is characterized by extremely high summer daytime temperatures, high degree of solar radiation and dryness.

The length of the growing period of 0–180 days is considered as a criterion for aridity by Food and Agriculture organization (FAO) (FAO, 2022). The ratio

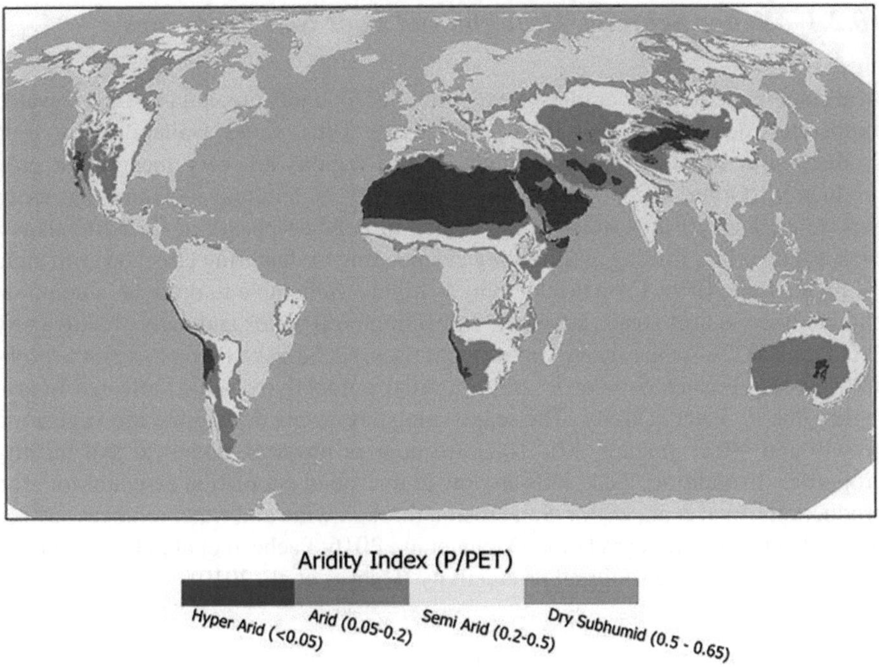

Fig. 10.1 Global dry land systems. *Source* Hanan et al. (2021), Fig. 1

of annual precipitation to potential evapotranspiration (P/PET), that varies between 0.05 and 0.65, is utilized by the United Nations Convention to Combat Desertification (UNCCD) (Rossi, 2020) for Aridity Index (Table 10.1), and used to present percent coverage of arid regions (Table 10.1) (UN, 2022; UNEP, 2000). The Asian arid region holding 33% of the global arid region experiences the magnitude of aridity to a greater degree (Naorem et al., 2023).

Table 10.1 Aridity index based classification of arid lands

S. no.	Arid region	Aridity index	% Area coverage
1	Hyper Arid	<0.05	6.6
2	Arid	0.05–0.2	10.6
3	Semi-arid	0.20–0.50	15.2
4	Sub humid zones	0.5–0.65	8.7

Source Safriel et al. (2005)

10.2.1 Water Scarcity, Drought and Poor Soil Properties

Arid zones experience poor moisture regimes due to scanty precipitation and elevated temperature leading to poor food productivity. Thus, water availability for food production is scanty. Annual rainfalls in arid regions are very meagre for crop production (Golla, 2021). Due to the availability of minimal moisture, the crops experience heavy water stress in the course of delicate phases of their life stages, such as flowering, fruiting, grain filling etc., leading to shattering effect on crop yield (Leng & Hall, 2019). Crop productivity is highly vulnerable to drought, a common phenomenon in arid zones, adversely impacting crop yields and native biodiversity (Madouh, 2022). As the severity of drought rises, so the risk of crop losses; the probability being greater. Similar to crops, animal productivity is also affected in arid region due to water scarcity. The scanty moisture levels deteriorate the vegetation growth and affect grazing. Arid soils are poor in physical, chemical and fertility properties. In addition, these soils are low in microbial population essential for crop production (Frost et al., 2007). Soil erosion through wind and water removes surface soil and causing soil degradation (Meng et al., 2016; Pacheco et al., 2018), and soil salinization affects agricultural productivity (Cuevas et al., 2019).

10.2.2 Vulnerability of Arid Lands to Climate Change

Climate change is chiefly accelerated by the emission of greenhouse gas (GHG) to the atmosphere. Dry arid lands are highly vulnerable to climate change effects (high temperature, prolonged droughts and low rainfall, floods, and sand and dust storms) (Doell, 2002; IPCC, 2008). Due to these complicated issues, the food security and agricultural production are under threat. In addition, climate change has a significant impact on arid rangelands and carrying capacity for grazing, and the rural people dwelling in the region are vulnerable to the effects of this phenomena. Climate change poses severe threats to the livelihoods of poor's depending on land and natural resources, and risking socioeconomic aspects. The sustainable management of arid lands and indigenous resources is essential in achieving food security and the conservation of biomass and biodiversity of global significance.

10.3 Climate Change and Food Security

Food systems are negatively affected by climate change and the greatest impact is experienced in the arid and semi-arid region of the globe (Nalau et al., 2018). Food and Agriculture Organization at 1996 World Food Summit defined the term "Food Security" as "when all people, at all times, have physical and economic access to sufficient, safe, nutritious food to meet their dietary needs and food preferences for an active and healthy life". The FAO recognizes four main extents of food security viz., availability, accessibility, utilization and stability (Peng & Berry, 2018). A deficit in food security is faced if any impairment is noticed in any of these dimensions affecting the population, ranging from a single person to global level. Developing new crops and learning to use wild plants creates the potential to diversify global food production and enable local adaptation to the diverse and changing environments humans inhabit.

The alarming climate change and the enormous human population have forced the world's agroecosystem for future survival and fitness demand. Feeding the globe with its ever-increasing population on healthy, quality, flavorful and nutritious food is an increasing demand and imposed a crucial task for humanity. Food security has never been a challenge as in the current century, witnessing extraordinary advances in technology in all sectors, where in the technology-imposed advantages on the ecosystem was countered by the leverages in technologies that mitigate food insecurity and safety tasks. Climate change is believed to reinforce other drivers and stressors to aggravate the susceptibility of agricultural systems (Aniah et al., 2019). The constant fluctuations experienced with precipitation and temperature modules and increased occurrence of extremities, have accounted for the decrease in crop productivity. It is predicted that the temperature rising scenario will further threaten the global food security. A frightening emergence of new pests and diseases due to changes in agro-ecological conditions, impacted by climate change, affects both crops and cattle production, further exaggerating the condition. At the forefront of this drive, the agricultural masterminds are obligated to enact food security schemes to produce enough food for all in the near future. The advancement made in the management of food security systems and their adaptation paths is insufficient to meet the current demands (Makonda & Thomas, 2018). Consequently, strategies that endorse collaborative thoughts and intrusions between farming and the scientific community to reinforce surviving tactics is an urgent need. A reasonably sensible solution is to focus on food plants grown in an environmentally sustainable manner, rather than requiring more resources. In this vision, the acquaintance, awareness and valuable experiences of indigenous and native plant communities in food security system justifies distinct consideration.

10.4 Indigenous and Native Plants of Arid Lands—A Solution to Food Security

Since the beginning of agriculture, roughly 10,000 of the 300,000 known edible plant species were utilized for human consumption, of which only 150–200 have been commercially cultivated, with rice, wheat, maize and potatoes supplying more than half of the global calorie intake (Thrupp, 1997). This indicates that several plants with the potential to improve food and nutritional security were not in the mainstream. Global crop production was enhanced and catalyzed by the green revolution in the middle of twentieth century, using elite varieties, expensive fertilizers and pesticides, that narrowed down the cultivation of indigenous plants and varieties, with bioaccumulation of harmful chemicals in food chain, introducing new disease to mankind, deteriorating soil and depleting water resources (John & Babu, 2021). The over dependence on wheat and rice has led to consuming more calories than nutraceuticals, without consuming the nutritional diversity required for a healthy life. This diversity can be achieved from other edible plants, underutilized, with specific advantages of being well adapted to local eco-settings, thrive well with limited water and nutrients in native soils, with the superiority of traditional usages by indigenous population. The context aids to achieve sustainable development goals (SDG), if rightly emphasized by developing mindsets and solutions to combat the impending climate change effects (FAO, 2019).

Globally, the indigenous underutilized plants are habitually consumed and utilized as the primary source of nourishment, particularly during periods of seasonal food scarcity and famine times. The utility of wild plants by diversified nations is confirmed by their national contributions to the recent "State of the Worlds Biodiversity of Food and Agriculture" (SOWFA) of the Food and Agriculture Organization (FAO) of the United Nations (2019). A report compiled from 91 countries stated that 69 nations reported 1955 wild plant species to contribute to the food security of their nations, in their respective countries, that contribute to food security and nutrition, supplementing to healthier diets. However, several unpublished reports demonstrate that the number is much higher and the species remain largely unreported in statistics, a vital contribution of biological resources to national economics in many parts of the world (Zulu et al., 2019).

Ancient humans survived by animal hunting and collection of indigenous wild plants for their food. Though agricultural cultivation of plants and livestock rearing for food emerged, consuming wild edible plants continued in tribal parts of the globe. Ethnic people carried the importance and traditional knowledge of indigenous plants as a heritage and transmitted to generations. The plant and human relation are considered as an ecological balance system, since ancient times to preserve living organisms on earth and the study 'ethno botany' aims to survey and document various indigenous traditional plants utilized as food by ethnic groups considering its nutrition and economics. Several ethno botanical researches resulted in the importance of wild edible plants in saving people during famine, drought and wars in different parts of the world (Abbasi et al., 2013). Though modern agricultural technology has evolved vigorously in the developing countries, attention by local people to the indigenous plants as an important food is due to aridity, drought, lack of rain and numerus abiotic and biotic constrains. Furthermore, the economic crisis is another major factor for the ethnic people to rely on the indigenous edible plants around the corner that thrive in the native region, withstanding all kinds of stresses.

Being educated to use the indigenous, traditional, underutilized, wild plants, generate the potential to broaden global food production and enable local adaptively to the changing ecosystems. The rising preference to replicate the indigenous populations in food systems can be ascribed to minimize the negative side of climate change impacts on food security and to acknowledge the acute role of indigenous traditional knowledge in food systems (Shelef et al., 2017).

10.5 Assessment of Indigenous and Native Plants of Arid Lands

The local traditional plants or wild edible plants (WEP), to a particular region is considered as a wild economic factor, combating hunger. They have been the part of diet in traditional food systems since ancient times serving as the main source of energy and nutrients for the indigenous people. Their importance cannot be underestimated that at least one in seven people worldwide rely on them as their food source. These indigenous traditional plants could be harvested in the lean seasons when the household food materials are depleted and new crops are still in the field or if there is a price hike of food materials in the market (De Beurs & Brown, 2013; Naorem et al., 2023). Though these indigenous plants make up to <1% of global calorie intake, these food materials are rich in essential nutrients and impact positively for improved dietary diversity and minimize the cost of nutritious diets (Sarfo et al., 2020). The greatest advantage of these plants is that most of them bear fruits, foliage, seeds, berries etc. year around that supplement the conventional foods available in the communities.

Apart being an inexpensive nutritional source, these plants are gaining importance for their role in disease resistance (Al-Fatimi, 2019). Most of the indigenous plants are being favored by traditional people, not only to appeal the hunger, but also to relieve themselves from pains and the effects. The phytochemical composition of these plants is the source of nutrition and constitute pharmaceutical effects. Plants possess two metabolite groups, the primary and secondary metabolites. The primary metabolites, including carbohydrates, proteins and fats are essential agents in the life of an organism (Madouh, 2022). The secondary metabolites include antioxidant agents such as vitamins, minerals, carotenoids, flavonoids, tannins, and polyphenols that protect plants from biotic and abiotic stresses (Madouh & Davidson, 2024). The indigenous edible plants are ideal and a richer natural source for inorganic metabolites particularly minerals, for balance of human and animal health. In ancient Arab-Islam Medicine, Abu Bakr Al-Razi, a Persian Physician (865–910 C.E.) recognized the affiliation between food and medicine for the foremost time. His renowned book about the benefits and to avoid their desired effect brings out the saying "Food is our medicine" and he is recognized as 'Father of Healthy Nutrition, or Nutraceutical Science' (Sanchez-Bel et al., 2018).

10.6 Importance of Indigenous and Native Plants as Food Source

Indigenous and native plants have been a vital food source for many communities throughout history. Recognizing their importance and addressing the challenges and opportunities surrounding their use can help promote cultural preservation, sustainable food systems, and biodiversity conservation. The sustenance of several ancient indigenous communities was from native wild plants, owing to their low cost (Seyoum et al., 2015), thus, rural communities across the globe, abide hunger through various indigenous traditional plants. They depend on these plants in diverse seasons to supplement both diets and income needs. These plants contribute to household food security and nutrition, chiefly in dry, arid lands, to meet seasonal shortages and emergency food during shortage periods. These plants shed light to the rural population with increased unemployment issues and large scale of single headed families, to alleviate themselves from poverty (Neudeck et al., 2012).

The reliance of rural poor's' on indigenous wild plants in the Zambezi valley and in Botswana were started from early 1960s (Shava, 2000). In the past, the wild indigenous plants serve as a food source providing a major food supply (Rodin, 1985). However, most local communities regarded the indigenous plants as primitive and substandard, and started to depend on a recent agrarian structure. The havoc of food shortages in the region impacts severely on the developing countries (Abraham, 1991).

The indigenous crop plants and fruit bearing trees in arid environment are extensively recognized as a food source for rural poor people. A comparative study on the usage of the fruits of indigenous tree, mahobohobo (*Uapaca kirkiana*), mufungu (*Anisophyllea* beohmi) and mobola plum (*Parinari curatellifolia*), to relieve hunger and as an income source at Copperbelt Province, Zambia exposed that 99% of the population undergoes starvation due to food unavailability during November to April and 97% of the population feed on the fruits of indigenous plants, *U. kirkiana, A. beohmii* and *P. curatellifolia* to subside hunger. About 31% of the population sell the fruits of *U. kirkiana, A. beohmii* and 46% process the fruits for juice and porridge, as an income source. The study concluded that the indigenous fruits were prioritized for relieving hunger than a revenue source (Kalaba et al., 2009). The indigenous leafy vegetables of cultivated pumpkins, melons and cowpeas and wild species, *Corchorus olitorius* (called Jew's mallow), and *Cleome gynandra* (called African cabbage), cat's whiskers, and stinkweed, perform a chief role in lessening starvation in Sub Saharan Africa (van Ransburg et al., 2004). The rural community processes the harvested leafy vegetables to preserve the perishable leaves by partial cooking, blanching and drying to meet hunger. Twenty-four species, belonging to 13 families have been utilized as a source of food and beverage in a rural village in Central Botswana (Dandy et al., 2015). The highest utilized plants among the countryside communities were *Grewia glava* (velvet raisin or wild currant), *Grewia bicolor* (white-leaved raisin or donkey berry), *Sclerocarya birrea* (jelly plum, cat thorn), *Amaranrhus thunbergia* (Thunberg's amaranthus or Thunberg's pigweed), *Cleome gynandra* (African cabbage or cat's whiskers) and *Corchorus olitorius* (Jew's mallow or Jute Mallow), serving as a food source for the rural poor. The distribution of wild indigenous plants in Western Nepal evidenced 72 wild plants belonging to 46 families and 61 genera supplementing food and nutrients to the rural poor, and also served as a source of cash income (Khakurel et al., 2021). A total of 50 indigenous, traditional, underutilized plants were documented in a rural village of Midakegn District, West Shewa Zone, Central Ethiopia, supplementing as staple food, emergency food for a chew and relief during drought, in times of poverty, and famine etc. Accordingly, indigenous traditional underutilized plants are indeed, a chief resource, empowering the poor, by reducing poverty (Guzo et al., 2023). However, underutilized plants are meant exclusively for poor communities in this globalized world, owing to their nutritional and therapeutic benefits.

10.7 Potential of Indigenous and Native Plants Underutilized for Food Systems

The indigenous and underutilized native plants can maintain and support the prevailing arrangements of food, as they are apt communally, economically and ecologically. The underutilized plants usually adapt to diverse environmental circumstances that are indispensable for resilient farming and sustainable food systems (Omotayo & Aremu, 2020). These plants are categorized as five groups viz., pseudocereals and millets, grain legumes, roots and tubers, leafy vegetables and fruits. The pseudo cereals and millets hold a protracted farming antiquity originated from the Americas, China and Africa. The crops to be domesticated initially before paddy and other important crops were the broomcorn (*Panicum miliaceum*) and foxtail millet (*Setaria italica*) (Alonso & Pérez-Jordà, 2023). Bambara groundnut (*Vigna subterraneae*), an ethnic legume native to Africa, is a potent crop suggested for hot region, where crops cannot withstand, able to fix nitrogen in soil, that lifts the rotational crops. The indigenous plants are not in main food stream due to deprived yield if cultivated other than their native niches, and scarcity of nationwide strategies to associate with small-scale business through their value-added products. Singh et al. (2022) summarized a list of ignored edible wild plants with massive nutritive prospective in global regions, including pseudo-cereals and millets: finger millet (*Eleusine coracona*), proso millet or broom corn (*Panicum milaceum*), fox tail millet (*Setaria italica*), banyard grass (*Echinochloa colonum*), kodo millet (*Paspalum scarbiculatum*), quinoa (*Chenopodium quinoa*), buck wheat (*Fagopurum esculentum*), Chia (*Salvia hispanica*), tef (*Eragrosrtis tef*), Job's tears (*Coix lacryma-jobi*), fanio (*Digitaria exilis*); grain legumes: groundnut (*Arachis hypogaea*), grass pea (*Lathyrus sativus*), bambara groundnut (*Vigna subbterranea*), horse gram (*Macrotonia uniflorum*), petai (*Parkia speciosa*), roots and tubers; yam (*Dioscorea alata*), mashua tuber (*Troeolum tuberosum*), yam bean (*Pachyrhizus erosus*); leafy vegetables: wild watermelon (*Citrullus lanatus*), chinese leek (*Allium tuberosum*), malabar spinach

(*Basella alba*), sparrow grass (*Asparagus officinalis*) and fruits: sugar apple (*Annona squamosa*), jujube (*Zizyphus mauritiana*) and bay berry (*Myrica escuenta*).

In Zambia, diversified cropping system with indigenous traditional crops has boosted yield efficiency and climate resilience (FAO, 2019). Rotating ignored crops with the prevailing crop system, replenishes the soil nutrients, diversify the pollinators, and disrupt disease and pest cycle. A cropping system of Indian mustard (*Brassica juncea*) with wild rocket (*Diplotaxis tenuifolia*), has considerably abridged the cucumber wilt disease initiated by the fungi, *Fusarium oxysorum* (Jin et al., 2019), with enhanced diversified soil beneficial microbes, including *Pseudomonas* sp., that yields an antibiotic product, 2,4-diacetylphloroglucinol. The indigenous native ethnic plants are regarded as "future food", for improving the nutritional prominence of mankind and increasing resilience of agriculture and farming food systems (Baldermann et al., 2016). Few important neglected crops with high nutritive values are discussed below.

10.7.1 Bambara Groundnut (Vigna Subterranea *Verdc*)

Bambara groundnut [an underutilized nut, Family: Leguminosae (Fabaceae), Subfamily Papilionoideae (Faboidea)], is an indigenous traditional legume plant wildly cultivated in African continent (Fig. 10.2). Bambara groundnut is a main crop among the rural livelihood of many Asian and African countries and has been incorporated in the farming systems as intercrops. The legume is hardy and has been documented as an imperative nourishing source of food, during times of food scarcity (Mbosso et al., 2020), due to climate smart features and adaptations to grow in poor soil, and withstand the extremities of drought. It is an intermediate annual crop, the foliage spreads close to the ground surface and produce the nuts underground (Gerrano et al., 2021). Bambara groundnut is well suitable for diverse agroecosystems being a highly adaptable legume and grows well under drought conditions and in different colors. Its water requirement is extremely low, suitable for arid and semiarid scenario and exhibit high lenience to soil salinity (Mayes et al., 2019). The existence of photoperiod variations, temperature sensitivity among the genotypes, make them a good indicator for improved varieties adapted for broad agro-ecological agricultural sectors (Kendabie et al., 2020). Consequently, Bambara groundnut has been recognized as an efficient next generation crop under the threatening global climate change (Khan et al., 2021).

Bambara groundnut is utilized by consumers in different ways, such as the green pods, dry nuts and grains and are utilized as processed and refined food. Bambara groundnut, as a drought resistant crop, apt for arid and semiarid region should be incorporated in cropping systems for benefitting the communities. However, its undesirable qualities viz., 'hard to cook' and the crop sensitivity to photoperiod during times of pod filling, demands scientific attention.

The nutrient contents of Bambara groundnut include approximately 64.4, 23.6, 6.6 and 5% of carbohydrates, protein, fat and fibres respectively, and rich in minerals

Fig. 10.2 Bambara groundnut. *Source:*Plant—Temegne et al. (2018), Fig. 2; Centre figure—Chibarabada et al. (2014), Fig. 1; Underground nuts, Nuts, Seeds, Seeds—El Bilali et al. (2024), Fig. A1

and micronutrients (Azman et al., 2019; Ndidi et al., 2014). Certain limitations, such as inadequate genetic progress, considerably long time for cooking and the existence of anti-nutritional factors in minute quantities, have minimized the wide utility and being categorized as an underutilized crop (Tan et al., 2020).

10.7.2 Quinoa (Chenopodium Quinoa *Wild.*)

Quinoa [a neglected pseudocereal (Family, Amaranthaceae)] is a promising food crop owing to its high nutrient content and ability to withstand abiotic stresses (Afzal et al., 2023). Quinoa is originated from the Andean region (Fig. 10.3) and has engrossed global attention due to its exceptional dietetic worth (Martínez-Villaluenga et al., 2000).

(a) (b)

Fig. 10.3 **a** Quinoa plants. **b** Quinoa grains. *Source* Lutz and Godoy (2017), Figs. 1 and 2

Quinoa possess high protein content (13.8–16.5%) than any cereal, rich in lysine (5.1–6.4%) and methionine (0.4–1%), with a better distribution of amino acids, suggested as milk protein (Pritham et al., 2021). The grain of quinoa is amylaceous with high carbohydrate content, especially starch, varying between 52 and 74% (dm) and a meagre portion of sugar (Antonio, 2017). Furthermore, higher levels of vitamins, minerals, essential fatty acids, dietary fibres and carbohydrates with hypoglycemic merits of lowering the blood sugar level and free from gluten, justifies quinoa as 'golden rice' (Angeli et al., 2020). Quinoa is a facultative halophyte, highly resilient to drought, and low levels to heat. Some quinoa cultivar's have been witnessed to grow in saline settings similar to sea water at 40 dSm^{-1} (Carrasco et al., 2016). Quinoa grain or flour is amalgamated with cereal grain or flour to advance the nutritive grade of conventional crops. Several quinoa-based confectioneries (pasta, bread), are popular due to superior nutritious quality and as a free nutritious routine dietary alternative (Nisar et al., 2018).

Quinoa research in African countries viz., Kenya yielded highly than in South American countries, displaying a boundless potential for productivity under harsh environmental stresses (Afzal et al., 2023). In nations with food shortages like Kenya and Columbia, quinoa is produced up to 4 tonnes/ha, making it a vital crop for food security (Jacobsen, 2003). In hot arid regions of South Africa, certain high yielding cultivars of quinoa with better grain quality yielded in a range of 1.23–2.8 t/ ha^{-1} (Mabrouka et al., 2023).

The quinoa greens (Fig. 10.4) endow an exceptional source of protein, amino acids, omega 3- fatty acids and other valuable nutrients, and can be grown year around due to short growth duration (Pradhan & Tamang, 2015; Pathan & Siddiqui, 2022). The prized Quinoa greens represent a promising food source to overcome malnutrition among rural livelihood. Nevertheless, assimilation of quinoa greens is not popularized for its practical utility. Further investigations are required to comprehend the dietary and practical implication of Quinoa microgreens in human health and to create awareness as 'new super vegetables'.

Fig. 10.4 Quinoa greens. *Source* Pathan and Siddiqui (2022), Fig. 1

10.8 Underutilized Indigenous and Native Plants in Arid and Semiarid Regions

Arid and semiarid regions are deliberated as the hot spot for abiotic traumas (temperature extremities, frequent droughts, intense solar radiation, salinity, poor soil fertility), where plants in general struggle for their potential performance (Madouh, 2023). In this climatic scenario, inclusion of the underutilized indigenous and native plants can afford an improved tactic for sustained and secured food production due to their external morphological, internal physiological, biochemical and anatomical features, that aid their performance in hot harsh weather (Madouh & Quoreshi, 2023b). Consequently, the adaptive traits upsurge the general flexibility and hostility to suboptimal ecological settings, and save the crops from forfeiting their yield productivity. It is usually presumed that the plant features that enhance adaptation, assure a stabilized yield and the performance being evaluated by the fertility and productivity of plant products. These traits comprise phenology alterations i.e., flowering/ripening in a precise period, morphological characteristics such as root/shoot ratio, characteristics of foliage at micro and macro level, presence of trichrome and other external plant adaptations that permit explicit genotypes to spurt ecological pressures, that aid in better productivity by escaping the harsh environmental conditions. To cope with the abiotic stress of aridity, the underutilized indigenous edible plants in the Asian arid region such as emblic (*Phyllanthus emblica*), bael (*Aegle marmlos*), Jamun (*Syzigium* spp.), jujube (*Zizyphus mauritiana*) and wood apple (*Feonia limonia*) have altered their body parts to ensure the physiological functions of the plant, amidst of aridity. The modified adaptations include strong and deep root systems, elevated shoot to root ratio, interaction with soil beneficial fungi such as Arbuscular mycorrhizal fungi (AMF) for more nutrient and moisture uptake from the deeper soils (Chundawat, 1990; Madouh & Quoreshi, 2023a). The promising indigenous underutilized plants from diverse regions of the world are listed out in Table 10.2.

The winged bean (Fig. 10.5) is an under exploited tropical legume with tuberous roots constituting 12–19% proteins. The bean is popularly known as "one species supermarket" for its nutrient rich green pods, leaves, tubers, immature and mature seeds (Tanzi et al., 2019). The prominent nodulation and root storage system of the

Table 10.2 Underutilized plants suitable for arid and semi-arid region

Crop type	Common name	Scientific name
Cereals	Tef	*Eragrostis tef*
	Broomcorn millet	*Panicum miliaceum*
	Foxtail millet	*Setaria italica*
	Oats	*Avicenia sativa, Avicenia nuda*
	Finger millet	*Eleusine coracana*
	Foxtail millet	*Setaria italica*
	Proso millet	*Panicum miliaceum*
Pseudocereal	Quinoa	*Chenopodium quinoa*
	Buckwheat	*Fagopyrum esculentum*
	Album seed	*Chenopodium album*
	Wattle seed	*Acacia* spp.
	Chia seed	*Salvia hispanica*
	Amaranth seed	*Amaranthus quitensis*
Legumes	Bambara groundnut	*Vigna subterannea*
	Lablab	*Lablab purpureus*
	Marama bean	*Tylosema esculentum*
	Word bean and jack bean	*Canavalia gladiata* and *C. ensiformis*
	Winged bean	*Psophocarpus tetragonolobus*
	African locust bean tree	*Parkia biglobosa*
	Broadbean, Faba bean,	*Vicia faba*
	Lima bean	*Phaseolus lunatus*
	Rice bean	*Vigna umbellata*
	Adzhuki bean	*Vigna angularis,*
	Cluster bean	*Cyamopsis tetragonoloba*
	Grass pea	*Lathyrus sativus*
	Horse gram	*Macrotyloma uniflorum*
	Moth bean	*Vigna aconitifolia*
	Narrow leafed lupin	*Lupinus angustifolius*
Nuts	Brazil nut	*Bertholletia excelsa*
	Sweet chestnut	*Castanea sativa*
	Sunflower seeds	*Helianthus annuus L*
Root and tubers	Cocoyam	*Colacasia esculenta*
	Foot yam	*Amorphophallus paeoniifolius*
	Arrowroot	*Maranta arundinacea*
	Aerial yam	*Dioscorea bulbifera*
	Water yam	*Dioscorea alata*
	Tannia	*Xanthosoma sagittifolium*

(continued)

Table 10.2 (continued)

Crop type	Common name	Scientific name
Leafy vegetables	Jews mallow	*Corchorusolitorius*
	Spider plant	*Cleome gynandra*
	Amaranth	*Amatranthus* sp.
	Night shade	*Solanum nigrum*
	Wild watermelon	*Citrullus lanatus*
	Sessile Joyweed	*Alternanthera sessilis*
	Alligator weed	*Alternenthera philoxeriodes*
	Red Joyweed	*Alternenthera polygonoides*
	Red amaranth	*Amaranthus cruentus*
	Spiny amaranth	*Amaranthus spinosus,*
	Diplazium fern	*Diplazium esculentum*
	Indian pennywort	*Hydrocotyle sibthorpioides*
	Chinese mallow	*Malva verticillata,*
	Creeping wood sorrel	*Oxalis corniculata*
	Knot weed	*Polygonum plebeium*
	Purslane	*Portulaca oleracea,*
	Indian Fenugreek	*Trigonella balansae*
	Bengal arum	*Typhonium trilobatum*
	Indian copperleaf	*Acalypha indica*
	Brazilian joy weed	*Alternanthera brasiliana*
	Red calico plant	*Alternanthera bettzickiana*
	Sessile joy weed	*Alternanthera sessilis*
	Cock's comb	*Celosia argentea*
	Common lambs quarter	*Chenopodium album*
	Ivy gourd	*Coccinia grandis*
	Asiatic pennywort	*Centella asiatica*
	Bird lime tree	*Pisonia alba*
	Sweet leaf	*Sauropus androgynus*
	Ceylon spinach	*Talinum fruticosum*
	Water spinach	*Ipomoea aquatica*
	Deccan hemp	*Hibiscus cannabinus*
Oilseeds	Sesamum	*Sesamum indicum*
	Mustard	*Brassica juncea*
Fruits	Indian gooseberry	*Emblica officinalis*
	Bael	*Aegle marmelos*
	Ber, Jujube	*Ziziphus mauritiana*
	Chironji	*Buchanania lanzan*

(continued)

Table 10.2 (continued)

Crop type	Common name	Scientific name
	Custard apple	*Annona squamosa*
	Jamun	*Syzygium cumunii*
	Jharber	*Ziziphus nummularia*
	Kair	*Capparis decidua*
	Karonda	*Carissa carandas*
	Khejri	*Prosopis cineraria*
	Khirni	*Manilkara hexandra*
	Lasora	*Cordia myxa*
	Mahua	*Madhuca longifolia*
	Manila tamarind	*Pithecellobium dulce*
	Mulberry	*Morus alba*
	Pilu	*Salvadora persica*
	Tamarind	*Tamarindus indica*
	Timroo	*Diospyros melanoxylon*
	Wood apple	*Feronia limonia*

Source Mabhaudhi et al. (2019), Rodríguez et al. (2020), Samal et al. (2023), Harbor and Aniedu (2021), Deb et al. (2018), Bhavithra et al. (2021), Meena et al. (2022), Rao and Poonia (2023)

plant recompense for the initial slow growth of foliage, and further accelerating the yield of highly digestible leaf protein and vitamins (Eagleton et al., 2023).

The thick, round-barked stems of fruit crops including asper (*Streblus asper*), bael (*Aegle marmlos*), lasora (*Cordia mixa*) and pilu (*Salvadora persica*) facilitate simple water storage and reduced cuticle transpiration. Due to the presence of moisture, crops such as kai (*Capparis decidua*), lasora (*Cordia mixa*), pilu (*Salvadora persica*) and aonla (*Phyllanthus emblica*) have synchronized flowering and quick fruit development during the season (Adhikary et al., 2019). Leaf shredding and dormancy are exhibited by crops including ber (*Zizyphus mauritiana*), phalsa (*Grewia asiatica*) and bael, to reduce water loss due to evapotranspiration in summer and to defend against frost in the winter (Saroj, 2018). The morphological characteristics of other underutilized crops are abundant as well; these include fur, sunken stomata, mucilaginous sap, tiny thick leaves, minimal foliage (as in ber), spiny cladodes (as in *Cactus*) and reduced transpiration rate (as in ber, phalsa, lasora and fig). In order to maintain cell turgor and allow for survival in arid and semiarid conditions, the features are linked to the accumulation of osmolytes, compatible organic and inorganic solutes (proline, phenolic, flavonoids, soluble sugars, glycine, betaine etc.), the biosynthesis of enzymatic and non-enzymatic antioxidants, heat shock proteins and drought responsive genes (Bhargava & Berwal, 2018).

Furthermore, the genetic basis of the adaptive traits justifies to be considered, for upcoming breeding programs involving genomes. These underutilized indigenous crop plants may exemplify the revolutionary crops for the upcoming generation, to

Plant Tender pod, seeds and tuberous roots

Fig. 10.5 Winged bean—*Psophocarpus tetragonobus*—an underutilized legume and vegetable. *Source* Eagleton et al. (2023), Fig. 1

augment the income of the rural farmers through ecological production systems even under climate change magnitudes.

10.9 Source of Balanced Food to Combat Malnutrition

The current research focus and international policies on food insecurity and malnutrition has shifted from amassed production to improved nourishment in diet by increasing access of nutritious food. Though not much investigated, underutilized plants are receiving attention by health-conscious people. Indigenous wild plants play a chief role in underdeveloped and developing countries in nutrient contribution, through their berries, nuts, roots, tubers, leaves etc. These neglected plants have greater nutritional content and a massive outlook for food and nutritional security and enhanced sustainability. It is a known fact that the neglected plants have the amazing power to improve human health, increase dietary diversity, decrease malnutrition and increase food and nutritional security. The indigenous wild edible leafy vegetables are an imperative area of food production throughout Arid regions of Africa, as they supplement vitamins, and micronutrients, particularly iron to the consumers (Mungofa et al., 2022). The consumption of underutilized plants has reduced iron, micro and macro nutrient deficiencies among Kenyan women and children (Kahane et al., 2013). Those plants have contributed to combate food insecurity and hidden hunger in Burkina Faso. A regular intake of three underutilized plants rich in vitamins and minerals as in Jackal berry (*Diospyros mespiliformis*), jujube (*Zizyphus mauritiana*), and desert date or soapberry (*Balanites aegyptiaca*) was evidenced to enhance the nutritive status of consumers in Burkina Faso (Lamien et al., 2009). The indigenous African plants have enhanced the life of people by their valuable indigenous nutraceutical properties in biodiversity conservation, ecosystem management and food production. Indigenous plants that are rich in minerals and vitamins such as Amaranth (*Amaranthus* sp.) African night shade (*Cleome gynandra*) and pumpkin seeds have the potential to significantly increase nutritional security. The enlargement of the territories of cultivation of underutilized indigenous plants for its beneficial trend toward human being is of supreme prominence. By preserving biodiversity and boosting the economics of rural inhabitants, these neglected plants have the potential to promote greater use in well-balanced, and healthful diet.

10.10 Prospective Role of Indigenous and Native Plants in Mitigating Food and Nutritional Insecurities in Different Regions

Indigenous wild edible plants are imperative modules of food systems that are allied to food security, as a chief source of nutrients for the rural community (Chitchumroonchokchai et al., 2017). In southern African countries, indigenous grains such as *Pennisetum* sp., *Panicum* sp. etc., pseudocereals such Bambara beans, *Vigna subterranean*, offers nutritional security. An investigation on the nutrition of few ignored fruits in Ethiopia *viz.,* simple thorned torchwood or desert date (*Balanites aegyptiaca),* Sudan teak *(Cordia africana)* and Christ's thorn jujube (*Ziziphus spina-christi*), revealed comparable level of proteins present in *C. africana* and *B. aegyptiaca fruits,* higher values of calcium in *B. aegyptiaca* and *Z. spina-christi*, higher fiber content in *B. aegyptiaca* and Z. *spina-christi.*, higher carbohydrate and energy content in *Z. spina-christi*, higher potassium, iron and zinc contents in *B. aegyptiaca* and *C. africana*, and proving food tree species are potential sources of macronutrients and minerals, suitable for landscape resilience to climate change impacts (Mokria et al., 2022). An investigation with eight edible indigenous plants in Thailand with regard to their nutritional factors states that shoot and young leaves of Jang (*Moerua siamensis*), Ta-Kuk (*Albizzia lebbeck*), (*Melienta suavis*) and Son Kob (*Hymenodictyon excelsum*) were rich in Vitamin C, fiber and protein content and bio-actives such as carotenoids and phenolic compounds (Sridonpai et al., 2022).

The only means for an assured food supply in arid and semi-arid region is encouraging the growth of the indigenous native plants that are well suitable for the region. Native wild plants of arid region have long been a staple of traditional meals providing a primary source of nutrition and energy. It should be noted that at least one in seven people in the world consume them on a global basis, that it should not be underestimated (FAO, 2019). Indigenous traditional plants are a rich source of phytochemicals such as terpenoids, flavonoids, phenols, tannin etc. in addition to several other secondary metabolites (Datta et al., 2019; Day, 2022). For instance, it is discovered that the fruit pulps of Christs thorn jujube, *Zizyphus spina-christi*, baobab (*Adansonia digitata*), duom (*Hyphaene thebaica*) and tamarind (*Tamarindus indica*) had increased phenolic compounds and were powerful antioxidants. The consumption of these fruits has reported a reduced risk of cardiovascular diseases due to their antioxidant effects, that act as metal chelating agents, radical scavengers, enzyme inhibitors, electron donors, etc. These indigenous plants, were widely utilized across the globe for their medicinal value (Tharmabalan, 2023).

10.10.1 Africa

The indigenous wild edible plants are the vital dietary components for the nomadic rural communities in most African countries. A survey of the local populations of Botswana, Kenya, Mali and South Africa, for the useful plant project managed by Royal Botanical Gardens, Kew, UK recognized 615 species used for food (Ulian et al., 2019). Among the plants are, morama bean (*Tylosema esculentum*), baobab (*Adansonia digitata*) and momgongo tree (*Schinziophyton rauntanenii*). Biodiversity International recorded 210 African leafy vegetables in Kenya alone in the 1990s; these are wild or semi-domesticated plants primarily grown for domestic use or informal trade. Of these, 23–42 are still consumed by the local communities; a few selected plants such as Amaranth, (*Amaranthus* sp.), spider plant (*Cleome gynandra*) and night shades (*Solanuum* sp.) are under further research (Borelli et al., 2020).

10.10.2 Asia Pacific

The native wild edible plants and fruit trees significantly improve human health in the Pacific region (Thomson et al., 2018). A list of wild plants is provided in the "*Food Plants of Papua New Guinea: A compendium*", and many of which are sourced from wild, regarded as 'minor food and flavorings', comprising root crops and staples, legumes, green leafy vegetables and other vegetables, fruits and nuts. The kernel of the prevalent wild plant in Papua New Guinea (PNG), Karuka (*Pandanus brosimos*), is consumed by roughly three in one share of the wild populace found at higher altitudes. The rural locals move to the higher altitudes during the fruit maturity seasons to gather the fruits and extract their nuts. The true sago palm, *Metoxylon sagu,* a staple meal rich in carbohydrates, is grown in Papua New Guinea and the adjacent areas (Lim et al., 2019). In Niue (Polynesia), the arrow root (*Tacca leonopetaloides*) is still traditionally processed. The starch produced from the root is processed as a local delicacy to make bread and puddings (Government of Niue, 2019). Sixty wild edible plants played a major role in famine or emergency foods, when climate extremities disrupt productions noticed in Malay islands (Tharmabalan, 2023).

10.10.3 The Mediterranean and the Arabian Peninsula

Wild edible plants are widely consumed in ethic indigenous cuisine and are extensively ingested locally in the Mediterranean region and the Arabian Peninsula (Taylor et al., 2019). About 2300 wild edible plants and fungal taxa were found in the Mediterranean region (Heinrich, 2006; Rivera et al., 2006), of which about 1000 are exclusively consumed by the indigenous populace. Out of the many known plants, the Mediterranean Biodiversity for food and nutrition (BFN) identified 42 wild edible plants for further study in Turkey (Dogan et al., 2013). In contrast, Morocco compiled a list of 246 wild plant species that are used extensively in Moroccan diets and culinary traditions (Nassif & Tanja, 2013). Aromatic edible herbs in the Moroccan diet that are commonly ingested are thyme (*Thymus* sp., *Mentha* sp.) and sage (*Salvia* sp.); however, their nutritional contribution is less as they are condiments (Hadjichambis et al., 2008). On the other hand, wild leafy vegetables are other edible group of plants widely consumed in the region, contributing to their nutritional needs (Powell et al., 2014). Few of the frequently eaten wild edible plants in the region were *Malva* sp., locally termed as "Khubbaizah", *Portulaca oleraceae*, "Barbir", *Chenopodium* sp., "Aifajan, Ghubayra, Khbaitha, Natna" based on the species, *Rumex* sp., locally called as "Hambasis, Hummayd" based on the species, *Foenicum vulgare*, commonly called "fennel", *Scolymus hispanicus* "golden thistle" and *Nasturtim officinale*, "water cress". These can be eaten as raw as salads or cooked as a side dish. Argan oil (*Sideoxylon spinosum*), capers (*Capparis spinosa* and *C. decidua*), acorns (*Quercus* sp.), strawberry tree (*Arbutus unedo*), jujube (*Zizyphus jujube*), mulberry (*Morus* sp.) and blackberry (*Rubus* sp.) are other common products derived from wild edible plants (Ellatifi, 2000). An assortment of 90 indigenous wild edible plants belonging to 40 families was used in the daily diet in the Khenifra region, Morocco. The study also details the experiences of older women who have managed to transfer and retain traditional knowledge on a wide range of local plant types and to utilize them in daily cuisine and ethnic therapy (Naciri et al., 2022). The medicinal plants, *Aerva javanica* (kapok bush), *Withania, Somnifera* (ashwagandha), *Zizyphus spina Christi* (Christ's thorn jujube), *Calotropis procera* (the apple of Sodom), and *Citrullus colocynthis* (bitter apple) locally called *as "Al-Sherry* or *Hanthal*", *Ruta chalepensis* (fringed rue), *Datura stramonium* (jimsonweed), *Aloe vera* (medicinal aloe) and *Tribulus terrestis* (puncture wine), colloquially termed as "Sharsheer" were the first to be documented as with ethnobotanical properties utilized by the local communities in the Jazan region of Saudi Arabia (Tounekti et al., 2019). In South Yemen, 58 wild edible indigenous plants were reported for the first, from 37 genera belonging to 21 families, the dominant being plants of the family Apocyanaceae, holding 18 species, with stem, leaves and fruits, being consumed as a food source. Herbs contribute more to food reserve than shrubs and trees, with 16 species present year around, 7 species in the dry spells and abundant species of 38 numbers present in rainy season and most of them (48) were consumed raw. Flowers of *Lantana camara*, commonly known as wild-sage and vernacularly in Arabic as 'Al-Darmaky', *Anisotus trisulcus* colloquially recognized among the natives as Madh", "Moze" and "Moddaid", *Cynanchum vanlessenii* locally called as "Qalab", *Monoluma solenophora* vernacularly known as "Khusmaa", stems of *Cynanchum viminale* sub sp. stipitaceum, commonly called by locals as "Radhaa or Milab", *Desmidorchis awdelianus* locally called by various names *viz.*, Uruz, Kuaaur, Muqrer, Kuaath, *Monolluma hexagona*, vernacularly called as "Mequrezh", and leaves of *Lavandula pubescens* (downy lavender), colloquially known as "Feheh" are among the widely consumed plants (Al-Fatimi, 2021). Traditional plant based indigenous food is an important source of Bahraini culture. The famous Bahraini dish, *Maghlaq* is made with wild oats, *Avena barbata* and dates, *Phoenix dactylifera* and mustard, *Brassica* sp. (Alalwan et al., 2017).

In Kuwait, old literature states tribal people consuming several indigenous wild plants *viz., Rumex vesicarius,* locally called as "Hummayd", *Malva parviflora,* "Khubbaizah", *Sonchus oleraceus,* "Khas Al Waz", *Suaeda aegyptiaca,* "Hartabel" and "Golleman" and *Cynomorium coccineum,* "Tarthouth" (Fig. 10.6) suggesting them as a nutritional food source. The Kuwaiti plant, *C. coccineum,* represents the species with maximum content of protein, fat and carbohydrate and the hydrolysate of its protein included eight necessary amino acids. The high nutritional content of these five plant species under investigation, especially with regard to proteins and amino acids, makes them ideal food sources for human consumption (El-Tantawy, 2002).

Numerous indigenous and native desert plants have been researched for a variety of purposes. In a more recent study, Madouh and Al-Sabbagh (2021) recognized five potential native desert fodder species *viz., Cenchrus ciliaris* widely known as foxtail

Fig. 10.6 Potential indigenous plants of Kuwait as food source

Fig. 10.7 Potential native fodder species for desert arid environment

buffalo grass, locally called in the Arabian desert regions as "Subbat" and" Thumum", *C. setigerus*, "Yarah", *Lasiurus scindus*, "Sbat", *Panicum turgidum*, "Thamam" and *Pennisetum divisum*, "Thumam" and "Sbatt", as an alternative animal feed to sustain livestock production in Kuwait and other desert arid and semi-arid regions (Fig. 10.7). The utility of substitutional fodder may well lessen the high dependency on imports and enhance self-sufficiency as regards animal feed and green fodders. More importantly, these native desert forages are known to have high nutritive value and biomass production coupled with several ecological significances which are recently gaining sight of many researchers. In addition, the native desert plant of Kuwait, *Farsetia aegyptia,* locally called as "Al labanah", is reported to be a traditional herb consumed for its therapeutic effects (Madouh & Davidson, 2024).

10.11 Research Gaps Pertaining to Indigenous Plants of Arid Regions for Food Security

Genotype selection, with low or high level of nutrients suitable for agricultural productivity should be considered for food production or pharmacological purposes (Ayilara et al., 2022). Indigenous traditional pulses are crucial in supplementing proteins and ensuring food safety for African rural communities. However, the greatest number of them are underutilized, as they are hard to cook. The Bambara groundnut demands very low precipitation can be utilized in the cropping system in rotation with groundnut and cowpea. Other physical mechanisms such as legume softening using softeners with phenolic compounds, pectins etc., influence of cell wall degrading enzymes, requires further appraisal. A native desert tree of arid region,

Prosopis cineraria, commonly known as ghaf, an ecologically and economically sound, slow growing tree, is often confronted by gall developments, in their inflorescence and flowers by a mite, *Eriophyes prosopidi,* that reduces pod development. In addition, it struggles for vegetative propagation and threatened by the competitive species, *P. juliflora* (Bhatnagar et al., 2019). The improvement strategies of this plant include disease resistance to flower galls and root rots, varietal development without spines for ease in handling the wood, and upsurge fruit nutrient quality (Rai et al., 2021), could be managed through traditional breeding programs and improved propagation methods. Choosing and enlightening the right genotype owing to its utility for food or drug is indispensable. The biochemical and nutritional based studies at diverse levels needs comprehension at molecular level through Randomly Amplified Polymorphic DNA (RAPD) analysis (Joshi et al., 2023). In addition, arid plants having difficulty to tolerate salinity, and these should be handled by genetic means. A salt tolerant germplasm of ber or jujube, *Zizyphus mauritiana,* using genome alterations and sequencing is suggested (Manzoor et al., 2023).

10.12 Future Research Ideas for Large Scale Production of Indigenous Plants for Food and Nutritional Securities

The present agri-food systems are susceptible to climate change, owing to industrialized production, less diversified plant selection, and high-water consumption (Perrings & Halkos, 2015). The annexation of indigenous traditional plants in cropping system has immense prospective to promote food and nutritional securities for the rural poor. However, the introduction, conservation, management and promotion of these plants, demands numerous issues to be sorted under various scientific grounds. The non-existence of data and literatures on the nutritional value of the indigenous traditional plant species is a key flaw for their expanded use and promotion. The rural local population feeds on them, as the ancestors traditionally pass on the nutritious value. Research arenas on the nutrient composition of these neglected plants to shed light on these natural treasures requires attention. In addition, the cost of production of these negligible plants demands further investigation to suggest its utility on a large scale. Though it is well known that the production of the wild indigenous plants is relatively low, information on their cost of production is limited for their promotion of large-scale production. Acquisition of nutrition in relatively low cost provides a great advantage to the rural poors' in the arid zones.

The key to more environmental changes is the decline in biodiversity in cropping systems, leading to problems of crop resilience against pest and disease pressures and climatic fluctuations. The ethnic neglected plants enrich a cropping system with diversity in gene. However, the conservation of these plants and the related farming technologies are scarce. In addition, the negative environmental impacts such as decline in pollinator populations due to vast usage of agrochemicals and land use changes, contribute more to weaken the expansion of indigenous plant production (Mubaiwa et al., 2016). The desert plants, chiefly legumes are consumed traditionally as a protein source. The plant-based proteins, lack in quality being structurally obstinate, devoid of amino acids such as lysine that are essential for calcium absorption and they cluster as well as aggregate resulting in lower digestibility. However, using the right processing technology, either physical (heating and ultra-sonication), chemical (deamination, acetylation and phosphorylation) and biochemical (hydrolysis or enzymatic crosslinking) can alleviate this matter (Day, 2022). Desert plant seeds should be explored for such strategies.

The organoleptic properties of the few desert seeds, fruits and products possessing unpleasant aroma and stringent taste, have been a deterrent to wide acceptability, despite of their nutraceutical benefits. Nano technology, an emerging branch of science can provide a solution to enhance the quality of food produces with regard to their flavor, taste, consistency, general outlook, nutritious significance and shelf-life (Singh et al., 2022). Nano encapsulation can facade aroma and taste and advance the biomolecules present in them and preserve them. Fruit bearing desert plants are mostly seasonal, and require storage for year-around consumption, which can be achieved through nanotechnology that requires further exploration. Additionally, the complex food matrix from indigenous plants can extend biopolymers, for developing nanomaterials for distribution of explicit bioactive compounds at specific target sites.

Value addition of the farm products and valorization of utmost waste is of paramount requisite in a cropping system, to ensure zero waste to achieve supreme monetary reimbursements for the growers. A strong research gap persists in this scenario, of the value addition of the ignored arid plant products, leading to ignorance in the farming community for active cultivation. Enhanced value chain logistics can greatly increase, the attractiveness of the products to a wide range of consumers. An in-depth awareness on value addition of ignored plant products, their marketing and distribution have not yet been promoted. A significant processing converts a food raw material into a value-added product that has year around availability, better shelf life, improved quality and barns, thus generates more revenue and more employments (Al Hinai et al., 2022). Value-added products include flour, pickle, jams, candies, bars, that require food processing techniques such as boiling, frying, steaming, baking and roasting, done to enhance their shelf life. However, such processes affecting the nutrients and secondary metabolites of the final food product are not well studied in most cases that necessitate further exploration. The knowledge gaps in cultivation, nutrition and processing are some of the challenges that can be mitigated through research and translation of academic knowledge in practice, at local, national and global level.

10.13 Conclusions and Recommendations

An environmentally sustainable food production system to meet the food, nutrition and health demands of the massive universal population is an important mission to be undertaken by the global scientists, particularly in the arid zones, highly vulnerable to climate change. The underutilized and indigenous desert wild edible plants perform an imperative part in assuring safety for food, feed and nourishments. A sum of prospective roles of the under researched food plants in the arid zones has been conferred in the current perspective. A wide ethnobotanical utility of these plants and their extensive culinary effects has long-established their safety and socioeconomic value to the rural communities in arid desert regions. Authentication of their nutritive and bioactive prospects scientifically would further aid in mobilizing funds, and channelizing the communal and private shareholders in their bio-conservation. The neglected and underutilized plants are indispensable for the livelihood of millions of poor farming communities in hot and dry regions of the globe. These biological assets if unnoticed, will be threatened, that requires further exploration in all possible directions. The underutilized desert plants have unlimited potential for processing and right technology is a requisite to popularize these plants. Intermediation through improved technologies such as artificial intelligence and nanotechnology would aid in preserving the superiority, ease anti-nutrients and augment the sensual features of the resulting produces, for ensured marketability and consumer acceptability. As a final note, streaming the wild indigenous plants from desert zone in to staple food production will aid in fulfiling the sustainable development goals of achieving food and nutrition security for global sustainable agricultural systems. The underutilized indigenous flora, prospering in the harsh topographical milieus, could be an alternative solution to create a climate smart and ready food on the table for future next generation human beings.

Acknowledgements The authors would like to express their gratitude to the Kuwait Institute for Scientific Research (KISR) for providing the resources and assistance required to obtain the material for this review chapter. The Kuwait Foundation for the Advancement of Sciences (KFAS) is acknowledged for supporting the scientific research. Furthermore, the authors would like to extend their appreciations to Dr. Niall P. Hanan, Dr. Carine Nono Temegne, Dr. Tendai P Chibarabada, Dr. Hamid El Bilali, Dr. Mariane Lutz, Dr. Safiullah Pathan and Dr. Graham Eagleton for granting permission to use the figures of native indigenous plants from their valuable publications. The figures are cited in the text and added in the reference section.

References

Abbasi, A. M., Khan, M. A., Khan, N., & Shah, M. H. (2013). Ethnobotanical survey of medicinally important wild edible fruits species used by tribal communities of Lesser Himalayas-Pakistan. *Journal Ethnopharmacol, 148*, 528–536.

Abraham, J. (1991). *Food and development. The political economy of hunger and the modern diet.* World Wide Fund for Nature & Kogan Page Ltd.

Adhikary, T., Kundu, S., Chattopadhayay, P., & Ghosh, B. (2019). Flowering intensity and sex ratio of ber (Ziziphus mauritiana Lamk.). *International Journal of Bio-resource and Stress Management, 10*, 107–112.

Afzal, I., Haq, M. Z. U., Ahmed, S., Hirich, A., & Bazile, D. (2023). Challenges and perspectives for integrating Quinoa into the agri-food system. *Plants (Basel), 22*, 12(19), 3361. https://doi.org/10.3390/plants12193361. PMID: 37836099; PMCID: PMC10574050.

Alalwan, T., Alalwan, A., Qaher, A., Mandeel, L., & Al-Sarhani. (2017). Traditional plant-based foods and beverages in Bahrain. *Journal of Ethnic Foods, 4*, 274–283, ISSN 2352-6181, https://doi.org/10.1016/j.jef.2017.10.003

Al-Fatimi, M. (2019). Ethnobotanical survey of medicinal plants in central Abyan governorate, Yemen. *Journal Ethnopharmacology, 241*, Article 111973. https://doi.org/10.1016/j.jep.2019.111973

Al-Fatimi, M. (2021). Wild edible plants traditionally collected and used in southern Yemen. *Journal of Ethnobiology and Ethnomedicine, 17*. https://doi.org/10.1186/s13002-021-00475-8

Al Hinai, H., Jayasuriya, P. B., Pathare, T., & Al Shukaili. (2022). Present status and prospects of value addition industry for agricultural produce—A review. *Open Agriculture, 7*(1), 207–216. https://doi.org/10.1515/opag-2022-0084

Almutawa, A., & Alfraih, A. (2023). Impact of climate change on agriculture, fisheries and livestock sectors in Kuwait. *Journal of Geoscience and Environment Protection, 11*, 141–166. https://doi.org/10.4236/gep.2023.1110011

Alonso, N., & Pérez-Jordà, G. (2023). The origins of millet cultivation (*Panicum miliaceum* and *Setaria italica*) along Iberia's Mediterranean Area from the 13th to the 2nd century BC. *Agronomy, 13*, 584. https://doi.org/10.3390/agronomy13020584

Angeli, V., Silva, Pedro, C. M., Danilo, K., Muhammad, H., Alicia, K., Forough, G., Simone, P., & Cinzia. (2020). Quinoa (Chenopodium quinoa Willd.): An overview of the potentials of the "Golden Grain" and socio-economic and environmental aspects of its cultivation and marketization. *Foods, 9*. https://doi.org/10.3390/foods9020216

Aniah, P., Kaunza-Nu-Dem, M. K., & Ayembilla, J. A. (2019). Smallholder farmers' livelihood adaptation to climate variability and ecological changes in the savanna agro-ecological zone of Ghana. *Heliyon, 5*(4). https://doi.org/10.1016/j.heliyon.2019.e01492

Antonio, M. M. F. (2017). Quinoa: Nutritional aspects. *Journal of Nutraceuticals and Food Science, 2*, 1(3), 1–5.

Ayilara, M. S., Abberton, M., Oyatomi, O. A., Odeyemi, A., & Babalola, O. O. (2022). Potentials of underutilized legumes in food security. *Frontiers in Soil Science, 2*, 60. https://doi.org/10.3389/fsoil.2022.1020193

Azman, R., Barkla, B. J., Mayes, S., & King, G. J. (2019). The potential of the underutilized pulse bambara groundnut (*Vigna subterranea* (L.) Verdc.) for nutritional food security. *Journal of Food Composition and Analysis, 77*, 47–59. https://doi.org/10.1016/j.jfca.2018.12.008

Baldermann, S., Blagojević, L., Frede, K., Klopsch, R., Neugart, S., Neumann, A., Ngwene, B., Norkeweit, J., Schröter, D., & Schröter, A. (2016). Are neglected plants the food for the future? CRC. *Critical Reviews in Plant Sciences, 35*, 106–119. https://doi.org/10.1080/07352689.2016.1201399

Bhargava, R., & Berwal, M. K. (2018). Adaptation for drought tolerance in arid horticultural crops; CIAH/Tech./Pub. No. 66; ICAR-Central Institute for Arid Horticulture: Bikaner, India, Techical Bulletin no 66 (p. 30).

Bhatnagar, S., Khan, A. U., Sharma, N., Tak, P. S., Singh, S., & Nirwan, B. (2019). *Eriophyes prosopidis* induced flower galls of *Prosopis cineraria. Journal of Entomology and Zoology Studies, 7*(2), 01–03.

Bhavithra, P., Balakumbahan, R., Kannan, P., & Sivakumar, V. (2021). Underutilized leafy vegetables in western ghats region for nutritional security. *Plant Archives, 21*(1), 32–39. https://doi.org/10.51470/Plantarchives.2021.v21.no1.005

Borelli, T., Gee, E., & Hunter, D. (2020). Neglected no more: Reframing the food systems narrative using agricultural biodiversity. In D. Hunter, T. Borelli, & E. Gee (Eds.,), *Biodiversity, food and nutrition. A new agenda for sustainable food systems.* Routledge.

Carrasco, K. B. R., Biondi, S., Martínez, E. A., Orsini, F., Antognoni, F., & Jacobsen, S. E. (2016). Quinoa—A model crop for understanding salt-tolerance mechanisms in halophytes. *Plant Biosystems, 150*(2), 1–48. https://doi.org/10.1080/11263504.2015.1027317

Chibarabada, T. P., Albert, T. M., & Mabhaudhi, T. (2014) Seed quality characteristics of a bambara groundnut (Vigna subterranea L.) landrace differing in seed coat colour. South African *Journal of Plant and Soil, 31*(4), 219–226. https://doi.org/10.1080/02571862.2014.966340

Chitchumroonchokchai, C., Diretto, G., Parisi, B., Giuliano, G., & Failla, M. L. (2017). Potential of golden potatoes to improve vitamin A and vitamin E status in developing countries. *PLoS One, 12*(11), e0187102. https://doi.org/10.1371/journal.pone.0187102

Chundawat, B. S. (1990). *Arid fruit culture.* Oxford and IBH Publishing Co. Pvt. Ltd.

Cuevas, J., Daliakopoulos, I. N., del Moral, F., Hueso, J. J., & Tsanis, I. K. (2019). A review of soil-improving cropping systems for soil salinization. *Agronomy, 9*(6), 295. https://doi.org/10.3390/agronomy9060295

Dandy, B., Joyce, L., & Demel, T. (2015). Utilization of edible wild plants and their contribution to household income in Gweta village, central Botswana. *African Journal of Food Science and Technology, 6*(7), 220–228. https://doi.org/10.14303/ajfst.2015.074

Datta, S., Sinha, B. K., Bhattacharjee, S., & Seal, T. (2019). Nutritional composition, mineral content, antioxidant activity and quantitative estimation of water soluble vitamins and phenolics by RP-HPLC in some lesser used wild edible plants. *Heliyon, 5*, e01431. https://doi.org/10.1016/j.heliyon.2019.e01431

Day, L., Cakebread, J. A., & Loveday, S. M. (2022) Food proteins from animals and plants: Differences in the nutritional and functional properties. *Trends in Food Science and Technology, 119*, 428–442. https://doi.org/10.1016/j.tifs.2021.12.020

De Beurs, K. M., & Brown, M. E. (2013). The effect of agricultural growing season change on market prices in Africa. In *Climate variability—Regional and thematic patterns.* https://doi.org/10.5772/56459

Deb, P., Prasad, B. V. G., Munsi, P., & Chakravorty, S. (2018). Underutilized leafy vegetables: Diversity, nutritional security and medicinal benefit in context of Eastern India. In J. Singh, R. Nigam, & W. Hasan (Eds.), *Advances in biodiversity conservation for sustainable development* (pp. 95–104). Parmar Publishers and Distributors.

Doell, P. (2002). Impact of climate change and variability on irrigation requirements: A global perspective. *Climatic Change, 54*, 269–293. https://doi.org/10.1023/A:1016124032231

Dogan, Y., Ugulu, I., & Durkan, N. (2013). Wild edible plants sold in the local markets of Izmir, Turkey. *Pakistan Journal of Botany, 45*(S1), 177–184.

Eagleton, G. E., Tanzi, A. S., Mayes, S., Massawe, F., Ho, W. K., Kuswanto, K., Stephenson, R. A., & Khan, T. N. (2023). Chapter 17—Winged bean (Psophocarpus tetragonolobus (L.) DC). In M. Farooq, H. M. Kadambot, & Siddique (Eds.), *Neglected and underutilized crops* (pp. 437–486). Academic Press. ISBN 9780323905374, https://doi.org/10.1016/B978-0-323-90537-4.00022-3

El Bilali, H., Kiebre, Z., Nanema, R. K., Dan, G. I., Rokka, V. M., Gonnella, M., Tietiambou, S. R. F., Dambo, L., Nanema, J., & Grazioli, F. (2024). Mapping research on Bambara Groundnut (Vigna subterranea (L.) Verdc.) in Africa: Bibliometric, geographical, and topical perspectives. *Agriculture, 14*, 1541. https://doi.org/10.3390/agriculture14091541

Ellatifi, M. (2000). The situation of non-wood forest products in Morocco. In *Proceedings of the Joint FAO/ECE/ILO Committee on Forest Technology, Management and Training Seminar Proceedings on Harvesting of Non-wood Forest Products*, Menemen-Izmir, Turkey, 2–8 October 2000. Food and Agricultural Organization of The United Nations, Rome.

El-Tantawy, H. (2002). The nutritional value of some desert plants in Kuwait, Arabian Pennisula. *Taeckholmia, 22*(1), 35–45.

FAO. (2019). The state of the world's biodiversity for food and agriculture. In J. Bélanger, & D. Pilling (Eds.), *FAO Commission on Genetic Resources for Food and Agriculture Assessments* (572 pp.). Rome. http://www.fao.org/3/CA3129EN/CA3129EN.pdf

FAO. (2022). World soil resource report No. 90: *Land resource potential and constraints at regional and country levels* (pp. 114). Retrieved November 17, 2022 from https://www.fao.org/documents/card/ru/c/15bf7a55-18ba-4c0b-95a3-fb63cb4849cb/

Frost, P., Campbell, B. M., Luckert, M., Mutamba, M., Mandondo, A., & Kozanayi, W. (2007). In search of improved rural livelihoods in semi-arid regions through local management of natural resources: Lessons from case studies in Zimbabwe. *World Development, 35*(11), 1961–1974. https://doi.org/10.1016/j.worlddev.2006.11.012

Gerrano, A. S., Eifediyi, E. K., Labuschagne, M., Ogedegbe, F. O., & Hassen, A. I. (2021). Production practices of bambara groundnut. In: S. A. Oyeyinka, & A.-O. Bio (Eds.), *Food and potential industrial applications of Bambara groundnut*. Springer. https://doi.org/10.1007/978-3-030-73920-1_2

Golla, B. (2021). Agricultural production system in arid and semi-arid regions. *International Journal of Agricultural Science and Food Technology, 7*(2), 234–244. https://doi.org/10.17352/2455-815X.000113

Government of Niue. (2019). *The State of Niue's biodiversity for food and agriculture*. Niue Country Report: FAO.

Guzo, S., Lulekal, E., & Nemomissa, S. (2023). Ethnobotanical study of underutilized wild edible plants and threats to their long-term existence in Midakegn District, West Shewa Zone, Central Ethiopia. *Journal of Ethnobiology Ethnomedicine, 19*(1), 30. https://doi.org/10.1186/s13002-023-00601-8

Hadjichambis, A. C., Paraskeva-Hadjichambi, D., Della, A., Giusti, M. E., De Pasquale, C., Lenzarini, C., Censorii, E., Gonzales-Tejero, M. R., Sanchez-Rojas, C. P., Ramiro-Gutierrez, J. M., Skoula, M., Johnson, C., Sarpaki, A., Hmamouchi, M., Jorhi, S., El-Demerdash, M., El-Zayat, M., & Pieroni, A. (2008). Wild and semi-domesticated food plant consumption in seven circum-Mediterranean areas. *International Journal of Food Sciences and Nutrition, 59*(5), 383–414. https://doi.org/10.1080/09637480701566495. PMID: 18979618.

Hanan, N. P., Milne, E., Aynekulu, E., Yu, Q., & Anchang. (2021). A role for drylands in a carbon neutral world? *Frontiers in Environmental Sciences, 9*, 786087. https://doi.org/10.3389/fenvs.2021.786087

Harbor, C. I., & Aniedu, C. (2021). Exploring some neglected and underutilized root and tuber crops for food security in Nigeria. In: O. O. Babalola (Ed.), *Food security and safety*. Springer. https://doi.org/10.1007/978-3-030-50672-8_6

Heinrich, M., Müller, W. E., & Galli, C. (Eds.). (2006). *Local mediterranean food plants and nutraceuticals*. Forum Nutrition.

IPCC. (2008). *Climate change and water*. Intergovernmental Panel on Climate Change Technical Report IV.

Jacobsen, S. E. (2003). The worldwide potential for quinoa (Chenopodium quinoa Willd.). *Food Review International, 19*(1), 167–177. https://doi.org/10.1081/FRI-120018883

Jin, X., Wang, J., Li, D., Wu, F., & Zhou, X. (2019). Rotations with Indian mustard and wild rocket suppressed cucumber *Fusarium* wilt disease and changed rhizosphere bacterial communities. *Microorganisms, 7*, 57. https://doi.org/10.3390/microorganisms7020057

John, D. A., & Babu, G. R. (2021). Lessons from the aftermaths of green revolution on food system and health. *Frontiers in Sustainable Food Systems, 5*, Article 644559. https://doi.org/10.3389/fsufs.2021.644559

Joshi, T., Deepa, P. R., Joshi, M., & Sharma, P. K. (2023). Matters of the desert: A perspective on achieving food and nutrition security through plants of the (semi) arid regions. *Journal of Agriculture and Food Research, 14*, 100725. ISSN 2666-1543. https://doi.org/10.1016/j.jafr.2023.100725

Kahane, R., Hodgkin, T., Jaenicke, H., Hoogendoorn, C., Hermann, M., Keatinge, J., d'Arros Hughes, J., Padulosi, S., & Looney, N. (2013). Agrobiodiversity for food security, health and income. *Agronomy for Sustainable Development, 33*, 671–693. https://doi.org/10.1007/s13593-013-0147-8

Kalaba, F. K., Chirwa, P. W., Prozesky, H., & Ham, C. (2009). The role of indigenous fruit trees in rural livelihoods: The case of communities in the Mwekera area, Copperbelt province, Zambia. *Acta Horticulturae, 806*, 129–136. https://doi.org/10.17660/ActaHortic.2009.806.14

Kendabie, P., Jørgensen, S. T., & Massawe, F. (2020). Photoperiod control of yield and sink capacity in Bambara groundnut (Vigna subterranea) genotypes. *Food and Energy Security, 9*(2). https://doi.org/10.1002/fes3.240CCBY4.0

Khakurel, D., Uprety, Y., Łuczaj, Ł., & Rajbhandary, S. (2021). Foods from the wild: Local knowledge, use pattern and distribution in Western Nepal. *PLoS One, 16*(10). https://doi.org/10.1371/journal.pone.0258905

Khan, M. M. H., Rafii, M. Y., Ramlee, S. I., Jusoh, M., & Al-Mamun, M. (2021) Bambara groundnut (Vigna subterranea L. Verdc): A crop for the new millennium, its genetic diversity, and improvements to mitigate future food and nutritional challenges. *Sustainability, 13*(10), 1–27. https://doi.org/10.3390/su13105530

Lamien, N., Lingani-Coulibaly, P., & Traore-Gue, J. (2009). Importance of local fruits consumption in diet balance in Burkina Faso, West Africa. *Acta Horticulturae, 806*(203), 208. https://doi.org/10.17660/ActaHortic.2009.806.24

Leng, G., & Hall, J. (2019). Crop yield sensitivity of global major agricultural countries to droughts and the projected changes in the future. *The Science of the Total Environment, 654*, 811–821. https://doi.org/10.1016/j.scitotenv.2018.10.434

Lutz, M., & Godoy, L. B. (2017). *The revival of quinoa: A crop for health.* https://doi.org/10.5772/65451. Chapter-3: 37–54.

Mabhaudhi, T., Chibarabada, T. P., Chimonyo, V. G. P., Murugani, V. G., Pereira, L. M., Sobratee, N., Govender, L., Slotow, R., & Modi, A. T. (2019) Mainstreaming underutilized indigenous and traditional crops into food systems: A South African perspective. *Sustainability, 11*, 72(1), 1–22. https://doi.org/10.3390/su11010172

Mabrouka, O., Smail, M., Tahar, H. M., & Chenchouni, H. (2023). Yield, growth development and grain characteristics of seven Quinoa (Chenopodium quinoa Willd.) genotypes grown in open-field production systems under hot-arid climatic conditions. *Scientific Reports, 13*(1). https://doi.org/10.1038/s41598-023-29039-4

Madouh, T. A. (2022). Eco-physiological responses of native desert plant species to drought and nutritional levels: Case of Kuwait. *Frontiers in Environmental Sciences, 10*, 1–13. https://doi.org/10.3389/fenvs.2022.785517

Madouh, T. A. (2023) The influence of induced drought stress on germination of Cenchrus ciliaris L. and Cenchrus setigerus Vahl.: Implications for rangeland restoration in arid desert environment of *Kuwait. Research in Ecology, 5*, 1–11. https://doi.org/10.30564/re.v5i1.5426

Madouh, T. A., & Al-Sabbagh, T. A. (2021). Innovating forage production system in Kuwait using locally adapted native desert plant species. *Journal of Agricultural and Horticultural Research, 4*(1), 53–61. https://doi.org/10.33140/jahr.04.01.01

Madouh, T. A., & Davidson, M. K. (2024). Phytochemical screening and potential effects of Farsetia aegyptia Turra seeds: A native desert herb, from Kuwait. *Journal of Tropical Agriculture, 62*(1), 81–96. https://jtropag.kau.in/index.php/ojs2/article/view/1428

Madouh, T. A., & Quoreshi, A. M. (2023a). The function of arbuscular mycorrhizal fungi associated with drought stress resistance in native plants of arid desert ecosystems: A review. *Diversity, 15*, 391. https://doi.org/10.3390/d15030391

Madouh, T. A., & Quoreshi, A. M. (2023b). Terrestrial habitats and ecosystems of Kuwait. In M. K. Suleiman, & S. A. Shahid (Eds.), *Terrestrial environment and ecosystems of Kuwait* (Ch. 10, pp. 247–263). Springer. https://doi.org/10.1007/978-3-031-46262-7_10

Makondo, C. (2018). Climate change adaptation: Linking indigenous knowledge with Western science for effective adaptation. *Environmental Science and Policy, 88*, 83–91. https://doi.org/10.1016/j.envsci.2018.06.014

Manzoor, M., Naz, S., Muhammad, H. M. D., & Ahmad, R. (2023). Smart reprogramming of jujube germplasm against salinity tolerance through molecular tools. *Functional and Integrated Genomics, 23*, 222. https://doi.org/10.1007/s10142-023-01140-x

Martínez-Villaluenga, C., Peñas, E., & Hernández-Ledesma, B. (2000). Pseudocereal grains: Nutritional value, health benefits and current applications for the development of gluten-free foods. *Food and Chemical Toxicology, 137*, 111178. https://doi.org/10.1016/j.fct.2020.111178

Mayes, S., Ho, W. K., Chai, H. H., Gao, X., Kundy, A. C., & Mateva, K. I. (2019). Bambara groundnut: An exemplar underutilised legume for resilience under climate change. *Planta, 250*, 803–820. https://doi.org/10.1007/s00425-019-03191-6

Mbosso, C., Boulay, B., Padulosi, S., Meldrum, G., Mohamadou, Y., Niang, A. B., Coulibaly, H., Koreissi, Y., & Sidibé, A. (2020). Fonio and bambara groundnut value chains in mali: Issues, needs, and opportunities for their sustainable promotion. *Sustainability, 12*, 4766. https://doi.org/10.3390/su12114766

Meena, V. S., Gora, J. S., Singh, A., Ram, C., Meena, N. K., Pratibha, R. Y., Basile, B., & Kumar, P. (2022). Underutilized fruit crops of Indian arid and semiarid regions: Importance, conservation and utilization strategies. *Horticulturae*, 171. https://doi.org/10.3390/horticulturae8020171

Meng, Q., Chen, X., Lobell, D. B., Cui, Z., Zhang, Y., Yang, H., & Zhang, F. (2016). Growing sensitivity of maize to water scarcity under climate change. *Scientific Reports, 6*(1), 19605. https://doi.org/10.1038/srep19605

Modarres, R. (2007). Rainfall trends in arid and semi-arid regions of Iran. *Journal Arid Environments, 70*(2), 344–355. https://doi.org/10.1016/j.jaridenv.2006.12.024

Mokria, M., Geretsadik, Y., Birhane, E., McMullin, S., Ngethe, E., Hadgu, K., Hagazi, N., & TewoldeBerhan, S. (2022). Nutritional and ecoclimatic importance of indigenous and naturalized wild edible plant species in Ethiopia. *Food Chemistry Molecular Sciences, 4*. https://doi.org/10.1016/j.fochms.2022.100084

Mubaiwa, J., Fogliano, V., Chidewe, C., & Linnemann, A. R. (2016). Hard-to-cook phenomenon in bambara groundnut (Vigna subterranea (L.) Verdc.) processing: Options to improve its role in providing food security. *Food Reviews International, 33*(2), 167–194. https://doi.org/10.1080/87559129.2016.1149864

Mungofa, N., Sibanyoni, J., Mashau, M. E., & Beswa, D. (2022). Prospective role of indigenous leafy vegetables as functional food ingredients. *Molecules, 27*(22), 7995. https://doi.org/10.3390/molecules27227995. PMID: 36432098.

Naciri, K., Aboukhalaf, A., Moujabbir, S., Essaih, S., Tbatou, M., Belahyana, A., & Belahsen, R. (2022). Ethnobotanical knowledge of wild food plants in Khenifra, a province in the Middle Atlas region of Morocco. *GSC Advanced Research and Reviews, 13*(02), 180–200. https://doi.org/10.30574/gscarr.2022.13.2.0306

Nalau, J., Becken, S., Loehr, J., Parsons, M., Brown, C., & Mackey, B. (2018). The role of indigenous and traditional knowledge in ecosystem-based adaptation: A review of the literature and case studies from the pacific islands. *Weather, Climate, and Society, 10*(4), 851–865. https://doi.org/10.1175/WCAS-D-18-0032.1

Naorem, A., Jayaraman, S., Dang, Y. P., Dalal, R. C., Sinha, N. K., Rao, C. S., & Patra, A. K. (2023). Soil constraints in an arid environment—Challenges, prospects, and implications. *Agronomy, 13*, 220. https://doi.org/10.3390/agronomy13010220

Nassif, F., & Tanja, A. (2013). Gathered food plants in Morocco: The long forgotten species in ethnobotanical research. *Life Sciences Leaflets, 3*, 17–54.

Ndidi, U. S., Ndidi, C. U., Aimola, I. A., Bassa, O., Mankilik, M., & Adamu, Z. (2014). Effects of processing (boiling and roasting) on the nutritional and antinutritional properties of Bambara groundnuts (Vigna subterranea [L.] Verdc.) from Southern Kaduna, Nigeria. *Journal of Food Processing, 472129*, 1–9. https://doi.org/10.1155/2014/472129

Neudeck, L., Avelino, L., Bareetseng, P., Ngwenya, B. N., Teketay, D., & Motsholapheko, M. R. (2012). The contribution of edible wild plants to food security, dietary diversity and income of households in Shorobe Village, northern Botswana. *Ethnobotanical Research and Applications, 10*, 449–462.

Nisar, M., More, D. R., Zubair, S., Sawate, A. R., & Hashmi, S. I. (2018). Studies on development of technology for preparation of cookies incorporated with Quinoa seed flour and its nutritional and sensory quality evaluation. *International Journal of Chemical Studies, 6*(2), 3380–3384.

Omotayo, A. O., & Aremu, A. O. (2020). Underutilized African indigenous fruit trees and food–nutrition security: Opportunities, challenges, and prospects. *Food and Energy Security, 9*(3), 3–16. https://doi.org/10.1002/fes3.220

Pacheco, F., Fernandes, L. F. S., de Valle Junior, R. F., Valera, C. A., & Pissarra, T. C. T. (2018). Land degradation: Multiple environmental consequences and routes to neutrality. *Current Opinion in Environmental Science and Health, 5*, 79–86. https://doi.org/10.1016/j.coesh.2018.07.002

Pathan, S., & Siddiqui, R. (2022). Nutritional composition and bioactive components in Quinoa (*Chenopodium quinoa* Willd.) greens: A review. *Nutrients, 14*(3), 558. https://doi.org/10.3390/nu14030558. PMID: 35276913; PMCID: PMC8840215.

Peng, W., & Berry, E. M. (2018). The concept of food security. In *Reference module in food science*. https://doi.org/10.1016/B978-0-08-100596-5.22314-7

Perrings, C., & Halkos, G. (2015). Agriculture and the threat to biodiversity in sub-saharan Africa. *Environmental Research Letters, 10*(9), Article 095015. https://doi.org/10.1088/1748-9326/10/9/095015

Powell, B., Ouarghidi, A., Johns, T., Ibn, T. M., & Eyzaguirre, P. (2014). Wild leafy vegetable use and knowledge across multiple sites in Morocco: A case study for transmission of local knowledge? *Journal of Ethnobiology and Ethnomedicine, 10*. https://doi.org/10.1186/1746-4269-10-34

Pradhan, S., & Tamang, J. P. (2015). Ethnobiology of wild leafy vegetables of Sikkim. *Indian Journal of Traditional Knowledge, 14*(2), 290–297.

Pritham, S. M., Revanna, M. L., Ravindra, U., Kalpana, B., Murthy, N., & Madhusudsnan. (2021). Quinoa (*Chenopodium quinoa*) grains processing and its value added products. *International Journal of Current Microbiology and Applied Sciences, 10*(04), 681–691. https://doi.org/10.20546/ijcmas.2021.1004.068

Quoreshi, A. M., & Madouh, T. A. (2023). Kuwait deserts and ecosystems in the context of changing climate. In M. K. Suleiman, & S. A. Shahid (Eds.), *Terrestrial environment and ecosystems of Kuwait*. (Ch. 14, pp. 341-359) Springer. https://doi.org/10.1007/978-3-031-46262-7_14

Rai, M. K., Shekhawat, J. K., Kataria, V., Phulwaria, M., & Shekhawat, N. S. (2021). Genomic and biotechnological interventions in *Prosopis cineraria*: Current status, challenges and opportunities. *Trees (Berl.), 35*, 1109–1121. https://doi.org/10.1007/s00468-020-02073-9

Rao, V., & Poonia, A. (2023). Protein characteristics, amino acid profile, health benefits and methods of extraction and isolation of proteins from some pseudocereals—A review. *Food Production Processing and Nutrition, 5*, 37(2–17). https://doi.org/10.1186/s43014-023-00154-z

Rivera, D., Obón, C., Heinrich, M., Inocencio, C., Verde, A., & Fajardo, J. (2006). Gathered Mediterranean food plants—Ethnobotanical investigations and historical development. *Bibliotheca Nutritio et Dieta, 59*, 18–74. https://doi.org/10.1159/000095207

Rodin, R. J. (1985). The ethnobotany of the Kwanyama Ovambos. In *Missouri: Missouri Botanical Gardens. Monograph in Systematic botany from the Missouri Botanical Garden* (Vol. 9).

Rodríguez, J. P., Rahman, H., Thushar, S., & Singh, R. K. (2020). Healthy and resilient cereals and pseudo-cereals for marginal agriculture: Molecular advances for improving nutrient bioavailability. *Frontiers in Genetics, 11*, 49. https://doi.org/10.3389/fgene.2020.00049

Rossi, F. (2020). Beneficial biofilms for land rehabilitation and fertilization. *FEMS Microbiology Letters, 367*(21). https://doi.org/10.1093/femsle/fnaa184

Safriel, U., Adeel, Z., Niemeijer, D., Puigdefabregas, J., White, R., Lal, R., Winslow, M., Ziedler, J., Prince, S., Archer, E., & King, C. (2005). Dryland systems. In R. Hassan, R. Scholes, & N. Ash (Eds.), *Ecosystems and human well-being: Current state and trends* (Vol. 1, pp. 623–662). Island Press.

Samal, I., Bhoi, T. K., Raj, M., Majhi, P. K., Murmu, S., Pradhan, A. K., Kumar, D., Paschapur, A. U., Joshi, D. C., & Guru, P. N. (2023). Underutilized legumes: Nutrient status and advanced breeding approaches for qualitative and quantitative enhancement. *Front in Nutrition, 18*(10), 1110750. https://doi.org/10.3389/fnut.2023.1110750

Sanchez-Bel, P., Romojaro, A., Egea, I., & Pretel, M. T. (2018). Wild edible plants as potential antioxidant or nutritional supplements for beverages minimally processed. *LWT-Food Science Technology, 62*, 830–837. https://doi.org/10.1016/j.lwt.2014.06.017

Sarfo, J., Keding, G. B., Boedecker, J., Pawelzik, E., & Termote, C. (2020). The impact of local agrobiodiversity and food interventions on cost, nutritional adequacy and affordability of women and children's diet in northern Kenya: A modeling exercise. *Frontiers in Nutrition, 7*, 129. https://doi.org/10.3389/fnut.2020.00129

Saroj, P. L. (2018). Global scenario for arid and semi-arid fruit production for combating hunger. In *Proceedings of the National Conference on Arid Horticulture for Enhancing Productivity and Economic Empowerment Held at ICAR-CIAH*, Bikaner, India, 27–29 October 2018 (pp. 1–14).

Seyoum, Y., Teketay, D., Dugo, G. S., & Wondafirash, M. (2015). Edible wild fruit trees and shrubs and their socioeconomic significance in Central Ethiopia. *Ethnobotany Research and Applications, 14*, 183–197.

Shava, S. (2000). *The use of indigenous plants as food by a rural community in the eastern cape: An educational exploration.* Thesis on Masters of Education submitted to Rhodes University.

Shayanowako, A. I. T., Morrissey, O., Tanzi, A. S., Muchuweti, M., Mendiondo, G. M., Mayes, S., Modi, A., & Mabhaudhi, T. (2021). African leafy vegetables for improved human nutrition and food system resilience in Southern Africa: A scoping review. *Sustainability, 13*(5), 2896. https://doi.org/10.3390/su13052896

Shelef, O., Weisberg, P. J., & Provenza, F. (2017). The value of native plants and local production in an era of global agriculture. *Frontiers in Plant Science, 8*. https://doi.org/10.3389/fpls.2017.02069

Shen, Y., Liu, D., Yin, J., Xiong, L., & Liu, P. (2020). Integrating hybrid runoff generation mechanism into variable infiltration capacity model to facilitate hydrological simulations. *Stochastic Environmental Research and Risk Assessment, 34*, 2139–2157. https://doi.org/10.1007/s00477-020-01878-x

Singh, R. K., Sreenivasulu, N., & Prasad, M. (2022). Potential of underutilized crops to introduce the nutritional diversity and achieve zero hunger. *Functional and Integrative Genomics, 22*(7), 1459–1465. https://doi.org/10.1007/s10142-022-00898-w

Sridonpai, P., Kongprapun, P., Sungayuth, N., Sukprasansap, M., Chimkerd, C., & Judprasong, K. (2022). Nutritive values and phytochemical compositions of edible indigenous plants in Thailand. *Frontiers in Sustainable Food Systems, 6*, Article 870147. https://doi.org/10.3389/fsufs.2022.870147

Tan, X. L., Azam-Ali, S., Goh, E. V., Mustafa, M., Chai, H. H., Ho, W. K., Mayes, S., Mabhaudhi, T., Azam-Ali, S., & Massawe, F. (2020). Bambara groundnut: An underutilized leguminous crop for global food security and nutrition. *Frontiers in Nutrition, 10*(7), Article 601496. https://doi.org/10.3389/fnut.2020.601496

Tanzi, A. S., Eagleton, G. E., Ho, W. K., Wong, Q. N., Mayes, S., & Massawe, F. (2019) Winged bean (Psophocarpus tetragonolobus (L.) DC.) for food and nutritional security: Synthesis of past research and future direction. *Planta, 250*(3), 911–931. https://doi.org/10.1007/s00425-019-03141-2. Epub 2019 Mar 25. PMID: 30911885.

Taylor, J., Sarkis, L., Hani, N., Abulaila, K., & Ulian, T. (2019). Conservation and sustainable use of wild edible plants in the Eastern Mediterranean region. In C. Mohanna (Ed.), *Role of mediterranean forests in the Paris agreement. Proceedings of the sixth mediterranean forest week*, Brummana, Lebanon, 1–5 April 2019 (pp. 293–300). Forêt Méditerranéenne.

Temegne, N. C., Gouertoumbo, W. F., Wakem, G. A., Nkou, F. T. D., Youmbi, E., & Ntsomboh-Ntsefong, G. (2018). Origin and ecology of Bambara Groundnut (Vigna subterranea (L.) Verdc: A review. *Journal of Ecology & Natural Resources, 2*(4), 000140. ISSN: 2578-4994.

Tharmabalan, R. T. (2023). Identification of wild edible plants used by the orang Asli, indigenous peoples of the Malay Peninsula. *Frontiers in Sustainable. Food Systems, 7*, 1–12. https://doi.org/10.3389/fsufs.2023.1036490

Thomson, L., Doran, J., & Clarke, B. (2018). *Trees for life in oceania: Conservation and utilisation of genetic diversity.* Australian Centre for International Agricultural Research Monograph.

Thrupp, L. A. (1997). *Linking biodiversity and agriculture: Challenges and opportunities for sustainable food security.* World Resources Institute.

Tounekti, T., Mosbah, M., & Khemira, H. K. (2019). Ethnobotanical study of indigenous medicinal plants of Jazan region, Saudi Arabia. In *Evidence-based complementary and alternative medicine.* Article ID 3190670, 45 p. https://doi.org/10.1155/2019/3190670

Ulian, T., Flores, C., Lira, R., Mamatsharaga, A., Mogotsi, K. K., Muthoka, P., Ngwako, S., Nyamongo, D. A., Omondi, W., Sanogo, A. K., Sanogo, S., & Mathana, E. (Eds.). (2019). *Wild plants for a sustainable future: 110 multipurpose species* (1st edn.). Kew Publishing.

UN. (2022). *UN decade for deserts and the fight against desertification.* Retrieved September 21, 2024, from https://www.un.org/en/events/desertification_decade/whynow.shtml

UNEP. (2000). UNEP-world conservation monitoring centre (Assessing forest integrity and naturalnessin relation to biodiversity. Cambridge, UK, 2000), p. 75.

van Ransburg, W. S. J., Venter, S. L., Netshiluvhi, T. R., den Heever, E. V., Vorster, H. J., de Ronde, J. A., & Bornman, C. H. (2004). Role of indigenous leafy vegetables in combating hunger and malnutrition. *South African Journal of Botany, 70*(1), 52–59.

Zulu, D., Ellis, R. H., & Culham, A. (2019). Collection, consumption, and sale of Lusala (Dioscorea hirtiflora)—A wild yam—By rural households in southern province, Zambia. *Economic Botany, 73*(2), 47–63. https://doi.org/10.1007/s12231-018-9433-3

Chapter 11
Agroforestry: A Nature Based Solution to Mitigate Climate Change Impacts and to Improve Crop Production Systems for Food Security in Arid Regions

M. Anisul Islam

Abstract Agroforestry is a land use management system in which combinations of trees or shrubs are grown with crops or livestock farming systems (pastures). Agroforestry combines agricultural and forestry technologies to create more diverse, productive, profitable, healthy, and sustainable land-use systems. Farmers in many countries and regions practice agroforestry to meet their demands for crops (grains) and forage (both roughage and grains). Research and practice of agroforestry demonstrated that it improves crop production diversity, soil quality through increasing organic matter, biological nitrogen fixation, soil microbial population and diversity to attain food security. Practice of agroforestry also helps in reducing wind erosion, reclamation of degraded land, enhance soil carbon sequestration, reduce the greenhouse gases and help in mitigating the global climate change effects. Since production of crops or trees can be arranged in spatial or temporal manner, agroforestry practice can also be modified according to changing climatic conditions. While agroforestry is practiced successfully in many parts of the world, it faces challenges such as prolonged drought, degraded soil with low fertility and often water logging and soil salinity in the arid and hyper-arid regions. Farmers in arid and hyper-arid region have been able to overcome these challenges by selecting species that are suited for local climate and meets farmers demand and mitigate the effects of global climate change through preventing land degradation, increased carbon sequestration, soil organic carbon, and decreased greenhouse gases emission. In addition, new crop and tree species have been used successfully to utilize salt affected areas of the arid region. Policy and institutional support and some form of incentives can attract and motivate farmers in the arid regions to adopt agroforestry in a larger scale. Future research should focus on innovative practices to promote agroforestry in the hyper-arid regions for enhancing soil fertility, undertaking trials for identification of salt tolerant tree species to be used to facilitate bio-drainage, sub-region specific best

---

M. Anisul Islam (✉)
Food Security Program, Environment and Life Sciences Research Center, Kuwait Institute for Scientific Research, Safat, Kuwait
e-mail: aislam@kisr.edu.kw

M. K. Suleiman and S. A. Shahid (eds.), *Fostering Arid Lands Agriculture in the Face of Climate Change*, Advances in Global Change Research 81,
https://doi.org/10.1007/978-3-031-98490-7_11

practices and cost–benefit analysis of small scale practices to ensure production and financial sustainability. This book chapter highlights the suitability of agroforestry systems for the arid regions and provide guidelines through which it can be practiced for improved crop production for food and nutritional securities, amelioration of local micro-climate and to mitigate the effects of global climate change.

Keywords Crop production · Forage · Soil improvement · Microbial diversity · Global warming

11.1 Introduction

Agroforestry is a land use management system in which combinations of trees or shrubs are grown around or among crops or pastureland. Agroforestry is defined as 'growing of trees and crops, and often involving livestock and their interactions (Nair et al., 2008). In a broader sense, agroforestry often refers to the cultivation of crops alongside growing trees in interacting combinations within a same unit of land in a spatial or temporal manner. Practice of agroforestry can also be viewed as nature based, ecologically friendly and dynamic and flexible (through choice of species based on geography and climate) natural resource management system that enhances crop productivity, diversity, carbon sequestration, soil ecosystem functions and services and arrest soil degradation and climate change related effects.

Practice of growing trees with agricultural crops were in existence for many centuries in different parts of the world. However, scientific and systematic studies to explore their production benefits and environmental sustainability came into attention during the seventies of the twentieth century. In today's world, practice of agroforestry through integration and utilization of trees and shrubs with cash crops ensures increased crop productivity, diversified products and services that offer sustainable solutions to food security and environmental degradations issues.

With increasing human population each year, there is a need to produce and provide more nutritious food mainly through agricultural practices. For centuries, increase in food productivity faced enormous challenges against rapid population growth, natural calamities, and resource degradation. In this context, practice of agroforestry has been viewed as a tool to increase agricultural productivity and diversity, as a strategy to primarily increase carbon sequestration (Ruruh et al., 2024) in the agricultural landscape, and as a mitigation measure to address the broader environmental issues in a world with changing climate scenarios. For example, research and practice of agroforestry demonstrated that it improves crop production diversity, soil quality through increasing organic matter, biological nitrogen fixation, soil microbial population and micro-climate (Bado et al., 2021; Islam et al., 2022). It has been demonstrated that practice of agroforestry also helps in reducing wind erosion. In addition, agroforestry has also been used for better weed control and as a pest management tool (Ramert et al., 2002). The significant advantage of practicing agroforestry is that it is best suited for resource (lack of water or degraded soil conditions) poor conditions across the world. Moreover, agroforestry is very flexible in its nature as it can be tailor made to the needs of farmers or needs that might arise due to effects of global climate change. Since the production of crops or trees can be arranged in spatial or temporal manner, agroforestry practice can also be modified according to changing climatic conditions through flexibility of crop and tree species selection suitable for local climatic conditions and need for grain and forage.

Although practice of agroforestry is quite common throughout the world, agroforestry practice and associated research in the arid regions has not been that widespread particularly in the arid areas of Gulf and Middle-Eastern region.

Despite all these positive attributes of practicing agroforestry can offer, it is logical to question whether the practice of agroforestry will be able to sustain the agricultural productivity to feed increasing global population or provide forage as well as mitigate the adverse effects of climate change effectively. This chapter provides an overview of existing agroforestry practices and the positive environmental and social impact it brings about. This chapter also aims to highlight the challenges of practicing agroforestry especially in the arid and hyper-arid region under a changing global environment and recommend on policy modifications and incentives from the respective governments to promote sustainable practice of agroforestry in the context of food security and deteriorating environment.

11.2 Background

Agriculture, the production of food through farming, is the principal source of food and nutrient supply for the entire population of the world. Agricultural production or form of farming started almost 11,000 years ago when people gradually transitioned from a hunter-gatherer lifestyle toward cultivating crops and raising animals for food (Montgomery, 2008). At that time only staple crops such as wheat or barley were cultivated for family consumption (d'Alpoim Guedes et al., 2015). A slightly larger scale cultivation of flax and papyrus (*Cyperus papyrus*) is also reported during the same time. Reports also indicate cultivation of wheat, barley and jujube in India some 9,000 years ago which was followed by integration of goat and sheep (Ghosal, 2021). The transition to agriculture is perceived to have taken place independently almost simultaneously in several parts of the world, including northern China, Central America, and the Fertile Crescent, a region in the Middle East. It is also believed that some 5000 years ago agriculture was practiced in every major continent except Australia (Bulliet et al., 2008). It is reported that currently 32% of the global land mass is used for agricultural production (Fuglie et al., 2024). Furthermore, it is roughly estimated that today the total global area under agroforestry system practices are about 1.6 billion hectares of which approximately 78% is located in the tropics and 22% in the temperate regions (Nair et al., 2021).

During the last three decades, increases in agricultural total factor productivity (the amount of output per unit of land, labor, capital and material inputs) has become the major driver of increased world agricultural output to cater to the need for humans and livestock. With ever increasing global population, the world currently is faced with two major challenges – food security and climate change effect. Broadly defined, food security refers to a situation/condition when all the people have both physical and economic access to safe and nutritious food to maintain a healthy life. Unfortunately, due to uneven agricultural production and economic condition of human population around the world, not all countries and regions can ensure food security adequately. Geographical conditions such as area under arable land, prevalent climatic challenges and local socio-economic conditions are often the biggest hinderance towards achieving food security. Climate change refers to long-term shifts in temperatures and weather patterns. Human activities have been reported to the main driver of climate change, primarily due to the burning of fossil fuels like coal, oil and gas. The concept of global warming or global climate change is a contentious issue. Some believe it to be true and some deny it. However, factual evidence from past decades indicate that global climate is generally changing and it is changing rapidly before mankind can come up with solutions to slow it down or mitigate the adverse effects of climate change. For example, the global atmospheric CO_2 level increase from 312.04 parts per million (ppm) in 1958 to 419.3 ppm in 2023 resulting an increase of 34.37% over 65 years (Lindsey, 2024). The overall rise of the global emission of CO_2 is attributed to fossil fuel burning (Yoro & Daramola, 2020; Ghosh and Ghosh, 2020). In addition, global temperature also increased in recent decades (Kraaijenbrink et al., 2017; Pascual & Bouma, 2009). As a consequence, sea water temperature also increased causing more frequent storm formation and resulting land inundation, waterlogging condition, habitat damage and crop loss (Hsiao et al., 2024; Sivakumar, 2018). Not only that, the ocean water during this time has absorbed enough carbon dioxide to lower its pH by 0.1 unit, resulting an 30% increase in acidity.

As a consequence of global warming, heated earth surface evaporates water quickly and results in infrequent and inadequate rain events. This in turn, results in many-rain free and very hot days (prolonged drought) which are far below the optimum farming condition – resulting in huge loss of crops (Asseng et al., 2015; Marion, 2015; Trenberth, 2011). Moreover, in many parts of the world, especially in the hyper-arid areas, salt builds up on the earth surface due to excessive evapotranspiration and slowly the land becomes unsuitable for cultivation of salt sensitive crop and tree species (Ali & Salem, 2024; Rengasamy, 2006).

It is well established that crop production through the practice of mono-cropping is vulnerable to environmental changes. In addition, single cropping of agricultural crops or forage crops, which is most often practiced by the farmers across the world and especially in the arid region including Kuwait, focuses on increased productivity through over use of fertilizers and pesticides (Malezieux et al., 2009). Monoculture is also linked to soil erosion, chemical contamination, and loss of biodiversity (Malezieux et al., 2009). In contrast, mixed species cropping or agroforestry is viewed as a viable tool to increase on-farm biodiversity in agriculture production systems and is a potentially important component of any sustainable cropping system (Ramert et al., 2002). Mixed cropping has also been studied as a pest management tool and disease management option (Ramert et al., 2002). Other benefits can also include improved soil structure, improved weed control, and microclimate amelioration (wind barrier) (Vandermeer, 1989). Therefore, integrating trees along with cultivation of other crops provides better protection against crop failure in changing global climatic scenario. Moreover, it ensures socio-economic benefits to the small farm holders along preserving biodiversity, improve soil and local micro-climate and other environmental services.

11.2.1 Agroforestry Practices: General Overview of Global Practices

Agroforestry combines agricultural and forestry technologies to create more diverse, productive, profitable, healthy, and sustainable land-use systems. Farmers in many countries and regions practice agroforestry to meet their demands for crops (grains) and forage (both roughage and grains). Agroforestry is widely practiced around the world. Nair et al. (2021) categorized two distinct regions tropical and temperate with reference to developing countries and industrialized regions of the world, respectively. Because of the inherent difference in climatic and socioeconomic conditions of these two regions, practice of agroforestry is also different in these regions. According to FAO classification and from agroforestry perspective, the arid and semiarid lands, and the subhumid and humid lowlands and highlands are considered as tropical developing regions, and the temperate and Mediterranean regions are considered as industrialized (Nair et al., 2021).

Gold and Garrett (2009) classified major agroforestry systems that are practiced in the temperate region into five practices. These include alley cropping, forest farming, riparian buffers, silvopasture, and windbreaks. The prominent agroforestry systems in the humid and subhumid tropics include slash and burn agriculture, forms of shifting cultivation and tree intercropping, small woodlot, and multilayer homegardens. In addition, various forms of silvopastoral (includes livestock) systems, tree intercropping, and windbreaks dominate the agroforestry scene in the semiarid and arid tropics. A detailed and updated description about the practice of agroforestry around the world can be accessed from Nair et al. (2021).

11.2.2 Agroforestry Systems in the Dry and Arid Regions

Agroforestry is widely practiced in drylands all over the world with successful examples found in northern Africa (Campos et al., 2009), sub-Saharan Africa (Lehmann et al., 1998), Asia (Zhaohua et al., 1991) and Latin America (Krishnamurthy et al., 2017). Agroforestry systems offer great flexibility as it is used based on the geographical location, socioeconomical conditions of the user and local consumer needs. For example, in arid regions of India livestock is integrated with the practice of agroforestry. The Central Arid Zone Research Institute (CAZRI) has developed agroforestry types of mixed cropping systems. In the first one, leguminous trees are emphasized and planted along with locally predominant grass species *Lasiurus sindicus*. In the second system, different species of drought tolerant Acacia species (*Acacia tortilis*, *Acacia senegal* and *Acacia seyal*) are planted with wide spacing so that farmers' preferred crops can be grown in between rows of trees (Harsh et al., 2003; Kavia & Harsh, 1991). The goal is to increase economic returns for the farmers and slow down the process of desertification.

Commercial silvopastoral systems and shaded tree-crop systems are the most prominent agroforestry systems in Latin America (Somarriba et al., 2012). In contrast, traditional oases represent the agroforestry systems in North African region through which farmers combine different crops such as date palms, fruit trees, vegetables, fodder and livestock to support their livelihood (Santoro, 2023). Traditional oases are an integral part for the livelihood of local farmers and communities since they still provide economic returns and many ecosystem services. However, practice of these agroforestry systems, despite their economic and cultural importance, are facing a number of environmental and social threats. For example, prolonged drought, salinization and desertification are perceived to be the main threats. Furthermore, for better economic returns, farmers are transitioning to the cultivation of commercial date varieties resulting in decreased agrobiodiversity which in turn, diminishing the use of local and traditional knowledge for farming and being considered as a cultural heritage loss (Santoro, 2023).

Three types of agroforestry are prevalent in sub-Saharan Africa: Agro-silvicultural systems (Taungya, alley cropping, home gardens and multipurpose trees on croplands, shelterbelts), silvi-pastoral systems (trees on rangelands, fodder banks, crops with pastures), and agro-silvi-pastoral systems (home garden with animals, multipurpose woody hedgerows, aquaforestry) (Awazi & Tchamba, 2019). As rainfall varies across the region so does the practice of agroforestry. Practice of small-scale agroforestry in sub-Saharan Africa not only contributes to the preservation of biodiversity (Jose, 2009), but also constitute the third largest carbon sink in Africa after primary forests and long-term fallows (Oke & Odebiyi, 2007), and gender parity as mostly women in rural areas are engaged in agroforestry which give them the opportunity to make income and be financially independent (Awazi & Tchamba, 2019). In addition, agroforestry-based farming systems provide various socio-economic benefits which facilitates the adaptation to climate change and enhance the sustainability of smallholder farming systems (Smith, 2010).

Moreover, in 2007 'The Great Green Wall or Great Green Wall for the Sahara and the Sahel' project was initiated by the African Union with the aim to restore most of the continent's degraded landscape by planting trees spanning a length of 8,000 km and a width of 15 km (UNCCD, 2024). Once completed across 22 participating countries, the overarching goal of this project is to capture 250 million tons of carbon dioxide. Already 18 million hectares of degraded land has been restored and different agroforestry systems are being in practice. For example, in Burkina Faso around 30,000 ha of land was restored and further 20,383 ha of land was reforested. The restoration efforts are focused on 'nutrition garden' where plants are being grown alongside trees that provide food and medicine to the local communities. Similarly, in Nigeria, an area of 2,801 hectares was restored, 373 hectares of multipurpose gardens created along with creation of 709 km long windbreak. Furthermore, in Ethiopia 792,711 ha of terraces were restored while 240 ha of multi-purpose gardens and 91 km long windbreaks were created (UNCCD, 2024). Additionally, there is potential for increasing bioenergy production through agroforestry involving short rotation nitrogen-fixing (leguminous) tree species throughout the arid region beyond few sub-Saharan countries such as Zambia (IRENA, 2019).

11.2.3 Agroforestry Systems in the Hyper-Arid Regions

Practice of agroforestry in some of the Gulf cooperation council (GCC) countries are very limited or non-existent except for Oman and Saudi Arabia. In northern Oman, fodder crops like alfalfa or wheat or vegetables are grown underneath date palms as an agroforestry system (Buerkert et al., 2021). In some places, farmers grow date palms along with citrus trees and grow grasses understory crops. However, there has been a shift from subsistence agroforestry practice to production of cash crops such as fodder in recent years due to changes in economic status of local population and resulting migration to city areas and declining rainfall (Buerkert et al., 2021). Similarly, in Saudi Arabia, being a hyper arid country, agroforestry is practiced in very smaller scale in areas such as Al-Jouf, Al-Asha, Tabuk and Al-Baha where date palm, olive and citrus tree are grown along with agricultural crops (Kouyate et al., 2022).

11.2.4 Agroforestry for Increased Crop Productivity and Food Security

Generally, increases in agricultural productivity competed against population growth throughout history. From 1961 to 2020, the global agricultural sector saw significant positive changes in production, productivity, and resource use mainly driven by widespread use of fertilizer, insecticide and development of newer varieties of crops and genetic modifications (called green revolution). While agricultural output increased nearly fourfold, so did global population by 2.6 times. This resulted an increase of just over 50% in agricultural output per capita. However, this growth in both agricultural productivity and global population was not evenly distributed across the world. Most of the world's agricultural output increase was in the Global South whereas agricultural output in the Global North remained approximately the same since the 1990s (Fuglie et al., 2024). Population growth saw huge jump in the Global South in mostly developing and least developing countries like China, India, Pakistan, Bangladesh, Indonesia, and Nigeria to name a few. According to the estimate by the Food and Agriculture Organization (FAO) world's population will reach 9.1 billion by 2050 (34% higher than today). In order to satisfy the food and nutritional demand of the increased population, the food production must increase by 70% (FAO, 2009).

Agroforestry is viewed as nature based, ecologically friendly and dynamic and flexible land use practice in which crop and tree species can be chosen and integrated based on geography and local climate to enhance crop productivity, on farm diversity, carbon sequestration, soil ecosystem functions and services and arrest soil degradation and climate change related effects. Studies have reported that agroforestry can increase agricultural productivity in the arid region (Amadu et al., 2020; Sileshi et al., 2008; Waldron et al., 2015) and enhances food and nutrition security (Rosenstock et al., 2019). For example, integrating ziziphus trees improved the productivity of millet (Bado et al., 2021). It is suggested that in agroforestry system, the tree-crop-environment interactions positively influence biomass production, nutrient uptake and availability, storage and availability of water in the soil, water uptake by trees and crops, and crop yields (Monteith et al., 1991). A recent meta-analysis in which 1106 observations were extracted from 126 peer-reviewed publications that fulfilled the selection criteria revealed that across ecological conditions in sub-Saharan Africa, agroforestry significantly increased crop yield along with total soil nitrogen, soil organic carbon, and available phosphorus compared to the control (Kuyah et al., 2019) while providing ecosystem services.

11.2.5 Agroforestry for Degraded Land Reclamation

Land degradation, which refers to the reduction in the capacity of soil to support agricultural production under a specified form of land management (FAO, 1999). Land degradation occurs especially in the arid regions when due environmental factors plant cover declines resulting in a deterioration of the physical, chemical and biological properties of soil (Cowie et al., 2018; Prăvălie, 2021). The main driver for land degradation in the arid and hyper-arid region is persistent severe aridity. Aridity is different from drought in the sense that aridity in nature is permanent whereas drought is somewhat temporary. The desert areas of the world have been divided into three zones according to aridity index which refers to the degree of dryness of the climate of any given location. Three arid zones are hyper-arid, arid and semi-arid. Of the total land area of the world, the hyper-arid zone covers 4.2 percent, the arid zone 14.6 percent, and the semiarid zone 12.2 percent. Therefore, almost one-third of the total area of the world is arid land (El-Ghani et al., 2017). Most of the GCC countries fall under hyper-arid zone where aside from extreme hot weather, soil salinity is the major factor that contributes to the scarcity of arable land.

Land degradation is a common phenomenon in the arid and hyper-arid region. According to GLASOD (1990) around 349.6 million hectares (ha) of land in arid zone are categorized as slightly to moderately degraded whereas some 42.8 million ha has been categorized as severely degraded. Agroforestry is mainly practised on degraded lands, which otherwise would have remained fallow or unutilized. The flexibility of species selection allows it encompass a wide range of approaches and technologies for restoring degraded lands. For example, alley cropping with trees and grasses are used for soil erosion control in sloping lands (Dagar et al., 2014). In addition, hedgerow intercropping involving trees such as *Leucaena leucocephala*, *Gliricidia sepium*, and *Flemingia congesta* with agricultural crops has also been used for soil erosion control (Catacutan et al., 2017).

Drifting sand often results in unstable sand dunes and by planting trees along with site specific grasses and other crops sand movement has been reduced in much of the western and central African regions. Approximately 18 million hectares of degraded land has been restored in sub-Saharan region through tree planting and different agroforestry practices. In countries like Burkina Faso, Nigeria and Ethiopia land degradation has been arrested and reclaimed through establishing multipurpose home gardens and windbreaks (UNCCD, 2024).

In Indian Thar desert areas *Prosopis cineraria*-based silvopastoral system has been found most suitable for sand dune stabilization (Gupta et al., 2020). In addition, agroforestry system which incorporated brushwood such as *Leptadenia pyrotechnica* and several local grass species such as *Cenchrus ciliaris* and *Panicum turgidum* have been successfully used for land stabilization and improving the livelihood of the farmers (Gupta et al., 2020).

11.2.6 Soil Salinity and Halophyte-Based Agriculture and Agroforestry

On a global basis, salinity damaged around 20% of cultivated land and nearly half of all irrigated fields to some extent. Saline soil covers 397 million hectares (Mandal et al., 2010) of which a significant area of land is located in the arid regions. Since soil salinity negatively impacts growth of most crops, practicing halophyte-based agriculture in some arid parts of the world most notably in UAE and India have received much attention from farmers and environmentalists. Halophytes not only grow in high saline soil, but they also act as effective carbon sink and contribute to the ecosystem services. Two species, Salicornia and Arthrocaulon, which are edible and have been grown extensively in Umm Al Quwain, UAE. These two species have been hailed as nature-based solution for food security and climate. In addition, halophytes have been grown in popularity as a source of fodder and biofuel (Dagar, 1995; Mathur et al., 2007; Shahid & Alkandari, 2024; Singh et al., 2009). Furthermore, halophytes such as *Haloxylon salicarnicum* has been successfully used for bioremediation of hydrocarbon contaminated soil (Al-Ateeqi, 2014).

There is potential to integrate salt tolerant tree species into halophyte-based agriculture to turn it into an agroforestry system. However, research is needed to find suitable native tree species so that it could potentially contribute towards greater carbon sequestration and lowering the groundwater level through bio-drainage (Heuperman et al., 2002; Khamzina et al., 2006). This will, in turn, facilitate to improve the growing conditions of crops particularly native grass species that are often used as forage and has very high market demand.

11.3 Agroforestry and Mitigation of Climate Change Effects

Agroforestry has been recognized as a land use tool to increase productivity and providing ecosystem services to benefit the livelihood of the farmers and environment. In addition, agroforestry is often viewed as an effective agro-ecological approach to mitigate the climate change effects because of the multitude of environmental benefits that different agroforestry systems provide.

11.3.1 The Drivers of Climate Change

Climate change commonly refers to the long-term shifts in temperatures and weather patterns which potentially have adverse consequences on global environment. Burning of fossil fuels accounts for over 75 per cent of global greenhouse gas (GHGs) emissions and nearly 90 per cent of all carbon dioxide emissions. Although climate varies naturally over a wide range of time but the warming trend since 1950s is somewhat strikingly high and it is extremely likely (> 95%) that human activities have been the dominant cause of that warming (IPCC, 2013).

Concentration of carbon dioxide, methane, and nitrous oxide (the greenhouse gases) have increased significantly in recent decades and these GHGs emissions have increased and accelerated the greenhouse effect and caused earth's surface temperature and ocean water temperature to rise (Fahey et al., 2017).

11.3.2 Agroforestry and Mitigation of Climate Change Effects

Agroforestry systems have attracted special attention in climate change mitigation and adaptations. Studies have shown that practice of agroforestry prevents soil degradation (Béliveau et al., 2017), improved soil structure (Fahad et al., 2022), soil microbial population and diversity (Islam et al., 2022), soil organic matter and nutrient recycling, increased carbon sequestration (Pan et al., 2025), improved local micro-climate (Zignol et al., 2023).

One of the primary benefits that agroforestry systems offer through integration of trees is increased carbon sequestration (Gupta et al., 2017; Lorenz & Lal, 2014; Nair,

2012; Ruruh et al., 2024). Boundary planting in agroforestry system has increased carbon sequestration rate by 35 to 50% in central Africa (Gupta et al., 2017). In addition, surface soil (0 – 30 cm depth) total and microbial stocks of carbon and nitrogen in the agroforestry system were significantly greater than single cropping system (Eddy & Yang, 2022). Soil carbon stocks increased from 62.7 Mega grams (Mg) C ha^{-1} in single cropping system to 72.1 Mg C ha^{-1} in agroforestry system (Eddy & Yang, 2022). Moreover, in a recent study, Pan et al (2025) reported global meta-analysis findings which suggest that compared to other land uses (cropland, forest and uncultivated land), agroforestry increased soil organic carbon (SOC) by 10.7%, dissolved organic carbon by 10.2% and CO_2 emissions by 10.2%. Strikingly, when compared with all land use practices across all climate zones, SOC of agroforestry increased the most in arid areas (18.7%). Furthermore, compared to practice of monoculture, agroforestry has more advantages in terms of microbial biomass, diversity and soil nutrient content (Islam et al., 2022).

Furthermore, agroforestry systems compared to monoculture systems, contribute more towards the reduction of GHGs especially CO_2, N_2O and CH_4 (Mutuo et al., 2005). Albrecht and Kandji (2003) found that N_2O emissions are much less in agroforestry systems due to the reduced application of supplementary nitrogen as inorganic nitrogen are recycled from tree leaf litter and can be taken up understory crops.

Several researchers have reported that agroforestry practices in smallholder farming systems contribute to climate change mitigation and adaptation. For example, farmers have been using various methodologies to overcome or cope up with the changing climate. One of the methods adopted by them is changing the species and agricultural production system used for agricultural practices (Zeratsion et al., 2023). In addition, Verchot et al. (2007) reported small-scale farmers in many parts of the world have undertaken improved farming practices such as not using the marginal lands for intensive agricultural production, restore degraded land, reduce tillage, decrease biomass burning, and introduce trees into agricultural landscapes. Furthermore, farmers adopted better crop management by choosing climate resilient cultivars, introduce forages into rotations, erosion control, leaf litter decomposing to increase nitrogen, animal waste recycling, and planting leguminous trees to augment soil nitrogen. These actions can potentially improve crop diversity and productivity and ensure sustainable farming systems and act towards climate change mitigation (Verchot et al., 2007).

Other studies (Harvey & Gonzalez-Villalobos, 2007; Bhagwat et al., 2008) have demonstrated that agroforestry plays a major role in conserving biodiversity which ultimately negate the adverse effects of global climate change specially in the arid regions. Traditional agroforestry practices have found to support biodiversity conservation in central south America, Africa and Asia (McNeely & Schroth, 2006). It was found that in human dominated landscapes, species diversity is well maintained if agroforestry was practiced (Bhagwat et al., 2008). Moreover, it is also reported that practice of agroforestry increases soil microbial diversity compared to cultivation of single crop (Islam et al., 2022).

11.3.3 Selection Criteria for Crops and Trees Specifically for Arid Regions

Selection criteria for crop and tree species for an agroforestry system especially in the arid region is dependent on existing local climate, intended purpose, farm size, farmers' own and market demand, and availability of climate resilient species. For example, in desert areas where drifting sand makes the land cover unstable, planting of native grass species is recommended along with drought tolerant native species for windbreaks. In parts of arid region where rain-fed agriculture or agroforestry is practiced in the form of multilayer home gardens, species should be selected which can offer food, nutrition, and medicinal benefits. Leguminous tree or shrub species can be used in agroforestry systems for the reclamation and increasing soil fertility of degraded arid lands. In saline arid areas halophytic grasses and shrubs can be used with native salt tolerant tree species which can help decrease the water table depth (bio-drainage) thereby facilitate growth of shallow rooted crops.

In the hyper-arid countries including Kuwait or Saudi Arabia where practice of agriculture or agroforestry is fully dependent on irrigation, alley cropping and windbreaks are the only options because of ease of setting up the irrigation system. Since soils in these areas have very low organic matter content, leguminous tree species such as *Leucaena leucocephala* or drought hardy variety of perennial alfalfa (*Medicago sativa*) can be planted for their ability to fix nitrogen (Islam et al., 2022). Tree species that can tolerate drought and salinity should be given more preference and long rotation tree can be planted for carbon sequestration. Similarly, same considerations should be given on selecting cereal and fodder crops due to lack of locally available fodder for the livestock. Native grass species such as *Panicum turgidum* and *Pennisetum divisum*, which are very well adapted to hyper-arid climate, can be grown in between tree rows which has proved to be grown well and preferred by the livestock as fodder (Madouh & Al-Sabbagh, 2020).

11.3.4 Economics of Crop and Forage Production and Local Market Demand

Farmers across the arid areas of sub-Sahara, Middle East and North Africa (MENA), and India mostly practice agroforestry in small scale for their subsistence and seldom sell excess crop to the local market for financial benefit. In these cases, very often, the cost of production does not play a huge role in their decision-making process on which crop to grow. Farmers, when practice agroforestry in medium to large scale, usually grow crops targeting to sell in local markets and look forward to the financial returns. Therefore, in such cases, farmers are somewhat mindful about the investment costs but they are not documented. Therefore, there is a lack of information on the cost–benefit analysis of small-scale operation of agroforestry practices across the arid region. Also, local market demand for agroforestry product varies from region to region and not well documented.

Farmers, especially in Gulf region, are transitioning towards growing economically viable agricultural crops. The scenario of fodder production is somewhat unclear as no local production data on country level is available. However, it is observed in Kuwait that some farmers sell a small quantity of fodder in roadside temporary spots which suggest there is local demand for locally produced fodder. In addition, locally grown vegetables crops are also being sold in mainstream supermarkets and some ethnic shops in Kuwait. However, demand for fodder (roughages) still remain high in Kuwait. According to Annual Bulletin of foreign trade statistics, 2020, Ministry of Planning, Kuwait imported over KD35,022,000/ worth of livestock feed during 2019 due to lack of local production and availability of forage crops (Ministry of Planning, 2020). It is worth noting here that when farmers grow crops, vegetables, and fodder in small-scale, usually the earning from selling them is often equal to or lesser than the cost of production. One study by Islam et al. (2022) showed that due to economies of scale, the unit cost for larger area (five ha) of fodder crop production was lower than in smaller area (0.25 ha). Therefore, it is recommended that local farmers adopt agroforestry system encompassing a sizeable area for the production of crops, vegetables (if applicable) and forage crops.

11.3.5 Future Approaches and Research Directions

Future research should focus on cost–benefit analysis of production especially at the small production scale to ensure sustainability. In addition, trial should be made to select salt tolerant tree species which could be used within halophyte-based agricultural system for greater carbon sequestration and lowering the groundwater level through bio-drainage. Research should also be carried out to identify best land use practices at different scales on various regions within arid environment.

Innovative practices should be adopted to promote agroforestry practices in hyper-arid regions especially for enhancing soil fertility and organic matter through the use of nitrogen fixing trees for crop and fodder production. More involvement of extension agencies is essential to promote best land use practices through the adoption of agroforestry in the MENA region.

11.4 Policy and Institutional Role on the Practice and Expansion of Agroforestry

It is reported (Ajayi & Place, 2012) that widespread adoption of agroforestry in Asia and Africa was strongly influenced by the policy and institutional context within which agroforestry is disseminated. One way of strengthening the agroforestry-support policies is when they are included as an important part of national conservation strategy. In sub-Saharan region, many international agencies and non-governmental organizations (NGOs) with the approval of respective governments are advocating practice of agroforestry and its financial and environmental benefits to the local farmers and communities.

In contrast, in some Gulf countries, the scenario is somewhat different as there are no subsistence farming. The land owners are motivated for economic benefit by taking up large scale farming which usually involves growing single crop in available land. In these circumstances, the government can encourage farmers towards biodiversity and environmental conservation by providing extra incentives on seed costs for desired and arid zone appropriate or climate resilient drought and salt hardy species.

11.4.1 Education, Extension and Awareness

The role of education, and awareness and extension cannot be undermined for the expansion of agroforestry practices in the arid region. The role of agriculture in securing food security, role of plants and trees in conserving and enhancement of biodiversity and environment should be included in school curriculum, broadcasted on TV and social media to create awareness among the population of all age groups. However, to be more effective, extension workers need to be involved to facilitate the information exchange and technology adoption (Scherr, 2012). Highlighting the benefits of agroforestry practice and how it aligns with sustainable crop production practice and mitigate climate change effects can also motivate more farmers to participate in agroforestry practices (Choi et al., 2022). Extension workers can educate and train farmers on site specific species selection and how to integrate trees with crops, livestock or even apiculture (honeybees), which can lead to improved environmental, economic, and social outcomes. By promoting these practices, agroforestry extension supports climate resilience, preserve biodiversity, increase crop production and ensure sustainable agricultural towards achieving food security.

11.4.2 Practice Model Farm Demonstration

Although different form and scales of agroforestry is practiced throughout the arid regions across the world, in some hyper-arid countries practice of agroforestry is still in the infancy stage. The main deterring factor for the farmers to practice agroforestry is the lack of knowledge for integrating appropriate trees with crop production system, management of crop rotation cycle for multiple crops and uncertainty about the outcomes. Although many agricultural and livestock farmers in the hyper-arid region have the financial means to start agroforestry practices, when it comes to initiating something on their own, they would rather wait and see if someone else is starting it first or setting the examples. In such context, it is believed that setting up an agroforestry demonstration production system which includes trees, vegetables, grains along with livestock will eliminate the psychological uncertainty barrier from the farmers to initiate agroforestry farming on their own. This can be achieved through collaboration between relevant ministry or extension agency and any enthusiastic farmer. On top of this, the relevant ministry or extension agency officials can also offer hands on training to the willing farmers on crop selection, soil and water requirements, crop production and management related issues. Above all, the concept of agroforestry and the practice of it in real should be publicized and be easily accessible for farmers and general public. It is often said 'seeing is believing' and believing something is the core driving inspiration and motivate people to do something good for them and for the environment. It is worth noting here that in Africa & India different NGOs and government agencies have partnership with local farmers for promoting and practicing agroforestry which is lacking in the MENA region.

11.5 Conclusions and Recommendations

In general, in most part of the world practice of agroforestry increases financial benefits for the farmer though diversified products, increase local employment, enhances self-sufficiency in food and fodder production, reduces reliance on importing grain and forage and use of fossil fuel for cooking especially for subsistence or small-scale farmers in least developing countries. From economic perspective, some agroforestry practices in drier regions of India and Latin America are linked with higher yields compared to the practice of single cropping. In some parts in North Africa and in Oman, practice of agroforestry is directly linked the local culture and heritage and plays an invaluable role for their livelihood. The potential benefits of agroforestry on improving environmental condition and adapting with changing global climatic condition cannot be overemphasized as cultivation of trees with crops acts as a carbon sequestration tool with mitigation benefits. Furthermore, the diversity of crops grown in any agroforestry systems offer the unique flexibility to the farmer from financial losses in situation when one specific crop may fail due to insect attack or weather-related issues.

The biogeography is unique and challenging in hyper-arid regions especially countries like Kuwait and surrounding countries in the Gulf Cooperation Council. There exists a huge local demand especially for grain and forage species for the sustainable production of livestock. However, local farmers, for higher economic returns, prefer to practice monocropping. Therefore, innovation is needed for crop diversification as well as formulation of value-added product (combining apiculture to produce honey) from traditional products derived from agroforestry can potentially benefit the farmers financially. Incentives in the form of policy support, increased awareness about the social and environmental benefits of agroforestry, and free direct guidance from extension works should be available for farmers practicing agroforestry. Wherever institutional framework and support is available in the arid region, farmers can practice agroforestry (carbon farming–to increase soil's organic matter and sequester carbon) and receive carbon credit. Furthermore, subsidies on seeds and crop insurance for sustainable agroforestry practice will enhance the possibility of wider agroforestry adoption in arid region which in turn will help achieving food security and mitigate the adverse effects of climate change.

References

Ajayi, O. C., & Place, F. (2012). Policy support for large-scale adoption of agroforestry practices: experience from Africa and Asia. In P. Nair, & Garrity D (Eds.), *Agroforestry–the future of global land use. Advances in agroforestry*, 9 Springer. https://doi.org/10.1007/978-94-007-4676-3_12

Al-Ateeqi, S. S. (2014). Phytoremediation of oil-polluted desert soil in Kuwait using native plant species. PhD thesis, University of Glasgow, UK.

Albrecht, A., & Kandji, S. T. (2003). Carbon sequestration in tropical agroforestry systems. *Agriculture, Ecosystems and Environment, 99*, 15–27.

Ali, A. M., & Salem, H. M. (2024). Salinity-induced desertification in oasis ecosystems: Challenges and future directions. *Environmental Monitoring and AssessMent, 196*, 696. https://doi.org/10.1007/s10661-024-12804-x

Amadu, F. O., Miller, D. C., & McNamara, P. E. (2020). Agroforestry as a pathway to agricultural yield impacts in climate-smart agriculture investments: Evidence from southern Malawi. *Ecological Economics, 167*, Article 106443.

Andres, R. J., Boden, T.A., Bréon, F. M., Ciais, P., Davis, S., Erickson, D., Gregg, J. S., Jacobson, A., Marland, G., Miller, J., Oda, T., Olivier, J. G. J., Raupach, M. R., Rayner, P., & Treanton, K. (2012). A synthesis of carbon dioxide emissions from fossil-fuel combustion, *Biogeosciences 9*, 1845–1871. https://doi.org/10.5194/bg-9-1845-2012

Annual Bulletin of Foreign Trade Statistics. (2020). Ministry of Planning Central Statistical Office, State of Kuwait.

Asseng, S., Zhu, Y., Wang, E., & Zhang, W. (2015). Chapter 20–Crop modeling for climate change impact and adaptation. In O. Victor & Sadras, & F. Daniel Calderini, (Eds.), *Crop physiology* (Second Edition), *Applications for genetic improvement and agronomy*, pp. 505–546.

Awazi, N. P., & Tchamba, M. N. (2019). Enhancing agricultural sustainability and productivity under changing climate conditions through improved agroforestry practices in smallholder farming systems in Sub-Saharan Africa. *African Journal of Agricultural Research, 14*(7), 379–388. https://doi.org/10.5897/AJAR2018.12972

Bado, B. V., Whitbread, A., & Sanoussi Manzo, M. L. (2021). Improving agricultural productivity using agroforestry systems: Performance of millet, cowpea, and ziziphus-based cropping systems in West Africa Sahel. *Agriculture, Ecosystems and Environment, 305*, Article 107175. https://doi.org/10.1016/j.agee.2020.107175

Béliveau, A., Lucotte, M., Davidson, R., Paquet, S., Mertens, F., Passos, C. J., & Romana, C. A. (2017). Reduction of soil erosion and mercury losses in agroforestry systems compared to forests and cultivated fields in the Brazilian Amazon. *Journal of Environmental Management, 203*(1), 522–532. https://doi.org/10.1016/j.jenvman.2017.07.037

Bhagwat, S. A., Willis, K. J., Birks, H. J. B., & Whittaker, R. J. (2008). Agroforestry: A refuge for tropical biodiversity? *Trends in Ecology and Evolution, 23*(5), 261–267.

Buerkert, A., Dix, B. A., Al Rawahi, M. N., & Schlecht, E. (2021). Agro-ecological landuse transformation in oasis systems of Al Jabal Al Akhdar, northern Oman. *Scientific Reports, 11*, 7709. https://doi.org/10.1038/s41598-021-85515-9

Bulliet, R. W., Crossley, P. K., Headrick, D. R., Johnson, L. L., & Hirsch, S. W. (2008). *The Earth and its peoples: A Global History* (Vol. I). Houghton Mifflin.

Campos, P., Daly-Hassen, H., Ovando, P., Chebil, A., & Oviedo, J. L. (2009). Economics of multiple use cork oak woodland: Two case studies of agroforestry systems. In A. RigueiroRodriguez, J. McAdam, & M. R. Mosquera-Losada (Eds.), *Agrofrestry in Europe: Current status and future prospects*. Springer.

Catacutan, D. C, van Noordwijk, M., Nguyen, T.H., Öborn, I., & Mercado, A.R. (2017). Agroforestry: Contribution to food security and climate-change adaptation and mitigation in Southeast Asia. White Paper. Bogor, Indonesia: World Agroforestry Centre (ICRAF) Southeast Asia Regional Program; Jakarta, Indonesia: *ASEAN-Swiss Partnership on Social Forestry and Climate Change.*

Choi, J., Wehde, W., & Maulik, R. (2022). Politics of problem definition: Comparing public support of climate change mitigation policies using machine learning. *Review of Policy Research, 41*(1), 104–134. https://doi.org/10.1111/ropr.12523

Cowie, A. L., Orr, B. J., Castillo Sanchez, V. M, Chasek, P., Crossman, N. D., Erlewein, A., Louwagie, G., Maron, M., Metternicht, G. I., Minelli, S., Tengberg, A., Walter, S., & Welton, S. (2018). Land in balance: The scientific conceptual framework for land degradation neutrality. *Environmental Science and Policy, 79*, 25–35. ISSN 1462–9011. https://doi.org/10.1016/j.envsci.2017.10.011

Dagar, J. C. (1995). Characteristics of halophyte vegetation in India. In: M. A. Khan, & I. A. Ungar (Eds.), *Biology of salt tolerant plants*, pp. 255–276. University of Karachi, Pakistan.

Dagar, J. C., Pandey, C. B., & Chaturvedi, C. S. (2014). Agroforestry: a way forward for sustaining fragile coastal and island agro-ecosystems. In Agroforestry systems in India: livelihood security & ecosystem services. In J. C. Dagar, A. K. Singh, & A. Arunachalam (Eds.), *Advances in agroforestry*, vol. 10. Springer, pp. 185–232.

d'Alpoim Guedes, J. A., Lu, H., Hein, A. M., & Schmidt, A. H. (2015). Early evidence for the use of wheat and barley as staple crops on the margins of the Tibetan Plateau. *Proc Natl Acad Sci U S A, 112*(18), 5625–30. https://doi.org/10.1073/pnas.1423708112

Eddy, W. C, & Yang, W. H. (2022). Improvements in soil health and soil carbon sequestration by an agroforestry for food production system. *Agriculture, Ecosystems & Environment, 333*. https://doi.org/10.1016/j.agee.2022.107945

El-Ghani, M. A., Huerta-Martinez, F. M., Liu, H., & Qureshi, R. (2017). Plant responses to hyperarid desert environments. Edition: 1st Publisher: Springer. https://doi.org/10.1007/978-3-319-59135-3.. ISBN: 978-3-319-59134-6.

Fahad, S., Chavan, S. B., Chichaghare, A. R., Uthappa, A. R., Kumar, M., Kakade, V., Pradhan, A., Jinger, D., Rawale, G., Yadav, D. K., Kumar, V., Farooq, T. H., Ali, B., Sawant, A. V., Saud, S., Chen, S., & Poczai, P. (2022). Agroforestry systems for soil health improvement and maintenance. *Sustainability, 14*(22), 14877. https://doi.org/10.3390/su142214877

Fahey, D. W, Doherty, S. J, Hibbard, K. A., Romanou, A., Taylor, P. C. (2017). Physical drivers of climate change. In Climate science special report: Fourth national climate assessment. In D. J. Wuebbles, D. W. Fahey, K. A. Hibbard, D. J. Dokken, B. C. Stewart, & T. K. Maycock (Eds.), *U.S. global change research program* (vol. 1, p. 80). Washington, DC, Figure 2.4. https://doi.org/10.7930/J0513WCR

FAO. (1999). Poverty alleviation and food security in Asia. In *Land degradation*, Chapter 3. *Food and Agricultural Organization of the United Nations (FAO).* Accessed 18 Oct 2020.

FAO. (2009). How to feed the world in 2050. https://www.fao.org/fileadmin/templates/wsfs/docs/expert_paper/How_to_Feed_the_World_in_2050.pdf

Friedlingstein, P., Houghton, R., Marland, G., Hackler, T. A., Boden, T. J., Conway, J. G., Canadell, M. R., Raupach, P., Ciais & Le Quéré, C. Update on CO_2 emissions. *Nature Geoscience, 3*, 811–812. https://doi.org/10.1038/ngeo1022

Fuglie, K. O., Morgan, S., & Jelliffe. (2024). World agricultural production, resource use, and productivity 1960–2020. Economic Research Service, USDA, February 2024. *Economic Information Bulletin*, No. (EIB-268), p. 61.

Ghosal, A. (2021). The shifting of natural farming to chemical farming in Indian agriculture. *Just Agriculture, 1*(9), e-ISSN 2582–8223, Article ID # 033.

Ghosh, S. K & Ghosh, B. K. (2020). Fossil fuel consumption trend and global warming scenario: energy overview. *Global Journal of Engineering Sciences 5*(2), 1–5. https://doi.org/10.33552/GJES.2020.05.000606.. ISSN: 2641-2039.

GLASOD. (1990). Global assessment of human-induced soil degradation (GLASOD) ISRIC–World soil information—Godert van Lynden (Deputy Director), PO Box 353, Wageningen, 6700AJ, Netherlands.

Gold, M. A., & Garrett, H. E. (2009). Agroforestry nomenclature, concepts and practices. In H. E. "Gene" Garrett (Eds.), *North American agroforestry: An integrated science and practice* (pp. 45–56). Madison, WI, USA. https://doi.org/10.2134/2009.northamericanagroforestry.2ed.c3

Gupta, R. K., Kumar, V., Sharma, K. R., Buttar, T. S., Singh, G., & Mir, G. (2017). Carbon sequestration potential through agroforestry: A review. *International Journal of Current Microbiology and Applied Science, 6*(8), 211–220. https://doi.org/10.20546/ijcmas.2016.501.029

Gupta, S. R., Dagar, J. C., & Teketay, D. (2020). Agroforestry for rehabilitation of degraded landscapes: achieving livelihood and environmental security. In J. C. Dagar, S. R. Gupta, & D. Teketay (Eds.), *Agroforestry for degraded landscapes. Recent advances and emerging challenges,* vol. 1. Springer.

Harsh, L. N., Khan, H. A., Bohra, M. D., & Tewari, J. C. (2003). Growth performance and gum production of Acacia Senegal in different landforms in hot arid zone of India. *Journal of Non-Timber Forest Products, 10*, 1–7.

Harvey, C. A., & Gonzalez-Villalobos, J. A. (2007). Agroforestry systems conserve species-rich but modified assemblages of tropical birds and bats. *Biodiversity and Conservation, 16*, 2257–2292.

Heuperman, A. F., Kapoor, A. S., & Denecke, H. W. (2002). Biodrainage-principles, experiences and applications. *International Programme for Technology and Research in Irrigation and Drainage (IPTRID) Secretariat, Food and Agriculture Organization of the United Nations (FAO)*, Italy, p. 79.

Hsiao, S., Fu, H., Wu, H., Liang, T., Chang, C., Chen, Y., Lin, L., & Chen, W. (2024) Impact assessment of sea level rise-induced high tide flooding and socioeconomic losses in a highly vulnerable coastal region. *Journal of Hydrology: Regional Studies, 55*, 101921. https://doi.org/10.1016/j.ejrh.2024.101921. ISSN 2214-5818

IPCC. (2013). Climate change 2013: The physical science basis. In T. F. Stocker, D. G. -K. Qin, M. Plattner, S. K. Tignor, J. Allen, A. Boschung, Y. Nauels, V. Xia, Bex, & P.M. Midgley (Eds.), *Working Group, I contribution to the fifth assessment report of the intergovernmental panel on climate change*. Cambridge University Press, p. 869.

IRENA. (2019). Sustainable harvest: Bioenergy potential from agroforestry and nitrogen-fixing wood crops in Africa, International Renewable Energy Agency, Abu Dhabi.

Islam, M. A., Suleiman, M. K., Quoreshi, A. M., Jallow, M., Abbas, S., AbdulMalik, N., Al-Mulla, L., Sivadasan, M. T., Jacob, S., Thomas, B., Manuvel, A. J., & Thomas, R. R. (2022). Performance and feed quality evaluation of selected forage species in mixed cropping agroforestry systems. Final Report, FA146C, KISR No. 17422, Kuwait Institute for Scientific Research (KISR), Kuwait.

Jose, S. (2009). Agroforestry for ecosystem services and environmental benefits: An overview. *Agroforestry Systems, 76*, 1–10.

Kavia, Z. D., & Harsh, L. N. (1991). Proven technologies of sand-dune stabilisation. In A. P. Dwivedi, & G. N. Gupta (Eds.), *Afforestation of arid lands. Scientific Publishers.*

Khamzina, A., Lamers, J. P. A., Martius, C., Worbes, M., & Vlek, P. L. G. (2006). Potential of nine multipurpose tree species to reduce saline groundwater tables in the lower Amu Darya River region of Uzbekistan. *Agroforestry SystEms.* https://doi.org/10.1007/s10457-006-9006-9

Köne, A. C., & Büke, T. (2010). Forecasting of CO_2 emissions from fuel combustion using trend analysis. *Renewable and Sustainable Energy Reviews, 14*(9), 2906–2915.

Kouyate, D., Abohassan, R. A., & Elfeel, A. A. (2022). Does arid land agroforestry contribute to achieving the Saudi green initiative goals? A review. *International Journal of Environment and Climate Change, 12*(11), 2707–2719. https://doi.org/10.9734/IJECC/2022/v12i1131259

Kraaijenbrink, P., Bierkens, M., Lutz, A., & Immerzeel, W. W. (2017). Impact of a global temperature rise of 1.5 degrees Celsius on Asia's glaciers. *Nature, 549*, 257–260. https://doi.org/10.1038/nature23878

Krishnamurthy, L., Krishnamurthy, K. P., Rajagopal, I., & Solares, A. P. (2017). Can agroforestry systems thrive in the drylands? Characteristics of successful agroforestry systems in the arid and semi-arid regions of Latin America. *Agroforestry Systems, 93*(1). https://doi.org/10.1007/s10457-017-0143-0

Kuyah, S., Whitney, C. W., Jonsson, M., Sileshi, G. W., Öborn, I., Muthuri, C. W., & Luedeling, E. (2019). Agroforestry delivers a win-win solution for ecosystem services in sub-Saharan Africa. *A Meta-Analysis. Agronomy Sustainable Development, 39*, 47. https://doi.org/10.1007/s13593-019-0589-8

Lehmann, J., Peter, I., Steglich, C., Gebauer, G., Huwe, B., & Zech, W. (1998). Below-ground interactions in dryland agroforestry. *Forest Ecology Management, 111*(2), 157–169.

Lindsey, R. (2024). Climate change: Atmospheric carbon dioxide based on annual report from National Oceanic and Atmospheric Administration Global Monitoring Lab. https://www.climate.gov/news-features/understanding-climate/climate-change-atmospheric-carbon-dioxide

Lorenz, K., & Lal, R. (2014). Soil organic C sequestration in agroforestry systems. *A Review. Agronomy for Sustainable Development, 34*, 443–454. https://doi.org/10.1007/s13593-014-0212-y

Madouh, T., & Al-Sabbagh, T. (2020). Enhancement of native forage species production as local feed resources for livestock in Kuwait-Phase II, Final Report FA130C, KISR No 16353, Kuwait.

Malezieux, E., Crozat, Y., Dupraz, C., Laurans, M., Makowski, D., Ozier-Lafontaine, H., Rapidel, B., De Tourdonnet, S., & Valantin-Morison, M. (2009). Mixing plant species in cropping system: Concepts, tools and models. A review. *Agronomy for Sustainable Development, 29*(1), 43–62.

Mandal, A. K., Sharma, R. C., Singh, G., & Dagar, J. C. (2010). Computerized database on salt affected soils in India. *Technical bulletin, 2.* Central Soil Salinity Research Institute, Karnal, India, p. 28.

Marion, A. (2015). Impacts of climate change on crop production. A case study of Tisai island, Ongino subcounty, Kumi district. *Dissertation School of Engineering and Applied Science.*

Mathur, B. K., Singh, J. P., Beniwal, R. K., & Singh, N. P. (2007). Utilization of salty shrub-lani (Salsola baryosma) of arid region as drought feed for goats. In *Abstracts of international tropical animal nutrition conference*, vol. 2, p. 33. National Dairy Research Institute, Karnal, India.

McNeely, J. A., & Schroth, G. (2006). Agroforestry and biodiversity conservation - traditional practices, present dynamics, and lessons for the future. *Biodiversity and Conservation, 15*, 549–554.

Ministry of Planning. (2020). Annual Bulletin of foreign trade statistics, Ministry of Planning, Kuwait.

Monteith, J., Ong, C., & Corlett, J. (1991). Microclimate interactions in agroforestry systems. *Ecology Management, 45*, 31–44. https://doi.org/10.1016/0378-1127(91)90212-E

Montgomery, D. (2008). *Dirt: The erosion of civilizations.* University of California Press.

Mullan, D. J., & Barrett-L. (2010). Breeding crops for tolerance to salinity, waterlogging and inundation. *Climate Change and Crop Production*, 92–114. https://doi.org/10.1079/9781845936334.0092

Mutuo, P. K., Cadisch, G., Alain, A., Palm, C. A., & Verchot, L. (2005). Potential of agroforestry for carbon sequestration and mitigation of greenhouse gas emissions from soils in the tropics. *Nutrient Cycling in Agroecosystems, 71*(1), 43–54.

Nair, P. K. R. (2012). Carbon sequestration studies in agroforestry systems: A reality-check. *Agroforestry System, 86*, 243–253. https://doi.org/10.1007/s10457-011-9434-z

Nair, P. K. R., Gordon, A. M., & Rosa Mosquera-Losada, M. (2008). Agroforestry. In S. E. Jørgensen, & B. D. Fath (Eds.), *Encyclopedia of ecology* (pp. 101–110). Academic Press. https://doi.org/10.1016/B978-008045405-4.00038-0

Nair, P. K. R., Kumar, B. M., & Nair, V. D. (2021). Global distribution of agroforestry systems. In An Introduction to agroforestry, four decades of scientific developments, Second Edition. Springer Nature. https://doi.org/10.1007/978-3-030-75358-0_4

Oke, D. O., & Odebiyi, K. A. (2007). Traditional cocoa-based agroforestry and forest species conservation in Ondo State, Nigeria. *Agriculture, Ecosystems and Environment, 122*, 305–311.

Pan, J., Chen, S., He, D., Zhou, H., Ning, K., Ma, N., Li, K., Liao, D., Mi, W., Wu, Q., Zhang, C., & Dong, Z. (2025). Agroforestry increases soil carbon sequestration, especially in arid areas: A global meta-analysis, *CATENA, 249*. https://doi.org/10.1016/j.catena.2024.108667

Pascual, M., & Bouma, M. J. (2009). Do rising temperature matter? *Ecology, 90*(40), 906–912.

Prăvălie, R., Patriche, C., Borrelli, P., Panagos, P., Roşca, B., Dumitraşcu, M., Nita, I., Săvulescu, I., Birsan, M., & Bandoc, G. (2021). Arable lands under the pressure of multiple land degradation processes. A global perspective, *Environmental Research, 194*, 110697. https://doi.org/10.1016/j.envres.2020.110697. ISSN 0013-9351.

Quadrelli, R., & Peterson, S. (2007). The energy–climate challenge: Recent trends in CO_2 emissions from fuel combustion. *Energy Policy, 35*(11), 5938–5952.

Ramert, B., Lennartsson, M., & Davies, G. (2002). The use of mixed species cropping to manage pests and diseases–theory and practice. In *Proceedings of the UK Organic Research Conference,* Organic Centre Wales, Institute of Rural Studies (pp. 207–210). University of Wales Aberystwyth, UK.

Rengasamy, P. (2006). World salinization with emphasis on Australia. *Journal of Experimental Botany, 57*(5), 1017–1023. https://doi.org/10.1093/jxb/erj108

Rosenstock, T. S., Dawson, I. K., Aynekulu, E., Chomba, S., Degrande, A., Fornace, K., Jamnadass, R., Kimaro, A., Kindt, R., Lamanna, C., Malesu, M., Mausch, K., McMullin, S., Murage, P., Namoi, N., Njenga, M., Nyoka, I., Paez Valencia, A. M., Sola, P., & Steward, P. (2019). A planetary health perspective on agroforestry in sub-Saharan Africa. *One Earth, 1*, 330–344.

Ruruh, A., Suma, Z. N. Y., & Usman, M. (2024). Carbon sequestration in agroforestry: A systematic review of global practices. *International Journal of Multidisciplinary Approach Sciences and Technologies, 1*(3), 158–168.

Santoro, A. (2023). Traditional oases in North Africa as multifunctional agroforestry systems: A systematic literature review of the provided ecosystem services and of the main vulnerabilities. *Agroforestry Systems, 97*, 81–96. https://doi.org/10.1007/s10457-022-00789-w

Scherr, S. J. (2012). The role of extension in agroforestry development: Evidence from Kenya. *Agroforestry Systems, 18*, 47–68.

Shahid, S. A., & Alkandari, A. J. (2024). Halophytic crops as a solution for food security, land rehabilitation, and mitigating future water crises by utilizing marginal quality waters. In J. C. Dagar, S.R. Gupta, & A. Kumar (Eds.), *Halophytes vis-à-vis Saline Agriculture.* Springer Nature, pp. 441–478. https://doi.org/10.1007/978-981-97-3157-2_17

Sileshi, G., Akinnifesi, F. K., Ajayi, O. C., & Place, F. (2008). Meta-analysis of maize yield response to woody and herbaceous legumes in sub-Saharan Africa. *Plant and Soil, 307*, 1–19.

Singh, J. P., Mathur, B. K., & Rathore, V. S. (2009). Fodder potential of Lana (Haloxylon salicornicum) in hot arid region. *Range Management and Agroforestry, 30*, 34–37.

Sivakumar, M. V. K. (2018). Chapter 23: Climate extremes and impacts on agriculture, In L. Jerry, Hatfield, V. K. Mannava, H. Sivakumar, John & Prueger Book (Eds.), *Agro climatology: Linking agriculture to climate*, vol. 60.

Smith, J. (2010). Agroforestry: reconciling productivity with protection of the environment. *A synopsis of research literature; The Organic Research Centre*. http://orgprints.org/18172/1/Agroforestry_synopsis.pdf

Somarriba, E., Beer, J., Alegre-Orihuela, J., Andrade, H. J., Cerda, R., DeClerck, F., Detlefsen, G., Escalante, M., Giraldo, L. A, Ibrahim, M., Krishnamurthy, L., Mena-Mosquera, V. E., Mora-Degado, J. R., Orozco, L., Scheelje, M., & Campos, J. J. (2012). Mainstreaming agroforestry in Latin America. In Agroforestry–the future of global land use. In P. Nair, & D. Garrity (Eds.), *Advances in agroforestry,* vol. 9. Springer. https://doi.org/10.1007/978-94-007-4676-3_21

Trenberth, K. E. (2011). Changes in precipitation with climate change. *Climate Services for Sustainable Development, 47*, 123–138.

UNCCD. (2024). Land degradation neutrality. Great Wall Initiative. Evaluation Reports; Independent review of the Great Green Wall accelerator. Accessed 23 Dec 2024. https://www.unccd.int/our-work/ggwi

Vandermeer, J. H. (1989). *The ecology of intercropping*. Cambridge University Press. https://doi.org/10.1017/CB09780511623523

Verchot, L. V., Noordwijk, M. V., Kandji, S., Tomich, T., Ong, C., Albrecht, A., Mackensen, J., Bantilan, C., Anupama, K. V., & Palm, C. (2007). Climate change: Linking adaptation and mitigation through agroforestry. *Mitigation and Adaptation Strategies for Global Change, 12*, 901–918.

Waldron, A., Justicia, R., & Smith, L. E. (2015). Making biodiversity-friendly cocoa pay: Combining yield, certification, and REDD for shade management. *Ecological Applications, 25*, 361–372.

Yoro, K. O., Daramola, M. O. (2020). Chapter 1-CO_2 emission sources, greenhouse gases, and the global warming effect. *Advances in carbon capture methods, technologies and applications.* Woodhead Publishing, pp. 3–28. https://doi.org/10.1016/B978-0-12-819657-1.00001-3. ISBN 9780128196571.

Zeratsion, B. T., Manaye, A., Gufi, Y., Tesfaye, M., Werku, A., & Anjulo, A. (2023). Agroforestry practices for climate change adaptation and livelihood resilience in drylands of Ethiopia. *Forest Science and Technology, 20*(1), 47–57. https://doi.org/10.1080/21580103.2023.2292171

Zhaohua, Z., Matang, C., Shiji, W., & Youxu, J. (1991). Agroforestry systems in China. Chinese Academy of Forestry, *People's Republic of China and International Development Research Center (IDRC), Canada.* Chinese Academy of Forestry, Beijing, China. http://hdl.handle.net/10625/11391

Zignol, F., Kjellström, E., Hylander, K., Ayalew, B., Zewdie, B., Rodríguez-Gijón, A., & Tack, A. J. M. (2023). The understory microclimate in agroforestry now and in the future–a case study of Arabica coffee in its native range. *Agricultural and Forest Meteorology, 340*, Article 109586. https://doi.org/10.1016/j.agrformet.2023.109586

Chapter 12
Economic Impacts Attributable to Climate Change on Arid Lands Agriculture and Food Security—Policy Development

Adil Naseeb, Ahmed Al-Khayat, Shaima Al-Shamali, Nora Abdulmalik, May Al-Asfour, Abdullah F. Al-Jaber, and Mariam S. Behbehani

Abstract After World War II, developed countries faced food security and economic collapses, leading to the development of the agricultural sector as a solution. Agriculture Economic Policies (AEP) were established to manage increasing food demands and population growth. These agricultural policies were highly successful for the developed countries, resulting in a prosperous economic growth for many decades. Therefore, many developing countries copied those AEP's and implemented them in their agricultural sectors. However, in the 1980s, the agricultural success of AEP for developed countries became an agricultural failure for many developing countries due to several reasons. The first was the insufficient choices and implementations conducted by the unexperienced government officials. The second was the static model designed specifically for the developed countries. The third was the agricultural environment's negative conditions, such as arid lands and climate changes. All

A. Naseeb (✉) · A. Al-Khayat · S. Al-Shamali · N. Abdulmalik · M. Al-Asfour · A. F. Al-Jaber · M. S. Behbehani
Techno-Economic Division, Kuwait Institute for Scientific Research, Safat, Kuwait
e-mail: ajnaseeb@kisr.edu.kw

A. Al-Khayat
e-mail: akhayat@kisr.edu.kw

S. Al-Shamali
e-mail: sshamali@kisr.edu.kw

N. Abdulmalik
e-mail: nmalik@kisr.edu.kw

M. Al-Asfour
e-mail: masfour@kisr.edu.kw

A. F. Al-Jaber
e-mail: ajaber@kisr.edu.kw

M. S. Behbehani
e-mail: mbehbehani@kisr.edu.kw

M. K. Suleiman and S. A. Shahid (eds.), *Fostering Arid Lands Agriculture in the Face of Climate Change*, Advances in Global Change Research 81,
https://doi.org/10.1007/978-3-031-98490-7_12

leading to the demise of the agricultural sector in the developing countries. Therefore, in the twenty-first century, many developing countries have developed a new approach using Dynamic Agricultural Economic Policy (DAEP) suitable for their agricultural, economic, and environmental conditions. This dynamic and customizable approach can be applied to any developing country based on its environmental and economic conditions. For example, Qatar's DAEP focused on education and innovation, while African regions focused on productivity and irrigation.

Keywords Developed countries · Developing countries · Agricultural policies · Agriculture economic policies · Dynamic agricultural economic policy

12.1 Introduction

After world war II, developed countries have been severely damaged from the negative impacts of the war. The majority of the countries have serious food security challenges as well as economic collapses. One of the main solutions to address World War II negative impacts is the development of the agricultural sector. This sector will alleviate the food security challenges and provides a strong platform for economic growth for many of the developed countries. Therefore, over several decades the developed countries have established an Agriculture Economic Policies (AEP) model, to manage the ever-increasing food demands and economic growth. As a result, in the sixties of the last century, agricultural commodities around the world were in surplus, because of the applied AEP's model in the agricultural sectors of the developed countries. Therefore, the developing countries have replicated that agricultural model as a benchmark for their agricultural independence and economic growth. This agricultural policy model, has successfully worked for the developing countries for several decades, but in the 80's of the last century an alteration in the agricultural outputs became noticeable. By way of explanation, the agricultural success of applying the model for the developed countries, has become an agricultural failure to a majority of the developing countries considering various external and internal factors (FAOC, 1957, 1960, 1983).

In addition, it was discovered that the primary cause responsible for the downfall of many of the agricultural sectors in the developing countries, is due to insufficient choices and implementations of agricultural policies conducted by the governments. These unwise choices and implementations are because of the selection of a static model that was designed specifically for the developed countries and not the developing countries and their unique agricultural environment (Arid Lands). Additionally, at the end of the last century, several problematic issues emerged which cause even further demise of the agricultural sector of the developing countries. These additional problematic issues, are climate change and their negative impacts on the agricultural sector as well as the food security challenges of many developing countries. Therefore, it was obvious that the conventional AEP's model was not effective for the developing countries, and a new approach was required (FAOC, 1983, 2000; Speece, 1997; El-Beltagy & Madkour, 2012).

Moreover, in the beginning of twenty-first century, many of the developing countries with the support of world organizations, have developed a new agricultural model that utilizes Dynamic Agricultural Economic Policy (DAEP) suitable for their agricultural, economic, and environmental conditions. This new agricultural model can be tailored to each country condition in order to increase or sustain its productivity to address the food security challenges, mitigate the climate change negative effects and improve disposable income of the farmer. Some countries have successfully applied this new approach and changed its dependency on agricultural imports into an exporting country such as the State of Qatar (Abubakar et al., 2021; FAO, 2024a; El-Beltagy & Madkour, 2012).

Therefore, the objective of this chapter is to illustrate the historical background of the AEPs used in the last centuries, as well as the conventional AEPs that were utilized in many of the developing countries in their agricultural sectors. Additionally, it is prudent to display the problematic issues that arose from implementing the static AEPs in the developing countries. Likewise, it is important as well to demonstrate the new AEPs approach used in the developing countries to address the food security challenges, climate change negative impacts, and economic growth in their agricultural sectors (El-Beltagy & Madkour, 2012).

12.2 Historical Background

Many consider the era of the sixties in the last century was the golden age of agriculture, as a result of the agricultural boom in many of the world's developed and developing countries. In fact, the agricultural sector in some of the developing countries is an essential sector, where it employs around two-thirds of the work force and generates around 50% of its Gross Domestic Product (GDP). Therefore, agricultural policies in that era were kept at minimum, and the utilization of the conventional policies methods was applied in many developing countries considering its successful implementation (FAOC, 1957, 1960; Ray, 1986; Abubakar et al., 2021; OECD, 2012).

However, in the period from 1970 to 1980, there were some disproportions in the agricultural sector of the world. A shift in income occurred between developed and developing countries, because of the unified agricultural policy implementations in different countries. This shift made the developed countries gain more yield in their agricultural commodities than the developing countries. Having said that, during the early years of the 1980's, some of the developing countries, in particular the African region have increased their agricultural yields. This outcome occurred because of two events, the first being the participation in the General Agreement on Tariffs and Trade (GATT) which facilitated international trade. The second event is because of the encouragement of some governments to change some of their conventional agricultural policies to tailor their needs (FAOC, 1983; Ray, 1986; WTO, 2024).

Furthermore, in the 1990's, two main calamities occurred, that affected the world's economy and overall safety. The first is the economic recession, and the second is the Gulf crisis, both of which contributed to affecting agricultural productivity in many regions, especially in developing countries. In addition, many agricultural policy-makers made some insufficient decisions, such as the reduction of subsidies in their agricultural sector as an attempt to reduce governmental expenditure. These agricultural policy adjustments aided the reduction of agricultural productivity even further in many developing countries. Therefore, a world concern has risen in different world organizations, such as the Food and Agriculture Organization (FAO) of the United Nations, regarding food security, the livelihood of rural farmers, and some aspects of environmental concern (FAOC, 1991; Kose et al., 2020).

Moreover, in this current century, agricultural countries around the world in association with world organizations such as the World Bank and FAO, have altered their focus towards the food supply chain and climate change. This is the result of many world predicaments that affected the agricultural sectors such as world economic crisis in 2008, COVID 19, Ukraine War, and other economic and environmental impacts. Therefore, to strengthen the world supply chain, many policies around the world focused on increasing agricultural productivity, sustainability, and accessibility. In addition, many agricultural policies have focused on mitigating the impacts of climate change on the farmland, as well as improving the livelihood of farmers, especially in the rural areas (FAOC, 2000, 2023; ERS-USDA, 2009, 2011; El-Beltagy & Madkour, 2012; WHO, 2024; Agriseta, 2024; Malhotra, 2024).

12.3 Conventional AEP's Approaches Applied in the Last Century

Figure 12.1 illustrates eight of the most Conventional AEP approaches utilized in the past century by many developed and developing countries. The First AEP, that is widely used by a majority of countries is *Irrigation*, where the main focus of this AEP is to develop and sustain the agricultural sector productivity, based on the countries environmental conditions. For example, Egypt in the middle of the last century, has initiated an irrigation policy that supervised the development of a water dam to be utilized in agricultural land irrigation process supporting the development of its agricultural sector (FAOC, 1957, 1983; Speece, 1997).

The Second AEP is *Technology*, consisting of *state-of-the-art* equipment that provided an efficient method of production, which in return lowers the overall cost in the agricultural sector. For instance, developed countries such as the United States of America (USA) promoted and supported the utilization of an innovative irrigation technology in their farms since the 1950's till 1990's. The applied irrigation system is the sprinkler technology, which has the ability to efficiently irrigate the land, lower irrigation cost, and boost productivity (FAOC, 1957, 1983, 1991; Evett et al., 2020; Ellis et al., 2024).

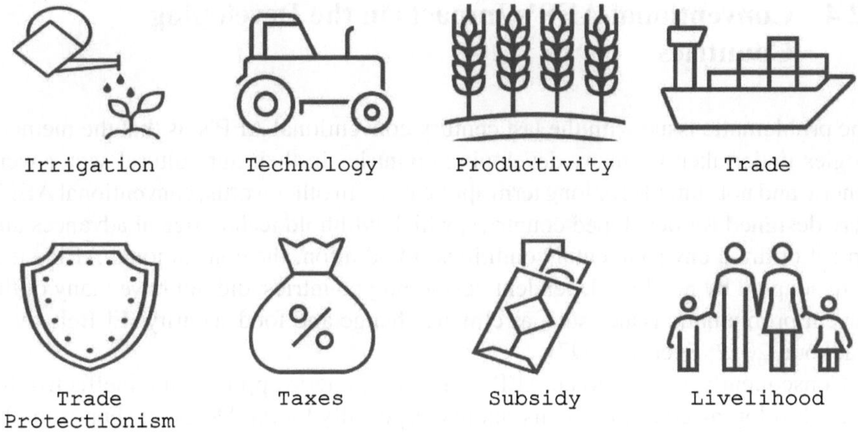

Fig. 12.1 Conventional agricultural economic policies utilized in the past century

The Third AEP is *Productivity*, the aim of this policy is to increase the productivity in order to supply the ever growing local and international demand of agricultural products. Additionally, another main purpose of this AEP is to increase the yield of the local farmer's farm land as well as the country's economy. The Fourth AEP is *International Trade*, here the decision makers developed the agricultural policies to facilitate exports of their agricultural commodities internationally. This process is considered to be an essential, because it will lead to increase production as well as economic growth and food security sustainability (FAOC, 1957, 1983, 1991; Ellis et al., 2024).

The Fifth AEP is *Trade Protectionism*, the objective here is the utilization of different trade protection tools such as tariffs, quotas, price margins, or any other governmental instrument to ensure market accessibility, sustainability, and economic growth. The Sixth AEP is *Taxes*, which is considered to be an important source of income in many developed and developing countries. Consequently, many governments apply taxes on their agricultural sector, as a source of additional income. In particular, the developing countries, since their agricultural productivity share in the GDP represents up to 50% in some countries (FAOC, 1957, 1983, 1991; OECD, 2012; Ellis et al., 2024; World Bank Group, 2024a; de Melo, 1979).

The Seventh AEP is *Subsidies*, where the purpose is to provide support to the agricultural sector for their commodities in order to sustain their production and maintain their access to the local market. For example, in the State of Kuwait local farmers are provided with subsidies by the Public Authority for Agriculture Affairs and Fish Resources (PAAFR) on their agricultural products, to support their productivity in Kuwait's harsh environment, as well as local market accessibility for their products. The eighth AEP is *Livelihood*, where the goal is to develop policies that sustain the livelihood of the rural farmers, maintain their lifestyles for future production sustainability (FAOC, 1957, 1983; de Melo, 1979; Ellis et al., 2024; Mesferlaw, 2024).

12.4 Conventional AEP's Impacts in the Developing Countries

The problematic issue with the last century conventional AEP's, is that the methodologies undertaken by many developing countries in their agricultural sector were generic and not suitable for long term application. In other words, conventional AEP's were designed for developed countries, which withhold technological advances and almost optimal environmental conditions. In addition, the conventional AEP's that were adapted by newly independent developing countries, did not have many of the current problematic issues such as climate change and food security (El-Beltagy & Madkour, 2012; Speece, 1997).

Consequently, conventional AEP's and their generic approach are ineffective for many developing countries in this century, especially for Arid Regions. For example, Egypt policy makers in the mid of the last century favored a short-term approach in solving their agricultural irrigation problem, by building the Aswan High Water Dam. Although the AEP approach was successful for many decades, in the long-run within this century it was revealed that it is very harmful to the environment of the land. This environmental damage was the result of extreme waterlogging in some areas which led to soil salinity in Egypt agricultural land. Additionally, the damage also included excessive water seepage losses in other areas from conventional irrigation methods which led to a shortage of water supply. Likewise, in Pakistan to solve their irrigation problems, they utilized the Indus river and its tributaries for agricultural land irrigation. This AEP utilized by the policy makers in Pakistan led to waterlogging and soil salinization to a major parts of the land (Speece, 1997; Zaman et al., 2018).

While in the Gulf region particularly, population growth led to a surge in demand for agricultural products. To meet this local demand surge, more land was cultivated which resulted in the excessive use of groundwater reservoirs. This excessive use for groundwater led to an increase in soil salinity. This problematic issue jeopardizes the production process of agricultural farming land, forcing the Gulf countries to use alternative methods in irrigation to mitigate the effect of soil salinity (Speece, 1997; Karanisa et al., 2021; GCC, 2024; Zaman et al., 2018).

In the case of the African region, many countries in the sub-Saharan depend on their agricultural export as a means of income. In fact, in some African countries the share of agriculture in their GDP is around 50%. Therefore, the government of the African region heavily taxed the agricultural exports, as a means of income generation. Thus, over the decades this taxation limited the growth of agricultural production capacity and increases rural poverty leading to raising concern of food security by world organizations (Abubakar et al., 2021; OECD, 2012; Ray, 1986).

12.5 Dynamic AEP's in the Developing Countries

Now that we have identified the ineffective approaches of the Conventional AEP's that were implemented for the developing countries, in the last century. It is apparent, to demonstrate what the developing countries with the support of FAO have done to mitigate the effect of the current world challenges on their agricultural sector. As it can be shown in Fig. 12.2, the Agricultural sector of the Arid Developing Countries (ADC) has transformed their static AEP's into a Dynamic Agricultural Ecosystem Policies (DAEP). This transformation provided the DAEP the ability to be flexible and rigid enough to manage their agricultural problematic issues, considering each countries circumstances.

Figure 12.2 presents the DAEP set by the ADC's and FAO to alleviate climate change negative effects and food security as well as economic challenges. This DAEP was adopted by the majority of the ADC's and currently oversees eight main DAEP's as well as their sub-categories. It is possible to include more AEP's but that depends entirely on the developing country prospective and objectives. The first DAEP is *Productivity*, where the main objective here for any ADC is to increase agricultural production (supply), in order to meet agricultural demands, sustain food security as well as boosting the country's economy (El-Beltagy & Madkour, 2012; Karanisa et al., 2021; FAO, 2024a; GCC, 2024).

The second DAEP is *Market*, where it's main focus is to provide a successful ecosystem in the agricultural sector, that interlinks between the suppliers and consumers. The third DAEP is *Trade*, where it is considered one of the most important goals of each country, to provide a successful accessibility to regional and international markets, facilitating food supply and the generation of wealth (Karanisa et al., 2021; FAO, 2024a; GCC, 2024).

The fourth DAEP is *Investment*, the objective here is to facilitate the financial resources to provide the ability to increase, improve, and sustain the agricultural sector either individually or as a whole. The fifth DAEP is *Education*, where it will provide future guidance to local farmers on new effective methods of agricultural productivity and continuity (El-Beltagy & Madkour, 2012; Karanisa et al., 2021; FAO, 2024a; GCC, 2024).

Fig. 12.2 Dynamic agricultural ecosystem policies (FAO, 2024a, b) (*Source* Authors' Design based on FAOC, 2023; FAO 2024a; Karanisa et al., 2021)

The sixth DAEP is *Innovation*, where its aim is to improve and boost productivity of the agricultural sector allowing future agricultural sustainability. The seventh DAEP is *Government Support*, this provides support from the government to the farmer in different forms and approaches so that the agricultural sector can withstand its obstacles (El-Beltagy & Madkour, 2012; Karanisa et al., 2021; FAO, 2024a; GCC, 2024).

The sixth DAEP is *Innovation*, where its aim is to improve and boost productivity of the agricultural sector allowing future agricultural sustainability. The seventh DAEP is *Government Support*, this provides support from the government to the farmer in different forms and approaches so that the agricultural sector can withstand its obstacles (El-Beltagy & Madkour, 2012; Karanisa et al., 2021; FAO, 2024a; GCC, 2024).

The eighth DAEP is Farmer's *Disposable Income*, the objective here is to sustain and improve the farmer's income over the difficulties of economic surges and climate change challenges. As stated in the above sentences, the agricultural sector in many developing countries is considered a great source of income generation. Where it not just provides income to the local farmer but it employs a large number of the work force of the developing country. Therefore, in sustaining and developing this sector, many developing countries will provide a strong platform for economic growth (FAOC, 2000, 2023; OECD, 2012; UKDFID, 2018).

12.5.1 DAEP–Productivity

Figure 12.3 demonstrates the first DAEP and its sub-categories labelled *Productivity*, containing three sub-categories all aiming to increase agricultural production and economic growth. The first DAEP-Productivity sub-category is agricultural farming methods, promoting the utilization of various agricultural approaches (new or old), fitting each country's environment and economic conditions. The second DAEP-Productivity sub-category is the irrigation methods, where it also promotes the employment of the most advanced and cost-effective irrigation systems that support agricultural productivity and infrastructure. The third DAEP-Productivity sub-category is Agricultural Equipment, here it promotes the use of the most advance and efficient equipment that boost and facilitate productivity at the lowest cost possible (Karanisa et al., 2021; Hubert et al., 2022; FAO, 2024a; GCC, 2024; SSCTK, 2024).

Placing the DAEP-Productivity into prospective, an actual example is required to show how the arid countries in Africa such as Kenya have utilized the DAEP, to increase productivity within the current climate change negative effects. The governments of these arid regions started supporting and encouraging farmers through DAEP to use alternative agricultural methods to boost and increase productivity. One of those methods, is the Conservation Agricultural (CA) method, where it promotes the utilization of three simple approaches. The first approach is soil conservation,

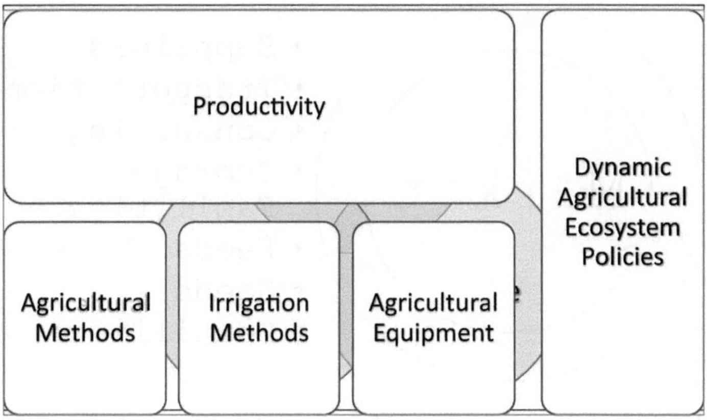

Fig. 12.3 Dynamic agricultural ecosystem policies–productivity

which aims to cover the agricultural soil using plants residue, in order to sustain moisture in the soil all year long. The second approach is minimal soil disturbance, where it aims to prevent ground turning as much as possible sustaining soil structure and moisture. The third approach is agricultural crop diversification, this approach has several beneficial outcomes, such as soil protection, insect resistance, preventing soil degradation, as well as negative climate impacts resistances. In addition, by applying this simple approach the farmer in these arid regions will have an excellent opportunity to increase his agricultural productivity and annual income. Moreover, this new agricultural approach will allow the farmers to have the ability to increase their farms productivities in an arid land by almost 30%, save irrigation water by almost 90%, and protect the soil from overuse of the pesticides. Additionally, in applying this conservation agriculture method, the government will successfully reach its desired objectives of increasing productivity, food security, and mitigating the climate change negative impacts (El-Beltagy & Madkour, 2012; FAO, 2024a; SSCTK, 2024).

12.5.2 DAEP–Market

Figure 12.4 presents the second DAEP and its sub-categories, representing the ecosystem of the local *Market*. The rural farmers of any country can only thrive when their agricultural community is homogeneous, meaning each stakeholder in the ecosystem provides a technical and economic support to each other within the local market boundaries, in order to achieve economic growth. Additionally, market accessibility is one of the key objectives for any developing arid country, enabling food sustainability and local market growth. Thus, to achieve the above objectives it is required by the market to establish a supply chain system that transports the farmers end products to the local market supplier in order to reach the end consumer. This supply chain cannot be established without the support of the government, in the

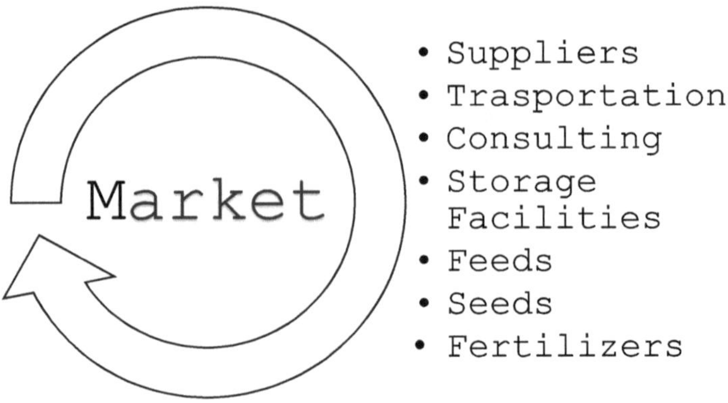

Fig. 12.4 Dynamic agricultural ecosystem policies–market

form of transportation infrastructure as well as agricultural policies. The agricultural supply chain has been successfully achieved in the State of Qatar, when it changed its overall objective to be self-sufficient in the agriculture sector, transforming the country into a producer of agricultural products (Karanisa et al., 2021; FAO, 2024a).

Additionally, fertilizers are considered key components to sustain and increase agricultural productivity in any developing country. One of the main methods to develop an organic fertilizer is composting, where agricultural residues as well as other food residues are utilized as a main source of ingredients to compose the end product. This usage of agricultural waste is a great example of market homogeneity and economic growth as well as mitigating environmental negative effect. Whereas compost is an organic fertilizer that is supplied in the market to improve soil's physical health, agricultural productivity, as well as increasing overall market yields. In addition, fertilizer-compost end product has the ability to decrease the carbon footprint in any developing country, through soil carbon sequestration, such as the case of State of Qatar (Karanisa et al., 2021; FAO, 2024a; GCC, 2024).

12.5.3 DAEP–Trade

Figure 12.5 shows the third DAEP which is *Trade* and its sub-categories, regional and international trade. The main focus of any developing country is to increase demand on its agricultural products and address the food security challenges. To illustrate, the government of Qatar as a method of expanding its agricultural exports, applied new exporting policies to facilitate the process, which resulted in the increase of its agricultural export earning around $ 9.0 Million in 2017 (OECD, 2012; Karanisa et al., 2021; FAO, 2024a; GCC, 2024).

Therefore, in applying DAEP-Trade to the developing countries, especially the arid regions, will benefit these countries in several economical and agricultural areas.

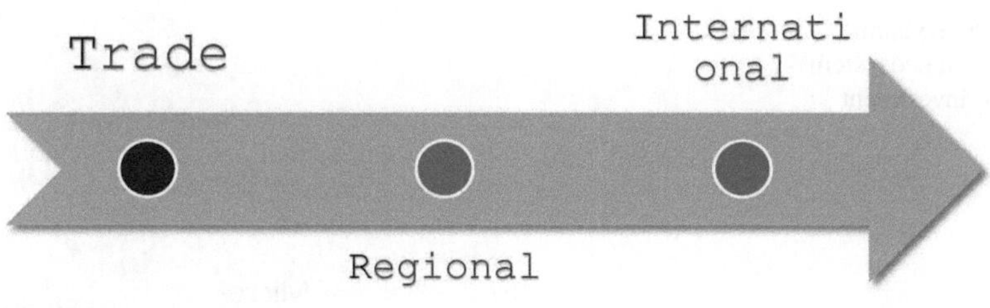

Fig. 12.5 Dynamic agricultural ecosystem policies–trade

Moreover, regional, and international Trades are an important source of mitigating the challenges of food security. This policy allows many developing countries to bridge the gaps of any shortages in their food supply, especially in time of world economic shocks or pandemic such as Covid-19. Additionally, it is considered as a great means of income generation for the rural farmer and has the ability to increase agricultural productivity of any developing country through high agricultural world demands. In addition, DAEP-Trade is a great tool in stabilizing the volatility of world agricultural commodity prices (ERS-USDA, 2009; OECD, 2012; Martin, 2017; Agriseta, 2024; FAOC, 2023; GCC, 2024; WHO, 2024).

12.5.4 DAEP–Investment

The fourth DAEP is *Investment* and its sub-categories, Finance, and Micro-Finance, as shown in Fig. 12.6. This DAEP represents a powerful instrument that facilitate production development and sustainability for agricultural sectors in developing countries. It was discovered by the world organizations and developing countries governments, that investment is a key instrument that will facilitate the reduction of negative effects caused by climate change. Moreover, investment into the agricultural sector will address food security challenges in many rural areas in the arid regions. As a method of explaining, implementing the DAEP- investment in a developing country helps the rural farmer to prevent or decrease post-harvest losses. In addition, the financial aid tool helps to build the rural farmer's infrastructure as well as adding value to his agricultural production. Moreover, this financial tool assists the rural farmer in accessing the markets (El-Beltagy & Madkour, 2012; OECD, 2012; Abubakar et al., 2021; FAO, 2024a; World Bank Group, 2024b).

Seeing the importance of this financial instrument, the World Bank in association with FAO and developing countries, has launched a financial agricultural program totaling around \$ 6.0 billion. This financial support fund comes in two forms, micro-financial loans, and agricultural financial loans. This agricultural fund will be distributed through each developing country's financial and agricultural organizations in an effort to support the local farmer and economic growth (World Bank Group, 2024c; FAO, 2024a; Malhotra, 2024).

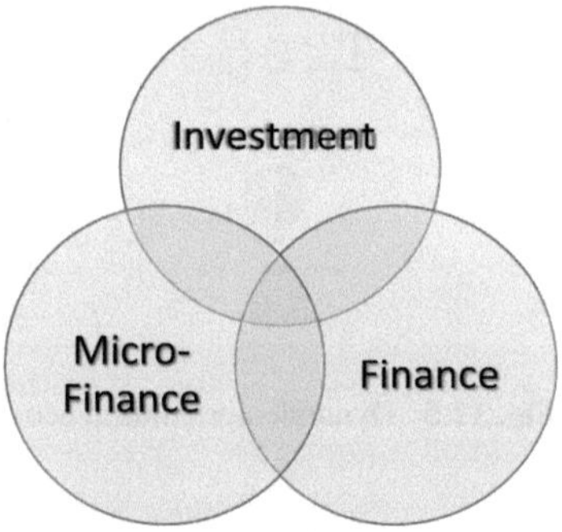

Fig. 12.6 Dynamic agricultural ecosystem policies–investment

12.5.5 DAEP–Education

Figure 12.7 exhibits the fifth DAEP of the main model entitled *Education*, containing significant sub-categories designated as Educational Program, Scientific Research, and Specialized Training. The DAEP-Education is a gateway to future development in the agricultural sector around the world, and an important tool in mitigating climate change negative effects as well as food security challenges. In many developing countries, one of the main reasons for success in their agriculture sectors is continuing the educational support system. For example, university of Qatar has established an educational programs focusing on the development of crop production including salinity tolerant crops, soil improvement, and water reuse. This educational program has opened an additional agricultural production line, utilizing the salt-tolerant plants as animal feed. This new line has the ability to increase the overall agricultural yields and at the same time preserve the consumption of agricultural water (El-Beltagy & Madkour, 2012; Karanisa et al., 2021; Agriseta, 2024; FAO, 2024a; GCC, 2024).

Furthermore, Kuwait Institute for Scientific Research (KISR) has conducted several research projects that aided the development of agricultural sector in the arid regions, such as “Controlled Environment Plant Factory” research project. Moreover, training programs provided by the government institutions or private sector, build the rural farmers technical, production, marketing, and institutional capacities leading to agricultural yield maximization (GCC, 2011; OECD, 2012; Agriseta, 2024; FAO, 2024a; GCC, 2024; KISR, 2024).

12.5.6 DAEP–Innovation

Figure 12.8 presents the sixth DAEP and its sub-categories, which is the *Innovation* of new technologies and climate smart agricultural (CSA) methods. This DAEP has a significant importance for the agricultural sector in the developing countries. Countries that will apply this DAEP will have the futuristic means to increase their agricultural production capacity and address the climate change challenges. As a way

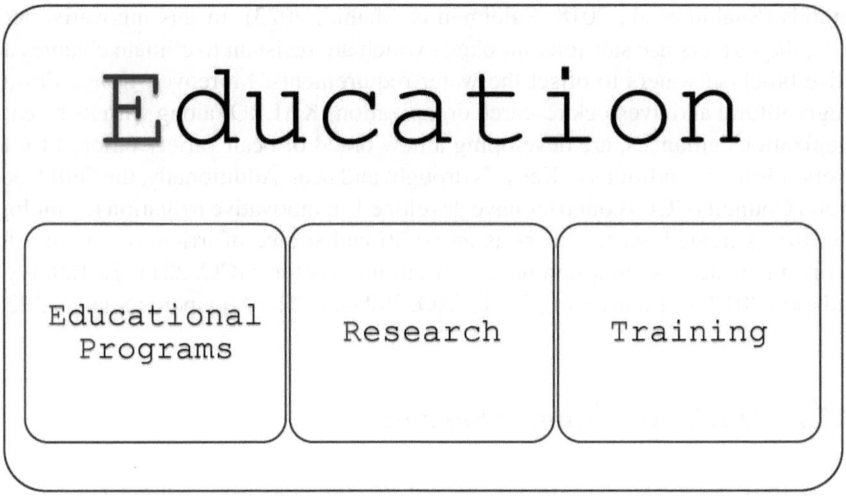

Fig. 12.7 Dynamic agricultural ecosystem policies–education

of illustration, Qatar through its educational and research programs has developed new innovative agricultural methods, such as urban and biosaline agriculture(El-Beltagy & Madkour, 2012; Karanisa et al., 2021; Hubert et al., 2022; FAO, 2024a, b; GCC, 2024; Malhotra, 2024).

For the urban agricultural method, it employs the technology of hydroponics and aquaponics techniques, which take the advantage of the symbiotic relationships between the plant and aquatic species, in a soilless production facility. While for the salt-tolerant plants agricultural method, it consists of the utilization of the climate smart agriculture technique in combination with biosaline agriculture

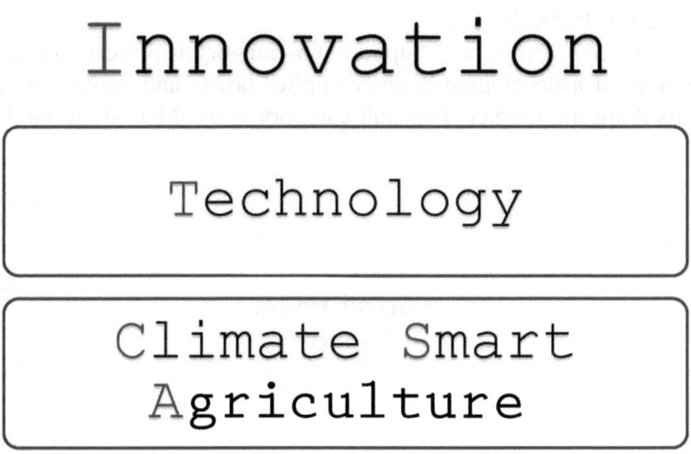

Fig. 12.8 Dynamic agricultural ecosystem policies–innovation

methods (Shahid et al., 2018; Suleiman & Shahid, 2023). In this innovative technique, the farmers use salt-tolerant plants which are resistant to climate changes and utilize brackish waters to offset the water requirements. Moreover, Kenya through its agricultural and livestock research organization (KALRO) along with its research organizations alliances, are developing a new breed of bean variety tailored for the adverse climate conditions in Kenya's drought and heat. Additionally, the Gulf Cooperation Council (GCC) countries have developed an innovative irrigation technology, that utilizes treated waste waters as an additional source of irrigation, in an effort to cope the water consumption in the agricultural sector (GCC, 2011; El-Beltagy & Madkour, 2012; Karanisa et al., 2021; FAO, 2024a, b; The Aquaponic Source, 2024).

12.5.7 DAEP–Government Support

Figure 12.9 shows the seventh DAEP and its sub-categories entitled *Government Support*, containing Subsidies, Small and Medium Entrepreneurs (SME's), Protection Policies, and Agricultural Infrastructure. The objective of this DAEP, is to improve and sustain productivity in the agricultural sector, as well as mitigating the climate change negative effects and food security challenges. The first DAEP sub-category is subsidies, an important financial tool that each developing country employ's it in the agricultural sector, in an effort to support the rural farmer. The governments of a developing country apply this DAEP sub-category in many agricultural classifications, such as Production, Energy (Water, Fuel, Electricity), Trade (Imports and Export), Lands, and Loans. All in an effort to maintain, improve, encourage, increase productivity and investment in the agricultural sector. The second DAEP- Government support sub-category is the SME's, where the government role is to facilitate the SME's start up and production process, in order to assist and expand productivity in the agricultural sector (El-Beltagy & Madkour, 2012; Karanisa et al., 2021; GCC, 2011; FAO, 2024a).

The third DAEP- Government Support sub-category is protection policies, here the government of a developing country applies tariffs and quotas, in an attempt to protect its domestic market. This sub-category is used based on the developing

Fig. 12.9 Dynamic agricultural ecosystem policies–government support

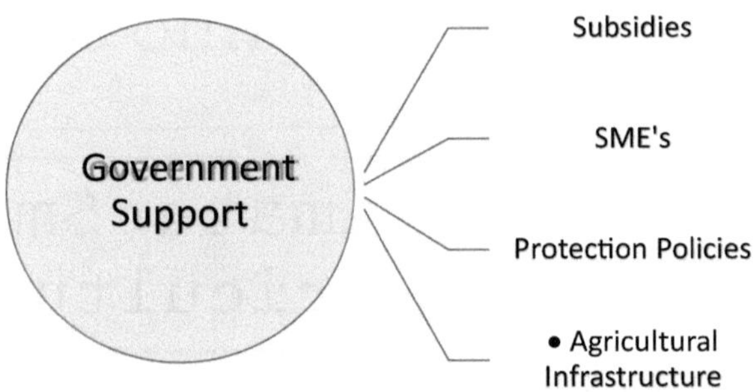

country goals, for instance it can be deployed as an agricultural dumping barrier or to boost agricultural competitiveness. The fourth DAEP- Government Support sub-category is the Agricultural Infrastructure, which is considered an important element in the development of the agricultural sector in any developing country. This sub-category provides the optimal operation essentials to the farmer to produce and deliver his agricultural products to the local market (GCC, 2011; Karanisa et al., 2021; FAO, 2024a).

12.5.8 DAEP–Farmer's Disposable Income

As demonstrated in Fig. 12.10, the eighth DAEP is *Farmers Disposable Income* and its sub-categories, the first is Income Growth and the second is Income Diversification. Agriculture in any developing country is considered to be an important and productive sector in the economy. This sector employs the majority of the workforce in the developing country and in some countries have a large share of the country's GDP. Additionally, the agricultural sector will assist in the reduction of poverty, addresses food security challenges, and promote economic growth (OECD, 2012; FAOC, 2000, 2023; UKDFID, 2018).

The main objective of the first sub-category is to facilitate income growth of the farmer by utilizing many of the above DAEP's. For example, a farmer can utilize innovative technology to increase his production in the harsh arid environment, as is the case of the state of Qatar and their technologically advanced greenhouses. Additionally, with the support of the government, the local farmer can export his agricultural products to other countries, allowing him to increase his income. The second sub-category in this DAEP is Income Diversification, where the farmer can

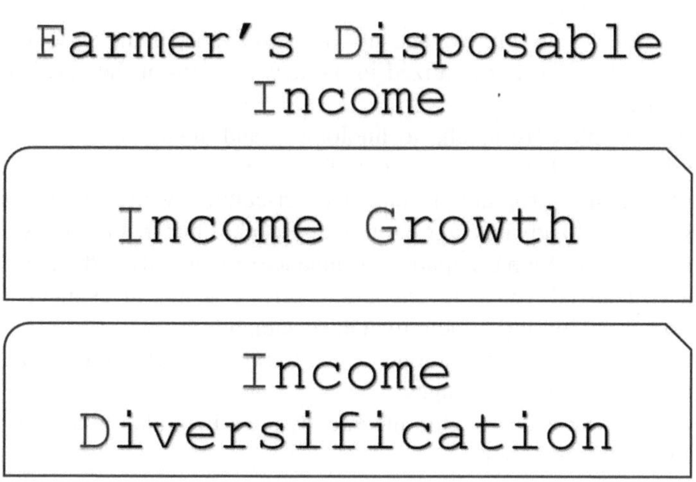

Fig. 12.10 Dynamic agricultural ecosystem policies–farmer's disposable income

utilize other production lines such as producing salt-tolerant plants as animal feeds, or plant multiple crops in different seasons, all in order to increase yield (OECD, 2012; UKDFID, 2018; FAOC, 2000, 2023; Shahid & Alkandari, 2024).

12.6 Conclusions and Recommendations

Agriculture is a great source of income generator for many developed and developing countries. In fact, some developing countries generate around 50% of their GDP from their agricultural sector. In the last century, many developing countries could not sustain agricultural productivity due to several environmental and economic factors. It was discovered that traditional static AEP's are the main cause of agricultural failure and could not sustain the progress and demands of an ever-changing developing country. Thus, a new approach was required to address food security and climate change challenges. Accordingly, many developing countries with the support of world organizations have developed a modern approach to the traditional static AEP's. This approach is dynamic, and customizable, which can be applied to any developing country based on its environmental and economic conditions. For example, in Qatar DAEP's were more oriented towards educational and innovation, which developed urban agriculture to increase its productivity to be more self-sufficient and mitigate food security challenges. Moreover, in the African regions the DAEP's is more oriented towards productivity and irrigation. Where the government supports the conservation agriculture method to increase productivity and preserve water irrigation consumption as well as the desegregation of the soil. Therefore, it is recommended that any developing country in the arid regions should at least have one of the following dynamic policies in their agricultural strategic plan, and they are as follows:

- Subsidies Policies, to improve and sustain productivity in the agricultural sector.
- Educational Programs, specialized in the development of the agriculture in the arid countries.
- Support Innovative research, technologies, and methodologies, to improve productivity in the harsh environment of arid regions.
- Establish Training programs to enhance agricultural workforce knowledge, in order to apply it to advance agricultural approaches and technologies.
- Support Private Sector participation, to enhance productivity and financial growth.
- Facilitate Financial Tools, to encourage private sector investment as well as improving agricultural productivity and sustainability.
- Facilitate regional and international Trade to boost agricultural productivity and mitigate food security challenges.
- Support the Farmers Disposable Income, by the utilization of the above DAEP's and income diversification.

In conclusion, to increase agricultural productivity, address food security challenges and mitigate climate change negative effects, the developing countries should create a dynamic policy structure tailored to their needs, for the purpose of agricultural development and economic growth.

References

Abubakar, A., Gambo, J., & Umar, S. (2021). An overview of the effects of some agricultural policies in Nigeria-1960-2020. *Nigerian Agricultural Journal, 52*(3), 151–162.

Agriseta (Agriculture Sector Education Training Authority). (2024). https://www.agriseta.co.za/wp-content/uploads/2022/09/AgriSETA-COVID-19-Report-2020.pdf. https://www.agriseta.co.za/about-us/

de Melo, M. H. (1979). Agricultural policies and development: A socioeconomic investigation applied to Sri Lanka. *Journal of Policy Modeling, 1*(2), 217–234.

El-Beltagy, A., & Madkour, M. (2012). Impact of climate change on arid lands agriculture. *Agriculture and Food Security, 1*(3). https://doi.org/10.1186/2048-7010-1-33

Ellis, F., Harvey, J., & Waites, T. (2024). History of agricultural policy. Retrieved December 19, 2024, from https://assets.publishing.service.gov.uk/media/57a08c5a40f0b64974001170/LRC_PLOW_LR_History_Agricultural_Policy.pdf

ERS-USDA (Economic Research Service-US Department of Agriculture). (2009). The 2008/2009 world economic crisis: What it means for US agriculture. https://www.ers.usda.gov/publications/pub-details/?pubid=40468

ERS-USDA (Economic Research Service-US Department of Agriculture). (2011). Impacts of higher energy prices on agriculture and rural economies. https://www.ers.usda.gov/webdocs/publications/44894/6806_err123_reportsummary.pdf?v=0#:~:text=Results%20are%20compared%20to%20the,income%20in%20the%20short%20run

Evett, S. R., Colaizzi, P. G., Lamm, F. R., O'Shaughnessy, S. A., Heeren, D. M., Trout, T. J., Kranz, W. L., & Lin, X. (2020). Past, present and future of irrigation on the US Great plains. *American Society of Agricultural and Biological Engineers, 63*(3), 703–729.

FAO (Food and Agriculture Organization). (2024a). Food and agriculture policy classification. Retrieved December 19, 2024, from https://www.fao.org/fileadmin/templates/fapda/docs/FAPDA_policy_classification_April2015.pdf

FAO (Food Agriculture Organization). (2024b). Climate smart agriculture sourcebook. https://www.fao.org/climate-smart-agriculture-sourcebook/enabling-frameworks/module-c3-policy/c3-overview/en/

FAOC (Food and Agriculture Organization Conference). (1957). *Report of the Conference of FAO Ninth Session 2–23 Nov 1957 Rome, Italy*. Retrieved December 19, 2024, from. https://www.fao.org/4/x5574E/x5574e03.htm

FAOC (Food and Agriculture Organization Conference). (1960). *Twenty-Fourth Session of FAO Conference 17–27 Oct 1960*. Retrieved December 19, 2024, from https://www.fao.org/4/AC969T/AC969T02.htm

FAOC (Food and Agriculture Organization Conference). (1983). *Twenty-Second Session of FAO Conference 14 Nov 1983*. Retrieved December 19, 2024, from https://www.fao.org/4/AK669E/ak669e14.htm

FAOC (Food and Agriculture Organization Conference). (1991). *Report of the Conference of FAO, Twenty-Sixth Session, 9–27 Nov 1991, Rome*. Retrieved December 19, 2024, from https://www.fao.org/4/x5587E/x5587e05.htm

FAOC (Food and Agriculture Organization conference). (2000). *Twenty-first FAO Regional Conference for Africa, Cameroon 21–25 Feb 2000*. Retrieved December 19, 2024, from. https://www.fao.org/4/X7467e/X7467e00.htm

FAOC (Food and Agriculture Organization Conference). (2023). *43rd session of the FAO Conference, 1–7 July 2023, Rome, Italy*. Retrieved December 19, 2024, from https://www.fao.org/governing-bodies/conference/C2023/en

GCC (Gulf Cooperation Council). (2011). Agricultural development in the Gulf Cooperation Council countries. In *Economic affairs*, (5th ed). Riyadh.

GCC (Gulf Cooperation Council). (2024). Cooperation in the field of agriculture. GCC achievements-economic cooperation. Retrieved December 19, 2024, from https://www.gcc-sg.org/en-cus/CooperationAndAchievements/Achievements/EconomicCooperation/CooperationintheFieldofAgriculture/pages/Home.aspx

Hubert, H., Li, F., Qiao, Y., Measho, S., Muhirwa, F., Tian, C., Leng, P., Ingabire, R., Itangishaka, A. C., Chen, G., & Turyasingura, B. (2022). Climate change–drylands–food security nexus in Africa: From the perspective of technical advances, challenges, and opportunities. *Frontiers in Environmental Science, 10*, 1–17. https://doi.org/10.3389/fenvs.2022.851249

Karanisa, T., Amato, A., Richer, R., Majid, S. A., Skelhorn, C., & Sayadi, S. (2021). Agricultural production in Qatar's hot arid climate. *Sustainability, 13*, 4059. https://doi.org/10.3390/su13074059

Kose, M. A., Sugawara, N., Terrones, M. E. (2020) Global recessions. World Bank Group, Policy Research Working Paper No.9172. pp. 71. https://documents1.worldbank.org/curated/en/185391583249079464/pdf/Global-Recessions.pdf

Kuwait institute for Scientific Research (KISR). (2024). Retrieved December 19, 2024, from https://www.kisr.edu.kw/en/program/14/

Malhotra, S. (2024). Reforming agricultural policies to sustainably transform food systems. Retrieved December 19, 2024, from https://www.agincentives.org/2024/04/09/reforming-agricultural-policies-to-sustainably-transform-food-systems/

Martin, W. (2017). Agricultural trade and food security. Asian Development Bank Institute, Working Paper, Tokyo, No.664. https://www.adb.org/publications/agricultural-trade-and-food-security

Mesferlaw. (2024). Law establishing the Public Authority for Agriculture Affairs and Fish Resources. Kuwait Today (1983), No. 1481. https://mesferlaw.com/archives/2238

OECD. (2012). Agricultural Policies for poverty reduction. *OECD Publishing*. https://doi.org/10.1787/9789264112902-en

Ray, A. (1986). Trade and pricing policies in the world of agriculture. International Monetary Fund. https://www.elibrary.imf.org/downloadpdf/journals/022/0023/003/article-A001-en.pdf

Shahid, S. A., & Alkandari, A. J. (2024). Halophytic crops as a solution for food security, land rehabilitation, and mitigating future water crises by utilizing marginal quality waters. In: Dagar, J. C., Gupta, S. R., & Kumar, A. (Eds.), *Halophytes vis-à-vis saline agriculture* (Ch. 17, pp. 441–478). Springer. https://doi.org/10.1007/978-981-97-3157-2_17

Shahid, S. A., Zaman, M., & Heng, L. (2018). Salinity and sodicity adaptation and mitigation options. In: Zaman, M. et al. (Eds.), *Guideline for salinity assessment, mitigation and adaptation using nuclear and related techniques* (Ch. 3, pp. 55–88). Springer. https://doi.org/10.1007/978-3-319-96190-3_3

Speece, M. (1997). Economic policy and desertification in arid and semi-arid developing countries. *The Pakistan Development Review, 36*(2), 191–201.

SSCTK (Small-Scale Conservation Tillage-Kenya). (2024). Retrieved December 19, 2024, from https://qcat.wocat.net/en/wocat/technologies/view/technologies_940/

Suleiman, M. K., & Shahid, S. A. (2023). Prospective of agriculture farming in Kuwait and energy-food-water-climate nexus. In: Suleiman, M. K., & Shahid, S. A. (Eds.), *Terrestrial environment and ecosystems of Kuwait* (Ch. 15, pp. 363–391). Springer. https://doi.org/10.1007/978-3-031-46262-7_15

The Aquaponic Source. (2024). https://www.theaquaponicsource.com/what-is-aquaponics/

UKDFID (UK Department for International Development). (2018). Economic growth in the arid and semi-arid lands of Kenya. Retrieved December 19, 2024, from https://assets.publishing.service.gov.uk/media/5c6fd72aed915d4a315f6552/482_Economic_Growth_in_the_Arid_And_Semi-Arid_Lands_of_Kenya.pdf

WHO (World Health Organization). (2024). Retrieved December 19, 2024, from https://data.who.int/dashboards/covid19/deaths

World Bank Group. (2024a). Agricultural policies in developing countries Exchange rates, prices, and taxation. Retrieved December 19, 2024, from https://openknowledge.worldbank.org/server/api/core/bitstreams/bfac7cac-fb29-54be-a1db-09ff10a34407/content

World Bank Group. (2024b). Agriculture and food. Retrieved December 19, 2024, from https://www.worldbank.org/en/topic/agriculture/overview

World Bank Group. (2024c). Climate-smart agriculture. Retrieved December 19, 2024, from https://www.worldbank.org/en/topic/climate-smart-agriculture

WTO (World Trade Organization). (2024). General agreement on tariffs and trade. Retrieved December 22, 2024, from. https://www.wto.org/english/docs_e/legal_e/gatt94_e.htm

Zaman, M., Shahid, S. A., & Heng, L. (2018). Irrigation water quality. In: Zaman, M. et al., (Eds.), *Guidelines for salinity assessment, mitigation and adaptation using nuclear and related techniques* (Ch. 5, pp. 113–131). Springer. https://doi.org/10.1007/978-3-319-96190-3_5

Part III
Mitigation and Adaptation Synergy

Chapter 13
Current Mitigation and Adaptation Progress: Gaps and Challenges in Arid Lands Agriculture

Majda Khalil Suleiman and Shabbir Ahmad Shahid

Abstract Despite significant progress, adaptation gaps exist between current levels of adaptation and levels needed to respond to impacts and reduce climate risks. Land restoration contributes to climate change mitigation and adaptation with synergies via enhanced ecosystem services and crops intensification for food security. Most observed adaptation practices are fragmented, small in scale, sector-specific, and focused more on planning rather than implementation. The challenges and barriers in the implementation of mitigation and adaptation include insufficient institutional and financial support, weak governance, insecure land ownership, low incomes and the risk of reversal. Some land-related adaptations have mitigation co-benefits, and these should be focused on. The arid lands in specific are affected by climate change, mainly due to rise in temperature, prolonged droughts, low and infrequent rainfall, and water scarcity. Climate-driven food insecurity and supply instability is likely to increase with rising temperature, competition for land between urban expansion and food production. In order to cope with the climate change impact, international scientific community and the organizations are making significant efforts to mitigate the climate change impact and adaptation efforts to deal the climate change holistically. The current efforts in dealing with climate change are significant but not sufficient to fully cope or reverse the impact of climate change, resultantly, the global surface temperature is rising. Therefore, there is a need to compile objective oriented and arid land specific tailor-made effective adaptation including but not necessarily limited to cultivar improvements (salt-tolerant, improved varieties), on-farm water management and storage, soil moisture conservation, diagnostics base reclamation of salt-affected soils, irrigation management, effective land use regulation and high adaptive capacity, farmers-based adaptation, farm level crops diversification, soil

M. K. Suleiman (✉) · S. A. Shahid
Food Security Program, Environment and Life Sciences Research Center, Kuwait Institute for Scientific Research, Safat, Kuwait
e-mail: mkhalil@kisr.edu.kw

S. A. Shahid
e-mail: sshahid@kisr.edu.kw

M. K. Suleiman and S. A. Shahid (eds.), *Fostering Arid Lands Agriculture in the Face of Climate Change*, Advances in Global Change Research 81,
https://doi.org/10.1007/978-3-031-98490-7_13

problems diagnostics based land management, and use of soil regenerative practices that work with the nature. Many agriculture and other land use options provide adaptation and mitigation benefits that could be up-scaled in arid lands.

Keywords Aridity · Drought · Gaps · Research · Extension · Policy development · Climate literacy · Climate action

13.1 Introduction

Climate change is a global phenomenon, but more evident in drylands including hyperarid, arid, semiarid and subhumid regions. Drylands are classified based on the Aridity Index (AI), which is less than 0.65 for drylands category (Middleton & Thomas, 1997). The aridity index is measured by dividing the net annual precipitation with potential evapotranspiration (P/PET), the losses from land surface (Jensen & Allen, 2016). The lower the AI, there are higher impacts of climate change on agriculture production for food security. The arid land region countries are more vulnerable to climate change impact than the cold region countries. Arid lands are plagued with biotic and abiotic stresses hindering optimum crop production for food security. Arid lands are part of drylands, which are characterized based on the aridity index (AI). About 47.2% global land area are classified into various types, hyper-arid ($AI < 0.05$) covering 7.5% global land area, arid ($0.05 < AI < 0.2$) covers 12.1% global land area, semiarid ($0.2 < AI < 0.5$) covering 17.7% of the global land area, sub-humid ($0.5 < AI < 0.65$) covering 9.9% of the global land area (Prăvălie, 2016). In drylands water scarcity, prolonged drought and soil salinization can limit crop production.

Therefore, concerted efforts are required to mitigate the climate change (reduction of greenhouse gases emission to reduce heat buildup) and adapt to climate change using multi-dimensional science based approaches. Agriculture sector is the most affected by climate change, especially the smallholder farmer. There are over 570 million farms of various sizes in 161 countries (FAO, 2013; Lowder et al., 2016), including 24% in India, 6% south Asia excluding India, 9% in Sub-Saharan Africa (SSA) and 35% in China (cumulatively 74% of total farms). These farms are occupying substantial area of world agriculture land. The small sized farm owners (<2 ha) are the most affected by climate change impact, thus, require to adopt well established adaptation technologies. Thus effective adaptation in agriculture takes place by farmers individually, by farmers and local institutions collectively, or through national level policy decisions which provide finance, research and development, knowledge transfer, and property rights or legal frameworks to enable individual or collective action (Wreford et al., 2010). It is to be noted that, although the arid land countries are located in specific climatic zones, however, within each country there are different agro-ecological zones, with different environments conducive for specific crops, and need different agriculture adaptations.

Despite significant progress made in different countries, adaptation gaps still exist between current levels of adaptation and levels needed to respond to impacts and reduce climate risks. In this chapter, the gaps between mitigation and adaptation in general and specific to countries in the arid land are presented and discoursed.

13.2 Climate and Climate Change

Climate is the average weather conditions, such as temperature, rainfall and wind usually taken over a 30 year-period, whereas the climate change (CC) is the long-term alteration in global weather patterns, especially increases in temperature and storm activity and the potential consequences of the greenhouse effect (Shahid, 2010). The impact of climate change can be realized in the form of increased level of GHGs and associated temperature rise.

13.2.1 Increase in Greenhouse Gasses

Due to increased pressure on natural resources to compensate unprecedented population growth and resultant urbanization and infrastructure development, the global greenhouse gas emissions have continued to increase. Other reasons of GHGs emission increase is unprecedented land use and land use change, for example conversion of prime arable land to urbanization development (housing projects and associated infrastructure development). IPCC (2023) has recently reported net global GHGs emission estimation as 59 ± 6.6 $GtCO_2$-eq (2019), a 12% increase higher than in 2010 (6.5 $GtCO_2$-eq) and 54% (21 $GtCO_2$-eq) higher than in 1990.

13.2.2 Greenhouse Gasses, Warming Potential and Concentration by 1900

The CO_2, CH_4 and NO_2 are the main GHGs causing global warming. Their relative global warming potential is 1, 23 and 296 for CO_2, CH_4 and NO_2 respectively. However, when the GHGs are reported as total carbon footprint, it includes all GHGs. The concentrations of GHGs have increased rapidly since 1850. In 2019, the concentrations of N_2O, CH_4 an CO_2 are recorded as 332 ppb, 1866 ppb and 414 ppm respectively (IPCC, 2023). The CO_2 level of 410 ppm has surpassed the safe limit of 350 ppm.

13.3 Crop Yield Prediction with Rising Temperature Scenarios

The current development rate and the population growth has increased the pressure on the natural resources and ecosystem services. The unsustainable human activities have exhausted the capacity of natural resources to a great extent to perform in healthy ways to provide ecosystem services and food security (FAO, 2009). These and many other activities have increased the GHGs emission, which has resulted into unequivocally caused global warming, with global temperature rise of 1.1 °C above 1850–1900 in 2011–2020 (IPCC, 2023).

Based on many studies covering a wide range of regions and crops, negative impacts of climate change on crop yields have been more common than positive impacts over the twenty-first century (IPCC 2014a, 2014b).

13.4 Mitigation and Adaptation Gaps with Respect to Arid Lands Countries

Despite significant progress on climate change action in many countries, adaptation gaps still exist between current levels of adaptation and levels needed to respond to impacts and reduce climate risks, and the largest adaptation gaps exist among lower income population groups (IPCC, 2023).

Most of the poor developing countries located in the arid lands are largely in Asia and Africa. In these countries a high level climate action is poorly defined and integrated to decision making, due to multiple reasons, the main one is lack of financial resources and climate action plan and literacy. The small scale farmers are using conventional technologies which have become weaker over time, and these farmers require climate literacy with respect to their farming services to produce food and improve livelihood. There are over 570 million farms of various sizes in 161 countries (FAO, 2014; Lowder et al., 2016), including 24% in India, 6% south Asia excluding India, 9% in Sub-Saharan Africa (SSA) and 35% in China (cumulatively 74% of total farms). Of 460 million farms in 111 countries, 84% are small size (72% farms are < 1 ha size and 12% are 1–2 ha size) (FAO, 2014). Farm sized less than 1 ha, 1–2 ha, 2–5 ha and larger than 50 ha control 8, 4, 7 and 65 percent of all agricultural land respectively. These farms manage the majority of the global land and produce major part of the world food. The small scale farmers are poor farmers and are highly vulnerable to climate change impacts. These farmers produce their food for themselves and small quantities are sold in the market.

In Pakistan almost half of its total land area (36 million ha) is dedicated to agriculture which remained relatively stable over the past four decades (FAO, 2017). Under such conditions when agriculture land is not increasing, the increased food production must come from crops intensification through adoption of CSA including cropping intensity and fertilizers use. Approximately 84% of land is classified as arable and 14% as permanent meadows or pastures. About 43% are smallholders holding less than one ha, and 22% hold more than 3 ha land, however, the average farm size is less than 2.6 ha (GoP, 2010). Pakistan has large rivers and canal systems, therefore, roughly 94% of the agricultural area in Pakistan is equipped for irrigation. In Pakistan the growth rate is positive throughout the years, however, it has been observed that in 2016, first time in the past 1.5 decade, the agriculture sector has experienced a negative growth rate of 0.2%, the reasons being primarily the impact of extreme events on key crops, a lack of access to key inputs, and a global downturn in commodity prices (GoP, 2016).

There are multiple gaps at various levels and in different regions and countries. However, the countries which are highly vulnerable to climate change are located in drylands (hyperarid, arid and semiarid) and are most affected. The adaptation gaps are less in high income countries and larger in low income poor developing countries. Each country is doing its best to reduce adaptation gaps within their available resources and know-how. For successful adaptation, poor countries need sufficient funds and climate action program for implementation over a period to see the significant impact of adaptation. However, there are financial barriers in these countries, due to limitations to redirect national capital to climate action, that may affect other essential developments in the country (education, health, transport, roads construction etc.). Following are the potential adaptation gaps hindering optimum crop production for food security in arid land countries.

13.4.1 Lack of Country Specific Climate Action Plan

Poor countries in the arid lands are plagued with multiple problems hindering their developments to full scale. These countries depend on significant loans on low–high interest rates from various countries and international organizations, such as International Monitory Fund (IMF), Islamic Development Bank, World Bank and Asian Banks etc.. The country specific climate action plan must reflect objectives oriented change needed to address climate change impacts holistically, and to observe through periodic monitoring the effect of change on agriculture production for food security and improvement in ecosystem services at large. For poor developing countries it is very difficult and may be risky to redirect the money to climate action program whose benefits can be seen after many years. Adaptation options often have long implementation times, long-term planning and accelerated implementation, it is important to close adaptation gaps, recognizing that constraints remain for some regions (IPCC, 2023). In addition, compared to adaptation more funds were floated on mitigation compared to effective adaptation. One of the best climate change mitigation is forestation/plantation on global scale. This plantation will improve the environment through carbon sequestration and lowering the temperature thus cooling the microenvironment. In order to have national action program on board, the governments to take the initiatives through lobbying with industrialist and business community to contribute to the climate action fund and the government should appreciate their contribution and provide some in-kind incentives for their continuous support. Therefore, it is pertinent that each country to develop climate action program and implement periodic monitoring to oversee the impact and exploit these benefits at international levels for appreciation to attract funds.

Here we share World Bank Group (2022) action plan for South Asia region countries, where, in case of Pakistan four priority themes are identified, (i) energy decarbonization, (ii) agriculture-water nexus, (iii) climate-resilient infrastructure and communities, and (iv) Macro-fiscal resilience. To address these four priority area, three notable activities are proposed including, (i) investments in floating PV systems and technical assistance for battery storage and wholesale power market development, (ii) investments in agricultural productivity and improved water resources management, (iii) support for climate-resilient and resource-efficient design features in the government's housing program actions are proposed (World Bank Group, 2022). The World bank group identified priority area for other South Asian countries (Bangladesh, Bhutan, India, Maldives, Nepal, and Sri Lana) which are country need specific. Other countries in the arid lands should follow similar strategies in their national climate action plans.

13.4.2 Lack of Climate Literacy

Significant number of the global population do not believe on climate change and its impacts. They believe that this is a natural region specific phenomenon, and there is no solution to mitigate climate change and to fight with the nature. They also believe that to reduce GHGs emission the industries are to be significantly cut down, which are producing substantial quantities of products of daily use, and to export to other countries to have national financial gains. So there is less knowledge based information to lobby international funding for climate action, which is true for Africa where severe climate data constraints and inequities in research funding and leadership reduced adaptive capacity exist (IPCC, 2023). However, the climate change phenomenon is well recognized by various development sectors and significant developments have been made to address this phenomenon.

13.4.3 Lack of Financial Resources to Enhance Investment in Agriculture

To comply national climate action plan and to enhance agriculture production, the world in general and the countries the most vulnerable to the impact of climate change in specific must have enough financial resources. The governments of the poor countries with high prevalence of hunger and famine (sub-Sahara Africa) should create funds locally and through international organization for a gradual increase of investment in agriculture sector. This will enhance agriculture production and rural infrastructure development to meet the demand to feed the hunger and ever growing population. Insufficient investment in climate change vulnerable poor developing countries may have severe detrimental impact on food security and food supply chain.

13.4.4 *Lack of Control on Population Growth*

United Nations projection about population growth has revealed that the world population will continue growing after 2050, although at a steadily decreasing rate, to peak at 10.4 billion in 2086, and then to start a slow decline to about 10.3 billion in 2100 with a growth rate at that time of −0.1% (UNDESA Population Division, 2022). Same report has shown a steady decline of total fertility rate (TFR) from 1950–2100, from over 2 to less than 2 in all region, but in Africa in1950 TFR was over six and in 2100 will be just over 2, the world average TFR was 5 in 1950 and is projected to be about 2 by 2100. The TFR is the average number of children that would be born to a woman over her lifetime. The population growth, food losses and waste, and shrinking resources for cultivation are the main reasons to sufficiently feed the growing population. The government should develop policies for population growth and food security for sustainability.

13.4.5 *Lack of Policy on Land Conversion and Uses–Urbanization*

The World Bank (2010) report has shown that about 4.4 billion (56% of world population) population is living in cities, and with the urban population more than doubling its current size by 2050, 70% (5.50 billion) population will live in cities (Jiang & O'Neill, 2017). The population growth and associated urbanization has serious impact on climate change (Satterthwaite, 2009). This makes sense that there will be significant land use and land change happening by 2050 (https://www.worldbank.org/en/topic/urbandevelopment/overview). The decisions on land use and land conversion (LULC) in the absence of science based backing can cause severe damages to ecosystem functions and provision of services. The policies should be developed to save arable lands for future expansion of agriculture to meet the demand of growing population. Simultaneous to secure the arable land, a parallel policy to be developed to identify areas to develop new cities and housing projects on non-arable lands, which will never be brought to agriculture production "*marginal lands*" due to severe marginality. To do it effectively, the government should involve national soil science societies to seek expert opinions for new developments. In Asia, 24 and 42 percent population was living in cities in the year 1970 and 2007 respectively, reaching 50% by 2022 (UN 2012). Urbanization is a serious threat affecting low- and middle-income nations. With high population growth rates, urbanization and economic development, India, like many other Asian and African countries faces the challenge to mitigate carbon emissions in its urban areas. With a total urban population of 377 million in India, urbanization increased from 27.81% in 2001 to 31.16% in 2011 (MoUD, 2011).

13.4.6 Reduction in Resource Capacity

The anthropogenic impacts on natural resources (soil, water) negatively impacted the agriculture land capacity to produce at higher level. Resource capacity is declining significantly due to short-term climate variability and longer-term climate change. This is the case of reduction of agricultural land due to salinization, where we are losing 2,000 ha irrigated agriculture land on daily basis in over 75 countries (IWMI, 2014). Currently 20% (62 million ha) of irrigated land is salinized (Qadir et al., 2014). In addition, desertification is causing severe impact on food production, as an example, erosion is causing 60% reduction in crop production (Shahid, 2023). In addition to anthropic impacts, the climate-change induced global warming has reduced the capacity of ecosystems to function to their optimum levels. An increase in the intensity and frequency of droughts, desertification, species depletion, soil degradation and crop failures has reduced livelihood options and the vulnerability of human beings (Singh, 2013). With regards to water resources in the face of climate change, both the renewable surface and groundwater is expected to reduce in the driest subtropical regions, where the rural areas are expected to experience major impacts on water availability and supply and food security (IPCC, 2007).

13.4.7 Crops Yield Gaps

Each crop has maximum potential to grow and gives high yield. The yield gap can vary between countries, the regions and at farmer fields. The progressive farmers are well equipped with agriculture machinery and other farm inputs (high yielding crop varieties, fertilizers, pesticides), well connected to markets and are making higher gains from their land. Whereas, the smallholder farmer lacks these facilities and show gap in crop yield compared to progressive farmers. This gap can be narrowed by empowering the farmers with new technologies and by the provision of timely inputs for use in their farms, economic incentives to adopt yield enhancing seeds or cropping techniques, to enable the improvised farmers to gain comparative yield to progressive farmers. The yield gap can be further reduced if farmers have easy access to information, extension services and technical skills. The extension services, especially to change farmer conventional crop management techniques to modern climate smart agriculture can also help closing yield gaps. It has been observed through working with several National Agricultural Research Systems (NARS) in Asia, Africa and Middle East and North Africa, that, in general there is a weak link between the research-extension-farmers, that should be strengthened for timely delivery of extension services in soil, water, plants and insects and pests management.

13.4.8 Water Scarcity and Poor Water Delivery System

In arid land countries, there are two main issues, (i) water scarcity and (ii) poor water storage and delivery systems. Water scarcity occurs when the demand for water exceeds available supplies, often in regions with low rainfall, leading to physical water scarcity.

Poorly developed countries either don't have or have poor infrastructure to harvest rain water and store in reservoirs or dams for future uses. Most of this water is wasted, and often causes flash floods, and erode huge quantities of soils and nutrients, leaving behind the unfertile exposed subsoil. In irrigated agriculture countries, the water conveyance systems (water transport from rivers to canals, water channels and farmers fields) are not lined to reduce water seepage. This leads to significant loss of water, causing salinization and water logging in nearby areas. These water losses can be reduced, provided the government enhances investments to line these water delivery systems, and simultaneously remove settled silt from the rivers and canals to increase their water storing capacities. In Pakistan, the government has taken serious actions with the help of USAID to line the water channels for a faster delivery of water from canals to small water channels leading to farmer fields.

13.4.9 Poor Insights of Farm Salinization and Mitigation

Soil salinization in irrigated fields has been a challenge for centuries, causing the collapse of ancient civilizations like Mesopotamia and the Viru Valley in Peru (Jacobson & Adams, 1958; Shahid et al., 2018). According to FAO (2021, 2022), about 424 million hectares of topsoil and 833 million hectares of subsoil globally are affected by salinity. This estimate is based on 73% of land mapped for soil salinization. Another estimate shows 20% (62 million hectares) of all irrigated land (310 million ha) is affected by soil salinization to various levels (Qadir et al., 2014). Recently, it has been reported that on daily basis 2,000 hectares (20 km^2) land has been abandoned due to soil salinity in over 75 countries (IWMI, 2014). This shows that our current interventions (gaps in salinity management) are not making the pace with the speed of salinization spread in agricultural farms.

13.4.10 Lack of Strategies to Promote Climate Smart Agriculture Practices

The climate smart agriculture (CSA) has been evolving over the last decade (FAO, 2013). It encompasses a variety of technologies refined for adapting to climate change, aiming to boost food security and reduce greenhouse gas emissions. Hundreds of technologies and approaches around the world fall under the heading of CSA, and now newly introduced term Regenerative Agriculture (RA) concept is also gaining recognition (Savills Research, 2021). The CSA concept is ambitious to integrate agriculture development and climate responsiveness, leading to improve food security and reduce GHGs emissions. The CSA requires a plan to address trade-offs and synergies between productivity, adaptation, and mitigation (FAO, 2010).

In recognition to the impact of climate change on agriculture and environment sectors, some countries have taken the initiatives and developed strategies to CSA practices (Rana & Gill, 2022). Some countries added climate change into their national agenda and either revised the names of the existing ministries by adding climate change or developed new authorities. In 2017, Pakistan introduced the Climate Change Act, establishing the Pakistan Climate Change Authority and Climate Change Fund to mobilize resources for climate adaptation, including CSA (GoP, 2024).

Recently World Bank took the initiative to develop South Asia roadmap as climate change action plan 2021–2025 (World Bank Group, 2022). While developing climate action strategies, it is important that each country to identify priority areas, where changes are needed, and develop theory of change (TOC) to achieve the objectives of the climate action plan. The objectives can be achieved by mobilizing essential resources (finances) through scaling up government grants, building capacity of the climate vulnerable communities and implementing potential country specific adaptation actions. This will ultimately reduce adaptation gaps, increase crop production for food security and improve ecosystem functions and services.

13.4.11 Integrated Water Resources Management Strategies, Water Use Efficiencies and Productivities

Arid countries face constraints like water scarcity, drought, poor soil fertility, and high temperatures, which require a national strategy integrating surface water, groundwater, and treated wastewater. Therefore, the water resources have to be combined together to develop national water use and conservation strategies. The need of such strategies is much important in capital rich and food insecure countries (where the countries are not able to grow sufficient crops within their national boundaries to meet population food demand). Such countries are net food importer, such as the case of GCC countries. Some Gulf Cooperation Council countries have taken the initiative to develop their water management and conservation strategies, such as Abu Dhabi water master plan (EAD, 2009), UAE water conservation strategy (UAMEW, 2010), a strategy for the use of recycled water (EAD-ICBA, 2010) and Oman salinity strategy (MAF, 2012).

The main water resources are, fresh water (surface and groundwater) and municipal waste water. In capital rich GCC nations, a high level of water treatment is used. The different levels of wastewater treatment start from raw sewage water to primary, secondary, tertiary and quaternary levels of treatments. The quality of water is improved at high level of treatments, and the choices for use also increased. It is noteworthy, that as the population will grow, the municipal waste water will increase. Therefore, this is a viable option for future use to mitigate water scarcity in different countries. In poor developing nations due to the lack of water transport system, modern irrigation systems, knowledge, significant volume of water is lost in various ways and decreasing the water use efficiencies and productivities. Improving the link between research-extension-farmers for technology transfer, and provisions of small scale funds to farmers can improve the water use efficiencies and productivities (crop per drop).

13.5 Conclusions and Recommendations

It is concluded that despite significant progress, adaptation gaps remain between current levels of adaptation and levels needed to respond to impacts and reduce climate risks. Countries must prioritize CSA practices tailored to agro-ecological zones, requiring coordinated efforts from policymakers, researchers, extension services, and farmers to mitigate climate risks. Climate change is recognized as serious threat to agriculture and food security. Reasons being the increase in greenhouse gasses emissions and associated temperature increase. The rising temperature has significant effects on crop production. Overall, The IPCC based on temperature rise scenarios over the twenty-first century has predicted more chances of crop yield decline than increase. The arid land agriculture is plagued with climate change effects, this is due to the lack of significant mitigation measures and lack of financial assistance to vulnerable smallholder farmers for climate smart agriculture adaptation. Many countries have included climate actions plan into their national strategies and are implemented using their domestic resources and international funds. The climate change impacts are of different nature in different countries and within the country have different agro-ecological zones. To address the climate change impact in specific country and agro-ecological zones, the impacts have to be identified, and addressed on priority basis, leading to mitigate climate change and agriculture adaptation to climate change. There is diversity of climate smart agriculture practices, and over hundred technologies exist under the CSA group, most suitable to particular location or agro-ecological zones are to be implemented. The successful implementation requires joint efforts of policy makers, the researchers, extension services and farmers.

References

EAD. (2009). *Abu Dhabi water resources master plan* (pp. 219). Environment Agency-Abu Dhabi, United Arab Emirates.

EAD-ICBA. (2010). A strategy for the use of recycled water–Abu Dhabi Emirate. Volume 1 (Main Report). Joint publication of Environment Agency-Abu Dhabi, UAE and International Center for Biosaline Agriculture Dubai UAE, pp. 92.

FAO. (2009). Proceedings of high-level experts forum on. "How to feed the World in 2050", 12–13 Oct 2009. Rome.

FAO. (2010). Climate-smart agriculture. Policies, practices and financing for food security, adaptation and mitigation. Food and Agriculture Organization of the United Nations, Rome.

FAO. (2013). Climate smart agriculture sourcebook. Food and Agriculture Organization of the United Nations (FAO). Rome. http://www.fao.org/docrep/018/i3325e/i3325e.pdf

FAO. (2014). The state of food and agriculture: Innovation in family farming. FAO Rome.

FAO. (2017). FAOSTAT Pakistan. Food and Agriculture Organization of the United Nations (FAO). http://www.fao.org/faostat/en

FAO. (2021). Global map of salt affected soils version 1.0. Retrieved December, 2021, from https://www.fao.org/soils-portal/data-hub/soil-maps-and-databases/global-map-of-salt-affected-soils/en/

FAO. (2022). Halt soil salinization, boost soil productivity—Proceedings of the global symposium on salt-affected soils, 20–22 Oct 2021.

GoP. (2010). Agricultural census. Pakistan Bureau of Statistics. Government of Pakistan. http://www.pbs.gov.pk/

GoP. (2016). Pakistan economic survey 2015-2016. Ministry of Finance, Government of Pakistan. http://www.finance.gov.pk

GoP. (2024). Climate smart agriculture in Pakistan. Joint report of UK-AID, World Bank Group, CCAF and CIAT (pp. 26).

IPCC. (2007). Climate change 2007: Synthesis report. In: Pachuari, R. K., & Reisinger, A. (Eds.), *Contribution of working group I, II and III to the fourth assessment report of the intergovernmental panel on climate change. Core writing team* (p. 104). IPCC.

IPCC. (2014a). Summary for policymakers. In: Field, C. B., Barros, V. R., Dokken, D. J., Mach, K. J., Mastrandrea, M. D., Bilir, T. E., Chatterjee, M., Ebi, K. L., Estrada, Y. O., Genova, R. C., Girma, B., Kissel, E. S., Levy, A. N., MacCracken, S., Mastrandrea, P. R., & White, L. L. (Eds.), *Climate Change 2014: Impacts, adaptation, and vulnerability. Part A: Global and sectoral aspects. Contribution of working group II to the fifth assessment report of the intergovernmental panel on climate change* (pp. 1–32). Cambridge University Press.

IPCC. (2014b). Climate change 2014: Synthesis report. In: Core Writing Team, Pachauri, R. K., Meyer, L. A. (Eds.), *Contribution of working groups I, II and III to the fifth assessment report of the intergovernmental panel on climate change* (p. 151). IPCC.

IPCC. (2023). Summary for policymakers. In: Climate change 2023: Synthesis report. In: Core Writing Team, Lee, H., & Romero, J. (Eds.), *Contribution of working groups I, II and III to the sixth assessment report of the intergovernmental panel on climate change* (pp. 1–34). IPCC. https://doi.org/10.59327/IPCC/AR6-9789291691647.001

IWMI. (2014). Press release: World losing 2,000 hectares of irrigated farmland to salt-induced degradation daily. Retrieved November 4, 2024, from https://www.iwmi.cgiar.org/news/press-release-world-losing-2000-hectares-irrigated-farmland-salt-induced-degradation-daily/

Jacobson, T., & Adams, R. M. (1958). Salt and silt in ancient Mesopotamian agriculture. *Science (New Series), 128*(3334), 1251–1258.

Jensen, M. E., & Allen, R. G. (Eds.). (2016). *Evaporation, evapotranspiration, and irrigation water requirements: Task Committee on Revision of Manual* (Vol. 70). American Society of Civil Engineers (ASCE). https://doi.org/10.1061/9780784414057

Jiang, L., & O'Neill, B. C. (2017). Global urbanization projections for the shared socioeconomic pathways. In: *Global Environmental Change* (Vol. 42, pp. 193–199). ISSN 0959-3780, https://doi.org/10.1016/j.gloenvcha.2015.03.008

Lowder, S. K., Skoet, J., & Raney, T. (2016). The number, size, and distribution of farms, Smallholder farms, and family farms worldwide. *World Development, 87*, 16–29.

MAF. (2012). Oman salinity strategy (Main report). Ministry of Agriculture and Fisheries, Sultanate of Oman.

Middleton, N. J., & Thomas, D. S. G. (1997). *World Atlas of Desertification* (2nd ed.).

MoUD. (2011). *India's urban demographic transition: The 2011 census results-provisional* (pp. 2–4). JNNURM Directorate and National Institute of Urban Affairs.

Prăvălie, R. (2016). Drylands extent and environmental issues. A global approach. *Earth-Sciences Review, 161*, 259–278.

Qadir, M., Quillérou, E., Nangia, V., Murtaza, G., Murari Singh, M., Thomas, R. J., Drechsel, P., & Noble, A. D. (2014). Economics of salt-induced land degradation and restoration. *Natural Resources Forum, 38*(4), 282–295. ff10.1111/1477-8947.12054ff.ffhal-01954413f

Rana, A. W., & Gill, S. (2022). *Country brief Pakistan: Strategy to promote climate smart agriculture practices*. International Food Policy Research Institute.

Satterthwaite. (2009). The implications of population growth and urbanisation for climate change'. In: Guzman, J. M., Martine, G., McGranahan, G., Schensul, D., & Tacoli, C. (Eds.), *Population dynamics and climate change* (pp. 45–63). UNFPA; IIED.

Savills Research. (2021). Regenerative agriculture. UK-Rural February. www.savills.co.uk/insight-and-opinion/

Shahid, S. A. (2010). Climate change–Its impact and biosaline agriculture. *Farming Outlook, 9*(2), 24–28.

Shahid, S. A. (2023). Desertification: A central problem to restoring ecosystems. In: Suleiman, M. K., & Shahid, S. A. (Eds.), *Terrestrial environment and ecosystems of Kuwait-assessment and restoration* (Ch. 5, pp. 119–137). Springer. https://doi.org/10.1007/978-3-031-46262-7_5

Shahid, S. A., Zaman, M., & Heng, L. (2018). Soil salinity: Historical perspectives and a world overview of the problem. In: Zaman, M., Shahid, S. A., & Heng, L. (Eds.), *Guideline for salinity assessment, mitigation and adaptation using nuclear and related techniques* (pp. 43–53). Springer. https://doi.org/10.1007/978-3-319-96190-3_2

Singh, A. (2013). Climate change, vanishing ecosystems and the challenge of achieving human prosperity. In: Huong, H., & Dhakal, T. N. (Eds.), *Governance approaches to mitigation of and adaptation to climate change in Asia* (pp. 13–37). Energy, Climate and the Environment Series. Palgrave Macmillan.

UAEMEW. (2010). *United Arab Emirates water conservation strategy* (p. 215). UAE Ministry of Environment and Water.

UNDESA Population Division. (2022). World population prospects, standard projections, archive, 2022 revision, United Nations Department of Economic and Social Affairs, Population Division. Retrieved November 6, 2024, from https://www.worldbank.org/en/topic/urbandevelopment/overview

World Bank. (2010). *Cities and climate change: An urgent agenda* (pp. 14–32). World Bank.

World Bank Group. (2022). World Bank Group Climate Change Action Plan 2021–2025. South Asia Roadmap. © World Bank. http://hdl.handle.net/10986/36321. (License: CC BY 3.0 IGO).

Wreford, A., Moran, D., & Adger, N. (2010). *Climate change and agriculture: Impacts, adaptation and mitigation*. OECD Publishing, Paris. https://doi.org/10.1787/9789264086876-en

Chapter 14
Mitigation and Adaptation Synergy in Arid Lands Agriculture for Food Security

Majda Khalil Suleiman

Abstract Arid lands due to their geographical location on the globe are highly vulnerable to climate change impact. The climate change might worsen existing national and regional disparities and reduce crop yield especially in arid lands located at lower latitudes where many developing countries are situated. Therefore, agricultural practices supporting climate change mitigation and facilitate adaptation to a changing climate are critical for arid land agriculture to ensure food security. Best agricultural practices must save water, improve soil health, reduce greenhouse gas emission to achieve synergies between mitigation and adaptation in agriculture. The arid lands have low aridity index (precipitation/potential evapotranspiration) classifying the drylands into four subtypes (hyperarid, arid, semi-arid and sub-humid). In the arid regions there is limited rainfall to meet the requirement of different crops for food security. The deficit in rainfall is offset by using supplemental water resources which may be marginal in quality and affecting the crop production. To have maximum crop production from arid lands, the soil properties are to be fully explored to mitigate soil effects on crop production and to formulate adaptation strategies to cope up the adverse effects both from the climate change and soil effects. The most important component of arid lands is to establish the land suitability classes for agriculture using established systems. The highly suitable soils for agriculture are to be included into the rationale land use policies where arable lands are to be designated for crop production and no other use. Marginal quality land which will never be used for agriculture to be recommended for other uses like housing projects to cope the pressure of growing population associated urbanization. It has been observed that land use change has increased greenhouse gases emission, and the farm sizes in Asia and Africa are decreasing. Globally there are 575 million farms of various sizes, distributed into different regions, and 500 million farms are less than 2-hectare size. In this chapter efforts have been made to propose multi-pronged mitigation-adaptation synergy to address climate change issues holistically to increase farm productivity for food security.

M. K. Suleiman (✉)
Food Security Program, Environment and Life Sciences Research Center, Kuwait Institute for Scientific Research, Safat, Kuwait
e-mail: mkhalil@kisr.edu.kw

M. K. Suleiman and S. A. Shahid (eds.), *Fostering Arid Lands Agriculture in the Face of Climate Change*, Advances in Global Change Research 81,
https://doi.org/10.1007/978-3-031-98490-7_14

Keywords Aridity index · Population · Marginal · Livelihood · Arable land · Shrinking · Synergy · Smallholder · Food losses and waste

14.1 Introduction

Arid lands are plagued with biotic and abiotic stresses hindering optimum crop production for food security. Arid lands are part of drylands, which are characterized based on the aridity index (AI). The AI is the ratio between precipitation (P) and potential evapotranspiration (P/PET), which is less than 0.65 for drylands (Middleton & Thomas, 1997; Schimel, 2010). In drylands water scarcity, prolonged drought and soil salinization can limit crop production (Shahid, 2022). Based on the aridity index (AI) four subtypes of drylands covering a total of 47.2% global land area are classified. Geographical distribution of drylands, delimited based on the aridity index (AI). The classification of AI is: Humid AI > 0.65, Dry sub-humid $0.50 < AI \leq 0.65$, Semi-arid $0.20 < AI \leq 0.50$, Arid $0.05 < AI \leq 0.20$, Hyper-arid $AI < 0.05$ (Abatzoglou et al., 2018).

Climate is the weather conditions prevailing in an area in general or over a long period, whereas the climate change (CC) is the long-term alteration in global weather patterns, especially increase in temperature, storm activity and the potential consequences of the greenhouse effect. Climate change is adding constraints to arid land agriculture crop production to meet the food demand of escalating population. In the last century, the global population was increased three times and now projected from the today 7.3 billion to an estimated 9.7 billion by 2050 and 11.2 billion by 2100, as the world population increases, more arable land needs to be converted for urban uses (Azadi et al., 2018; UN, 2017). It has long been realized that to feed the escalated population of 10 billion by 2050, 56% more food will be needed (Ranganathan et al., 2018).

The main concerns about climate change are multidimensional, increased level of greenhouse gasses (GHGs) emissions, associated global warming, prolonged drought, water scarcity, increasing level of natural resources marginality, salinization and sifting trend of soil biodiversity including insects and pests. The main GHGs in Earth's atmosphere are carbon dioxide (CO_2), methane (CH_4), and nitrous oxide (N_2O). The global warming is the biggest concern influencing economic growth. Therefore, the GHGs have far reaching effects through trapping heat and causing climate change. All these factors need a strong synergy development between multiple interconnected sectors to understand insights of climate change and ultimate consequences in terms of natural resources, vulnerable climate change and degradation trend. Well cultivated land or densely forested land can play significant role in cycling and sequestering GHGs (methane-CH_4; carbon dioxide-CO_2 and nitrous oxides NO_2) and similarly deforested land due to land use change can lead to release GHGs to the atmosphere, and the most significant spatial effect of land use change was recorded in African and Asian countries (Huang et al., 2023). Agricultural land conversion (ALC) to urban development (housing, infrastructure development) due to burgeoning population growth is occurring at rigorous level around the globe with a higher rate in emerging economies (Azadi et al., 2018). Therefore, the conversion of productive agriculture land into urban areas may be crucial obstacle to world food production, and a constraint to achieve the UN Sustainable Development Goal-2 (Zero hunger) by 2030, through doubling food production (Pedersen, 2018).

Mitigation and adaptation are two different terms commonly used in publications. Mitigation practices are more likely related to human intervention to reduce anthropic impacts on climate system through GHGs emission reduction and increasing GHGs sink "*soil carbon sequestration*" using multidisciplinary approaches. Whereas adaptation (IPCC, 2007) is about the adjustment in natural or human systems in response to actual or expected climatic stimuli or their effects, which moderates harm or exploits beneficial opportunities.

Adaptation can be multifaceted and can be divided in three types, that is anticipatory or proactive (which takes place well before the impact of climate change). The *autonomous or spontaneous* (adaptation that does not constitute a conscious response to climatic stimuli but is triggered by ecological changes in natural systems and by market oriented or welfare based changes in human systems). The third type is *planned adaptation*, which includes adaptation that is resulting from a deliberate policy decision, based on awareness and actions to maintain, or achieve the desired state.

Atmospheric temperature and rainfall are linked to climate change impact (CCI) in terms of melting glaciers, sea level rise (SLR), hurricanes, floods and drought. It is wiser to state that the CC is an environmental issue of global concern but has recently reached the realm of high politics. Thus, it is pertinent to visualize that the arid land countries are highly vulnerable to CCI, and these are located in the category of water scarce developing countries, and even countries of the Arab region (Shahid & Behnassi, 2014). The Arab region, especially the countries in the Gulf Cooperation Council (GCC) are associated with temperature fluctuations, decline in precipitation, increase in evapotranspiration, change in water cycle and depleting groundwater resources, which are considered crucial for the survival of agriculture farming and provision of balanced ecosystem services, survival of biodiversity (living beings-plants, native vegetation, wildlife of different nature, and microbial population) and also to combat desertification.

Water is crucial for both farming and survival of nature, living beings and biodiversity, and in combating the desertification. It is noteworthy that insignificant mitigation efforts are made in poor countries located in the arid environments due to the fact that these countries are developing and need to progress for economic growth. Scientifically determined, climate patterns record barely exists in arid region countries in general, and economic considerations are merely ignored.

This study is part of insight designed to present CC and CCI and briefly describe potential mitigation and adaptation in relation to arid lands and then try to develop synergy among different sectors. This will lead for a better managing natural resources for their sustainable uses to cope the CCI shocks and adoption of adaptation measures to optimize the resource use to improve farm lands productivities for food security.

14.2 Why Mitigations Are the Global Concerns?

The major global concerns of the twenty-first century are food security, soil degradation by land misuse, soil management, and anthropogenic increase in atmospheric GHGs as well as land use change from agriculture land towards other use, that are significantly affecting the GHGs emission (Huang et al., 2023). In addition to these concerns, soil salinity in arid lands countries is also increasing to an unprecedented levels and the farmers are not able to exploit their agricultural lands to their full production potential capacity, leading to poor livelihood and at extreme cases abandonment of farm land which were once highly productive (FAO, 2022). The CC impact is not only limited to agriculture lands but also affecting the ecosystems on broader scale. The misuse and over-exploitation of terrestrial ecosystems in arid land countries, like over grazing (anthropic effect) has also caused degradation, and due to less recovery potential under drought conditions. The extent to which any ecosystem can withstand land-use pressure or its resilience or potential to recover from a degraded state is a function of the ecosystem in question, land degradation level and drylands being particularly vulnerable (Abdelrahman, 2023; Olsson et al., 2019).

14.3 Small, Medium and Large Farms in Arid Lands

In poor arid land countries, the farmers owned mostly small, medium and sometimes large size farm holdings. Overtime due to inheritance the farm land is distributed among families and the farm sizes continued to shrinking. In few cases where the large farms are owned by progressive farmers, their ownership is expanding due to high profit compared to the subsistence famer. The shrinking arable land per capita (Smil, 2005; Jaggard et al., 2010; FAO, 2022) is an indication of future food impeding problems, as the finite natural resources (land, water and genetic potential) are unlikely to provide food to escalating population, especially in the poor developing countries (Asia and Africa). Cropland area per capita decreased in all regions between 2000 and 2021 as population increased faster than the cropland area, the world average declined by 18 percent (FAO, 2023).

There are a large number of subsistence farmers in poor and developing arid land countries such as in Asia and Africa. These subsistence farmers grow crops, raise livestock to maintain their family and very little, if any, surplus for sale. The climate change is impacting the livelihood of small scale farmers more than the large scale farmers. Therefore, the government supports to small scale farmers is essential to reduce poverty and continuation of farm operation as subsistence farming, this will compliment UN-SDG-2 (Zero hunger) to be achieved by 2030 (Pedersen, 2018). This is also supported by agricultural economists and other development specialists that investment in agriculture is an effective strategy for reducing poverty, inequality and hunger. This is especially in countries where the agriculture sector employed a large share of the population (FAO, 2012; World Bank, 2007), such as the areas in Asia and Africa where poor farmers are facing the vagaries of climate change, which is hindering them to have full production from their improvised farms. Investing in small scale farming makes sense as a large share of the world's food production (e.g., Fairtrade International, 2013)—or that a large share of the food consumed in Africa and Asia is produced by smallholders in those regions (HLPE, 2013; IFAD & UNEP, 2013).

14.3.1 Global Distribution of Small, Medium and Large Farms

A comprehensive information on the sizes and global distribution of various sized farms has been focus of a recent publication (Lowder et al., 2016). The study has used agricultural census data from 167 countries and territories. It is considered a ready source of information to develop strategies on a wider range of issues related to climate change adaptation and poverty eradication as envisioned in SDG-2. Following are the inferences observed from the published literature to track the distribution of different size farms (Tables 14.1 and 14.2).

The majority of farms are found in lower-middle or upper-middle income countries representing respectively 36 and 47% of the 570 million farms worldwide, whereas in low income countries they are 13%, and 4% in high-income countries (Lowder et al., 2016). A comprehensive analyses of farms distribution is shown in Table 14.2 (Lowder et al., 2016 and other sources as cited).

The Table 14.2 has clearly shown an increasing or decreasing trend of farm sizes. Overall on the global scale, it appears that farm sizes are decreasing, especially in Asia and Africa regions. This is very alarming, because there must be strong reasons behind this decreasing trend, including increasing marginality of the farms due to soil degradation (erosion and soil salinization), and impact of climate change in terms of

Table 14.1 Global distribution of agricultural farms

World/countries	Number of farms (millions)	Percent on world basis
World	570.0	100.00
China	200.0	35.00
India	137.0	24.00
Sub-Saharan Africa	51.0	9.00
Europe in Central Asia	40.0	7.00
Latin America and Caribbean	23.0	4.00
Middle East and North Africa (MENA)	17.0	3.00
Rest of the world	102.0	18.0

Source Lowder et al. (2016)

Table 14.2 Global distribution of small and medium sized farm distribution

Geographical coverage	Inferences
America and Europe	Between 1950 and 1990 average farm size increased
Asia	Average size decreased during 1950 to 1990
Africa	Average size decreased during 1970 to 1990 (Eastwood et al., 2010)
Africa, Asia and Latin America	An analyses of 16 countries revealed median farm size decreased during 1970–2000 (Hazell et al., 2010)
Latin America, Eastern Europe and Central Asia, Southeast Asia and Sub-Saharan Africa (selected land abundant countries)	There has been an increase in large scale farming *Southeast Asia*—increased large plantation due to expansion of palm oil industry; but rubber plantation increased from large plantation to production by smallholders (Deininger and Byerlee, 2012) *Sub-Saharan Africa*: Between 1970 and 1980's moves towards large farm agriculture failed, whereas, small farming operations have persisted, with an increase in large-scale investments in more recent years (Deininger and Byerlee, 2012)
Global coverage (114 countries)	Analyses of 114 countries between 1930 to 2000, has revealed that the average of mean farm size and the median of mean farm sizes decreased from 1930 to 2000 in most regions except Europe. Whereas, in African region a decrease in average farm size was recorded from 1960 to 1980 followed by a slight increase from 1980 onward. In other countries (Australia, New Zealand, the United States, Canada, Argentina, and Uruguay), an increasing trend is reported (FAO, 2013)
Global (91 countries)	Analyses of 91 countries during 1930–2000 has revealed average farm size has decreased for the African region as a whole and it has also decreased in China About 73% of farms are smaller than 1 ha and 85% are smaller than 2 ha. On average for 14 African countries 80% of holdings are smaller than 2 ha in size; they operate about 25% of the agricultural land In the European Union 50% of farms are smaller than 2 ha in size and operate only about 2% of the agricultural land (HLPE, 2013)

(continued)

Table 14.2 (continued)

Geographical coverage	Inferences
Africa and Asia	Since 1950, in the entire Africa and Asia, an average farm sizes are decreased. However, recently a rising trend is observed in entire Asia, but continue decreasing in Africa for quite some time (Masters et al., 2013)
Global (63 countries)	In general, it is observed that average farm size is larger in countries with higher average per capita GDP. Whereas, in countries with high average incomes farms larger than 20 ha operate 70% of land, while in the poorest countries 70% of land is operated by farms smaller than 5 ha (Adamopoulos and Restussia, 2014)
Africa (12 countries)	Africa is land abundant region (Deininger et al., 2011). The majority of the uncultivated arable land is located in few countries. Where, in all land-constrained countries average farm size has decreased. Most but not all land abundant African countries have shown an increase in average farm size (Jayne et al., 2014)

increasing temperature, prolonged droughts, salinization and water scarcity etc. In order to achieve, SDG-2 by 2030 a robust synergy is to be developed to mitigate the impact of climate change and other constraints to smallholder's farmers and adoption of strong adaptation options to sustain their farms for long term food production to meet the demand of escalating population.

14.4 Mitigation and Adaptation Synergy

On a global scale a total of 525 million farms of various sizes have been reported (Nagayets, 2005). The search of the literature (Hazell et al., 2010; Wiggins et al., 2010; IFAD, 2011; HLPE, 2013), has further clarified that there are about 500 million farms with size of less than 2 ha (<20,000 m^2). In order to have maximum crop production of high number of small farms a comprehensive mitigation-adaptation synergy is the need of the time.

14.4.1 Why Soil and Land Evaluation as a Means of Mitigation Synergy?

It is noted that different kinds of land use may have different requirements. The land itself and the land use are equally fundamental to land suitability evaluation. Likewise, the land use includes the broader context of agriculture production system and its biophysical and socio-economic environment. The designation of one land type for specific use must be based on comparison of benefits obtained and the inputs needed on different types of lands in addition to the provision of healthy ecosystem services.

Soil and land qualities can be assessed at various levels, that is from farm to national scales. At the national level, usually the soil assessment is done through science based soil surveys using international soil classification systems (Soil Science Division Staff, 2017; Soil Survey Staff, 2022) and soil and thematic maps for various uses are produced. Arid lands have huge barren lands which are uncultivated. These areas may have good arable land, and some areas may be not suitable for cultivation, due to occurrence of near surface hardpan, topography and highly gravelly nature. Therefore, for science based land use planning, soil mapping is the viable option to delineate highly suitable soils for agriculture (Dornik et al., 2022). Soil map will delineate the land into different parcels called Soil Map Units, which may be highly pure (one type of soil) called *soil consociation*, or mixture of soils in different proportions called "*soil associations*" or mixture of soils which cannot be separated at certain scales "*soil complexes* and *undifferentiated groups*" (Soil Science Division Staff, 2017). Using the component soils information in each soil map unit and making evaluation for suitability of map units using land use evaluation criteria (FAO, 1976), the soil map can be divided into different land suitability classes for agriculture (FAO, 1976; Table 14.3) as well as suitability for wildlife habitats and recreation purposes (Omar et al., 2002). Based on the land suitability classification, soil map units classified as S_1 and S_2 categories can be brought into production system, which will ultimately lead to improve food security, increase soil carbon sequestration (SCS), and mitigate GHGs emission. It has been recognized that agriculture as a sector offers several opportunities to mitigate the portion of global GHGs emissions that are directly dependent upon land use, land-use change, and land-management techniques (Rosenzweig & Tubiello, 2007). Thus conversion of uncultivated arable land to cultivated land will be considered an important synergy to mitigation climate change impacts in Asia and Africa where uncultivated land is abundant. Some views, that all fertile lands are already under cultivation, and this suggests that to expand irrigated agriculture on degraded land with lower yielding capacity are to be brought into cultivation, this can adversely affect food and livelihood production (AbdelRahman, 2023), which needs high inputs for crop production.

The other soils (N_1 and N_2) which are either currently or permanently unsuitable can be used for other uses, such as construction of industries, and new housing schemes to accommodate population growth to cope the pressure of urbanization.

Table 14.3 Irrigated land suitability classification (FAO, 1976)

Land suitability type	Class	Description of land suitability category
Highly suitable	S1	Highly suitable land with no significant limitations to the specified use
Moderately suitable	S2	Moderately suitable land with moderate limitations to the specified use
Marginally suitable	S3	Marginally suitable land with severe limitations to the specified use
Currently unsuitable	N1	Currently unsuitable land with severe limitations which cannot be corrected with existing knowledge and technology at currently acceptable costs
Permanently unsuitable	N2	Permanently unsuitable land with severe limitations which cannot be corrected

14.4.2 Assessing Inherent and Dynamic Soil Properties to Mitigate Climate Change Impacts

Soil health is the capability to sustain the productivity, diversity, and environmental services of the terrestrial ecosystems including food security. For a better assessment of soil health, it is appropriate to consider soils physical, chemical, mineralogical, and biological properties and their interaction in the soil system during the operation of management practices. These soil properties can be grouped into two main categories.

14.4.2.1 Inherent Soil Properties

Soils in general deliver multiple functions, however, different soils vary inherently in their capacity to function; therefore, quality is specific to each kind of soil. The soils possessing good inherent properties are basically productive. For example, soil texture and mineralogy are determined by the nature of the parent material the soil is established. The soil texture and mineralogy are considered *inherent* soil property and are determined by factors of soil formation i.e., parent material, climate, topography, vegetation, and time. These, almost permanent soil properties determine different uses of soils and differentiate other soils for specific use. Inherent properties (loamy texture) and nutrient retention clay with high cation exchange capacity are considered inherently productive soils. Such soils present dense vegetation covers and high farm land productivity, thus, leading to high soil carbon sequestration and reduce GHGs emission.

14.4.2.2 Dynamic Soil Properties

Dynamic soil qualities are important to be considered when the soils are used for specific purpose. The *dynamic* properties change with the level of management for a specific use; therefore, the user should forecast the consequences of management practice. As an example, if farm lands are irrigated with salty water, this may create soil salinity in the root zone which will hinder plant growth, and if the salts levels are developed to a toxic level, it may significantly affect the plant growth and yield decline. The dynamic soil salinity which effects plant growth must be measured at farm level scale (Shahid et al., 2008) and timely decision made to reverse the hiked salinity for optimum crop production, and soil carbon sequestration (Shahid et al., 2018a, 2018b; Zaman et al., 2018a, 2018b).

14.5 Potential of Protected Agriculture in Arid Lands Countries

The level of aridity index determines the nature of climate (hyperarid, arid, semiarid and subhumid). In general, the aridity indices of arid lands are low due to low precipitation and high potential evapotranspiration. In hot climatic conditions, where there is prolonged drought and insufficient and infrequent rainfall throughout the year, and water scarcity at the peak, the open field agriculture can be on stake. Under such circumstances it is wiser to state that open field agriculture to be shifted to Protected Environment Agriculture (PEA). This way production can be achieved throughout the year, such as growing vegetables (cucumber, lettuce, tomato, capsicum etc.). However, PEA may not be viable options to grow cereals for large scale production. The greenhouse facilitated the controlled environmental conditions and justified the importance of controlled environmental conditions for the maximum growth characters in tomato production (Chacha et al., 2023). Desert-greenhouse farming principally has a high potential for enhancing regional, national and global food security (Goddek et al., 2023).

14.6 Mitigation and Adaptation Options

IPCC (2023) has highlighted effective adaptation options which includes, (i) cultivar improvements, (ii) on-farm water management and storage, (iii) soil moisture conservation, (iv) irrigation, (v) agroforestry, and community-based adaptation (farm and landscape level diversification in agriculture, sustainable land management approaches, use of agro-ecological principles and practices and other approaches that work with natural processes).

Projections show that while overall global food production in the coming decades may keep pace with the food requirements of a growing world population, climate change might worsen existing regional disparities because it will reduce crop yields mostly in lands located at lower latitudes where many developing countries are situated.

14.6.1 Mitigation and Adaptation Options and Synergies

Scientists have different opinions about food security for coming years, some views that current practices may cope the pressure of population growth and meet the food demand, however, climate change impact may disturb the food production by 2050 (Alexandratos & Bruinsma, 2012). Therefore, locality based adaptation options are to be adopted to keep the pace if food production to meet the food demand of growing population and at the same time to keep the resources healthy and productive through adoption of climate smart agriculture and regenerative agriculture practices (Savills Research, 2021). Adaptation and mitigation synergy to be implemented for crop production should be aimed to minimize or alleviate the potential negative effects of climate change. Therefore, in the light of this, key synergies need to be debated locally and then finalized which is locality specific. There may be danger that some mitigation practices may compete with modifications to local agricultural practices aimed at maintaining production and income.

Strategies to enhance local adaptation capacity are therefore needed to minimize climatic impacts, alleviate negative impacts of climate change, and to maintain national, regional and global stability of food production. The potential mitigation and adaptation synergies in arid lands agriculture are summarized in Table 14.4, which can be taken as guideline and used on case by case basis, where appropriate and suitable for farming conditions.

14.7 Global Food Production and Potential Immediate Threat

Globally 9.2 billion metric tons (BMT) crop is produced, constituting 34% (3.128 BMT) produced in irrigated agriculture (competing water demands) and 66% (6.07 BMT) rainfed agriculture (stressed by climate change, drought and erratic weather) as reported by World Resource Institute https://www.wri.org/insights/growing-water-risks-food-crops). It is noteworthy that only ten countries (China, India, United States, Pakistan, Brazil, Egypt, Mexico, Vietnam, Indonesia and Thailand) produce 72% of the world's irrigated crops (FAO, 2023). It is to be noted that of 310 million ha irrigated land, 20% (62 million ha) is affected by soil salinization to various levels (Qadir et al., 2014). It has been further highlighted that, daily 2000 ha land is lost

Table 14.4 Proposed mitigation and adaptation synergies for Arid Land Agriculture for food security

Climate change impact	Mitigation strategy	Adaptation strategy	Synergy for arid lands
Global warming	Reduce GHGs	• Increase soil carbon sequestration • Rehabilitate deserts • Control overgrazing • Control conversion agriculture land to housing and infrastructure projects • Introduce heat resisted (drought) crop varieties • Adopt greenhouse production	• Strong science based policy development and implementation • Greenhouse farming to be integral part of climate-change mitigation and adaptation strategies (Goddek et al., 2023)
Prolonged drought	Drought resistant crop varieties	• Evolve drought resistant varieties • Increase soil moisture retention by using organic soil amendments for long term water retention, such as recalcitrant biochar (pyrogenic carbon). Biochar remains in soil for decades and not decomposed by soil microorganisms • Add farm based compost to agriculture fields and increasing soil carbon sequestration	• Introduce farm level biochar production kilns through empowering the farmers with the technology • Empower the famer with farm level composting technology to convert their farm waste into resource
Warming induced sea level rise	Shift coastal farms to inland	• Build dikes to protect farms from sea level rise • Coastal land farms to be abandoned and relocated to fertile inland	• Governments to develop policy to shift coastal sites agriculture farms to inlands
Water scarcity	Alternate water sources	• Use alternate water resources including treated waste water (Khan et al., 2022) • Use brackish water to grow salt-tolerant crops • Use desalinated water to grow cash crops (vegetables)	• Government to build infrastructures to collect and treat waste water • Construct pipelines to transport treated waste water to farming areas

(continued)

Table 14.4 (continued)

Climate change impact	Mitigation strategy	Adaptation strategy	Synergy for arid lands
Increased soil salinization	Reverse salinization	• Scientific diagnostics of salinity extent at country level • Regular salinity diagnostics at farm levels • Laser land leveling of farms for uniform distribution of water • Crop selection based on salt-tolerance level • Selection of suitable modern irrigation system • Integrated soil reclamation using combination of physical, chemical, hydrological and biological methods	• Establish soil salinity research institute to oversee the salinity problems and provide guidance to potential stake holders • Educate farmers with modern technologies to address salinity problem in their farms
Increased soil sodicity	Reverse sodification	• Analyze soil for sodicity level (exchangeable sodium percentage) • Apply gypsum based on gypsum requirement of sodic soil • Grow leguminous crops and use as green manuring to improve soil health • Improve in-depth soil properties through chiseling or subsoiling	• Government to provide financial support to farming community to purchase gypsum for sodic soils, establish suitable irrigation system, and provide agriculture machinery for subsoiling
Floods causing erosion	Water harvesting and storage, soil carbon sequestration	• Stabilization of sloppy lands • Soil stabilization • Revegetation of degraded lands • Water harvesting and diversion to small, medium and large scale storage facilities (dams) and use in agriculture lands	• Governments to fund revegetation projects • Government to support building water storage facilities at various scales • Construction of farm level mini dams

(continued)

Table 14.4 (continued)

Climate change impact	Mitigation strategy	Adaptation strategy	Synergy for arid lands
Deforestation for agriculture and urbanization	• Stop deforestation • Stop agriculture land conversion to develop new cities	• Increase forestation to reduce GHGs emission and increase soil carbon sequestration • Accommodate increased population in marginal lands (permanently unsuitable for agriculture) to build new cities and housing projects • Increase forestation to mitigate GHGs emission and increase soil carbon sequestration	• Governments to make informed decisions on land use based on their suitability for specific use • Governments to involve national soil science societies for advisory services for land use planning
Reduced farm sizes	• Reverse climate change impact to improve soil health	• Improve agriculture farm productivities • Strengthen the link between research-extension-farmers to enhance extension services • Shift from open field to protected agriculture (cash crops)	• Enhance extension services for fast transfer of technologies to farmers • Governments to provide funds for greenhouse establishment to grow cash crops
Livelihood crash	• Develop crop insurance policies	• Introduce crop insurance policies to recover damages to crops due to climate induced calamities	• Governments to secure farmers loses due to climate change induced damages (hurricanes, floods, drought etc.) • Provide crop insurance to the farmers
Reduced farm productivities	• Mitigate climate induced impacts on plant growth	• Precision agriculture • Adoption of climate smart agriculture practices • Adoption of regenerative agriculture practices • Educate farming community for new technologies	• Share education material with farming community in the form of easy to use friendlier brochures, pamphlets in local languages

(continued)

Table 14.4 (continued)

Climate change impact	Mitigation strategy	Adaptation strategy	Synergy for arid lands
Food loss and waste	• Improve storage facilities and educate food users	• Help farmers to improve storage facilities to reduce postharvest losses • Reach to food users through electronic media to educate how to reduce food loss and waste	• Governments to take initiative to mobilize financial resources to help farmers to improve storage facilities • Governments to buy agriculture commodities at farm gate rather than selling in open market • Governments to create large scale hygienic storage facilities for long term food store
Crop yield decline	• Crops intensification	Use seven sustainable intensification sub-types: integrated pest management, conservation agriculture, integrated crop and biodiversity, pasture and forage, trees, irrigation management and small or patch systems (Pretty et al., 2018)	• Strong commitment by government agriculture sector to support famers with timely inputs, subsidies and extension services on modern technologies

only due to salinity in over 75 countries (IWMI, 2014). At the current rate of farmland degradation and abandonment (2,000 ha daily) by the year 2100, 1/4 of the total irrigated land (310 million ha) will be out of production (Shahid and Alkandari, 2024). This is serious threat to food production in irrigated farms, thus, it is timely to mitigate the salinity spread and continue cultivating for food production, following science base diagnostics of the problems, and the implementation of integrated soil reclamation program using a holistic approach including a combination of physical, chemical, hydrological and biological methods (EAD, 2018; MAF, 2012; Shahid, 2010; Shahid & Rahman, 2011; Shahid et al., 2018a, 2018b; Zaman et al., 2018a, 2018b). It is to be noted that, economic damages from climate change have been detected in climate-exposed sectors, such as agriculture, forestry and fishery (IPCC, 2023), therefore, there is urgency to address potential mitigation options to continue producing food even from marginal natural resources by using science based best management practices and tested technologies. On the other hand, if salinity mitigation options are not adopted on priority basis, and business as usual will lead us to reduce current food production (9.2 billion metric tons) from irrigated agriculture to 6.9 billion tons by 2100. At that time global population with more than 10 billion, resulting into poor African countries to be on UN food aid policies.

14.8 Conclusions and Recommendations

Temperature rise scenarios during twenty-first century projected reduction in crop production more than increasing trend. About 34% crop production is from irrigated agriculture, where the climate change is likely to worsen the crop production. Especially salinity if not addressed holistically, at the current rate of farm land degradation, it may reduce global food production from irrigated agriculture by ¼ by 2100. Thus it is desirable to adopt climate smart and regenerative agriculture practices to maintain the pace of crop production and also to increase crop production to achieve UN-SDG-2 to zero hunger by 2030. To mitigate climate change (reduce GHGs emission) and to cope up the impact of climate change, there is dire need to develop region and country specific mitigation and adaptation synergies to cope the impact, and to continue production food for the current population and to feed the escalating population. However, the mitigation and adaptation synergy should have designed which should not affect land use which are environmental friendly and providing sustainable ecosystem services. It has been observed that in low income developing countries farm sizes are reduced whereas in high income counties the size is increased. The most of the reduction in sizes is recorded in Asia and China. It is noteworthy that in low income countries there is weak link between research, extension and farmers, which is leading to low crop production due to poor knowledge about latest technologies developed in crop production, water saving and nutrient management.

References

Abatzoglou, J. T., Dobrowski, S. Z., Parks, S. A., & Hegewisch, K. C. (2018). TerraClimate, a high-resolution global dataset of monthly climate and climatic water balance from 1958–2015. *Science Data, 5*, Article 170191. https://doi.org/10.1038/sdata.2017.191

AbdelRahman, M. A. E. (2023). An overview of land degradation, desertification and sustainable land management using GIS and remote sensing applications. *Rendiconte Lincei Scienze Fisechee Naturalli, 34*, 767–808. https://doi.org/10.1007/s12210-023-01155-3

Adamopoulos, T., & Restuccia, D. (2014). The size distribution of farms and international productivity differences. *The American Economic Review, 104*(6), 1667–1697.

Alexandratos, N., & Bruinsma, J. (2012). World agriculture towards 2030/2050: The 2012 revision. *ESA Working paper No.* 12-03. Rome, FAO.

Azadi, H., Keramati, P., Taheri, F., Rafiaani, P., Teklemariam, D., Gebrehiwot, K., Hosseininia, G., Passel, S. V., Lebailly, P., & Witlox, F. (2018). Agricultural land conversion: Reviewing drought impacts and coping strategies. *International Journal of Disaster Risk Reduction, 31*, 184–195. https://doi.org/10.1016/j.ijdrr.2018.05.003

Chacha, J. M., Thirumalai, M., Idawa, O. K., Chilwea, J. P., Kilamba, C. J., Hussy, B. D., Ishaq, H., & Rajendran, K. (2023). Greenhouse and open-field farming: A comparison through yield and growth parameters investigated in Dar es Salaam, Tanzania. *Innovation in Agriculture* (Vol. 6). https://doi.org/10.25081/ia2023-02

Deininger, K., & Byerlee, D. (2012). The rise of large farms in land abundant countries: Do they have a future? *World Development, 40*(4), 701–714.

Deininger, K., Byerlee, D., Lindsay, J., Norton, A., Selod, H., & Stickler, M. (2011). *Rising global interest in farmland: Can it yield sustainable and equitable benefits?* World Bank.

Dornik, A., Cheţan, M. A., Drăguţ, L., Iliuţă, A., & Daniel, D. D. (2022). Importance of the mapping unit on the land suitability assessment for agriculture. *Computers and Electronics in Agriculture, 201*, 107305, ISSN 0168-1699. https://doi.org/10.1016/j.compag.2022.107305

EAD. (2018). Farm soil salinity. In: *Groundwater atlas of Abu Dhabi emirate* (pp. 104–112). Environment Agency–Abu Dhabi, United Arab Emirates.

Eastwood, R., Lipton, M., & Newell, A. (2010). Farm size. In Pingali, P. L., & Evenson, R. E. (Eds.), *Handbook of agricultural economics,* Elsevier.

Fairtrade International. (2013). Powering up smallholder farmers to make food fair: A five point agenda. Retrieved January 4, 2025, from https://re-emergingworld.com/news/powering-up-smallholder-farmers-to-make-food-fair-a-five-point-agenda/

FAO. (1976). A framework for land evaluation. In *FAO soils bulletin* (Vol. 32). Food and Agriculture Organization, Rome Italy.

FAO. (2012). The State of Food and Agriculture 2012: Investing in agriculture for a better future, Rome Italy.

FAO. (2013). *2000 world census of agriculture methodological review (1996–2005): FAO statistical development series 14.* FAO.

FAO. (2022). Halt soil salinization, boost soil productivity. In *Proceedings of the Global Symposium on Salt-affected Soils.* 20–22 Oct 2021. Rome. https://doi.org/10.4060/cb9565en

FAO. (2023). Land use. In: *FAOSTAT*. Rome. Retrieved October, 2023, from https://www.fao.org/faostat/en/#data/RL. https://doi.org/10.4060/cc8166en-fig03

Goddek, S., Körner, O., Keesman, K. J., Tester, M. A., Lefers, R., Fleskens, L., Joyce, A., van Os, E., Gross, A., & Leemans, R. (2023). How greenhouse horticulture in arid regions can contribute to climate-resilient and sustainable food security. *Global Food Security, 38*, Article 100701. https://doi.org/10.1016/j.gfs.2023.100701

Hazell, P., Poulton, C., Wiggins, S., & Dorward, A. (2010). The future of small farms: Trajectories and policy priorities. *World Development, 38*(10), 1349–1361.

HLPE. (2013). Investing in smallholder agriculture for food security. In *A report by The High Level Panel of Experts on Food Security and Nutrition* (Vol. 6). Rome: FAO.

Huang, S., Ghazali, S., Azadi, H., Moghaddam, S. M., Viira, A.-H., Janečková, K., Sklenička, P., Lopez-Carr, D., Köhl, M., & Kurban, A. (2023). Contribution of agricultural land conversion to global GHG emissions: A meta-analysis. *Science of the Total Environment, 876*, Article 162269. https://doi.org/10.1016/j.scitotenv.2023.162269

IFAD. (2011). IFAD Proceedings of conference on new directions for smallholder agriculture, 24–25 Jan 2011, IFAD, Rome Italy.

IFAD and UNEP. (2013). Smallholders, food security and the environment. Rome Italy.

IPCC. (2007). Intergovernmental Panel on Climate Change Report. Fifth Assessment Report.

IPCC. (2023). Intergovernmental Panel on Climate Change Report. Sixth Assessment Report.

IWMI. (2014). Press release: World losing 2,000 hectares of irrigated farmland to salt-induced degradation daily. Retrieved November 4, 2024, from https://www.iwmi.cgiar.org/news/press-release-world-losing-2000-hectares-irrigated-farmland-salt-induced-degradation-daily/.

Jaggard, K. W., Qi, A., & Ober, E. S. (2010). Possible changes to arable crops yields by 2050. *Philosophical Transactions of the Royal Society, 365*, 2835–2851. https://doi.org/10.1098/rstb.2010.0153

Jayne, T. S., Chamberlin, J., & Headey, D. D. (2014). Land pressures, the evolution of farming systems, and development strategies in Africa: A synthesis. *Food Policy, 48*, 1–17.

Khan, M. M., Siddiqi, S. A., Farooque, A. A., Iqbal, Q., Shahid, S. A., Akram, M. T., Rahman, S., & Al-Busaidi, W. (2022). Towards sustainable application of wastewater in agriculture: A review on reusability and risk assessment. *Agronomy, 12*, 1397. https://doi.org/10.3390/agronomy12061397

Lowder, S. K., Skoet, J., & Raney, T. (2016). The number, size, and distribution of farms, smallholder farms, and family farms worldwide. *World Development, 87*, 16–29.

MAF. (2012). Oman salinity strategy (Main report). Ministry of Agriculture and Fisheries, Sultanate of Oman.

Masters, W. A., Djurfeldt, A. A., De Haan, C., Hazell, P., Jayne, T. S., Jirström, M., & Reardon, T. (2013). Urbanization and farm size in Asia and Africa: Implications for food security and agricultural research. *Global Food Security, 2*(3), 156–165. https://doi.org/10.1016/j.gfs.2013.07.002

Middleton, N. J., & Thomas, D. S. G. (1997). *World Atlas of desertification* (2nd ed.). Hodder Education Publishers. (15 Sept, 1997).

Nagayets, O. (2005). Small farms: Current status and key trends. In *The future of small farms: Proceedings of a Research Workshop, Wye, UK*, 26–29 June 2005. International Food Policy Research Institute.

Olsson, L., Barbosa, H., Bhadwal, S., Cowie, A., Delusca, K., Flores-Renteria, D., Hermans, K., Jobbagy, E., Kurz, W., Li, D., Sonwa, D. J., & Stringer, L. (2019). Land degradation. In: P. R. Shukla, J. Skea, E. Calvo Buendia, V. Masson-Delmotte, H.-O. Pörtner, D. C. Roberts, P. Zhai, R. Slade, S. Connors, R. van Diemen, M. Ferrat, E. Haughey, S. Luz, S. Neogi, M. Pathak, J. Petzold, J. P. Pereira, P. Vyas, E. Huntley, K. Kissick, M. Belkacemi, & J. Malley (Eds.) *Climate Change and Land: an IPCC special report on climate change, desertification, land degradation, sustainable land management, food security, and greenhouse gas fluxes in terrestrial ecosystems.* https://doi.org/10.1017/9781009157988.006

Omar, S. A. S., Grealish, G., Shahid, S. A., King, P., Al-Shatti, K., Abo-Rezq, H., Madouh, T., & Roy, W. (2002). Land-use evaluation of arid lands using soil information and GIS for wildlife habitats and recreational camping. *Arabian Gulf Journal of Scientific Research, 29*(4), 219–227.

Pedersen, C. S. (2018). The UN sustainable development goals (SDGs) are a great gift to business! *Procedia CIRP, 69*, 21–24. https://doi.org/10.1016/j.procir.2018.01.003

Pretty, J., Benton, T. G., Bharucha, Z. P., Dicks, L. V., Flora, C., Godfray, C., Goulson, D., Hartley, S. E., Lampkin, N., Morris, C., Pierzynski, G., Prasad, P. V. V., Rockstrom, J., Smith, P., Thorne, P., & Wratten, S. D. (2018). Global assessment of agricultural system redesign for sustainable intensification. *Nature Sustainability, 1*, 441–446. https://doi.org/10.1038/s41893-018-0114-0

Qadir, M., Quillerou, E., Nangia, V., Murtaza, G., Singh, M., Thomas, R. J., Drechsel, P., & Noble, A. D. (2014). Economics of salt-induced land degradation and restoration. *Natural Resources Forum, 38*(4), 282–295. https://doi.org/10.1111/1477-8947.12054

Ranganathan, J., Waite, R., Searchinger, T., & Hanson, C. (2018). How to sustainably feed 10 billion people by 2050, in 21 Charts. World Resource Institute. Retrieved January 2, 2025, from https://www.wri.org/insights/how-sustainably-feed-10-billion-people-2050-21-charts

Rosenzweig, C., & Tubiello, F. N. (2007). Adaptation and mitigation strategies in agriculture: An analysis of potential synergies. *Mitigation Adaptation Strategies Global Change, 12*, 855–873. https://doi.org/10.1007/s11027-007-9103-8

Savills Research. (2021). Regenerative agriculture. UK-Rural February. www.savills.co.uk/insight-and-opinion/

Schimel, D. S. (2010). Drylands in the earth system. *Science, 327*, 418–419.

Shahid, S. A. (2010). Climate change–Its impact and biosaline agriculture. *Farm Outlook, 9*(2), 24–27.

Shahid, S. A. (2022). Innovative thinking and use of salt-affected soils in irrigated agriculture. In *Proceedings of the Global Symposium on Salt-affected Soils.* 20–22 Oct 2021 (pp. 292–293). Rome. https://doi.org/10.4060/cb9565en

Shahid, S. A., & Alkandari, A. J. (2024). Halophytic crops as a solution for food security, land rehabilitation, and mitigating future water crises by utilizing marginal quality waters. In J. C. Dagar, S. R. Gupta, & A. Kumar (Eds.), *Halophytes vis-à-vis saline agriculture* (Ch. 17, pp. 441–478). Springer. https://doi.org/10.1007/978-981-97-3157-2_17

Shahid, S A., & Behnassi, M. (2014). Climate change impact in the Arab region: Review of adaptation and mitigation potential and practices. In: M. Behnassi, M. S. Mteng'e, G. Ramachandra, K. Shelat (Eds.), *Book–Vulnerability of agriculture, water and fisheries to climate change–Toward sustainable adaptation strategies* (Ch. 2, pp. 15–38). Springer.

Shahid, S. A., & Rahman, K. (2011). Soil salinity development, classification, assessment and management in irrigated agriculture. In M. Pessarakli (Ed.), *Handbook of plant and crop stress* (3rd ed., Ch. 2, pp. 23–39). CRC Press.
Shahid, S. A., Dakheel, A., Mufti, K. A., & Shabbir, G. (2008). Automated in-situ soil salinity logging in irrigated agriculture. *European Journal of Scientific Research, 26*(2), 288–297.
Shahid, S. A., Zaman, M., & Heng, L. (2018a). Introduction to soil salinity, sodicity, and diagnostics techniques. Chapter 1. In M. Zaman, S. A. Shahid, L. Heng (Eds.), *Guideline for salinity assessment, mitigation and adaptation using nuclear and related techniques* (pp. 1–42). Springer, Cham. https://doi.org/10.1007/978-3-319-96190-3_1
Shahid, S. A., Zaman, M., & Heng, L. (2018b). Salinity and sodicity adaptation and mitigation options. Chapter 3. In M. Zaman, S. A. Shahid, L. Heng (Eds.), *Guideline for salinity assessment, mitigation and adaptation using nuclear and related techniques* (pp. 55–88). Springer, Cham. https://doi.org/10.1007/978-3-319-96190-3_3
Smil, V. (2005). Do we need higher farm yield during the first half of the 21st century? In Yields of farmed species.
Soil Science Division Staff. (2017). Soil survey manual (pp. 437). USDA Handbook no 18. United States Department of Agriculture, U.S. Government Printing Office, Washington, DC 20402.
Soil Survey Staff. (2022). *Keys to soil taxonomy* (13th ed.) USDA-NRCS.
UN. (2017). World Population Projected to Reach 9.8 Billion in 2050, and 11.12 Billion in 2100 [WWW Document]. UN DESA U. N. Dep. Econ. Soc. Aff. Retrieved June 7, 2018, from https://www.un.org/development/desa/en/news/population/world-population-prospects-2017.ht ml
Wiggins, S., Kirsten, J., & Lambi, L. (2010). The future of small farms. *World Development, 38*(10), 1341–1349.
World Bank. (2007). World Development Report 2008: Agriculture for development. World Bank. Retrieved November 30, 2024, from https://www.wri.org/insights/growing-water-risks-food-crops
Zaman, M., Shahid, S. A., & Heng, L. (2018a). Irrigation systems and zones of salinity development. Chapter 4: In M. Zaman, S. A. Shahid, L. Heng (Eds.), *Guideline for salinity assessment, mitigation and adaptation using nuclear and related techniques* (pp. 91–111). Springer. https://doi.org/10.1007/978-3-319-96190-3_4
Zaman, M., Shahid, S. A., & Heng, L. (2018b). Irrigation water quality. Chapter 5. In M. Zaman, S. A. Shahid, L. Heng (Eds.), *Guideline for salinity assessment, mitigation and adaptation using nuclear and related techniques* (pp. 113–131). Springer. https://doi.org/10.1007/978-3-319-96190-3_15

[illegible]

Chapter 15
Mitigation and Adaptation in Arid Lands Agriculture to Achieve Sustainable Development Goal 2–End Hunger

Habibah Al-Menaie and Merlene Ann Babu

Abstract Arid land agriculture faces severe challenges due to non-conducive environmental conditions, including low precipitation, high evapotranspiration, water scarcity, high temperature, prolonged drought, soil salinity, land degradation, and increased frequency of extreme weather events. Thus, climate change possesses significant impacts on arid land agriculture. Nearly a billion people around the world are going hungry, and achieving food security amidst unpredictable climate changes are increasingly challenging. In order to address hunger holistically, the United Nations set up 17 Sustainable Development Goals (SDGs) covering various aspects, including SDG 2 (end hunger, achieve food security and improved nutrition and promote sustainable agriculture). The SDG 2 has various targets to be achieved by 2030. The Target 2.3 aims to double the agricultural productivity and income of small-scale food producers by 2030. Additionally, Target 2.4 seeks to progressively enhance land and soil quality by 2030, through implementing sustainable food production systems and resilient agricultural practices. There are few years to reach 2030, under current circumstances when soil and water resources are continuously degraded due to natural (climatic affects) and anthropogenic effects (secondary salinization), the SDG-2 appears to be very ambitious. However, if proper mitigation and adaptation practices are followed and various countries take necessary actions, the SDG2 can be achieved partially by 2030. In this chapter the constraints to arid land agriculture are addressed and mitigation (reduction in greenhouse gases) and adaptation (soil, crops, water, fertility etc.) efforts suitable for arid land agriculture are amassed to improve arid land agriculture for food security.

H. Al-Menaie (✉) · M. A. Babu
Food Security Program, Environment and Life Sciences Research Center, Kuwait Institute for Scientific Research, Safat, Kuwait
e-mail: hmanaie@kisr.edu.kw

M. A. Babu
e-mail: mbabu@kisr.edu.kw

M. K. Suleiman and S. A. Shahid (eds.), *Fostering Arid Lands Agriculture in the Face of Climate Change*, Advances in Global Change Research 81,
https://doi.org/10.1007/978-3-031-98490-7_15

Keywords Climate change · Mitigation · Adaptation · Sustainable development goal · Food security · Hunger · Climate smart · Regenerative · Conservation agriculture

15.1 Introduction

Climate change poses significant impacts on agriculture drastically affecting the crop productivity, food security and rural livelihoods owing to its increased temperatures, extreme weather events and altered precipitation patterns (Bibi & Rahman, 2023). Mitigation and adaptation constitutes the two major strategies adopted to combat climate change. Greenhouse gases (GHGs) contribute critically to climate change due to the ability to warm the planet by trapping heat in earth's atmosphere (Sovacool et al., 2021). Therefore, addressing the GHGs emission plays a significant role in reducing the impact on society and ecosystems. Mitigation constitutes different strategies intending to decrease GHGs emission and thus slow the progress of global warming. Crop management practices such as conservation tillage enhances soil carbon sequestration by reducing soil disturbance and minimizing organic matter decomposition (Zhang et al., 2022). Cover crops and crop rotations enhance soil health by increasing soil fertility which in turn reduces the use of synthetic fertilizers (Acharya et al., 2024). Agroforestry integrates trees and shrubs into crop production systems, thereby enhancing biodiversity, soil and water quality as well as carbon sequestration (Yasin et al., 2023). Over application of consumable resources such as water and fertilizer for crops drastically reduces the resource use efficiency. Precision agriculture techniques optimizes fertilizer use and mitigates climate change by reducing the nitrous oxide emissions due to fertilizer overdose (Hassan et al., 2022). Effective livestock management practices such as improved feed practices and manure management reduces methane emissions via enteric fermentation in ruminants, thereby contributing to climate change mitigation (Ayal & Mamo, 2024). In addition, use of renewable energy sources in agricultural operations reduces the use of fossil fuels and decrease GHG emission (Majeed et al., 2023). In contrast to mitigation, adaptation ensures recovery of agricultural systems from climatic shocks through modifying practices, technologies and policies that reduce risks while exploiting any benefits of climate change. Development of climate resilient crop varieties is one of the major adaptation strategies (Cabusora, 2024). Improved soil management techniques that include application of organic amendments such as compost and biochar enhance soil health as well as water retention and nutrient availability (Al-Kilani, 2024). Drip and sprinkler irrigation as well as construction of water storage facilities enhance water security especially during dry periods (Zhao & Boll, 2022). Crop diversification and livestock provides an excellent platform for developing alternative sources of income and nutrition. In addition, providing farmers with timely climate information, early warning systems for extreme weather events and strengthening agricultural policies to support climate-resilient farming practices contributes greatly to climate change adaptation. Thus, a multidimensional approach that integrates mitigation and adaptation strategies is critical to ensure sustainable agricultural development and global food security through reducing GHG emissions and improving resilience to climate change.

15.2 Arid Land Countries

The arid region occupies nearly 43% of the lands surface (Abdelhak, 2022) and is described by aridity index (AI) which is calculated by dividing the precipitation amount an area receives by the amount lost to evapotranspiration (Gamo et al., 2013). It gives a fraction less than 0.50 in arid regions (FAO, 1989). It is characterized by extreme temperatures, long dry periods, occasional intense storms and highly variable rainfall with annual precipitation lesser than 250 mm. The prevailing high temperatures in the region leads to high evaporation rates that results in significant loss of water. Based on the AI values the global arid zone is divided into four categories namely hyperarid, arid, semiarid and dry sub humid areas. Majority of the regions possess soil which is low in organic matter and nutrient contents. Sparse vegetation is found in arid lands as only species that are highly adapted to the arid climate with the ability to conserve water, deep roots, small or waxy leaves could exist (Abdelhak, 2022).

15.2.1 Asia

Around 32% of global arid zone is occupied in Asia (Thomas, 2011). The arid land climate in Asia is defined by limited and variable rainfalls, extreme temperature variations and limited vegetation. The predominantly arid regions include Arabian Peninsula (Saudi Arabia, Oman, UAE, Yemen), Thar desert (India and Pakistan), Central Asian desert (Kazakhstan, Uzbekistan, Turkmenistan), Gobi Desert (Mongolia and China), and Iranian Plateau (Iran, parts of Afghanistan, Pakistan) (Dennell, 2013). The countries experience a unique climate where the summer season temperature often exceed 40 °C, whereas winter temperatures vary with some areas exhibiting cold winters while others dropping below freezing (Dennell, 2013). Agriculture productivity is severely affected by extreme weather events, scarce water resources, changing precipitation patterns, low percentage of arable land, desertification and climate change effects which results in decreased yield, and loss of land productivity.

15.2.2 Middle East

The Middle east region is vast and generally experiences hot and arid climate characterized by extreme dryness, low and erratic rainfall and high temperature, less fertile soil with low nutrient content and lack of biodiversity. It covers vast area which includes countries such as Kuwait, Bahrain, the United Arab Emirates, Oman, Qatar, Yemen, Iraq, Saudi Arabia, Jordan, and some parts of Iran and Israel (Francis & Fonseca, 2024). The temperature in summer season often exceeds 45 °C reaching up to 50 °C (Francis & Fonseca, 2024). Humidity is low, though coastal areas experience higher levels especially during summer. The arid zones record an annual rainfall generally less than 200 mm. At times it occurs as short intense bursts causing flash floods. With scarce water resources and in general lacking permanent rivers or lakes, the region depends on groundwater and desalinated sea water to cater water needs. The region experiences dust storms that is most prevalent between May and September (Al-Delaimy, 2020). Poor and improper land management practices in addition to overgrazing and deforestation leads to desertification which in turn reduce the land productivity. With the view to overcome these challenges raised by extremely high temperatures and limited water resources, different adaptation as well as mitigation programs which include improved soil, water and nutrition management practices, conservation agriculture, use of renewable energy sources are adopted to improve climate change resilience.

15.2.3 Sub-Saharan Africa

Above 15% of Earth's land surface is occupied by the Sub-Saharan Africa (SSA), an incredibly diverse region (Charles et al., 2018). Global warming is a serious issue and recent studies state that temperature is increasing over time in almost all SSA countries. It is predicted to experience greater rise in temperature and warming than rest of the world in the coming years (Umeh & Gil-Alana, 2024). The arid climate zones are mainly found in Sahel region, a semi-arid belt towards south of the Sahara Desert. During the hottest months high temperatures often exceed 40 °C. It is characterized by limited precipitation, high temperatures, extreme dryness, less fertile soil with low nutrient contents, high species endemism, and sparse vegetation cover etc. (Deng, 2000). Highly variable and unpredictable nature of precipitation in these areas records a very low annual rainfall that is less than 250 mm. These limitations drastically reduce agricultural productivity, thereby affecting the health and livelihood of the people in the region due to food and water shortages. Desertification is worsened by anthropic activities such as overgrazing and deforestation. Different adaptation and mitigation strategies are adopted to improve water management, implement sustainable agricultural techniques and enhance the resilience of community to climate change.

15.2.4 Others

The major countries which possess regions characterized by arid land features such as low precipitation, high temperature, limited water resources other than Asia, Middle East and SSA includes Australia, Mexico, Chile, United States of America (USA), Argentina and Peru. Owing to their arid environment, it is challenging to manage agriculture and biodiversity in these countries. Australia is the driest inhabited continent on the planet. The Great Victoria and the Simpson deserts are characterized by arid and semiarid climate (Hesse et al., 2004). More than 52% of Mexico is occupied by arid and semiarid climates. The northern Mexico regions such as Chihuahuan Desert and Sonoran Desert display low rainfall and arid conditions (Perez-Aguilar et al., 2021). The driest desert on Earth, Atacama Desert in northern Chile is characterized by extreme dryness, high UV radiation levels, saline and oxidizing soils (Azua-Bustos et al., 2022). Similarly, southwestern part of the USA which includes Mojave Desert and parts of Great Basin as well as Pantagonian region and areas near the Andes in Argentina experience arid conditions. The coastal area of Peru which includes Sechura desert exhibits hyperarid climate marked by high temperatures under the influence of cold Humbolt current. It supports little vegetation where the inland dunes are occupied by woody species and coastal dunes are covered by salt grass as well as woody species (Jury, 2024).

15.3 Climate and Climate Change

Climate and climate change phenomenon is wide spread and is not limited to drylands only as in general perceived. Climate change presents variations in climatic conditions over a long period of time. In the following section, climate change aspects are briefly described with specific reference to arid land countries.

15.3.1 Arid Land Countries

The long-term average, frequency, and extremes of several meteorological variables, determines the climate of a region of which temperature and precipitation holds a significant role (Giddings et al., 2005). The long term mean precipitation forms the simplest index to classify climate regions where three major types of climate such as Mediterranean, tropical and continental climates exist in arid zones. Summers are hot without rains while winter season records mild temperatures in the Mediterranean climate. Rain is prevalent during autumn and winter seasons. However, with tropical climate, rainfall commonly occurs during the summer and it possess long and dry winters. In case of continental climate, rainfall is dispensed evenly across the year, with comparatively greater rainfall during summers (FAO, 1989). Arid regions generally experience high temperature that results in higher evaporation from water bodies, exacerbating the adverse effects of climate change. The arid zone climate is featured by a cool dry season followed by hot dry season and moderate rainy season. It also fluctuates both seasonally and daily. Temperature range of 15–22 °C is common in arid regions (Vaughn, 2005) and during hot dry season, the daytime temperature could reach up to 45 °C and drops to 15 °C at night (Juang et al., 2012). A previous study on surface air temperature trends over global land from 1901 to 2009 noted that semi-arid regions contribute 44.46% to global annual mean land surface temperature (Juang et al., 2012). Climate change worsens arid climate resulting in even hotter conditions leading to higher evaporation rates, lower soil moisture content and severe water scarcity. A warming climate imposes serious implications which includes health risk to local population and reduced agricultural production in the region. Previous studies have reported significant expansion of global dry lands in the past 60 years which will continue expanding in the future (Feng & Fu, 2013). A recent study has warned even drier future for arid lands with 10% expansion by the end of twenty-first century under high GHG emission scenario (Allan & Douville, 2024). Nevertheless, unpredictable precipitation patterns along with shrinking glaciers due to altered weather patterns reduce the availability of water exacerbating water scarcity in the region. More intense, prolonged and frequent droughts leads to fewer harvests and reduced agricultural yield. Over time, climate change combined with anthropogenic activities leads to loss of biodiversity and land degradation affecting the native species. Changes in weather patterns leads to increased flooding risk after short intense rainfall due to poor water absorption capacity of the soil and sparse vegetation in arid areas. Besides, increased frequency of dust storms could impair the air quality and transport pathogens and pollutants over large distances. Therefore, the direct and indirect effects of climate change on agriculture and livelihood in arid regions could affect food security, imposing serious economic and social impacts.

15.3.2 Impact on Agriculture and Food Security

The impact of climate change on agriculture constitutes direct and indirect effects which possesses significant risks to food security, livelihoods, and economic stability (Alotaibi, 2023). Integrated assessment models have noted that the indirect effects were much larger than the direct effects as the climate effects range from −1% to + 6% in 2040, whereas the indirect effects of climate range from −12% to + 15% (Calvin & Fisher-Vanden, 2017). Changes in temperature alter the traditional planting and harvesting schedules significantly affecting the crop yield. Higher temperatures under climate change induced higher rates of evapotranspiration that results in shortened growing seasons and reduced crop yields (Mahmood et al., 2012). A study conducted using a new integrated climate-hydrological-economic model in Iran, an arid region has predicted 5.88% and 6.05% increase in minimum and maximum temperatures respectively and a 30.68% decrease in precipitation (Shayanmehr et al., 2022). Climate change has significantly reduced the quantity and quality of rice and wheat productivity in Asia (Din et al., 2022). The crop model DSSAT has predicted a yield reduction of 15.2% in rice and 14.1% in wheat, whereas APSIM model has shown a yield reduction of 17.2% in rice and 12% in wheat in a study conducted in Pakistan (Habib-ur-Rahman et al., 2022). A study on wheat conducted in Egypt North Nile Delta has shown that a decrease in wheat yield up to 17.6% could happen for a 1–4 °C increase in temperature (Kheir et al., 2019). Besides, a significant reduction in the protein content, sugar and starch percentage was also noted in a study conducted on impact of climate change on wheat (Asseng et al., 2019). In addition, the water stress caused by climate change exacerbates water scarcity issue in the arid region and reduced the area under cultivation (Kirby et al., 2016) switching to less water intensive, and less profitable crops. Nevertheless, the heat stress on plants brought by extreme heat events damages the crops productivity and quality. Altered precipitation patterns, another major consequence of climate change could lead to prolonged droughts or unexpected heavy rains in arid regions. It was shown that drought stress reduced the protein, starch as well as sugar content of the wheat (Rao et al., 2021). Droughts limit the water available for irrigation and cause soil degradation, while soil erosion and nutrient losses are caused due to heavy rainfalls. Soil degradation caused by soil erosion, desertification and salinization turns the arable land less productive, and difficult to sustain crop productivity over time (Kumar & Pani, 2011). Besides, increased carbon dioxide levels in the atmosphere affect the crop nutritional quality (Ziska, 2022). A direct correlation was drawn between disease infestation and high temperature as well as carbon dioxide levels associated with climate change (Tonnang et al., 2022). Thus, through its effects on agriculture, food production and distribution systems, climate change imposes both direct and indirect effects on food security.

Addressing the impact of climate change on food security requires a multi-faceted approach which involves international cooperation, and community-level engagement to ensure availability, access, safety, stability, utilization of food. Food availability is one of the critical components in ensuring food security (Capone et al., 2014). Temperature rise, unpredictable precipitation patterns and harsh weather conditions brought about by climate change exhibit yield declines. Besides, it can lead to proliferation of pests and plant diseases which reduces the crop yield further. Food accessibility constitutes another important factor that underlines food security (Capone et al., 2014). Negative effects of climate change on crop production affects communities largely dependent on agriculture, especially the subsistence farmers. The reduced household income affects the ability of the families to purchase food and daily used accessories, health and children education. Access to food is also affected due to disrupted local as well as global food supply chains caused due to extreme weather events and political issues. A reduction in protein, zinc and iron content is noted in rice and other staple crops like wheat under climate change induced elevated carbon dioxide levels (Victor et al., 2022). Food safety is very important in determining food security, and higher temperatures can result in the proliferation of food borne pathogens and mycotoxins (Duchenne-Moutien & Neetoo, 2021). Seasonal variability caused by climate change leads to irregular food production cycles that affects the stability of food supplies. Thus, ensuring food security in the face of climate change via combining different mitigation and adaptation strategies is crucial to build resilience against the adverse impacts of climate change on food production critical for global stability and human well-being.

15.4 UN Sustainable Development Goals

The United Nations Sustainable Development Goals (SDGs) are a set of 17 and are adopted by all UN Member States in 2015 as part of the 2030 Agenda for Sustainable Development (https://sdgs.un.org/goals). The aim is to create a better and more sustainable future for everyone by the year 2030 through addressing a wide range of issues such as poverty, inequality, climate change, environmental degradation, peace, and justice.

15.4.1 Sustainable Development Goals Related to Food Security

Seventeen SDGs aimed toward sustainable development are formulated with its core focus on five points (i) people's well-being, (ii) planet with protection of earth's ecosystem, (iii) prosperity with eradication of poverty and inequality, (iv) peace, and (v) partnership (Tremblay et al., 2020; WHO, 2015). Among 17 SDGs, number 2, 5, 12, 13 and 15 are related to food security are briefly given below:

SDG 2—Zero Hunger: End hunger, achieve food security and improved nutrition, and promote sustainable agriculture.

SDG 5—Gender Equality: Achieve gender equality and empower all women and girls.

SDG 12—Responsible Consumption and Production: Ensure sustainable consumption and production patterns.

SDG 13—Climate Action: Take urgent action to combat climate change and its impacts.

SDG 15—Life on Land: Protect, restore, and promote sustainable use of terrestrial ecosystems, manage forests sustainably, combat desertification, and halt and reverse land degradation and halt biodiversity loss.

15.4.2 SDG 2–End Hunger, Achieve Food Security and Improved Nutrition, and Promote Sustainable Agriculture

Hunger eradication, improved nutrition and enhanced food security are important factors that underline overall human health and well-being especially children's development and productivity (Ritchie et al., 2018). Around 690 million people and

155 million children worldwide suffer from hunger and chronic malnutrition respectively. A comprehensive approach needs to be adopted to enable provision of reliable, sufficient, and nutritious food for all the people all year round. The SDG 2 adopted by UN in 2015 aimed to end hunger, attain food security and improve nutrition, and promote sustainable agriculture by 2030. The SDG2 aims at improving food production while preserving natural resources under reduced environmental footprints, thereby ensuring environmental sustainability. Thus, it promotes sustainable agriculture, maintains ecosystem services, enhance biodiversity conservation, mitigate climate change and ensure long term viability. It also contributes to economic growth by boosting economic stability, promotes social stability and alleviates social inequalities through addressing root causes of hunger and malnutrition. However, the main challenges in achieving SDG2 are rapidly growing population, climate change, food waste, malnutrition, inadequate infrastructure, economic downturns such as pandemic etc. The deleterious impacts of climate change on agricultural productivity makes it really challenging for arid regions with limited resources to achieve SDG2 within set time frame of 2030. The consequences of climate change such as reduced crop yields, decreased nutrition values, increased pest and disease incidence, water scarcity, livestock health impacts result in reduced agricultural productivity and increased hunger. Despite its collective efforts, the world is not in line yet to attain zero hunger by 2030. Global cooperation facilitated through collective actions sharing technologies and best practices via international partnerships is vital to secure SDG 2 which is key for human well-being, environmental sustainability and economic development. The key targets of SDG 2 Zero Hunger are as follows.

1. End Hunger (2.1)
2. End Malnutrition (2.2)
3. Double Agricultural Productivity and Incomes (2.3)
4. Sustainable Food Production Systems (2.4)
5. Maintain Genetic Diversity (2.5)
6. Increase Investment (2.a)
7. Correct and Prevent Trade Restrictions and Distortions (2.b)
8. Ensure Proper Functioning of Food Commodity Markets (2.c)

Among eight targets of SDG-2, target number 2.3 and 2.4 are directly related to agriculture and food security.

15.4.2.1 SDG 2: Target 2.3 - Double Agricultural Productivity and Incomes

Target 2.3 seeks to double the agricultural productivity and income of small-scale food producers and other groups in the society including women, indigenous peoples, family farmers, pastoralists, and fishers. This could be achieved through secure and equal access to land, productive resources, inputs, knowledge, financial services, markets, opportunities for value addition and non-farm employment (https://globalgoals.org/goals/2-zero-hunger). With the view to increase the productivity while maintaining environmental sustainability, equal access to sustainable practices and resourceful inputs such as land, seeds, tools and technology is necessary. Capacity building programs that empower small scale producers through theoretical as well as hands-on-training on sustainable farming techniques and practices are to be organized at farm gate level. These training programs will strengthen technical and entrepreneurial skills of farmers as well as social protection programs to render support to local producers. Access to different types of financial services is provided to support investments in agriculture and mitigate risks. Infrastructure such as storage facilities, transportation networks, and market information systems are developed and access to local, national, and international markets is facilitated to ensure fair prices and stable income for small-scale producers. Employment and value addition opportunities are generated through processing, packaging and branding of products. To achieve the target, precision agriculture, digital tools, organic farming, agroforestry and climate smart agricultural practices are to be promoted. In addition, public–private partnerships must be fostered to provide resources, knowledge, and market access investing more in the research and development of innovative technologies to enhance agriculture productivity. Thus SDG 2- Target 2.3 contributes towards economic growth, food security, social equity and environmental sustainability through doubling agricultural productivity and incomes.

15.4.2.2 SDG 2: Target 2.4—Sustainable Food Production Systems

Target 2.4 of SDG 2 seeks to develop sustainable food production systems and promote resilient agricultural practices that improve productivity and production. The main goal is to maintain ecosystems, enhance adaptability to climate change, harsh weather, drought, flooding and other disasters while progressively improving land and soil quality (https://globalgoals.org/goals/2-zero-hunger). This includes the implementation of resilient agricultural practices like crop diversification, agroforestry, conservation agriculture and integrated pest management. Crop rotation, inter cropping, reduced tillage, organic farming are various ways to manage land in a sustainable manner and at the same time combat soil degradation and desertification. These practices result in better soil fertility, water retention and increased biodiversity that contribute to higher ecosystem services. It bolstered productivity and output hence building resilience to climate change, perils such as extreme weather, drought or flooding, and other disasters. It also encourages sustainable agricultural practices to improve climate resilience by encouraging the use of climate-smart crops, which are the most suitable for the local environmental context and fit ecosystem profiles and the promotion of organic agriculture. A comprehensive approach involving innovation, policy support, community engagement, and international collaboration is required to achieve Target 2.4 by 2030. Investment needs to be intensified through public–private partnerships in agricultural research and development for the dissemination of innovative technologies in sustainable farming practices. The policies made for sustainable agriculture can increase agricultural productivity and use of services while conserving natural resources and also can contribute to global food security. To support farmers to adopt sustainable practices and improve sustainable food production and mitigate environmental degradation, they need financial backing through appropriate subsidies as well as crops insurance policies. International cooperation in sustainable agriculture allows the sharing of knowledge, technology, and resources. Introduction of global initiatives connecting farmers to markets as well as certification schemes and labels for sustainably produced goods promotes sustainable food production systems and climate resilience.

15.5 Mitigation and Adaptation to Climate Change Impacts

Different mitigation and adaptation strategies adopted in the agricultural management practices could improve crop production as well as restore ecosystem in the face of climate change. Implementation of water management practices such as installation of efficient irrigation systems, water conservation and storage infrastructure, rainwater harvesting as well as aquifer recharge, enhance water use efficiency and reduce water wastage (Roldan-Canas & Moreno-Perez, 2021). Development of mutant crop varieties tolerant to different biotic and abiotic stress factors prominent in the arid regions can improve crop yield under changing climate conditions. Reforestation and afforestation (Cunningham et al., 2015) using drought tolerant species enhances biodiversity and ecosystem resilience. In addition, conservation agricultural practices which includes reduced tillage, mulching and crop rotation enhance soil health, crop productivity and resource use efficiency (Al-Menaie et al., 2024a; Sadiq et al., 2024). Sustainable energy solutions such as solar and wind energy reduces the dependence on fossil fuels which in turn lower GHGs emission (Majeed et al., 2023). Introduction of different policies on climate resilience as well as strengthening cooperation on transboundary resource sharing contributes to sustainable resource management during climate change (FAO, 2010, 2012). Involvement of local communities raising awareness on sustainable practices and climate resilience is another critical factor in the implementation of adaptation strategies. Besides, research and development focused on innovative technologies tailored to specific impacts of climate change could improve the performance of agricultural sector in arid region. Thus, a holistic approach that takes into account water management, sustainable agricultural practices, renewable energy sources and robust policy frameworks in arid regions can enhance their resilience to climate change.

15.6 Climate Smart Agricultural Practices Specific to Arid Land Agriculture

There is a set of climate smart agricultural practices used globally, which may be specific to specific locations and climatic conditions. Not all CSA practices are applicable everywhere, therefore, objectives oriented, location and climate specific CSA are to be selected for their successful use to cope climate change impacts, improve farm productivities leading the crops intensification for food security.

15.6.1 Soil, Crop, Water, Fertility Management Aspects

Climate-smart agricultural practice is an approach that aims to increase productivity, enhance climate change resilience and reduce GHG emissions to manage agricultural systems. It focusses on adaptation of different farming practices to climate variability as well as mitigation strategies that slow down the onset of climate change impacts. Through enhancing soil health, reducing erosion, optimizing water use, enhancing water and nutrient use efficiencies, improving water quality, reducing irrigation related costs, improving crop yield, and resilience, enhancing biodiversity conservation, enhancing carbon sequestration, reducing GHG emissions and promoting integrated pest management (Zheng et al., 2024), the CSA plays an inevitable role in sustaining productivity and ensuring long-term viability in arid lands. Conservation tillage is one of the key CSA practices that ensures minimum soil disturbance by planting seeds into the individual seed furrows through direct drill or planter (Rahman et al., 2021; Shahid & Alshankiti, 2013). The different forms of conservation tillage are no till farming, strip till, ridge till, mulch till, reduced tillage, zone tillage, vertical tillage, conservation chiseling etc. These practices help to maintain soil structure, reduces erosion, improves soil moisture, reduce evaporation losses and increase organic matter content. Similarly, the use of organic mulches such as plant residues, straw, wood chips as well as inorganic mulches on soil surface facilitates soil health improvement, erosion control, soil temperature regulation, enhanced soil moisture content and nutrient availability, reducing evaporation losses, increased yield, and weed suppression etc. (Riaz et al., 2022). The long-term effect of tillage on soil organic matter content was studied by comparing the effect of continuous tillage and no tillage in Mediterranean regions, and it was found that no tillage significantly increased soil organic carbon and soil total nitrogen by 22% in the shallow layers of soil. In contrast, the continuous tillage significantly reduced soil organic carbon by 3% and soil total nitrogen (Mazzoncini et al., 2016). A previous study conducted on desert oasis farmland has shown that reduced tillage and straw mulching has significantly increased soil carbon and nitrogen accumulation up to 59 and 130% respectively and decreased carbon footprint for crop production in comparison to conventional farming techniques. In addition, the crop yield was shown to increase up to 14% under rotational cropping (Yang et al., 2017). No-tillage significantly decreased the soil carbon dioxide emission and rice straw mulching significantly enhanced rice yield (Dossou-Yovo et al., 2016). Another study conducted in Japan displayed similar increase in the soil organic matter content and soil fertility upon addition of organic matter in upland fields of different types of andosols (Iwasaki et al., 2017). In contrast, reduced tillage did not significantly enhance the yield (Koga, 2017). It was shown that the tied ridging improved water use efficiency by 75%, 48%, 17%, and 85% for cotton, sorghum, groundnut, and wheat, respectively, compared to that of control treatment (Ali et al., 2017). A two-year study conducted in maize reported increased nitrogen use efficiency, nitrogen harvest index, nitrogen remobilization, and nitrogen remobilization efficiency under no-till tillage system in a continuous cover cropping system (Habbib et al., 2016). Contour farming technique reduces soil erosion and helps in water conservation by planting along natural contours of the land retains the top soil rich in nutrients; while terrace farming technique involves creating stepped levels on slopes (Scherr et al., 2012). Incorporation of organic amendments like composting, green manure and well decomposed animal manure possess great impact in improving the overall soil health via enhancing the soil structure, organic matter content, fertility and water retention capacity. Agroforestry is another CSA practice which offers numerous benefits by integrating trees and shrubs in to agricultural systems (Sileshi et al., 2023), which is described in a later section. Similarly, cover crops increase soil moisture retention, reduce erosion and upgrade the soil structure (Koudahe et al., 2022). A study conducted in West Africa Sahel, has noted an increased crop productivity as well as income for low input mono-cropping of millet or millet/cowpea intercropping upon planting ziziphus trees at the density of 80 plants ha $^{-1}$ along with farmyard manure (Bado et al., 2020).

In arid regions with scarce irrigation water resources, intense heat and harsh environmental conditions, precise determination of actual crop water requirements and irrigation scheduling is essential to ensure water use efficient crop production (Al-Menaie et al., 2021). Soil moisture sensors could be used to determine the soil moisture content, and water can be delivered directly to the root zone based on actual crop water requirement via drip irrigation system to reduce wastage of water. The use of efficient micro-irrigation systems such as drip irrigation avoids overuse of water as it delivers water directly to the roots reducing the evaporation losses (Kumar et al., 2023). It was shown that drip irrigation increased crop yield by 28.92%, 14.55%, 8.03% in comparison to flood, border, and furrow irrigation respectively (Yang et al., 2023). Another study conducted for a period of 5 years in arid region of north-west China noted that drip irrigation decreased soil evaporation by 30 mm (~23%) in average, shortened the growth days by 13 days (~9%) and decreased evapotranspiration by 7% when compared with the traditional border irrigation (Wang et al., 2023). Likewise, a significant increase in the plant height, leaf area index, yield, water use efficiency and economic benefit was noted in maize under drip irrigation when compared to conventional border irrigation (Liu et al., 2023). Similarly, sprinkler irrigation reduces water wastage and run off by distributing water evenly across the fields. In addition to irrigation systems, rainwater harvesting systems such as water collection systems can be introduced in areas with rainfall to store rain water for use in dry periods (Belay et al., 2023). The micro catchments capture and retain rainwater in the soil.

Soil salinization has a long history and not a recent phenomenon affecting the crops (Shahid et al., 2018a). Soil salinity is a major issue that drastically affects crop production and through leaching which flushes out extra salts from root zone using water, soil health as well as crop yield can be improved (Hoshan, 2022; Shahid et al., 2018b; Zaman et al., 2018). In addition to water, nutrients play a crucial role in improving crop productivity and its efficient management is essential to optimize soil fertility while maintaining soil health (Al-Menaie et al., 2024b). Precise amounts of organic and inorganic fertilizers required by the crops could be determined and applied directly to the root zone to avoid overuse of fertilizers (Penuelas et al., 2023). Biochar, a stable carbon rich form of charcoal, also improves soil health by enhancing soil moisture and nutrient availability (Iqbal et al., 2023). The selection of crop and cropping techniques is another important factor to be considered in enhancing crop production in arid regions. The use of drought and salinity tolerant crop varieties maintain productivity under drought and saline conditions respectively (Al-Ragam et al., 2023). Cropping techniques such as cover cropping, mixed cropping as well as crop rotation with legumes also helps to enhance soil moisture conservation. This reduces the use of water and contributes greatly to increased ecosystem services and community resilience (Al-Menaie et al., 2024c). Thus, the numerous benefits offered by different CSA practices contribute to sustainable agricultural systems under climate change through increased crop yields, greater climate resilience and improved environmental health.

15.7 Regenerative Agricultural Practices

Regenerative agricultural (RA) practice is a conservation and rehabilitation approach aimed at enhancing food and farming systems. Through regenerating natural resources and restoring ecosystems, particularly focusing on soil health, biodiversity, and water cycles, it aims to create long term sustainability. Although both RA and CSA aim to create a sustainable and resilient farming system (Table 15.1). The major emphasis of CSA practices is on productivity, adaptation and mitigation. However, overlaps exist between the two as both utilize methods such as agroforestry, cover cropping, crop rotation and no-till farming, that contribute to soil health, water management, biodiversity, and carbon storage.

The RA encompasses a range of sustainable integrated farming techniques aimed at enhancing topsoil regeneration, biodiversity, ecosystem services, water cycle bio sequestration, resilience to climate change, and farm soil health and vitality (Khangura et al., 2023). The key RA practices include the following practices.

Table 15.1 Climate smart agriculture and regenerative agriculture

	Climate-smart agriculture	Regenerative agriculture	Overlap
Primary Objective	Adapt agricultural practices to climate change, enhance productivity, and reduce greenhouse gas emissions	Ecosystem, soil health and biodiversity restoration	Build sustainable and resilient farming systems
Soil Health	Secondary to productivity and resilience objectives	Central focus and primary objective	Application of practical methods like cover cropping and no-till farming to improve soil health
Carbon Sequestration	Promotes practices that reduce GHGs emissions and sequester carbon where possible	Core principle is to actively encourage carbon capture in soil	Carbon sequestration through agroforestry, crop rotation
Productivity Focus	Importance for sustainable increase in crop yields alongside resilience to climate variability	Accords priority to ecosystem health, at times at the cost of short-term yields	Both work on sustainable yields but differ in priorities
Water Management	Practices for improving water use efficiency,	Concentrates on natural processes to improve soil water retention	Use drip irrigation, rainwater harvesting, and to enhance soil water retention
Biodiversity	Not a major goal, often considered as a beneficial by-product	One of the primary goals to enhance biodiversity	Implement agroforestry and integrated pest management (IPM) to enhance biodiversity

15.7.1 *Cover Cropping*

It is an agricultural practice which involves planting crops in soil to cover the soil as well as produce agriculture commodities. Diverse cover crops such as buckwheat, alfalfa, clovers, and various cereals, are grown during off-seasons to reduce soil erosion, increase water retention, improve soil health, reduce weed growth, and enhance biodiversity (Quintarelli et al., 2022). It protects the soil from erosion by wind and water, reducing soil loss and maintaining its structure. It also plays a key role in soil quality management where it adds organic matter to the soil upon decomposition. It improves soil health and fertility resulting in a productive growing environment (Adetunji et al., 2020).

Cover crops offers a multitude of benefits that promote a more sustainable and environmentally friendly farming approach enhancing soil health, biodiversity, and overall ecosystem resilience. It provides a natural source of fertility for subsequent crops, suppresses weed growth and thus reduces the need for synthetic fertilizers and herbicides (Mukherjee & Sen, 2021). Some cover crops can reduce disease pressure in agricultural systems and serve as hosts for beneficial insects that prey on pests (Gerhards & Schappert, 2020). In addition, cover crops enhance water infiltration into the soil and reduce water erosion, thereby mitigating the impacts of heavy rainfall events and droughts (Delgado et al., 2021). The cover crops also play a vital role in transforming farmland into carbon sinks through carbon sequestration (Lal, 2021). It creates additional value for the community through generating potential income sources for farmers, and supporting sustainable agricultural practices.

15.7.2 Holistically Managed Grazing

Holistically managed grazing is a sustainable approach that allows the movement of livestock regularly between small paddocks or sections of pasture to maximize forage utilization while allowing adequate rest and recovery periods for vegetation. Also known as intensive rotational grazing, this practice supports the overall health and vitality of the land mimicking natural grazing patterns (Savory, 2013). It enhances soil health by enhancing the growth of beneficial microorganisms, increasing soil organic matter, and improving soil structure (Nordborg, 2016). This leads to improved water retention, nutrient cycling, and overall soil fertility. It also increases the biodiversity and plays an important role in maintaining healthy ecosystems via growth of diverse plant species and attracting a variety of insects, birds and other wildlife (Janzen, 2011). Besides, it improves water infiltration and reduce runoff leading to increased water retention in the soil (Dobert et al., 2021). The enhanced growth of grasses and plants under rotational grazing leads to carbon sequestration as it utilizes carbon dioxide from the atmosphere for photosynthesis. Thus, it helps mitigate climate change by soil carbon sequestration in the soil and vegetation, reducing GHGs emissions (Dowhower et al., 2020). Thus it results in improved farm profitability, environmental stewardship, and resilience to climate change.

15.7.3 No-Till Farming

Under no-till farming, seeds are sowed into the untilled soil subjecting it to minimal disturbance. Also called low till, or zero-till, or direct drilling. This technique is fundamental to regenerative farming and possess number of benefits. It preserves soil structure, retains organic matter, reduces soil erosion, improve water retention and water infiltration (Blanco-Canqui & Ruis, 2018). The increased carbon sequestration along with reduced need of chemical fertilizers and mechanical operations results in reduced emission of GHGs thereby contributing greatly to climate change mitigation (Ogle et al., 2019). Nevertheless, it enhances plant growth and biodiversity supporting a diverse range of soil organisms which contribute to nutrient cycling and soil health such as earthworms, insects, and microorganisms (Froslev et al., 2022). Thus, no till farming fosters climate-resilient farming practices facilitating sustainable and resilient agricultural systems.

15.7.4 Composting

Composting involves conversion of organic waste into organic matter generating nutrient-rich humus that benefits both soil as well as plants and increase soil microbial population. It supplies essential nutrients slowly and steadily, thereby generating valuable inputs for farming. The enhanced soil biodiversity improves plant growth, helps control pests and improves nutrient cycling thereby alleviating the need for synthetic fertilizers. It improves soil structure and fertility that result in enhanced biological activity and nutrient availability (Martinez-Blanco et al., 2013). Thus, it reduces the need for synthetic fertilizers and lowering the cost of production and chemical runoff into water bodies. The improved water-holding capacity of the soil reduces the irrigation requirements of plants and helps to withstand drought conditions (Gould, 2015). It also contributes significantly to climate change mitigation, through soil carbon sequestration (Dijkstra & Keitel, 2024). In addition, it minimizes the methane emissions from decomposing waste in anaerobic conditions (Nordahl et al., 2023). Thus, by adopting composting, farmers can transition to more sustainable and environmentally friendly agricultural methods supporting the ecological balance sought in RA.

15.7.5 Agroforestry

Agroforestry (AF) is an RA practice that combines agriculture and forestry integrating trees, crops, and livestock in a mutually beneficial manner (Ntawuruhunga et al., 2023). Different modes of agroforestry such as alley cropping, silvopasture and forest can be adopted to improve the sustainability and productivity of farming systems. It optimizes ecological interactions and offers numerous benefits which includes improved soil health, biodiversity, climate change mitigation, and sustainable land management (Ali et al., 2024; Rioux, 2018). The organic matter developed from the decay of leaves and roots of trees and shrubs in agroforestry improves overall soil health owing to the enhanced soil organic carbon levels. It serves as carbon sinks storing the captured atmospheric carbon dioxide in biomass and soils (Raj et al., 2020). The AF practice helps to bind the soil, prevents nutrient runoff, and retain moisture, leading to increased soil fertility and decreased dependency on synthetic fertilizers. The AF also reduces water runoff and erosion via improving water infiltration and retention (Udawatta & Gantzer, 2022). The AF regulates water cycle by enhancing ground water which is very useful in regions prone to drought or heavy rainfall, and enhances biodiversity as trees and shrubs creates habitats for different wild life including birds, insects and soil organisms (Santos et al., 2022). The AF also provides diversified income streams producing numerous products like fruits, nuts, medicinal plants resulting in enhanced the food security. Furthermore, the AF regulates the microclimate (van Noordwijk et al., 2021) by creating a more favorable environment for crops and livestock, reducing heat stress, and wind speed, and maintaining humidity levels. Overall, the AF results in improved ecosystem resilience, pest and disease control promoting sustainable agricultural systems.

15.7.6 Conservation Crop Rotation

By alternating crops with varying root structures and nutrient requirements, crop rotation facilitates efficient use of nutrients and ensures natural restoration of the soil nutrients. Thus, it improves soil fertility reducing the need for synthetic fertilizers (McDaniel et al., 2014). It also aids in the conservation of water allowing more efficient use of water resources reducing runoff (Bordovsky et al., 1994). It contributes to mitigate climate change by reducing GHGs emissions through enhanced carbon sequestration (Yu et al., 2022). The use of different crops enhances biodiversity as it creates habitats for various beneficial insects, pollinators and other organisms (Beillouin et al., 2021). Thus, it promotes a healthy and biologically active soil ecosystem. Weed populations could be managed efficiently as different crops compete differently with weeds. Alternating crops could also break the life cycles of pests and diseases contributing greatly to pest and disease management (Yin et al., 2010).

These RA practices aim to restore soil health, increase resilience to climate change, reduce environmental impact, and promote sustainable farming methods that benefit both the ecosystem and agricultural productivity.

15.8 Mutation Breeding

Yield and nutritional quality are two other major factors drastically affected by climate change as unpredictable weather events often results in crop failures, yield variability and reduced nutrient content. Different stress factors such as heat, cold, drought, salinity, nutrients, arise as a consequence of climate change and developing stress tolerant varieties is crucial in climate change adaptation. Plant mutation breeding plays a significant role in climate change adaptation and mitigation where it harnesses the potential of genetic diversity and targeted trait improvements to develop climate resilient crops via induced genetic mutations (Forster & Shu, 2012; Penna et al., 2023). It accelerates the conventional breeding process in developing new crop varieties through thorough screening and evaluation process. Development of mutant crop varieties with enhanced yield and better quality could maintain the crop productivity under climate change (Shabani et al., 2022; Wanga et al., 2023). Around 3332 mutant cultivars are produced from agricultural important 228 crop species through induced mutagenesis (IAEA, 2024; http://mvd.iaea.org; Manjaya & Gupta, 2023). Plants that can combat extreme temperatures through maintaining productivity and quality were developed through induced mutation via physical or chemical mutagens. Heat tolerant mutant varieties could avoid land conversion which can otherwise release stored carbon by producing stable yield at rising temperatures. Research on different factors that regulate tolerance to heat were conducted in different crops such as wheat (Comastri et al., 2018; Draeger & Moore, 2017), rice (Takai et al., 2020; Xu et al., 2020), maize (Momcilovic & Ristic, 2007; Ristic et al., 2004) barley (Sakata et al., 2000; Savin et al., 1996), tomato (Ding et al., 2018; Raja et al., 2017) etc. to prevent yield losses in regions under extremely high and low temperatures. Similarly, development of deep-rooted plants via mutation breeding results in enhanced carbon sequestration that could mitigate climate change. Mutant crops with deeper root systems (Rizi & Mohammadi, 2023) and improved water use efficiency were also developed to adapt to drought conditions (Kell, 2011). With the view to establish sustainable agriculture, salinity tolerant mutant crops that thrive in saline soils were developed to improve crop production in salt affected lands (Mwando et al., 2020; Thabet & Alqudah, 2023). Changes in weather patterns results in change in duration of growing seasons. Rapidly maturing plants were developed to cope with the shorter growing seasons (National Research Council, 1996; Gautam et al., 2021).

Mutation breeding accelerates the development of climate resilient crop varieties and reduce GHGs emission, thereby contributing to climate change mitigation. Mutant crop varieties with improved water use efficiency and drought tolerance reduced the energy utilized in irrigation systems and the associated carbon foot print (Human et al., 2011; Wanga et al., 2023). Yield losses due to disease and pest incidence associated with climate change could be alleviated through the introduction of mutant crop varieties resistant to emerging pathogens and prevailing pests. These disease and pest resistant mutant varieties eventually reduce the usage of chemical pesticides thereby reducing emission of GHGs (Nanhapo et al., 2024). Likewise, mutation breeding technique was harnessed to develop bioenergy crops (Sihono et al., 2020) as well as crop varieties with improved nitrogen use efficiency to reduce the emission of GHGs. Mutant crops with enhanced photosynthetic capacity also contribute to climate change mitigation owing to the increased carbon capture due to higher yields (Hong et al., 2022). Thus, mutation breeding plays a crucial role in climate change adaptation and mitigation and helps agriculture adapt to its challenges through enhancing resilience, reducing GHGs emissions, and promoting sustainable agricultural practices.

15.9 Policy Development to End Hunger in Poor African Countries

A multifaceted approach that addresses the root causes of food insecurity, including poverty, agricultural productivity, infrastructure, governance, and climate resilience are important in developing policies intended to end hunger in poor African countries. A large part of agricultural workforce in Africa is smallholder farmers. Therefore, providing financial support, training and access to seeds, fertilizers and technology to small holder farmers leads to significant agricultural development (Runge et al., 2004). Introduction of subsidies reduces the financial burden on famers encouraging increased investment on inputs required for improving and sustaining agricultural productivity. This enables the farmer to continue with their agricultural operations in the time of unexpected crises. Similarly, introduction of crop insurance schemes could offer compensation to farmers during pandemics and other climate shocks that drastically affect the crop production. Educating the farmers with the relevant knowledge and skills on climate resilient farming methods that enhance resilience to climate variability and extreme weather events through agricultural extension services can improve food security. The implementation of sustainable farming methods, such as crop diversification, agroforestry, conservation agriculture, and efficient water and fertilizer management systems could be promoted to establish climate resilient agriculture in the region (Chakrabarty & Suri, 2023). In addition, nutrition education programs to be promoted to addresss malnutrition and improve dietary practices. Farmers access to market needs to be ensured to facilitate sale of their produce at fair prices. Towards this end, local, regional and international market linkages as well as rural infrastructure such as roads, storage facilities need to be strengthened to provide farmers with a broader access to market. Social protection schemes provide conditional financial assistance to households under poverty line and food assistance programs address immediate food needs and support vulnerable population, ensuring food security (Regmi & Meade, 2013). Capacity building is an important factor in addressing the food security issues and strengthening of capacity of local and national institution plays important roles in coordinating food security policies effectively (USAID, 2008). Development of more inclusive and effective policies ensuring the participation of stakeholders, farmers and private sectors facilitates the design of comprehensive strategies to fight hunger in these countries. Trade and economic policies are to be improved that reduce the trade barriers in transporting food across borders to improve food security. Economic diversification needs to be encouraged in rural areas to avoid sole dependency on agriculture and create alternative income (Fan et al., 2013). Equitable access to land needs to be promoted and long-term investments in land and sustainable agricultural practices need to be encouraged. International cooperation is critical in the implementation of global partnerships that mobilize resources, support capacity building programs to help enhancement of food security (Otekunrin, 2021).

15.9.1 Future Research Direction to Achieve SDG2

Achieving SDG2 to end hunger, ensure food stability, enhance nutrition, and foster sustainable farming by 2030, demands coordinated and interdisciplinary research efforts across several areas. Advanced biotechnology and genetics research need to be conducted to develop more number of mutant climate smart crop breeds with higher yield, resistance to various stress factors, improved nutrition, and adaptability to the specific region. Socio-economic research needs to be conducted to develop different poverty alleviation strategies to increase house hold incomes and access to safe, healthy and adequate food. Impact of existing policies on improving food security needs to be evaluated followed by identification of policy gaps and best practices. Nutrition and food safety contributes greatly to improving food security. Therefore, different nutrition enhancing programs such as fortification, biofortification, diet diversification as well as methods to improve food safety standards and practices need to be researched. Comprehensive studies on the impacts of climate change on agriculture in arid regions need to be conducted followed by development of context specific climate change adaptation and mitigation strategies. Agricultural practices that can sequester carbon such as AF and RA needs to be researched to mitigate climate change. Introduction of digital agricultural techniques such as remote sensing, Internet of Things, big data analytics could improve monitoring and decision making which in turn improves productivity (Balasundram et al., 2023). The incorporation of renewable energy sources such as solar powered irrigation and bioenergy needs to be investigated to promote sustainability. Gender and social equity need to be considered to address gender disparities and ensure equal access for marginalized groups. In addition, installation of vertical farming, rooftop gardens, and community gardens along with efficient distribution networks could improve urban food production and increase access to fresh produce. Thus, focused, holistic and innovative research in collaboration with the local government, research institutions, private sector and local communities is essential to design innovative solutions to antagonize the challenges of food security with the view to achieve SDG2.

15.9.2 Conclusions and Recommendations

Adoption of climate change mitigation and adaptation strategies is crucial in achieving 'SDG-2 "end hunger' in arid regions marked with unique vulnerabilities which includes scarce water resources, high temperatures, less fertile soil, and harsh environmental conditions. Climate change impacts worsen the agriculture scenario in these regions with intense and frequent droughts, altered heat events and precipitation patterns. Adaptation strategies such as cultivation of crop varieties tolerant to various biotic and abiotic stress factors, application of improved soil, water and nutrient management practices can enhance yield as well as improve soil health significantly. Besides, mitigation strategies such as reduced tillage, mulching, crop rotation, agroforestry, renewable energy adoption enhance crop production under improved resource use efficiency and reduced emission of greenhouse gases. Thus adaptation and mitigation strategies are important in minimizing adverse impacts of farming practices on environment and building climate change resilience in these fragile ecosystems.

Sustainable agricultural practices that enhance productivity while reducing the environmental foot print are needed to be strengthened. The use of climate-resilient crops and application of farming techniques that aid carbon sequestration and reduce greenhouse gas emissions should be promoted with the view to mitigate and adapt to climate change. Ecosystem restoration using reforestation and afforestation and utilization of renewable energy sources such a solar and wind energy contribute significantly towards sustainable agriculture. Capacity building programs which facilitate knowledge and technology transfer through training and workshops needed to be promoted to educate the farming community with the latest and the most effective climate smart agricultural practices. Subsidies and crop insurance programs are proposed to be introduced for farmers allowing them to invest in inputs required for improved and sustained crop productivity. An efficient support system which includes improved infrastructure, easy access to the climate information and other resources such as quality seeds, modern farming technologies, for small holder farmers could help them to cope with climate change consequences. Through strengthening food supply chains, food loss and waste could be reduced enhancing food security. Thus, promoting climate-resilient agriculture, strengthening food supply chains and markets, enhancing social protection, supporting vulnerable populations, fostering global cooperation and knowledge sharing, strengthening different policies and research on climate resilience and sustainable agricultural practices, food security could be achieved promising a healthy and sustainable future to all sections of the society.

An awareness on the diversified diets and nutrition ensuring access for all especially in vulnerable regions will help in reducing malnutrition and hunger significantly. Gender equality must be taken into consideration and women empowerment needs to be encouraged by providing women farmers with equitable property access, credit and other resources. Different programs that empower women and girls in rural regions need to be conducted to enhance their engagement in crop production and food security. Collaboration is to be developed between governments, private sector, society, international organization and involvement of local communities and stakeholders in decision analysis, facilitate adoption of different cutting-edge techniques to ensure food availability, food access, and food nutrition and food safety.

Acknowledgements We would like to thank Kuwait Institute for Scientific Research for the valuable support and resources that made this work possible. We would also like to extend our deep appreciation to the editors of the book, for their valuable support and feedback to enhance the quality of the manuscript.

References

Abdelhak, M. (2022). Soil improvement in arid and semiarid regions for sustainable development. In: Jhariya, M.K., Meena, R.S., Banerjee, A., & Meena, S.N. (Eds.), *Natural resources conservation and advances for sustainability* (pp. 73–90). Elsevier. https://doi.org/10.1016/B978-0-12-822976-7.00026-0

Acharya, P., Ghimire, R., & Acosta-Martínez, V. (2024). Cover crop-mediated soil carbon storage and soil health in semi-arid irrigated cropping systems. *Agriculture Ecosystems & Environment, 361*, Article 108813. https://doi.org/10.1016/j.agee.2023.108813

Adetunji, A. T., Ncube, B., Mulidzi, R., & Lewu, F. B. (2020). Management impact and benefit of cover crops on soil quality: A review. *Soil Tillage Research, 204*, Article 104717.

Al-Delaimy, W. K. (2020). Vulnerable populations and regions: Middle East as a case study. In: Al-Delaimy, W., Ramanathan, V., & Sánchez Sorondo, M. (Eds.), *Health of people, health of planet and our responsibility*. Springer, Cham. https://doi.org/10.1007/978-3-030-31125-4_10

Ali, A. B., Elshaikh, N. A., Hong, L., Adam, A. B., & Haofang, Y. (2017). Conservation tillage as an approach to enhance crops water use efficiency. *Acta Agricul Scandinavica, Section B—Soil and Plant Science, 67*, 252–262.

Ali, M. M., Ahmad, B., Bari, M. S., Pal, A. C., Rahman, M. L., & Sarmin, I. J. (2024). An assessment of agroforestry as a climate-smart practice: Evidences from farmers of northwestern region of Bangladesh. *Agrosystems Geosciences and Environment, 7*, E20501. https://doi.org/10.1002/gg2.20501

Al-Kilani, M. R. (2024). Agricultural land measures for climate change adaptation in arid regions: Can the farmers do it alone? *Journal of Arid Land Agriculture, 10*, 82–93. https://doi.org/10.25081/jaa.2024.v10.7945

Allan, R. P., & Douville, H. (2024). An even drier future for the arid lands. *Proceedings of the National Academy of Sciences, 121*, Article e2320840121. https://doi.org/10.1073/pnas.2320840121

Al-Menaie, H. S., Al-Ragam, O., Al-Shatti, A., McCann, I., Naseeb, A., El-Hadidi, M. A., & Babu, M. A. (2021). Impact of different irrigation and nitrogen treatments on barley yield, yield components and water use efficiency. *Asian Journal of Agriculture Research, 15*, 7–19.

Al-Menaie, H., Al-Ragam, O., Al-Shatti, A., Al-Hadidi, M., & Babu, M. A. (2024a). Enhanced forage yield and water productivity with cereal-legume mixed cropping and deficit irrigation under arid conditions. *Legume Research, 47*(4), 583–589. https://doi.org/10.18805/LRF-769

Al-Menaie, H., Al-Ragam, O., Al-Shatti, A., Al-Hadidi, M. A., & Babu, M. A. (2024b). Effect of tillage and mulching on growth and yield performance of barley under different nitrogen and irrigation application rates. *Indian Journal of Animal Research, 58*(8), 1410–1417. https://doi.org/10.18805/IJAR.BF-1757

Al-Menaie, H. S., Al-Ragam, O., Al-Shatti, A., Al-Hadidi, M.A., & Babu, M. A. (2024c). Optimizing nitrogen fertilization for barley crop at full and deficit irrigation in the arid region. *Indian Journal of Animal Research, 58*(3), 517–524. https://doi.org/10.18805/IJARe.AF-823

Alotaibi, M. (2023). Climate change, its impact on crop production, challenges, and possible solutions. *Notulae Botanicae Horti Agrobotanici Cluj-Napoca, 51*, 13020 https://doi.org/10.15835/nbha51113020

Al-Ragam, O., Al-Menaie, H. S., Al-Shatti, A., & Babu, M. A. (2023). Enhancing food security through identification of best performing barley cultivars under arid conditions. *Indian Journal of Animal Research, 57*(6), 731–736. https://doi.org/10.18805/ijare.AF-784

Asseng, S., Martre, P., Maiorano, A., Rötter, R. P., O'Leary, G. J., Fitzgerald, G. J., Girousse, C., Motzo, R., Giunta, F., Babar, M. A., Reynolds, M. P., Kheir, A. M. S., Thorburn, P. J., Waha, K., Ruane, A. C., Aggarwal, P. K., Ahmed, M., Balkovič, J., Basso, B., ... Ewert, F. (2019). Climate change impact and adaptation for wheat protein. *Global Change Biology, 25*(1), 155–173. https://doi.org/10.1111/gcb.14481

Ayal, D. Y., & Mamo, B. (2024). Farmer's climate smart livestock production adoption and determinant factors in Hidebu Abote District, Central Ethiopia. *Scientific Reports, 14*(1), 10027. https://doi.org/10.1038/s41598-024-59967-8

Azua-Bustos, A., Carlos, G., & Fairén, A. G. (2022). The Atacama Desert in northern Chile as an analog model of Mars. *Frontiers in Astronomy and Space Science, 8*. https://doi.org/10.3389/fspas.2021.810426

Bado, B. V., Whitbread, A., & Manzo, M. L. S. (2020). Improving agricultural productivity using agroforestry systems: Performance of millet, cowpea, and ziziphus-based cropping systems in West Africa Sahel. *Agriculture Ecosystems & Environment, 305*, Article 107175. https://doi.org/10.1016/j.agee.2020.107175

Balasundram, S. K., Shamshiri, R. R., Sridhara, S., & Rizan, N. (2023). The role of digital agriculture in mitigating climate change and ensuring food security: An overview. *Sustainability, 15*(6), 5325. https://doi.org/10.3390/su15065325

Beillouin, D., Ben-Ari, T., Malezieux, E., Seufert, V., & Makowski, D. (2021). Positive but variable effects of crop diversification on biodiversity and ecosystem services. *Global Change Biology, 27*, 4697–4710.

Belay, A. D., Kebede, W. M., & Golla, S. Y. (2023). Determinants of climate-smart agricultural practices in smallholder plots: Evidence from Wadla district, northeast Ethiopia. *International Journal of Climate Change Strategies and Management, 15*(5), 619–637. https://doi.org/10.1108/IJCCSM-06-2022-0071

Bibi, F., & Rahman, A. (2023). An overview of climate change impacts on agriculture and their mitigation strategies. *Agriculture, 13*(8), 1508. https://doi.org/10.3390/agriculture13081508

Blanco-Canqui, H., & Ruis, S. J. (2018). No-tillage and soil physical environment. *Geoderma, 326*, 164–200.

Bordovsky, J. P., Lyle, W. M., & Keeling, J. W. (1994). Crop rotation and tillage effects on soil water and cotton yield. *Agronomy Journal, 86*, 1–6.

Cabusora, C. C. (2024). Developing climate-resilient crops: Adaptation to abiotic stress-affected areas. *Technology in Agronomy, 4*, Article e005. https://doi.org/10.48130/tia-0024-0002

Calvin, K., & Fisher-Vanden, K. (2017). Quantifying the indirect impacts of climate on agriculture: An inter-method comparison. *Environmental Research Letters, 12*, Article 115004. https://doi.org/10.1088/1748-9326/aa843c

Capone, R., El Bilali, H., Debs, P., Cardone, G., & Driouech, N. (2014). Food system sustainability and food security: Connecting the dots. *Journal of Food Security, 2*(1), 13–22. https://doi.org/10.12691/jfs-2-1-2

Chakrabarty, M., & Suri, S. (2023). Ending Africa's chronic food insecurity: What the G20 can do? Occasional Paper No. 407, Observer Research Foundation.

Charles, R. G., Davies, M. L., Douglas, P., & Hallin, I. L. (2018). Sustainable solar energy collection and storage for rural Sub-Saharan Africa. In: Letcher, T. M., Fthenakis, V. M. (Eds.), *A comprehensive guide to solar energy systems*. (pp. 81–107). Academic Press. https://doi.org/10.1016/B978-0-12-811479-7.00005-1

Comastri, A., Janni, M., Simmonds, J., Uauy, C., Pignone, D., Nguyen, H. T., & Marmiroli, N. (2018). Heat in wheat: Exploit reverse genetic techniques to discover new alleles within the *Triticum durums* Hsp26 family. *Frontiers in Plant Science, 9*, 1337. https://doi.org/10.3389/fpls.2018.01337

Cunningham, S. C., Nally, R. M., Baker, P. J., Cavagnaro, T. R., Beringer, J., Thomson, J. R., & Thompson, R. M. (2015). Balancing the environmental benefits of reforestation in agricultural regions. *Perspectives in Plant Ecology, Evolution Systematics, 17*(4), 301–317. https://doi.org/10.1016/j.ppees.2015.06.001

Delgado, J. A., Mosquera, V. H. B., Alwang, J. R., Villacis-Aveiga, A., Ayala, Y. E. C., Neer, D., Monar, C., & López, L. O. E. (2021). Potential use of cover crops for soil and water conservation, nutrient management, and climate change adaptation across the tropics. In: Sparks, D. L. (Eds.), *Advances in agronomy*, (Vol. 165, pp. 175–247). Academic Press.

Deng, L. A. (2000). *Integrating drylands issues into country assistance strategies: A review* (p. 67). World Bank.
Dennell, R. (2013). Hominins, deserts, and the colonisation and settlement of continental Asia. *Quaternary International, 300*, 13–21. https://doi.org/10.1016/j.quaint.2012.12.002
Dijkstra, F. A., & Keitel, C. (2024). Maximising carbon sequestration through mixing compost in moist soil. *Soil Biology and Biochemistry, 191*, Article 109330. https://doi.org/10.1016/j.soilbio.2024.109330
Din, M. S. U., Mubeen, M., Hussain, S., Ahmad, A., Hussain, N., Ali, M. A., El Sabagh, A., Elsabagh, M., Shah, G. M., Qaisrani, S. A., Tahir, M., Javeed, H. M. R., Anwar-ul-Haq, M., Ali, M., & Nasim, W. (2022). World nations priorities on climate change and food security. In: *Building Climate Resilience in Agriculture* (pp. 365–384). Cham: Springer. https://doi.org/10.1007/978-3-030-79408-8_22
Ding, H., He, J., Wu, Y., Wu, X., Ge, C., Wang, Y., Zhong, S., Peiter, E., Liang, J., & Xu, W. (2018). The tomatomitogen-activated protein kinase SlMPK1 is as a negative regulator of the high-temperaturestress response. *Plant Physiology, 177*(2), 633–651.
Dobert, T. F., Bork, E. W., Apfelbaum, S., Carlyle, C. N., Chang, S. X., Khatri-Chhetri, U., Sobrinho, L. S., Thompson, R., & Boyce, M. S. (2021). Adaptive multi-paddock grazing improves water infiltration in Canadian grassland soils. *Geoderma, 401*, Article 115314. https://doi.org/10.1016/j.geoderma.2021.115314
Dossou-Yovo, E. R., Bruggemann, N., Jesse, N., Huat, J., Ago, E. E., & Agbossou, E. K. (2016). Reducing soil CO_2 emission and improving upland rice yield with no-tillage, straw mulch and nitrogen fertilization in northern Benin. *Soil Tillage Research, 156*, 44–53.
Dowhower, S. L., Teague, W. R, Casey, K. D., & Daniel, R. (2020). Soil greenhouse gas emissions as impacted by soil moisture and temperature under continuous and holistic planned grazing in native tallgrass prairie. *Agriculture Ecosystems & Environment, 287*. 1 06647.https://doi.org/10.1016/j.agee.2019.106647
Draeger, T., & Moore, G. (2017). Short periods of high temperature during meiosis prevent normal meiotic progression and reduce grain number in hexaploid wheat (Triticum aestivum L.). *Theoretical and Applied Genetics, 130*(9), 1785–1800. https://doi.org/10.1007/s00122-017-2925-1
Duchenne-Moutien, R. A., & Neetoo, H. (2021). Climate change and emerging food safety issues: A Review. *Journal of Food Protection, 84*(11), 1884–1897. https://doi.org/10.4315/JFP-21-141
Fan, S., Brzeska, J., Keyzer, M., & Halsema, A. (2013). From subsistence to profit: Transforming smallholder farms. Food policy report, International Food Policy Research Institute.
FAO. (1989). *Arid zone forestry: A guide for field technicians*. Food and Agriculture Organization, Rome.
FAO. (2010). Climate-smart agriculture: policies, practices and financing for food security, adaptation and mitigation. Technical input for the first conference on agriculture, food security and climate change in the Netherlands. Food and Agriculture Organization.
FAO. (2012). Developing a climate-smart agriculture strategy at the country level: Lessons from the recent experience. Technical input for the second conference on agriculture, food security and climate change in Vietnam. Food and Agriculture Organization.
Feng, S., & Fu, Q. (2013). Expansion of global drylands under a warming climate. *Atmospheric Chemistry and Physics, 13*, 10081–10094. https://doi.org/10.5194/acp-13-10081-2013
Forster, B. P, & Shu, Q. Y. (2012). Plant mutagenesis in crop improvement: basic terms and applications. In: Shu, Q. Y, Forster, B. P, Nakagawa, H. (Eds.), *Plant mutation breeding and biotechnology* (pp. 9–20). Wallingford, CABI. https://doi.org/10.1079/9781780640853.0009
Francis, D., & Fonseca, R. (2024). Recent and projected changes in climate patterns in the Middle East and North Africa (MENA) region. *Scientific Reports, 14*, 10279. https://doi.org/10.1038/s41598-024-60976-w
Froslev, T. G., Nielsen, I. B., Santos, S. S., Barnes, C. J., Bruun, H. H., & Ejraes, R. (2022). The biodiversity effect of reduced tillage on soil microbiota. *Ambio, 51*(5), 1022–1033. https://doi.org/10.1007/s13280-021-01611-0

Gamo, M., Shinoda, M., & Maeda, T. (2013). Classification of arid lands, including soil degradation and irrigated areas, based on vegetation and aridity indices. *International Journal of Remote Sensing, 34*(19), 6701–6722. https://doi.org/10.1080/01431161.2013.805281

Gautam, V., Swaminathan, M., Akilan, M., Guruswamy, A., Suresh, M., Kaithamalai, B., & Joel, A. J. (2021). Early flowering, good grain quality mutants through gamma rays and EMS for enhancing per day productivity in rice (Oryza sativa L.). *International Journal of Radiation Biology, 97*(12), 1716–1730. https://doi.org/10.1080/09553002.2021.1987563

Gerhards, R., & Schappert, A. (2020). Advancing cover cropping in temperate integrated weed management. *Pest Management Science, 76*, 42–46.

Giddings, L., Soto, M., Rutherford, B., & Maarouf, A. (2005). Standardized precipitation index zones for Mexico. *Atmosfera, 18*, 33–56.

Gould, C. (2015). *Compost can increase the water holding capacity of droughty soils*. Michigan State University Extension.

Habbib, H., Verzeaux, J., Nivelle, E., Roger, D., Lacoux, J., & Catterou, M. (2016). Conversion to no-till improves maize nitrogen use efficiency in a continuous cover cropping system. *PLoS ONE, 11*, Article e0164234.

Habib-ur-Rahman, M., Ahmad, A., Raza, A., Hasnain, M. U., Alharby, H. F., Alzahrani, Y. M., Bamagoos, A. A., Hakeem, K. R., Ahmad, S., Nasim, W., Ali, S., Mansour, F., & EL Sabagh, A. (2022). Impact of climate change on agricultural production; Issues, challenges, and opportunities in Asia. *Frontiers in Plant Science, 13*, 925548. https://doi.org/10.3389/fpls.2022.925548

Hassan, M. U., Aamer, M., Mahmood, A., Awan, M. I., Barbanti, L., Seleiman, M. F., Bakhsh, G., Alkharabsheh, H. M., Babur, E., Shao, J., Rasheed, A., & Huang, G. (2022). Management strategies to mitigate N_2O emissions in agriculture. *Life, 12*(3), 439. https://doi.org/10.3390/life12030439

Hesse, P. P., Magee, J. W., & van der Kaars, S. (2004). Late quaternary climates of the Australian arid zone: A review. *Quaternary International, 118*(119), 87–102. https://doi.org/10.1016/S1040-6182(03)00132-0

Hong, M. J., Kim, D. Y., Jo, Y. D., Choi, H. I., Ahn, J. W., Kwon, S. J., Kim, S. H., Seo, Y. W., & Kim, J. (2022). Biological effect of gamma rays according to exposure time on germination and plant growth in wheat. *Applied Sciences, 12*(6), 3208. https://doi.org/10.3390/app12063208

Hoshan, M. N. (2022). Review of reclamation of salinity affected soils by leaching and their effect on soil properties and plant growth. *Tikrit Journal Agriculture Science, 22*(1), 149–168. https://doi.org/10.25130/tjas.22.1.14

Human, S., Andreani, S., Sihono, S., & Indriatama, W. M. (2011). Stability test for sorghum mutant lines derived from induced mutations with gamma-ray irradiation. *Atom Indonesia, 37*(3), 102. https://doi.org/10.17146/aij.2011.76

IAEA mutant variety database (mvd). (2024). International Atomic Energy Agency, Vienna. Retrieved April 24, 2024, from http://mvd.iaea.org

Iqbal, S., Xu, J., Khan, S., Worthy, F. R., Khan, H. Z., Nadir, S., & Ranjitkar, S. (2023). Regenerative fertilization strategies for climate-smart agriculture: Consequences for greenhouse gas emissions from global dry lands. *Journal of Cleaner Production, 398*, Article 136650. https://doi.org/10.1016/j.jclepro.2023.136650

Iwasaki, S., Endo, Y., & Hatano, R. (2017). The effect of organic matter application on carbon equestration and soil fertility in upland fields of different types of Andosols. *Soil Science and Plant Nutrition, 63*, 200–220.

Janzen, H. H. (2011). What place for livestock on a re-greening earth? *Animal Feed Science and Technology, 166–167*, 783–796. https://doi.org/10.1016/j.anifeedsci.2011.04.055

Juang, J., Guan, X., & Ji, F. (2012). Enhanced cold-season warming in semi-arid regions. *Atmospheric Chemistry and Physics, 12*, 5391–5398.

Jury, M. R. (2024). Regional controls on climate and weather variability on the southwest coast of Peru. *Coasts, 4*, 49–62. https://doi.org/10.3390/coasts4010004

Kell, D. B. (2011). Breeding crop plants with deep roots: Their role in sustainable carbon, nutrient and water sequestration. *Annals of Botany, 108*(3), 407–418. https://doi.org/10.1093/aob/mcr175

Khangura, R., Ferris, D., Wagg, C., & Bowyer, J. (2023). Regenerative agriculture: A literature review on the practices and mechanisms used to improve soil health. *Sustainability, 15*, 2338. https://doi.org/10.3390/su15032338

Kheir, A. M., El Baroudy, A., Aiad, M. A., Zoghdan, M. G., Abd El-Aziz, M. A., Ali, M. G. M., & Fullen, M. A. (2019) Impacts of rising temperature, carbon dioxide concentration and sea level on wheat production in North Nile delta. *The Science of the Total Environment, 651*, 3161–3173. https://doi.org/10.1016/j.scitotenv.2018.10.209

Kirby, J. M., Mainuddin, M., Mpelasoka, F., Ahmad, M. D., Palash, W., Quadir, M. E., Shah-Newaz, S. M., & Hossain, M. M. (2016). The impact of climate change on regional water balances in Bangladesh. *Climate Change, 135*, 481–491.

Koga, N. (2017). Tillage, fertilizer type, and plant residue input impacts on soil carbon sequestration rates on a Japanese Andosol. *Soil Science and Plant Nutrition, 63*, 1–9.

Koudahe, K., Allen, S. C., & Djaman, K. (2022). Critical review of the impact of cover crops on soil properties. *International Soil and Water Conservation Research, 10*(3), 343–354. https://doi.org/10.1016/j.iswcr.2022.03.003

Kumar, H., & Pani, P. (2011). Effects of soil erosion on agricultural productivity in semi-arid regions:The case of lower chambal valley. *Journal of Rural Development, 32*(2), 165–184.

Kumar, S., Yadav, A., Kumar, A., Hasanain, M., Shankar, K., Karan, S., Rawat, S., Sinha, A., Kumar, V., Gairola, A., Prajapati, S. K., & Dayal, P. (2023). Climate smart irrigation practices for improving water productivity in India: A comprehensive review. *International Journal of Environment and Climate Change, 12*(23), 333–348. https://doi.org/10.9734/ijecc/2023/v13i123689

Lal, R. (2021). Soil management for carbon sequestration. *South African Journal of Plant and Soil, 38*(3), 231–237.

Liu, M., Liang, F., & Li, Q. (2023). Enhancement growth, water use efficiency and economic benefit for maize by drip irrigation in Northwest China. *Scientific Reports, 13*, 8392. https://doi.org/10.1038/s41598-023-35611-9

Mahmood, N., Ahmad, B., Hassen, S., & Baskh, K. (2012). Impact of temperature and precipitation on rice productivity in rice-wheat cropping system of Punjab province. *The Journal of Animal and Plant Sciences, 22*(4), 993–997.

Majeed, Y., Khan, M. U., Waseem, M., Zahid, U., Mahmood, F., Majeed, F., Sultan, M., & Raza, A. (2023). Renewable energy as an alternative source for energy management in agriculture. *Energy Reports, 10*, 344–359. https://doi.org/10.1016/j.egyr.2023.06.032

Manjaya, J. G., & Gupta, S. K. (2023). Mutation breeding for adaptation to climate change in seed propagated crops. In: Raina, A., Wani, M. R., Laskar, R. A., Tomlekova, N., Khan, (Eds.), *Advanced crop improvement* (Vol. 2). Springer. https://doi.org/10.1007/978-3-031-26669-0_8

Martinez-Blanco, J., Lazcano, C., Christensen, T., Munoz, P., Rieradevall, J., Moller, J., Anton, A., & Boldrin, A. (2013). Compost benefits for agriculture evaluated by life cycle assessment. A review. *Agronomy for Sustainable Development*, 12. https://doi.org/10.1007/s13593-013-0148-7

Mazzoncini, M., Antichi, D., di Bene, C., Risaliti, R., Petri, M., & Bonari, E. (2016). Soil carbon and nitrogen changes after 28 years of no-tillage management under Mediterranean conditions. *European Journal of Agronomy, 77*, 156–165.

McDaniel, M. D., Grandy, A. S., Tiemann, L. K., & Weintraub, M. N. (2014). Crop rotation complexity regulates the decomposition of high- and low-quality residues. *Soil Biology and Biochemistry, 78*, 243–254

Momcilovic, I., & Ristic, Z. (2007). Expression of chloroplast protein synthesis elongation factor, EF-Tu, in two lines of maize with contrasting tolerance to heat stress during early stages of plant development. *Journal Plant Physiology, 164*(1), 90–99.

Mukherjee, R., & Sen, S. (2021). Bi-Annual sustainable agriculture & agricultural sustainability through nitrogen fixation: Approaches and techniques. *Harvest, 6*, 48–55.

Mwando, E., Angessa, T. T., Han, Y., & Li, C. (2020). Salinity tolerance in barley during germination- homologs and potential genes. *Journal of Zhejiang University. Science. B, 21*(2), 93–121. https://doi.org/10.1631/jzus.B1900400

Nanhapo, P. I., Valombola, J. S., Wanga, M. A., Elungi, K., Awala, S. K., & Horn, L. N. (2024). Optimum gamma radiation doses to enhance genetic diversity in selected cowpea (Vigna unguiculata) genotypes. *Reproduction and Breeding, 4*(2), 83–87. https://doi.org/10.1016/j.repbre.2024.01.001

National Research Council. (1996). *Lost crops of Africa* (Vol. I). Grains. The National Academies Press. https://doi.org/10.17226/2305

Nordahl, S. L., Preble, C. V., Kirchstetter, T. W., & Scown, C. D. (2023). Greenhouse gas and air pollutant emissions from composting. *Environment Science and Technology, 57*(6), 2235–2247. https://doi.org/10.1021/acs.est.2c05846

Nordborg, M. (2016). Holistic management–A critical review of Allan Savory's grazing method. SLU/EPOK–Centre for Organic Food and Farming and Chalmers. ISBN 978-91-576-9424-9.

Ntawuruhunga, D., Ngowi, E. E., Mangi, H. O., Salanga, R. J., & Shikuku, K. M. (2023). Climate-smart agroforestry systems and practices: A systematic review of what works, what doesn't work, and why. *Forest Policy and Economics, 150*, Article 102937. https://doi.org/10.1016/j.forpol.2023.102937

Ogle, S. M., Alsaker, C., Baldock, J., Bernoux, M., Breidt, F. J., McConkey, B., Regina, K., & Vazquez-Amabile, G. G. (2019). Climate and soil characteristics determine where no-till management can store carbon in soils and mitigate greenhouse gas emissions. *Scientific Reports, 9*, 11665. https://doi.org/10.1038/s41598-019-47861-7

Otekunrin, O. A. (2021). Is Africa ready for the SDG 2 (zero hunger) target by 2030? *Current Agriculture Research Journal, 9*(1), 1–3. https://doi.org/10.12944/CARJ.9.1.01

Penna, S., Bidabadi, S. S., & Jain, S. M. (2023). Mutation breeding to promote sustainable agriculture and food security in the era of climate change. In: Penna, S., & Jain, S. M. (Eds.), *Mutation breeding for sustainable food production and climate resilience*. Springer. https://doi.org/10.1007/978-981-16-9720-3_1

Penuelas, J., Coello, F., & Sardans, J. (2023). A better use of fertilizers is needed for global food security and environmental sustainability. *Agriculture and Food Security, 12*, 5. https://doi.org/10.1186/s40066-023-00409-5

Perez-Aguilar, L. Y., Plata-Rocha, W., Monjardin-Armenta, S. A., Franco-Ochoa, C., & Zambrano-Medina, Y. G. (2021). The identification and classification of arid zones through multicriteria evaluation and geographic information systems-Case study: Arid regions of northwest Mexico. *ISPRS International Journal of Geo-Information, 10*, 720. https://doi.org/10.3390/ijgi10110720

Quintarelli, V., Radicetti, E., Allevato, E., Stazi, S. R., Haider, G., Abideen, Z., Bibi, S., Jamal, A., & Mancinelli, R. (2022). Cover crops for sustainable cropping systems: A Review. *Agriculture, 12*(12), 2076. https://doi.org/10.3390/agriculture12122076

Rahman, M. D. M., Aravindakshan, S., Hoque, M. A., Rahman, M. A., Gulandz, M. A., Rahman, J., & Islam, M. T. (2021) Conservation tillage (CT) for climate-smart sustainable intensification: Assessing the impact of CT on soil organic carbon accumulation, greenhouse gas emission and water footprint of wheat cultivation in Bangladesh. *Environmental Sustainability Indicators, 10*(11). https://doi.org/10.1016/j.indic.2021.100106

Raj, A., Jhariya, M. K., Yadav, D. K., & Banerjee, A. (2020). Climate change and agroforestry system: Adaptation and mitigation strategies. *CRC-Apple Academic Press*. https://doi.org/10.1201/9780429286759

Raja, V., Majeed, U., Kang, H., Andrabi, K. I., & John, R. (2017). Abiotic stress: Interplay between ROS, hormones and MAPKs. *Environmental and Experimental Botany, 137*, 142–157.

Rao, D. S., Raghavendra, M., Gill, P., Madan, S., & Munjal, R. (2021). The effect of drought stress treatment on some grain quality traits in wheat (*Triticum aestivum* L.) Varieties). *International Journal of Plant and Soil Science, 33*, 117–123.

Regmi, A., & Meade, B. (2013). Demand side drivers of global food security. *Global Food Security, 2*(3), 166–171. https://doi.org/10.1016/j.gfs.2013.08.001

Riaz, A., Ali, S., Inayat, M., Safi, A., Imtia, M., Shabbir, Q., Yasin, M. T., & Shafiq, M. T. (2022). Effect of mulching on crop production and soil health in dry land region: An overview. *International Research Journal of Modernization in Engineering Technology and Science, 4*(10): 774–780. https://doi.org/10.56726/IRJMETS30482

Rioux, J. (2018). Opportunities and challenges of promoting agroforestry for climate change mitigation: A case-study of the mitigation of climate change in agriculture (MICCA) pilot project in Tanzania. *Nature and Faune, 26*(2), 63–68.

Ristic, Z., Wilson, K., Nelsen, C., Momcilovic, I., Kobayashi, S., Meeley, R., Muszynskib, M., & Habben, J. (2004). A maize mutant with decreased capacity to accumulate chloroplast protein synthesis elongation factor (EF-Tu) displays reduced tolerance to heat stress. *Plant Science, 167*(6), 1367–1374.

Ritchie, H., Roser, M., Mispy, J., & Ortiz-Ospina, E. (2018). Measuring progress towards the sustainable development goals. SGD-Rracker.org. United Nations Sustainable development goals. https://www.un.org/sustainabledevelopment

Rizi, M. S., & Mohammadi, M. (2023). Breeding crops for enhanced roots to mitigate against climate change without compromising yield. *Rhizosphere, 26*, Article 100702. https://doi.org/10.1016/j.rhisph.2023.100702

Roldan-Canas, J., & Moreno-Perez, M. F. (2021). Water and irrigation management in arid and semiarid zones. *Water, 13*(17), 2446. https://doi.org/10.3390/w13172446

Runge, C., Senauer, B., Pardey, P. G., & Rosegrant, M. W. (2004). Ending hunger in Africa. Prospects for the small farmer. International Food Policy Research Institute (IFPRI), Issue briefs 16, Washington DC, USA.

Sadiq, M., Rahim, N., Tahir, M. M., Alasmari, A., Alqahtani, M. M., Albogami, A., Ghanem, K. Z., Abdein, M. A., Ali, M., Mehmood, N., Yuan, J., Shaheen, A., Shehzad, M., El-Sayed, M. H., Cheng, G., & Li, G. (2024). Conservation tillage: A way to improve yield and soil properties and decrease global warming potential in spring wheat agroecosystems. *Frontiers in Microbiology, 15*, 1356426. https://doi.org/10.3389/fmicb.2024.1356426

Sakata, T., Takahashi, H., Nishiyama, I., & Higashitani, A. (2000). Effects of high temperature on the development of pollen mother cells and microspores in barley Hordeum vulgare L. *Journal of Plant Research, 113*(4), 395–402.

Santos, M., Cajaiba, R. L., Bastos, R., Gonzalez, D., Petrescu-Bakis, A. L., Ferreira, D., Leote, P., Barreto da Silva, W., Cabral, J. A., Goncalves, B., & Mosquera-Losada, M.R. (2022). Why do agroforestry systems enhance biodiversity? Evidence from habitat amount hypothesis predictions. *Frontiers in Ecology and Evolution, 9*, 630151. https://doi.org/10.3389/fevo.2021.630151

Savin, R., Stone, P. J., & Nicolas, M. E. (1996). Response of grain growth and malting quality of barley toshort periods of high temperature in field studies using portable chambers. *Australian Journal Agricultural Research, 47*(3), 465–477.

Savory, A. (2013). Response to request for information on the "science" and "methodology" underpinning holistic management and holistic planned grazing. 25 August 2018. http://www.savoryinstitute.com

Scherr, S. J., Shames, S., & Friedman, R. (2012). From climate-smart agriculture to climate-smart landscapes. *Agriculture and Food Security, 1*(1), 12. https://doi.org/10.1186/2048-7010-1-12

Shabani, M., Alemzadeh, A., Nakhoda, B., Razi, H., Houshmandpanah, Z., & Hildebrand, D. (2022). Optimized gamma radiation produces physiological and morphological changes that improve seed yield in wheat. *Physiology and Molecular Biology of Plants, 28*(8), 1571–1586. https://doi.org/10.1007/s12298-022-01225-0

Shahid, S. A., & Alshankiti, A. (2013). Sustainable food production in marginal lands – Case study GDLA member countries. *International Soil and Water Conservation Research, 1*(1), 24–38.

Shahid, S. A, Zaman, M., & Heng, L. (2018a). Soil salinity: Historical perspectives and world overview of the problem. In: Zaman, M., Shahid, S. A., Heng, L. (Eds.), *Guideline for salinity assessment, mitigation and adaptation using nuclear and related techniques* (Ch. 2, pp. 43–53). Springer. https://doi.org/10.1007/978-3-319-96190-3_2

Shahid, S. A., Zaman, M., & Heng, L. (2018b). Salinity and sodicity adaptation and mitigation options. In: Zaman, M., Shahid, S. A., & Heng, L. (Eds.) *Guideline for salinity assessment, mitigation and adaptation using nuclear and related techniques.* (Ch. 3, pp. 55–88). Springer. https://doi.org/10.1007/978-3-319-96190-3_3

Shayanmehr, S., Porhajasova, J. I., Babosova, M., Sabouni, M. S., Mohammadi, H., Henneberry, S. R., & Foroushani, N. S. (2022). The impacts of climate change on water resources and crop production in an arid region. *Agriculture, 12*, 1056. https://doi.org/10.3390/agriculture12071056

Sihono, S., Human, S., Indriatama, W., Maryono, M. Y., Puspitasar, W., & Wahyono, T. (2020). Exploring the potential of sorghum for food, animal feed and bioenergy through induced mutation breeding. In: *Proceeding of International Conference on Nuclear Capacity Building, Education, Research and Applications (I-Concern'19)* (Vol. 1, pp. 22–27), 6–7 Sept 2019, Yogyakarta, Indonesia.

Sileshi, G. W., Dagar, J. C., Nath, A. J., & Kuntashula, E. (2023). Agroforestry as a climate-smart agriculture: Strategic interventions, current practices and policies. In: Dagar, J. C., Gupta, S. R., & Sileshi, G. W. (Eds.), *Agroforestry for sustainable intensification of agriculture in Asia and Africa.* (pp. 589–640). Springer. https://doi.org/10.1007/978-981-19-4602-8_18

Sovacool, B. K., Griffiths, S., Kim, J., & Bazilian, M. (2021). Climate change and industrial F-gases: A critical and systematic review of developments, sociotechnical systems and policy options for reducing synthetic greenhouse gas emissions. *Renewable and Sustainable Energy Reviews, 141*(3), Article 110759.

Takai, T., Lumanglas, P., & Simon, E. V. (2020). Genetic mechanism of heat stress tolerance at anthesis among three different rice varieties with different fertilities under heat stress. *Plant Production Science, 23*(4), 529–538.

Thabet, S. G., & Alqudah, A. M. (2023). New genetic insights into improving barley crop with salt stress via regulating mineral accumulation, cellular ion homeostasis, and membrane trafficking. *Environmental and Experimental Botany, 208*, Article 105252. https://doi.org/10.1016/j.envexpbot.2023.105252

Thomas, D. S. G. (2011). *Arid Zone Geomorphology: Process, form and Change in Drylands* (pp. 656). John Wiley & Sons Ltd.

Tonnang, H. E., Sokame, B. M., Abdel-Rahman, E. M., & Dubois, T. (2022). Measuring and modelling crop yield losses due to invasive insect pests under climate change. *Current Opinion in Insect Science, 50*, Article 100873. https://doi.org/10.1016/j.cois.2022.100873

Tremblay, D., Fortier, F., Boucher, J. F., Riffon, O., & Villeneuve, C. (2020). Sustainable development goal interactions: An analysis based on the five pillars of the 2030 agenda. *Sustainable Development, 28*, 1584–1596. https://doi.org/10.1002/sd.2107

Udawatta, R. P., & Gantzer, C. J. (2022). Soil and water ecosystem services of agroforestry. *Journal Soil Water Conservation, 77*(1), 5A-11A. https://doi.org/10.2489/jswc.2022.1028A

Umeh, S. C., & Gil-Alana, L. A. (2024). Trends in temperatures in Sub-Saharan Africa. Evidence of global warming. *Journal of African Earth Sciences, 213*, Article 105228. https://doi.org/10.1016/j.jafrearsci.2024.105228

USAID (United States Agency for International Development). (2008). Initiative to end hunger in Africa. Tackling the root causes of food insecurity. Annual report, Unites States.

van Noordwijk, M., Coe, R., Sinclair, F. L., Luedeling, E., Bayala, J., Muthuri, C. W., Cooper, P., Kindt, R., Duguma, L., Lamanna, C., & Minang, P. A. (2021). Climate change adaptation in and through agroforestry: Four decades of research initiated by Peter Huxley. *Mitigation and Adaptation Strategies for Global Change, 26*(5), 18. https://doi.org/10.1007/s11027-21-09954-5

Vaughn, D. M. (2005). Arid climates. In: Oliver, J. E. (Eds.), *Encyclopedia of world climatology, encyclopedia of earth sciences series.* Springer. https://doi.org/10.1007/1-4020-3266-8_16

Victor, O., Chiza, K., Beatrice, E., Megan, P. E., Laina, E., Nanna, R., Warren, L. T., & Daniel, T. (2022). The impact of climate change on food systems, diet quality, nutrition, and health outcomes: A narrative review. *Frontiers in Climate, 4*, Article 941842. https://doi.org/10.3389/fclim.2022.941842

Wang, F., Harindintwali, J. D., Wei, K., & Tiedge, J. M. (2023). Climate change: Strategies for mitigation and adaptation. *The Innovation, 1*(1), 100015. https://doi.org/10.59717/j.xinn-geo.2023.100015

Wanga, M. A., Shimelis, H., Mashilo, J., Horn, L. N., & Sarsu, F. (2023). Responses of elite sorghum (Sorghum bicolor [L.] Moench) lines developed via gamma-radiation for grain yield, component traits and drought tolerance. *Reproduction and Breeding, 3*(4), 184–196. https://doi.org/10.1016/j.repbre.2023.10.005

WHO. (2015). Africa can rely on United Nations Development System in decade of action to fulfil 2030 Agenda, Deputy Secretary-General tells Regional Forum 2020. Retrieved February 28, 2021, from https://www.un.org/press/en/2020/dsgsm1390.doc.htm

Xu, J., Henry, A., & Sreenivasulu, N. (2020). Rice yield formation under high day and night temperatures-A prerequisite to ensure future food security. *Plant Cell and Environment, 43*(7), 1595–1608.

Yang, R., Su, Y., & Kong, J. (2017). Effect of tillage, cropping, and mulching pattern on crop yield, soil C and N accumulation, and carbon footprint in a desert oasis farmland. *Soil Science and Plant Nutrition, 63*, 599–606.

Yang, P., Wu, L., Cheng, M., Fan, J., Li, S., Wang, H., & Qian, L. (2023). Review on drip irrigation: Impact on crop yield, quality, and water productivity in China. *Water, 15*, 1733. https://doi.org/10.3390/w15091733

Yasin, G., Nawaz, M. F., Zubair, M., Azhar, M. F., Mohsin Gilani, M., Ashraf, M. N., Qin, A., & Ur Rahman, S. (2023). Role of traditional agroforestry systems in climate change mitigation through carbon sequestration: An investigation from the semi-arid region of Pakistan. *Land, 12*(2), 513. https://doi.org/10.3390/land12020513

Yin, C., Jones, K. L., Peterson, D. E., Garrett, K. A., Hulbert, S. H., & Paulitz, T. C. (2010). Members of soil bacterial communities sensitive to tillage and crop rotation. *Soil Biology and Biochemistry, 42*, 2111–2118.

Yu, T., Mahe, L., Li, Y., Wei, X., Deng, X., & Zhang, D. (2022). Benefits of crop rotation on climate resilience and its prospects in China. *Agronomy, 12*, 436. https://doi.org/10.3390/agronomy12020436

Zaman, M., Shahid, S. A., Heng, L. (2018). Irrigation systems and zones of salinity development. In Zaman, M., Shahid, S. A., & Heng, L. (Eds.), *Guidelines for salinity assessment, mitigation and adaptation using nuclear and related techniques* (Ch. 4, pp. 91–111). Springer. https://doi.org/10.1007/978-3-319-96190-3_4

Zhang, H., Hobbie, E. A., Feng, P., Niu, L., & Hu, K. (2022). Can conservation agriculture mitigate climate change and reduce environmental impacts for intensive cropping systems in North China Plain? *Science of the Total Environment, 806*, Article 151194. https://doi.org/10.1016/j.scitotenv.2021.151194

Zhao, M., & Boll, J. (2022). Adaptation of water resources management under climate change. *Frontiers in Water, 4*, Article 983228. https://doi.org/10.3389/frwa.2022.983228

Zheng, H., Ma, W., & He, Q. (2024). Climate-smart agricultural practices for enhanced farm productivity, income, resilience, and greenhouse gas mitigation: A comprehensive review. *Mitigation and Adaptation Strategies for Global Change, 29*(28). https://doi.org/10.1007/s11027-024-10124-6

Ziska, L. H. (2022). Rising carbon dioxide and global nutrition: Evidence and action needed. *Plants, 11*(7), 1000. Retrieved December 19, 2024, from https://doi.org/10.3390/plants11071000. https://sdgs.un.org/goals. https://globalgoals.org/goals/2-zero-hunger

[illegible]

Chapter 16
Climate Change-Induced Variation in Ecosystem Structure in Arid Lands Countries

Amal Jaafer Alkandari, Mohammad A Alsayegh, and Ali A Abdulraheem

Abstract Ecosystems are geographic areas where various groups of plants, animals, and other organisms coexist, and can range in size from vast deserts to tiny ponds. These ecosystems include living organisms and non-living components such as rocks, sand, and moisture. The structure of an ecosystem is a complex and dynamic interplay between its biotic and abiotic components. Climate plays a fundamental role in determining the distribution of ecosystems, species ranges, and rates of processes on Earth. Consequently, changes in ecosystem functionality due to climate change have significant potential to influence the climate itself, serving as either positive or negative feedback mechanisms. Climate change is speeding up ecological shifts, with changes projected for 2100 happening sooner than anticipated. The increased aridity caused by climate change is expected to significantly impact the structure and functionality of drylands, which make up about 45% of the Earth's land surface. This rising aridity is likely to push more than 20% of these areas past critical thresholds, potentially leading to extensive land degradation and desertification. Over 2,000 million hectares of land have become degraded, resulting in a loss of agrobiodiversity, increased water scarcity, and the destruction of natural resources. To sustain ecosystems, both public and private sectors, along with international bodies, must integrate climate change considerations into their strategies and actions. This approach can empower vulnerable communities in dryland regions to manage natural resources sustainably, safeguarding the environment for future generations. By investing in research, adopting sustainable practices, and promoting international cooperation to

A. J. Alkandari (✉) · M. A. Alsayegh · A. A. Abdulraheem
Food Security Program, Environment and Life Sciences Research Center, Kuwait Institute for Scientific Research, Safat, Kuwait
e-mail: ajkandari@kisr.edu.kw

M. A. Alsayegh
e-mail: msayegh@kisr.edu.kw

A. A. Abdulraheem
e-mail: aabdulraheem@kisr.edu.kw

M. K. Suleiman and S. A. Shahid (eds.), *Fostering Arid Lands Agriculture in the Face of Climate Change*, Advances in Global Change Research 81,
https://doi.org/10.1007/978-3-031-98490-7_16

achieve sustainable development goals, we can mitigate adverse effects on vulnerable communities and ensure the preservation of ecosystems for the future. The well-being of our planet depends on the collective efforts and wisdom of all nations.

Keywords Climate impact · Drylands · Sustainability · Biome · Wildfire · Hydrological cycle · Desertification · Protected areas

16.1 Introduction

This chapter examines the relationship between ecosystems, climate change, and human activities in arid lands countries. Ecosystems are defined by the interactions among living organisms and their physical, chemical, and biological environments. While ecosystems can vary greatly in size, the essential principles of their dynamics remain unchanged, underscoring the interconnection between life and the environment. These systems comprised of diverse components that includes plants, animals, fungi, and bacteria, which interact with each other as well as with abiotic factors like land, water, and weather conditions. This interplay creates a complex web of relationships essential for maintaining ecosystem health. Gaining insight into these interactions allows to appreciate the fragile balance necessary for sustaining ecosystems and to understand the profound effects of human activities on these natural systems (Grimm et al., 2013). The Intergovernmental Panel on Climate Change (IPCC) stated in 2001 that human activities are the leading causes of global climate change, mainly resulting from increased emissions of greenhouse gases (IPCC, 2023). Climate change affects ecosystems worldwide, subsequently impacting human life. Variables like temperature and rainfall, which are influenced by climate change, alter water cycles and resource availability, necessitating adaptation strategies, especially in arid and semi-arid regions (Wang & Liu, 2023). These regions face low land productivity and high-water scarcity, making them particularly vulnerable to climate change, which threatens agricultural productivity and biodiversity (Singh & Chudasama, 2021). Since the 1970s, a 1.15 °C increase in temperature has affected wind patterns and contributed to changes in the physical and biological properties of mountain lakes (Michelutti et al., 2016). Biodiverse areas, such as thickets, are susceptible to climate-induced temperature rises and droughts, increasing the risk of fires, especially in hot, dry seasons (McKenzie et al., 2004). Effects of climate change on agriculture and crops can lead to an expansion of the geographic distribution of crops, as well as pests. Warmer temperatures can increase the survival rates of pests during the winter. The changing climate can lead to an increased number of pest generations per year. Altered timing in the life cycles of plants and pests can disrupt their interactions. There is an increased risk of invasion by migratory pests due to climate changes. The incidence of diseases transmitted by insects is likely to increase. The effectiveness of biological control methods, especially natural enemies of pests, may be reduced. There is a serious risk of economic losses in crop production. Human food security is challenged by the increased vulnerability of crops to pests (Skendži´c et al., 2021).

16.2 Climate and Climate Change

Since the 1800s, anthropogenic activities have led to the release of greenhouse gases (GHGs) such as carbon dioxide (CO_2), methane (CH_4), and nitrous oxide (N_2O) into the atmosphere, primarily due to the burning of coal, oil, and gas for energy production. This has been identified as a major driver of climate change. In the late 1850s, scientists began to focus on GHGs like CO_2 and their role in absorbing heat from solar radiation, with the understanding that these gases could influence global warming. By the late 1950s, scientific consensus had emerged supporting the idea that CO_2 and other GHGs were significantly contributing to climate change. During this time, the first global data and studies began to emerge in support of the theory of global warming. In 1896, Swedish chemist Svante Arrhenius notably argued that elevated CO_2 levels in the atmosphere could result in global warming. His calculations predicted that a doubling of CO_2 could increase global temperatures by about 5 °C. In 1800s, human activities have contributed to the release of GHGs into the atmosphere, primarily due to the combustion of coal, oil, and gas for energy production. This has been recognized as a major factor driving climate change. In the late 1850s, scientists began examining GHGs like CO_2 and their capacity to absorb heat from solar radiation. In the early 1970s, scientists began to explore the possibility that atmospheric pollutants, effects on climate. However, by 1979, the focus shifted as understanding of climate change progressed. At the First World Climate Conference in Geneva, a declaration was made urging governments worldwide "to foresee and prevent potential man-made changes in climate that might adversely affect humanity's well-being." This marked a crucial step in acknowledging the need for international action on climate change.

In 1988, the United Nations established the Intergovernmental Panel on Climate Change (IPCC) to provide a scientific assessment of climate change and its repercussions. The IPCC's first comprehensive report, released in 1990, raised serious concerns about global warming, projecting severe outcomes such as more frequent heatwaves, increased resource scarcity, and intensified hurricanes, all associated with rising sea surface temperatures. This report was a landmark moment in solidifying the scientific foundation for climate change and energizing international policy initiatives (IPCC, 2023; Nwaerema, 2020). The main source for GHGs in the world is human activities and since 1900 global surface temperature increases above 1.1 °C.

Fig. 16.1 A typical hyperarid desert landscape in Kuwait

16.3 Arid Lands

Arid land, also referred to as dryland or desert, is characterized by water scarcity, low nutrient availability, low rainfall, sparse vegetation, limited biodiversity, harsh living conditions, and fragile ecosystem (Fig. 16.1). Land degradation and erosion from wind or water are significant features of these regions. Arid lands make up about one-third (41%) of the Earth's land surface (Fig. 16.2), and the primary drivers of degradation in these areas include human activities, such as deforestation and unsustainable agricultural practices, overgrazing in some regions, as well as climate change. All these interacting conditions will affect the land sustainability (Abdelhak et al., 2022; Roeder & Hill, 2009).

16.4 Terrestrial Ecosystems of Arid Lands Countries

Arid land environments are highly fragile and face numerous environmental challenges. Human activities in these regions often lead to soil degradation and salinization, further exacerbating the already challenging conditions. Climate change intensifies these issues by raising temperatures and reducing rainfall, which in turn decreases soil moisture. In arid lands, the rate of evaporation far exceeds precipitation, leading to severe drought conditions and further degradation of the land, making the soil less fertile (Huang et al., 2016).

Desertification, a critical and severe problem in arid regions, results in the land becoming less productive than anticipated. This process depletes the terrestrial

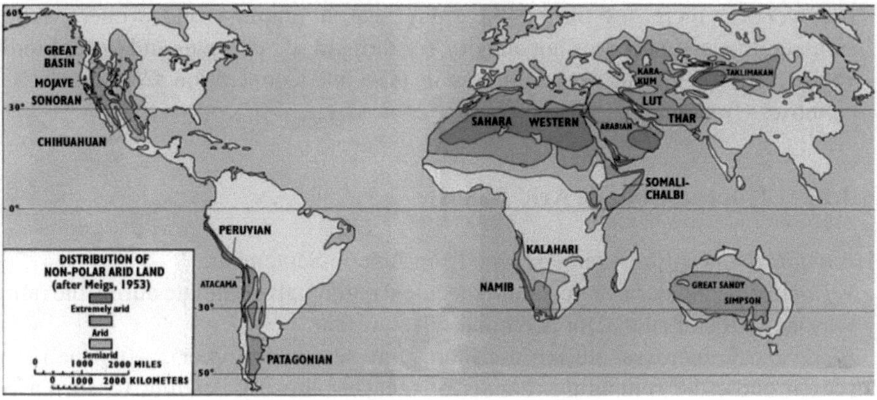

Fig. 16.2 World arid and semiarid countries (Roeder & Hill, 2009)

ecosystem, making it poor, fragmented, vulnerable, and increasingly unsuitable for sustaining life. Over time, the ecosystem loses its ability to maintain sustainability.

Arid and semi-arid lands cover approximately 6,150 million hectares, characterized by low biodiversity among plants and animals (Boutaj et al., 2022; Gupta et al., 2018). Deserts within these regions can vary widely—they can be hot or cold, foggy coastal deserts, or sunny continental deserts. Rainfall patterns also vary, with some deserts receiving rain in the winter and others in the summer. Additionally, deserts may be composed of sand or rocks, further influencing their ecological characteristics (Barakat, 2011).

16.4.1 Types of Arid Land Ecosystems

Arid land ecosystem in general is classified depending on high temperature and scarce of water in rain fall season. So, type of arid land ecosystem depends on the climate (Barakat, 2011).

16.4.1.1 Moderate Climate Arid Lands

- *Rainfall Pattern*: Rainfall occurs in two seasons, typically from October to May (autumn and winter).
- *Temperature*: Mild.
- *Vegetation*: Includes ephemeral annuals and succulent perennials, with non-succulent perennials being less common.

- *Biodiversity*: Increases during the rainy season (autumn and winter). After summer, the land becomes hot and dry, resulting in scarce water and lower biodiversity. This is similar to conditions in the Gulf Cooperation Council (GCC) countries (Boutaj et al., 2022).

16.4.1.2 Tropical Climate Arid Lands

- *Rainfall Pattern*: Short rainy season from June to September.
- *Vegetation*: Ephemeral annuals and succulent perennials dominate during the rainy season, with non-succulent perennials also present.
- *Biodiversity*: Increases during the short rainy season (summer). After the rainy season ends, the remaining months of the year are dry, leading to decreased biodiversity, similar to conditions in parts of India.

In both climates, the patterns of rainfall and temperature significantly impact vegetation types and biodiversity throughout the year.

16.4.1.3 Continental Climate Arid Lands

Precipitation is often distributed fairly evenly throughout the year. The observation that each monthly precipitation is less than twice the corresponding mean monthly temperature suggests a dry climate. This means that despite fluctuations in temperature, the amount of rain received is generally low compared to the temperature. However, there can be variations, with a tendency for increased rainfall in the summer months. In Alice Springs, precipitation is relatively low throughout the year and doesn't necessarily follow the typical continental pattern. The dry season in Alice Springs extends throughout the year due to the overall low precipitation. Even though there might be some seasonal variations in rainfall, it remains quite dry relative to the temperature. So, Alice Springs experiences a dry climate year-round with relatively consistent low precipitation, and this aligns with its classification as a desert or semi-arid region rather than a continental one, this condition can be seen clearly in Northern Territory of Australia.

- The plants grow in these areas either Ephemeral annual or Succulent perennials.– Non-Succulent perennials
- Biodiversity is low over the year

16.5 Type of Deserts

Deserts can be classified based on the amount of annual rainfall. Here's a detailed breakdown (Barakat, 2011; Cooke & Warren, 1973):

16.5.1 Very Dry Desert

- *Annual Rainfall*: Less than 100 mm (mm).
- *Characteristics*: Extremely arid with very sparse vegetation. These deserts have the harshest conditions and include well-known examples like the Atacama Desert in Chile, and parts of the Sahara and GCC countries (Fig. 16.3).

Fig. 16.3 Desert views from GCC Countries, **a** Kuwait, **b** Saudi Arabia, photo courtesy Dr. Naser-aldeen B. Asadalla, **c** Qatar (photo courtesy Dr. Mohammed bin Batti Al-Abdulla), **d** Bahrain (photo courtesy Khalid Mustafa), **e** United Arab Emirates, (photo courtesy Shaikha Mohamed AlKetbi), and **f** Oman (photo courtesy Ali Bin Nasser Al Rasbi)

16.5.2 *Semi-deserts*

Annual rainfall is between 100 and 300 mm. Semi-deserts receive more rainfall than very dry deserts but still experience significant dryness. Vegetation may include hardy shrubs and grasses adapted to low water availability. Examples include parts of the Kalahari Desert and the southwestern United States.

16.5.3 *Woodland Savanna*

Annual Rainfall is between 300 and 600 mm. This region receives more rainfall, supporting a mix of grasses and scattered trees or shrubs. The climate is still dry but supports a more diverse range of plant life compared to the drier desert types. Examples include parts of the African savanna and some regions in Australia.

16.5.4 *Evergreen Scrub*

Annual Rainfall is in excess of 500 mm. This type of desert has enough rainfall to support evergreen shrubs and denser vegetation compared to the lower rainfall deserts. The climate is somewhat more temperate and can support a greater variety of plant species. Examples include certain Mediterranean regions and parts of the coastal desert of southern California.

Each category reflects how rainfall influences the types of vegetation and overall ecological conditions in desert regions.

16.6 Threats to Ecosystems

Arid ecosystems, which cover about one-third of the Earth's land surface, face numerous challenges that threaten their biodiversity (Bernardino et al., 2020). Factors affecting these ecosystems can be broadly classified into natural and anthropogenic.

16.6.1 Natural Factors

16.6.1.1 Climate Variability and Climate Change

Arid areas experience sharp temperature changes and erratic, limited rainfall, making it hard for many species to thrive, which leads to reduced biodiversity. Moreover, extended periods of drought can significantly impact plant and animal populations.

Rising global temperatures and changing precipitation patterns due to climate change are intensifying the stress on arid ecosystems. This can lead to shifts in species distribution, loss of biodiversity, and increased vulnerability to invasive species. These factors make arid and semi-arid ecosystems particularly sensitive and vulnerable, these regions.

16.6.1.2 Soil Erosion

The lack of dense vegetation in arid regions often leads to greater soil erosion, which may be exacerbated during rare, intense rainfall events. Such erosion can damage habitats and reduce the land's capacity to support diverse organisms.

16.6.1.3 Desertification

Fertile land is becoming more arid and desert-like as a result of human activities encroaching on ecosystems and climate change. This process of desertification increases stress on existing ecosystems, leading to a decline in both diversity and sustainability.

16.6.2 Anthropogenic Factors

16.6.2.1 Overgrazing

Overgrazing, a significant contributor to develop landscape to look like desert and occurs when livestock grazing is poorly managed. This practice depletes vegetation, compacts the soil, and heightens erosion, diminishing the land's capacity to support a variety of plant and animal species.

16.6.2.2 Land Use Change

Unsustainable human activities such as urbanization, mining, and deforestation can exacerbate desertification in semi-arid and arid regions. Additionally, these unpopular

Fig. 16.4 Land use change, new settlement for increased population

actions can change the landscape of arid areas, disrupt local climates and water cycles in nature resulting in habitat loss and further strains the ecosystem (Fig. 16.4).

16.7 Key Climate Change Impacts on Ecosystem Structure and Functions

Climate change significantly impacts ecosystem structure and functions, leading to various alterations that affect biodiversity, ecosystem services, and overall environmental stability. Ecosystem state change in species distribution and composition by shift in habitats: As temperatures rise, species are forced to migrate toward cooler areas, such as higher altitudes or latitudes. This can lead to changes in community composition as some species are outcompeted or unable to migrate quickly enough. Species that cannot adapt or migrate may face extinction, particularly those with limited ranges or specialized habitat needs.

16.7.1 Biome Shift

The shift in biomes due to climate change is particularly evident in arid regions, where the impacts can be severe and transformative. Temperature and precipitation patterns are critical factors in defining and maintaining ecosystems, or biomes. In arid lands, these changes can lead to desertification. Major biomes in the world are shown in Fig. 16.5. Increased temperatures and reduced precipitation can lead to the expansion of deserts into semi-arid and even previously fertile regions. This process reduces the availability of water and soil fertility, leading to a decline in vegetation cover. As climate conditions change, these types of plants that can survive in a region also change. For instance, grasslands might become dominated by shrubs, or woodlands might transform into savannas. This shift in plant communities can have cascading effects on the local fauna and overall biodiversity. In arid regions, water is already a limiting resource. Climate change can exacerbate this by reducing the amount of rainfall or by increasing evaporation rates due to higher temperatures. This leads to a reduction in available water for plants and animals, further stressing the ecosystem (Brown et al., 1997; Bonannella et al., 2023).

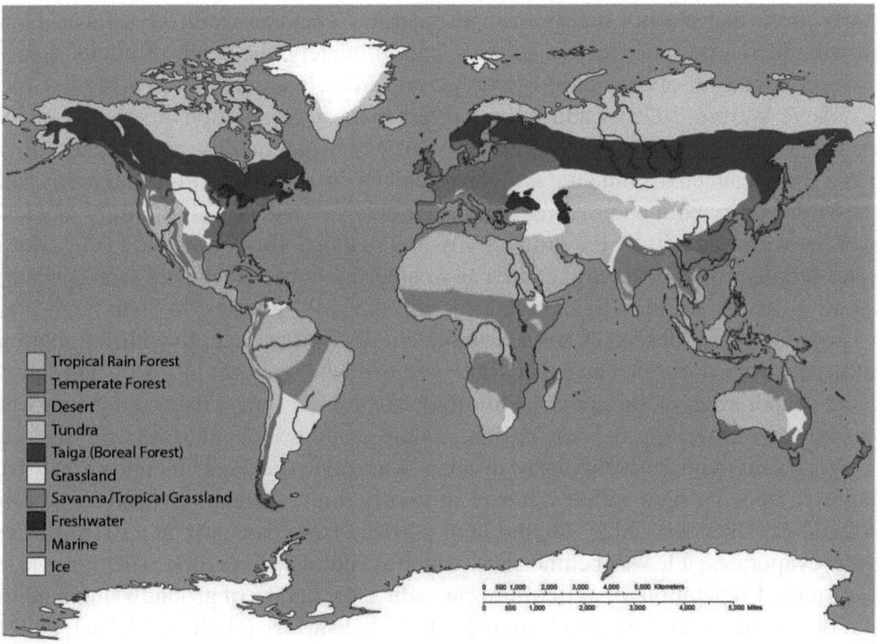

Fig. 16.5 Major Biomes land in the world. *Source* Arizona State University, https://askabiologist.asu.edu/explore/biomes

16.7.2 Drylands Desert Vegetation and Ecosystems—Loss and Gain

Desert vegetation dynamics in the context of climate change involve both losses and gains, which can have profound impacts on desert ecosystems and their sustainability (Brown et al., 1997). Loss of desert vegetation varied from place to place depending on climate temperature and rainfall. Land degradation, one of the most significant losses of desert vegetation is due to desertification, which is often accelerated by climate change. Land degradation involves the degradation of land in arid, semi-arid, and dry sub-humid areas, primarily due to human activities and climatic variations. The loss of vegetation cover leads to soil erosion, reduced soil fertility, and the eventual collapse of the ecosystem.

Increased temperature and prolonged droughts can stress plant species, particularly those that are not highly drought-tolerant. This can result in reduced plant growth, decreased reproductive success, and ultimately, the death of plants. Sensitive species may disappear entirely from certain areas, leading to a loss in biodiversity (Cooke & Warren, 1973). Land salinity refers to the accumulation of soluble salts in the soil to levels that adversely affect plant growth, water quality, and land usability. It is a significant environmental issue, particularly in arid and semi-arid regions, and can be driven by both natural processes and human activities. Land salinity increase over the world and effect the productivity and sustainability of the land (Fig. 16.6). Land salinity could be natural causes as weathering of rocks releases salts (primary salinity), which are then transported through the soil by water. In areas with high evaporation rates, these salts can accumulate in the soil surface. In addition, marine influence in coastal areas can experience salinity due to seawater intrusion and subsequent evaporation, or surface deposition of salts by sea sprays from sea. Moreover, groundwater upwelling in some regions, saline groundwater can rise to the surface, bringing salts with it, particularly in areas with poor drainage. Inappropriate irrigation, especially with saline water or in poorly drained soils, can lead to salinity (Shahid & Alkandari, 2024; Shahid et al., 2018, 2024; Zaman et al., 2018). When water evaporates, it leaves behind salts, which accumulate over time. The removal of deep-rooted vegetation for agriculture can reduce the uptake of groundwater, leading to a rise in the water table and bringing salts to the surface (Hailu & Mehari, 2021).

Salinity can impair plant growth by affecting water uptake, leading to reduced agricultural productivity. Some crops are particularly sensitive to salts, resulting in poor germination, stunted growth, and lower yields. Increased salinity can lead to a decline in biodiversity as species that cannot tolerate high salt levels are displaced or die out. Decrease in the biodiversity could facilitate the spread of invasive species that outcompete native desert plants. These invasives often have different water and nutrient needs, altering the balance of the ecosystem and leading to the decline of native vegetation (Hailu & Mehari, 2021; Singh, 2009).

Grazing, particularly when unmanaged or excessive, can lead to significant vegetation loss, which has far-reaching impacts on ecosystems, when animals graze intensively in one area, they consumed vegetation faster than it can regrow. This reduces

Fig. 16.6 Soil salinity induced by shallow water table due to restricted drainage (*Photo curtsey* Dr. Shabbir Shahid)

plant cover, leaving the soil exposed to erosion and less capable of retaining moisture. With reduced vegetation cover, the soil becomes more susceptible to wind and water erosion. Erosion not only removes fertile topsoil but also displaces seeds and young plants, which further hampers the recovery of vegetation. As plant life diminishes, the habitat for various wildlife species also shrinks, leading to a decline in animal biodiversity, especially among those species that rely on dense vegetation for food, shelter, and breeding grounds.

Vegetation is essential for regulating local climate conditions, such as temperature and humidity. Losing the vegetation cause more extreme temperatures and drier conditions, intensifying stress on the remaining plants. In arid and semi-arid regions, overgrazing is a major contributor to desertification, as vegetation is depleted and soil health deteriorates, the land becomes increasingly barren and unable to sustain plant life. Degraded lands produce less forage for livestock, leading to a decline in livestock health and productivity, creating a vicious cycle of overgrazing and further vegetation loss (Mata-González et al., 2007) (Fig. 16.7).

Compacted soil significantly contributes to desertification (Fig. 16.8), particularly in arid and semi-arid regions. Desertification refers to the degradation of land in these areas, turning once-productive land into barren deserts. Soil compaction is a significant factor that contributes to desertification, especially in arid and semi-arid regions. Compacted soil leads to decrease the infiltration rate and making it increasingly difficult for vegetation to grow (D'Odorico et al., 2019).

Hot weather with strong winds accelerates erosion. In arid regions, where vegetation is sparse, this can lead to significant soil loss and further reduce plant cover. The ultimate effect is often a cycle of increasing desertification. For example, State of Kuwait is one of the hottest deserts in the world, where wind erosion is common (Alkandari, 2023; D'Odorico et al., 2019; Shahid, 2023a, 2023b).

(a) (b)

Fig. 16.7 Overgrazing in the desert environments (**a**, **b**)

Fig. 16.8 Compacted soil devoid of vegetation cover

Soil plays a crucial role in determining the productivity of arid and semi-arid lands. The ability of soil to support vegetation is significantly affected by its physical and chemical properties, especially in semi-arid and arid regions where water is limited and conditions are tough. Soils with high sand content consist of larger particles and have lower water-holding capacity, leading to poor nutrient retention. This makes it difficult for plants to access the moisture they need, ultimately hindering their growth and reducing productivity. Similarly, gravelly or loose soils often share this poor structure, which limits their capacity to retain water and nutrients and provides inadequate root support, complicating plant establishment and survival. While clay soils are effective at retaining water, they can become compacted, negatively affecting aeration and root penetration.

In many arid and semi-arid areas, soils are shallow due to underlying rock or hardpan layers. Shallow soils offer limited space for root growth and have low water and nutrient storage capacity, restricting plant development. Soils that are rocky or contain a high percentage of calcium carbonate (caliche) layers are particularly challenging for vegetation. These soils are often impermeable to water, preventing roots from accessing moisture and nutrients, and making it nearly impossible for most vegetation to thrive.

Most plants prefer soils that are neutral to slightly alkaline (pH 6.5–8). In arid regions, soils tend to be more alkaline, which can limit the availability of phosphorus and micronutrients (iron, copper, manganese, and zinc). Soils with very high pH levels (due to exchangeable sodium percentage) can become too alkaline for most plants, leading to nutrient deficiencies and poor vegetation growth. Soils rich in essential nutrients like nitrogen, phosphorus, and potassium support healthy plant growth. In arid and semi-arid lands, however, soils are often nutrient-poor due to low organic matter content and slow nutrient cycling (Fig. 16.9) (Chigani et al., 2012; Huntley, 2023).

Organic matter is crucial for improving soil fertility and structure. It enhances water retention, nutrient availability, and microbial activity, all of which are vital for supporting vegetation (Chigani et al., 2012; Huntley, 2023). In arid and semi-arid land, the highest rainfall variability ranging between 100 and 300 mm. Increase in rainfall and water supply in arid and semi-arid lands can indeed significantly boost vegetation growth, leading to a range of ecological and economic benefits. Water is a crucial factor in seed germination. With increased rainfall or water supply, more seeds can germinate, resulting in denser vegetation cover and greater plant biomass. This abundance of plants also offers more resources for herbivores and other wildlife. The

Fig. 16.9 Loss of fertile surface soil due to wind erosion

adequate water, native plants that are well-adapted to local conditions can flourish and outcompete invasive species, contributing to a more stable and resilient ecosystem.

In some instances, increased water supply can effectively flush out salts from the soil, lowering salinity levels and making the land more conducive to plant growth. This enhanced plant diversity supports a wider range of animal species, including insects, birds, and mammals, potentially leading to the return of species that had vanished due to habitat loss. A rise in vegetation cover resulting from increased water availability helps to stabilize the land and reduces the risk of desertification. Additionally, plants capture carbon dioxide from the atmosphere and store it as biomass, meaning that more vegetation leads to greater carbon sequestration, which aids in mitigating climate change (Panagos et al., 2017; Ragab & Prudhomme, 2002).

Moist soils foster the growth of soil microorganisms, including bacteria that play key roles in nutrient cycling and enhancing soil fertility. Typically, beneficial bacteria known as rhizobacteria colonize plant roots, promoting plant health. Additionally, various nitrogen-fixing bacteria, such as *Allorhizobium, Azospirillum, Azorhizobium, Bradyrhizobium, Enterobacter, Klebsiella, Mesorhizobium, Pseudomonas, Rhizobium, and Sinorhizobium,* contribute to soil fertility. Each type of bacteria has distinct functions, but all thrive under favorable environmental conditions. In semi-arid and arid regions, however, these bacteria can be impacted by climate change, and their distribution may shift depending on the availability of water and soil humidity (Jansson & Hofmockle, 2020).

Restoring desert vegetation is vital for enhancing ecosystems and supporting wildlife. Rehabilitating the degraded areas, reintroducing native plant species, and improving soil health, can yield numerous ecological benefits. For instance, plant root systems stabilize the soil, preventing erosion from wind and water, which is especially important in arid environments where soil can be easily displaced. Additionally, vegetation enhances the soil's ability to retain moisture, a crucial factor in desert areas where water is limited. Plants also reduce surface runoff, allowing more water to infiltrate the soil, thereby promoting further plant growth and maintaining soil health.

Restored vegetation provides essential habitats for wildlife, offering food, shelter, and breeding grounds. Diverse plant species create various microhabitats, supporting a wide range of animals, including insects, birds, reptiles, and mammals. Moreover, vegetation plays a significant role in regulating local microclimates by providing shade, lowering ground temperatures, and increasing humidity. These conditions can foster a more favorable environment for other plant and animal species to thrive. Even in desert settings, plants contribute to capturing and storing carbon dioxide from the atmosphere. Although on a smaller scale compared to temperate ecosystems, restored vegetation can still help mitigate climate change by sequestering carbon.

The restoration of deserts by growing native plants attracts and supports vital pollinators such as bees and butterflies, essential for the reproduction of many plant species. This enhances plant diversity and productivity, which in turn supports wildlife. Moreover, restoring vegetation helps reinstate natural nutrient cycles. Native plants, particularly those suited for arid environments, improve the ecosystem's ability to withstand drought through their deep roots and survival adaptations. Diverse and healthy plant communities are more resilient to invasions by non-native species, enabling restored native vegetation to effectively compete with invasive plants that threaten to dominate.

16.7.3 Revitalization of Degraded Ecosystems Through Protected Areas

Protecting the degraded ecosystems, and establishing reserves in arid regions is a highly effective approach to boost vegetation and improve ecosystem productivity. The Protected Planet Report (IUCN, 2024; UNEP-WCMC, 2024) reveals that 17.6% of land and inland waters and 8.4% of the ocean and coastal areas globally are within documented protected and conserved areas. The world database on protected areas (WPDA) present a total of 305,195 protected areas including 293,259 polygons and 11,936 points covering 244 countries and territories.

According to the data provided by the World Database on Protected Areas (UNEP-WCMC, 2024), in Kuwait there are one National Park, 3 Marine Parks, 14 Nature Reserves, one Scientific Reserve, one Wildlife Nature Reserve, two Parks, and four other Protected Areas. Terrestrial and inland waters protected area coverage is17.13% of the 17,818 km^2 reported area. Various studies and local initiatives, like those currently underway in Kuwait and Qatar, demonstrate that designating land as a protected area allows it to recover from degradation, fosters diverse plant communities, and ultimately enhances productivity. By alleviating grazing pressure, we enable natural processes to rejuvenate soil and promote vegetation growth. Furthermore, protected areas restrict harmful human activities that can damage vegetation cover, leading to issues such as soil erosion and compaction. When land is protected from disturbances, native plant species often begin to naturally regenerate (McNeely, 2003; Sharma et al., 2018). These plants are well-adapted to local conditions and can help rebuild a healthy, resilient ecosystem. Protected areas conserve the habitats of native plants and animals, which helps maintain biodiversity. A diverse plant community supports a wider range of wildlife and contributes to a more stable and productive ecosystem. With increased vegetation cover, the soil is better protected from erosion and further supports vegetation growth and finally insure the sustainability (Fig. 16.10).

In 2024, under the issue of "The global threat of Drying lands: regional and global aridity trend and future projection" the United Nation held COP 16 in Saudi Arabia agreement to bring the issue of Desertification (UNCCD) to the global voice for land and as one of the three major UN treaties known as the Rio Conventions, alongside climate and biodiversity. The aim of COP is to reduce land degradation for food security and environment sustainability by creating scientific policies to gain more understandings for the causes of dry land and find scientific solutions to restore land for sustainability.

Fig. 16.10 Dense vegetation cover in a protected area in Kuwait

16.7.4 Hydrological Cycles

In hydrologic cycle the water moves within atmosphere, or known as a movement of the water from the earth to the atmosphere and back to the earth (Shahid & Alkandari, 2024). Climate change significantly affects the hydrological cycle, which in turn influences various environmental components such as rivers, land, and streams. As the climate warms, changes in precipitation patterns, evaporation rates, and snowmelt timing can alter river flow, soil moisture, and groundwater levels. These changes impact water availability, ecosystem health, and agricultural productivity. The climate system comprises key components like the atmosphere, cryosphere (frozen water part of the Earth), and biosphere (global sum of all ecosystems) and play key role in regulating temperature and precipitation, while the cryosphere affects sea level and freshwater availability. The biosphere interacts with these components, and impact biodiversity and ecosystem services (Davies & Simonovic, 2005). Rising atmospheric temperatures can increase evaporation rates, leading to changes in rainfall patterns. Population growth increases the demand for resources, leading to more deforestation and land use changes that release GHGs. These factors create a feedback loop where climate change impacts water systems, and demonstrating the complex and interdependent nature of Earth's environmental systems.

16.7.5 Insects and Pests

One third of the animal kingdom is Insect and pests. Climate change has significant impacts on insect and pest populations, which can have cascading effects on ecosystems and human activities. Variations in precipitation and humidity can also influence insect distribution (Peace 2020). For instance, increased rainfall can create more breeding sites for mosquitoes, while drought conditions might reduce the availability of water sources for other insects. Higher temperatures can accelerate the development and reproduction rates of insects. For some species, this means more generations per year, potentially leading to population explosions. Changes in temperature and other climate variables can affect insect behaviour, such as feeding patterns, mating rituals, and migration routes. These changes can disrupt ecological balances and agricultural practices. Climate change can affect the timing of flowering and the availability of nectar and pollen (Fig. 16.11), which in turn impacts pollinators like bees and butterflies. Mismatches between the timing of insect activity and plant blooming can reduce food availability for insects. Insects that are crucial food sources for other animals may experience changes in population dynamics, potentially affecting the species that rely on them. This can lead to shift in food webs and ecological relationships. Changes in climate can alter interactions between insect species and their competitors or predators. Overall, understanding and addressing the effects of climate change on insects and pests is vital for maintaining ecological integrity, ensuring agricultural productivity, and protecting human health (Karuppaiah & Sujayanad, 2012).

Fig. 16.11 The interaction between climate change, plants and insect behaviour

16.8 Ecosystem Structuring

The structure of an ecosystem is a fundamental concept in ecology, characterized by the organization and interactions of its biotic (living) organisms—such as plants, algae, and some bacteria that produce energy and form the basis of the food web—and its abiotic (non-living) components. The climate of an ecosystem, which includes factors like temperature, humidity, and precipitation, plays a crucial role in shaping its structure by influencing which species can survive and thrive in that environment. Thus, the structure of an ecosystem is a complex and dynamic system formed by the interactions between its biotic and abiotic elements, along with energy flow and climatic conditions (Upadhyay, 2020). An increase in global temperatures can significantly impact the structure and biodiversity of both marine and terrestrial ecosystems. In terrestrial ecosystems, rising temperatures may prompt species to migrate to higher or lower altitudes or latitudes. However, not all species are capable of adapting or relocating quickly enough, resulting in mismatches in ecosystem functions and potential local extinctions. Climate changes can disrupt the timing of natural events, such as flowering, breeding, and migration. Additionally, higher temperatures, coupled with drought conditions, can lead to more frequent and severe wildfires. These fires not only destroy habitats but also alter soil composition and nutrient cycling, contributing to long-term ecosystem degradation. Warmer climates can also expand the range of pests and pathogens, increasing their effects on both plants and animals.

The extinction or migration of keystone species (those crucial for ecosystem stability) can lead to the collapse of ecosystems, affecting all other species within the system. As species adapt, migrate, or perish due to temperature changes, food webs can become imbalanced, leading to cascading effects throughout the ecosystem. Both marine and terrestrial ecosystems may experience a reduction in biodiversity, leading to less resilient ecosystems that are more vulnerable to further environmental changes and disturbances. Overall, the increase in global temperatures by 2–4 °C is likely to cause significant and possibly irreversible changes to the structure and function of ecosystems across the planet, with widespread consequences for biodiversity and the services these ecosystems provide.

16.9 Ecosystem Degradation

Climate change highly impacts on environmental degradation, particularly in arid and semi-arid regions. These areas are already suffering from poor soil and water scarcity, and will face even harsher conditions as temperatures rise. Climate change affects the timing of seasonal events like flowering, breeding, and migration. Plants may bloom earlier in the spring, disrupting the synchronization between species, such as pollinators and flowering plants. These shifts can lead to mismatches between species that depend on one another, such as predators and prey or plants and pollinators, potentially reducing survival and reproductive success. Ecosystems that sequester carbon, such as forests and wetlands, could become less effective as carbon sinks if they face increased disturbances like wildfires, pest infestations, or drying conditions. Elevated temperatures and drier environments can alter ecosystem structure, diminish biodiversity, and disrupt nutrient cycles.

Intensified storms and flooding can result in habitat loss, erosion, and shifts in species diversity and composition. Climate change impacts the availability of essential resources such as fresh water, food, and medicinal plants. For example, variations in precipitation and temperature can diminish crop yields. Certain ecosystems may reach critical thresholds, leading to sudden and irreversible transformations, such as the shift from forest to savanna or desert. Global climate patterns can drastically influence and cause melting permafrost releases stored GHG like methane, which is driving global warming and exacerbating its effects on ecosystems. Collectively, these changes jeopardize the stability and resilience of ecosystems, underscoring the urgency of addressing climate change to safeguard the services and functions they offer. This environmental strain may lead to significant displacement of both human and animal populations due to food scarcity and health challenges (Warner et al., 2010).

Human-induced climate change impacts ecosystems by altering climate variability, which poses direct threats to biodiversity, ecological systems, and food security. Ecosystems, especially in arid and semi-arid areas, are particularly susceptible to these changes, leading to significant degradation. The rising frequency and severity of extreme weather events, along with shifting precipitation patterns and increasing temperatures, destabilize these delicate environments, making it harder to maintain biodiversity and ensure food security (Malhi et al., 2020). The Millennium Ecosystem Assessment (MEA Board, 2005) highlights the decline of ecosystem services, indicating that 15 out of 24 services are either degraded or unsustainably exploited. This degradation disproportionately affects impoverished, resource-dependent communities. The MEA report also noted that agricultural expansion driven by population growth has resulted in the loss of 55% of biomes, leading to considerable biodiversity decline. Over the past 50 years, climate change has been a significant factor driving changes in biodiversity.

To sustain or enhance ecosystems in the context of climate change, mitigation strategies must focus on reducing human environmental impact. Effective approaches emphasize adaptive management techniques that tackle the effect of climate change on various aspects, such as crop production, in large the biodiversity, reduce water scarcity, and increase food security. Prioritizing research on the effects of climate change on ecosystems enables more informed decision-making. Additionally, ongoing monitoring of climate impacts is essential for adapting management practices to evolving conditions.

16.10 Ecosystem Management Options

Climate change affects various components of the climate system, including humidity, wind patterns, rainfall, and moisture levels. The greenhouse effect is a natural phenomenon where certain gases in the Earth's atmosphere trap heat, preventing it from escaping into space, and these GHGs drive to climate change. As the climate warms, weather patterns can become more extreme and unpredictable, resulting in more frequent and severe storms, heatwaves, and heavy rainfall events. In polar regions, the thawing of permafrost releases stored methane, a powerful GHG that can further accelerate global warming. Modifying public policies is essential to mitigating the long-term impacts of climate change. By understanding these processes, scientists can better predict and manage the effects of climate change. Effectively addressing climate change necessitates a comprehensive strategy that includes reducing GHGs emissions, improving energy efficiency, and adapting to ongoing changes. Ultimately, effective management and a clear understanding of this global issue are crucial for ensuring the sustainability of natural resources (Weiskopf et al., 2020).

Changes in temperature and variations in precipitation—whether increased or decreased—can significantly impact ecosystems and habitats. Species may move to new locations, adapt to altered conditions, or face extinction if they cannot adjust. For instance, rising temperatures can shift the distribution of many species, influencing biodiversity and the overall health of ecosystems. Additionally, higher global temperatures lead to the melting of polar ice caps and glaciers, which contributes to rising sea levels. This can result in coastal erosion, increased flooding, and the displacement of both human and animal populations (Upadhyay, 2020). The interplay of various climate change factors endangers the sustainability of natural systems, hindering their ability to support life, control climate conditions, and deliver crucial resources for future generations.

16.11 Change the Policy of the Environment to Ensure the Sustainability

Developing and implementing policies that promote adaptive management and sustainable land use in semi-arid and arid regions is essential for maintaining long-term ecosystem health. These policies should seek to harmonize human needs with environmental conservation, ensuring responsible use of natural resources while protecting and restoring ecosystems. Key policy approaches include:

16.11.1 Adaptive Management Policies

Establish policies that encourage adaptive management, enabling flexibility and responsiveness to evolving environmental conditions. This approach includes ongoing evaluation and modification of practices in light of new scientific findings and observed changes in the environment.

16.11.2 Sustainable Agricultural Practices

Formulate policies that promote sustainable soil management techniques, such as incorporating clay into sandy soils, increasing soil carbon sequestration by adding organic fertilizers, utilizing various irrigation technologies, improved crop varieties, and planting tall native trees to minimize soil erosion. These practices help lessen environmental impact, enhance soil health, and boost biodiversity.

16.11.3 Land Use Planning

Enact policies that support sustainable land use practices, preventing the excessive exploitation of natural resources in semi-arid and arid areas and ensuring a balance between land development and conservation activities.

16.11.4 Conservation Incentives

Introduce economic incentives, such as subsidies or tax reductions, for businesses that adopt conservation practices. Additionally, Payment for Ecosystem Services (PES) programs can be utilized to compensate individuals and organizations that conserve or restore ecosystem services.

16.11.5 Climate Resilience

Incorporate climate resilience into all environmental policies to ensure ecosystems can endure the effects of climate change. This involves investing in research and infrastructure that facilitate ecosystem adaptation.

16.11.6 Community Engagement

Promote the participation of local communities in the creation and implementation of environmental policies. Policies that involve stakeholder engagement are generally more effective and sustainable over the long term. By focusing on these policy areas, governments and organizations can create a framework that supports adaptive management, promotes sustainable practices in semi-arid and arid land, and ensures the long-term sustainability of ecosystems.

16.12 New Technologies and Opportunities to Reduce the Climate Change Impact

Emerging technologies and opportunities to combat climate change present promising solutions for mitigating environmental harm and advancing sustainability. These innovations aim to lower GHGs emissions, improve resource efficiency, and adapt to evolving environmental conditions (Alkandari, 2023) as:

- ***Solar Power***: Advanced photovoltaic systems and concentrated solar power (CSP) are becoming more efficient and cost-effective, allowing for cleaner energy production.
- ***Wind Energy***: Wind turbines, especially offshore wind farms, are expanding and provide a significant reduction in carbon emissions compared to fossil fuels.
- ***Hydroelectric and Tidal Power***: Innovations in hydroelectric power, including small-scale and low-impact designs, as well as tidal and wave energy, harness natural water flows for clean electricity.
- ***CCS*** technologies capture CO_2 emissions from industrial processes and store them underground, preventing them from entering the atmosphere. This is a critical technology for reducing emissions from sectors like cement, steel, and chemical manufacturing.
- ***Precision Agriculture***: Uses different devices as drones, and satellite imagery to monitor crops, optimize water use, and reduce the need for fertilizers and pesticides, minimizing the environmental footprint of farming.
- ***Vertical Farming***: Indoor farming techniques, using LED lighting and hydroponics, reduce water usage and land footprint, helping to produce food more sustainably in urban environments.
- ***Drought-Resistant Crops***: Advances in biotechnology have led to the development of crops that can withstand extreme weather conditions, reducing the risk of food shortages in climate-vulnerable regions.
- ***Smart Transportation Systems***: Technologies such as electric public transit, bike-sharing, and smart traffic management can reduce emissions from the transportation sector.

- ***Smart Grids:*** These grids use digital technologies to manage electricity distribution more efficiently, integrating renewable sources and reducing energy waste.
- ***Utilizing Clean Energy in Buildings:*** Implementing energy-efficient designs and technologies, such as green roofs and solar panels, can significantly lower a building's carbon footprint and decrease GHGs emissions
- ***3D Printing for Construction:*** This technology allows for the creation of energy-efficient, affordable housing with less material waste.
- ***Waste-to-Energy***: Modern waste-to-energy plants convert non-recyclable waste into electricity or heat with reduced emissions compared to traditional incineration.
- ***Solar Radiation Management***: Techniques like stratospheric aerosol injection aim to reflect a portion of solar radiation back into space, temporarily reducing global temperatures.
- ***Developing Advanced Monitoring and Forecasting Tools***: For creating best solution to reduce the adverse effect from climate change on ecosystem.
- ***Investigating the Impacts of Climate Change***: On natural enemies and improving biological control methods.
- ***Breeding and Genetically Engineering Crops***: To be more resilient to changing climatic conditions.
- ***Enhancing Water-use Efficiency***: Establishing sustainable water management practices is essential in semi-arid and arid regions, which are vulnerable to water scarcity. This involves techniques such as rainwater harvesting, implementing efficient irrigation systems, and safeguarding water sources from contamination.
- ***Educating Population and Stakeholders***: Awareness about climate change impacts and adaptive strategies.

By utilizing these technologies, governments, industries, and communities can make meaningful progress in reducing the effects of climate change while fostering more sustainable and resilient systems for the future. Furthermore, by incorporating these strategies, human activities can be more effectively aligned with environmental needs, aiding in the mitigation of climate change's negative impacts on ecosystems in semi-arid and arid regions.

16.13 Conclusions and Recommendations

It has been observed that climate change has significant consequences on ecosystem degradation, provision of ultimate services and sustainability. Simultaneous to climate change impacts, anthropic activities are also causing degradation of ecosystem by declining vegetation cover and accelerating erosion leading to poor soil quality of arid and semi-arid land countries. These fragile and degraded ecosystems should be prioritised for their protection to assure long term and sustainable existence and healthy services. Few activities, among others, which can cause significant impact on the recovery of degraded ecosystems are, (i) controlled grazing based on the recurrence of vegetation after grazing cycle, (ii) increasing the protected areas, nationally and globally, (iii) rehabilitation of degraded ecosystems using locally adapted indigenous and native plant species and, (iv) regular patrolling to assure there is no external effects. The creation of protected areas must have objective oriented policy, in terms of conservation of flagship and nearly extinct plants and wildlife species. As has been observed that countries in the drylands (especially hyperarid, arid and semi-arid) are highly vulnerable to climate change and anthropic impacts, stringent policy development and implementation is essential to reap the rewards from ecosystems rehabilitation efforts. The population in these countries to be educated to assure they understand the effects of climate change and anthropic effects on ecosystems, and make efforts to reverse the ecosystems degradation trend. To address these challenges, it is crucial to adopt adaptive management strategies, such as sustainable land use practices, biodiversity conservation, efficient water management, and the restoration of degraded lands. Furthermore, incorporating climate change considerations into policy-making and planning is essential for the long-term sustainability of ecosystems in arid regions. Ongoing research is also necessary to deepen our understanding of climate change and its effects on the environment.

References

Abdelhak, M., Jhariya, M., Meena, R., Banerjee, A., & Meena, S. (2022). Soil improvement in arid and semiarid regions for sustainable development. *Natural Resources Conservation and Advances for Sustainability*. (pp. 73–90). Elsevier. ISBN 9780128229767. https://doi.org/10.1016/B978-0-12-822976-7.00026-0

Alkandari, A. (2023). Climate and climate change aspects of Kuwait. In M. K. Suleiman, S. A. Shahid (Eds.), *Terrestrial environment and ecosystems of Kuwait* (Ch. 3, pp. 57–91). Springer. https://doi.org/10.1007/978-3031-46262-7_3.

Barakat. H. N. (2011). Arid lands: Challenges and hopes. In *Earth system: History and natural variability* (Vol. III, pp. 8). ©Encyclopedia of Life Support Systems (EOLSS).

Bernardino, P., De Keersmaecker, W., Fensholt, R., Verbesselt, J., Somers, B., & Horion, S. (2020). Global-scale characterization of turning points in arid and semi-arid ecosystem functioning. *Global Ecology and Biogeography, 29*(7), 1230–1245.

Bonannella, C., Hengl, T., Parente, L., & de Bruin, S. (2023). Biomes of the world under climate change scenarios: Increasing aridity and higher temperatures lead to significant shifts in natural vegetation. *Peer J, 11*, Article e15593. https://doi.org/10.7717/peerj.15593

Boutaj, H., Moumni, A., Nassiri, O., & Aitouna, A. O. (2022) Climate change impacts on biodiversity in arid and semi-arid areas: Biodiversity under climate change. In *Research anthology on environmental and societal impacts of climate change* (Vol. 25). https://doi.org/10.4018/978-1-6684-3686-8.ch028

Brown, J. H., Valone, T. J., & Curtin, C. G. (1997). Reorganization of an arid ecosystem in response to climate change. *Proceedings of the National Academy of Sciences of the United States of America, 94*(18), 9729–9733. https://www.pnas.org/doi/epdf/. https://doi.org/10.1073/pnas.94.18.9729

Chigani, K., Khajeddin, S., & Karimzadehi, H. (2012). Soil-vegetation relationship of three arid land plant species and their use in rehabilitating degraded sites. *Land Degradation and Development, 23*(1), 92–101. https://doi.org/10.1002/ldr.1057

Cooke, R. U., & Warren, A. (1973). *Geomorphology in deserts*. University of California Press.
Davies, E. G. R., & Simonovic, S. P. (2005). Climate change and the hydrological cycle. In *17th Canadian hydrotechnical conference* (pp. 47–58), 17–19 Aug 2005, Edmonton, Alberta.
D'Odorico, P., Rosa, L., Bhattachan, A., & Okin, G. S. (2019). Desertification and land degradation. In P. D'Odorico, A. Porporato, R. C. Wilkinson (Eds.), *Dryland ecohydrology*. Springer. https://doi.org/10.1007/978-3-030-23269-6_21
Grimm, N. B., Chapin, I. I. I. F. S., Bierwagen, B., Gonzalez, P., Groffman, P. M., Luo, Y., Melton, F., Nadelhoffer, K., Pairis, A., Raymond, P. A., Schimel, J., & Williamson, C. E. (2013). The impacts of climate change on ecosystem structure and function. *Frontiers in Ecology and the Environment, 11*(9), 474–482. https://doi.org/10.1890/120282
Gupta, S., Seth, R., & Sharma, A. (2018). Sundry of PGPR as a potential source of plant growth promotion in arid and semi-arid regions. *International Journal of Current Microbiology and Applied Sciences, 7*(9), 455–461. https://doi.org/10.20546/ijcmas.2018.709.055
Hailu, B., & Mehari, H. (2021). Impacts of soil salinity/sodicity on soil-water relations and plant growth in dry land areas: A review. *Journal of Natural Sciences Research, 12*(3), 1–10.
Huang, J., Ji, M., Xie, Y., Wang, S., He, Y., Ran, J. (2016). Global semi-arid climate change over last 60 years. *Climate Dynamics, 46*(3), 1131–1150. https://doi.org/10.1007/s00382-015-2636-8
Huntley, B.J. (2023). The Namib desert biome. In *Ecology of Angola: Terrestrial biomes and ecoregions*, Book Springer International Publishing, (Ch. 16, pp. 361–382). https://doi.org/10.1007/978-3-031-18923-4_16
IPCC. (2023). Climate change 2023 Synthesis report. In Core Writing Team, H. Lee, & J. Romero (Eds.) *Contribution of Working Groups I, II and III to the Sixth Assessment Report of the Intergovernmental Panel on Climate Change*. (pp. 35–115). IPCC, Geneva, Switzerland. https://doi.org/10.59327/IPCC/AR6-9789291691647
IUCN. (2024). https://iucn.org/press-release/202410/world-must-act-faster-protect-30-planet-protected-and-conserved-areas-need
Jansson, J. K., & Hofmockel, K. S. (2020). Soil microbiomes and climate change. *Nature Reviews Microbiology, 18*(1), 35–46. https://doi.org/10.1038/s41579-019-0265-7
Karuppaiah, V., & Sujayanad, G. (2012). Impact of climate change on population dynamics of insect pests. *World Journal of Agricultural Sciences, 8*(3), 240–246.
Malhi, Y., Franklin, J., Seddon, N., Solan, M., Turner, M. G., Field, C. B., & Knowlton, N. (2020). Climate change and ecosystems: Threats, opportunities and solutions. *Philosophical Transanctions, B, 375*(1794), 20190104. https://doi.org/10.1098/rstb.2019.0104
Mata-González, R., Figueroa-Sandoval, B., Clemente, F., & Manzano, M. (2007). Vegetation changes after livestock grazing exclusion and shrub control in the southern Chihuahuan desert. *Western North American Naturalist, 67*(1), 63–70. https://doi.org/10.3398/1527-0904(2007)67[63:VCALGE]2.0.CO;2
Mckenzie, D., Gedalof, Z., Peterson, D. L., & Mote, P. (2004). Climatic change, wildfire, and conservation. *Conservation Biology, 18*(4), 890–902. https://doi.org/10.1111/j.1523-1739.2004.00492.x
McNeely, J. (2003). Biodiversity in arid regions: Values and perceptions. *Journal of Arid Environments, 54*(1), 61–70. https://doi.org/10.1006/jare.2001.0890
MEA Board (Millennium Ecosystem Assessment Board). (2005). Ecosystem and human wellbeing. The MA Board represents the users of the findings of the MA process. https://www.greenfacts.org/en/biodiversity/index.htm
Michelutti, N., Labaj, A., Groom, C., & Smol, J. P. (2016). Equatorial mountain lakes show extended periods of thermal stratification with recent climate change. *Journal of Limnology, 75*(2), 403–408. https://doi.org/10.4081/jlimnol.2015.1444
Nwaerema, P. (2020). Impact of climate change on insects, pest, diseases and animal biodiversity. *International Journal of Environmental Sciences and Natural Resources, 23*(5), 151–153.
Panagos, P., Borrelli, P., Meusburger, K., Yu, B., Klik, A., Lim, K., Yang, E., Ni, J., Miao, Ch., Chattopadhyay, N., Sadeghi, H., Hazbavi, Z., Zabihi, M., Larionov, G., Krasnov, S., Gorobets, V., Levi, Y., Erpul, G., Birkel, Ch., Hoyos, N., Naipal, V., Oliveira, P., Bonilla, C., Meddi,

M., Nel, W., Al Dashti, H., Boni, M., Diodato, N., Oost, K., Nearing, M., & Ballabiol, C. (2017). Global rainfall erosivity assessment based on high-temporal resolution rainfall records. Scientific Reports. https://doi.org/10.1038/s41598-017-04282-8

Ragab, R., & Prudhomme, C. (2002). SW—Soil and water: Climate change and water resources Management in arid and semi-arid regions: Prospective and challenges for the 21st century. *Biosystems Engineering, 81*, 3–34. https://doi.org/10.1006/bioe.2001.0013

Roeder, A., & Hill, J. (Eds.). (2009). *Recent advances in remote sensing and geoinformation processing for land degradation assessment* (1st ed., pp. 418). CRC Press. https://doi.org/10.1201/9780203875445

Shahid, S. A. (2023a). Potential threats to soil functions and mitigation options for sustainable use. In M. K. Suleiman, & S. A. Shahid (Eds.), *Terrestrial environment and ecosystems of Kuwait, (Ch. 1, 3-20)* Springer. https://doi.org/10.1007/978-3-031-46262-7_1

Shahid, S. A. (2023b). Desertification: a central problem to restoring ecosystems. In M. K. Suleiman, S. A. Shahid (Eds.), *Terrestrial environment and ecosystems of Kuwait, (Ch. 5, 119-137)* Springer. https://doi.org/10.1007/978-3-031-46262-7_5

Shahid, S. A., & Alkandari, A. J. (2024). Halophytic crops as a solution for food security, land rehabilitation, and mitigating future water crises by utilizing marginal quality waters. In J. C. Dagar, S. R. Gupta, & A. Kumar (Eds.), *Halophytes vis-à-vis saline agriculture* (Ch. 17, pp. 441–478). Springer. https://doi.org/10.1007/978-981-97-3157-2_17

Shahid, S. A., Zaman, M., & Heng, L. (2018). Salinity and sodicity adaptation and mitigation options. In Zaman et al. (Eds.), *2018 Guideline for salinity assessment, mitigation and adaptation using nuclear and related techniques* (Ch. 3, pp. 55–88). Springer. https://doi.org/10.1007/978-3-319-96190-3_3

Shahid, S. A., Alkandari, A., Baron, J., Othman, A., Burezq, H., Al-Najdi, F., Abdulraheem, A., Al-Jeri, M., Thomas, M., & Quoreshi, A. (2024). Farm land salinity risk assessment to develop a management strategy for food security. Annual Progress Report No 1, Kuwait Institute for Scientific Research, Kuwait (KISR # 18906).

Sharma, M., Samanta, A., & Sharma, K. R. (2018). Degraded lands and their management for rehabilitation and improving crop productivity. *Journal of Soil and Water Conservation, 17*(1), 25–33. https://doi.org/10.5958/2455-7145.2018.00004.8

Singh, G. (2009). Salinity-related desertification and management strategies: Indian experience. *Land Degradation and Development*, *20*, 367–385. Wiley Inter Science. www.interscience.wiley.com. https://doi.org/10.1002/ldr.933

Singh, P. K., & Chudasama, H. (2021). Pathways for climate change adaptations in arid and semi-arid regions. *Journal of Cleaner Production, 284*(6), 124–744. https://doi.org/10.1016/j.jclepro.2020.124744

Skendžić, S., Zovko, M., Živković, I. P., Lešić, V., & Lemic, D. (2021). The impact of climate change on agricultural insect pests. *Insects, 12*(5), 440. https://doi.org/10.3390/insects12050440

UNEP-WCMC. (2024). Protected area profile for Kuwait from the World Database on Protected Areas, December 2024. www.protectedplanet.net

Upadhyay, R. K. (2020). Markers for global climate change and its impact on social, biological and ecological systems: A review. *American Journal of Climate Change, 9*(03), 159. https://doi.org/10.4236/ajcc.2020.93012

Wang, X., & Liu, L. (2023). The impacts of climate change on the hydrological cycle and water resource management. *Water, 15*(13), 2342. https://doi.org/10.3390/w15132342

Warner, K., Hamza, M., Oliver-Smith, A., Renaud, F. G., & Julca, A. (2010). Climate change, environmental degradation and migration. *Natural Hazards, 55*(33), 689–715. https://doi.org/10.1007/s11069-009-9419-7

Weiskopf, S. R., Rubenstein, M. A., Crozier, L. G., Gaichas, S., Griffis, R., Halofsky, J. E., Hyde, K. J. W., Morelli, T. L., Morisette, J. T., Muñoz, R. C., Pershing, A. J., Peterson, D. L., Poudel, R., Staudinger, M. D., Sutton-Grier, A. E., Thompson, L., Vose, J., Weltzin, J. F., & Whyte, K. P. (2020). Climate change effects on biodiversity, ecosystems, ecosystem services, and natural

resource management in the United States. *Science of the Total Environment, 733*, Article 137782. https://doi.org/10.1016/j.scitotenv.2020.137782
Zaman, M., Shahid, S. A., & Heng, L. (2018) Irrigation systems and zones of salinity development. In M. Zaman, S. A. Shahid, & L. Heng (Eds.), *Guideline for salinity assessment, mitigation and adaptation using nuclear and related techniques* (Ch. 4, pp. 91–111). Springer. https://doi.org/10.1007/978-3-319-96190-3_4

Part IV
Poultry and Aquaculture

Chapter 17
Effect of Rising Temperatures on Overall Health, Productive and Reproductive Performance of Chickens—A Review

Hanan Sulaiman Al-Khalaifah and Afaf Yaqoob Al-Nasser

Abstract The poultry sector is of significant importance in ensuring global food safety, as it is the primary source of consumption of meat which accounts for nearly two-thirds of all meat consumed worldwide. To optimize poultry production, it is essential to manage ecological factors such as temperature, humidity, light, nutrition, air, and sound at specific intervals. However, the poultry industry is susceptible to the negative influences of climate change, which can result in physiological stress and reduced production and profits for farmers. Several strategies can be employed to mitigate the consequences of high temperatures and humidity in poultry production. These include passive cooling, insulation of poultry houses, nutritional factors, genetic selection, shading, ventilation, cool water, effective management during the summer, and appropriate housing systems. This review manuscript is dedicated to examining the effects of increasing temperatures on the well-being, productivity, and reproductive capabilities of chickens. Additionally, it presents various methods to mitigate the effect of high temperatures and humidity in poultry farming.

Keywords Climate change · Eggs · Meat · Poultry · Temperature · Mitigation · Ventilation

17.1 Introduction

Poultry farming, an important factor in agriculture sector includes the raising of bird species such as chickens, ducks and turkeys, is widely recognized as a fundamental feature of animal husbandry with global significance. The chicken products like

H. S. Al-Khalaifah (✉) · A. Y. Al-Nasser
Food Security Program, Environment and Life Sciences Research Center, Kuwait Institute for Scientific Research, Safat, Kuwait
e-mail: hkhalifa@kisr.edu.kw

A. Y. Al-Nasser
e-mail: anasser@kisr.edu.kw

M. K. Suleiman and S. A. Shahid (eds.), *Fostering Arid Lands Agriculture in the Face of Climate Change*, Advances in Global Change Research 81,
https://doi.org/10.1007/978-3-031-98490-7_17

eggs and meat, are rich in high-quality protein sources for human consumption, therefore poultry farming plays a vital role on the food industry. The rapid growth and strengthening of this sector can be attributed to advancements in areas of animal husbandry such as breeding, diet, housing, and disease control. As global population growth is increasing, there will be a rising demand for poultry products, which requires the industry's adaptation and expansion to meet the rising demand for food. Hence the importance and benefits of food supply chain is very crucial.

The two major challenges in poultry farming which show a significant threat to food security are the fall in poultry production and the rise in the cost of production. The major environmental factors affecting the decrease in poultry production includes change in climate, environmental instability, and extreme variations in temperature. These changes in climate will lower the poultry yields which adversely increases the cost of production. As a result, the poultry industry should find suitable methods to adapt and expand to encounter the mounting demand for food, while addressing the challenges brought on by a changing climate (Lacetera, 2019).

The profit and a long term success in poultry industry is quite modest due to the major climatic issues the industry has to face. It is important to note that all domesticated animals, including poultry, enable them to maintain their body temperature which is essential to optimize the range for biological behaviors and actions through homeothermy. Due to the limited level of tolerance of temperature, poultry is exceptionally susceptible to heat stress. Achieving thermal stability without consuming excessive nutrients is crucial for optimizing animal welfare, nutrition, and health in systems where high production is desired. Heat stress is a devastating condition in the overall poultry system and results in significant economic losses for every country. It happens when the amount of heat energy produced by the animal is less than the total amount of heat energy lost by the animal (Renaudeau et al., 2012). According to its duration, heat stress is classified into two types, long term stress known as chronic heat stress, and short term stress known as acute heat stress. The effect of heat stress can be intensified when there is relatively higher humidity (Safi et al., 2022).

The reaction of animals to environmental challenges is not limited to any specific species, but poultry appears to be mainly susceptible to environmental challenges related to temperature, particularly heat stress. This happens due to the higher metabolic activity in modern poultry genotypes that leads to increased body heat production. Successful poultry management and welfare depends on the ability to understand and control environmental conditions (Wasti et al., 2020).

17.2 Status of Poultry Production and Consumption

Livestock systems are vital for meeting the nutritional requirements of the projected 9.8 billion people by 2050 (UNDESA, 2017). Livestock products (milk, meat, and eggs) provide around 30% of the global resource of proteins, with notable regional differences (Godde et al., 2021; FAOSTAT, 2022). To meet the expected rise in demand, meat production around the world is anticipated to increase by 14% by the end of this era, compared to the present status (Carlin et al., 2023). United Nations (UN) estimated the global population to surpass nine billion by 2050, continues to grow, feeding the planet has become a crucial priority (Shrestha & Mahat, 2022).

Poultry production is a subsector of livestock farming focused on raising domesticated birds such as chickens, ducks, turkeys, guinea fowls and ostriches for meat, eggs, and other related products. Over the past 40–50 years the production and consumption of poultry products have increased and it is estimated to increase further, mainly in the developing countries. Global demand for food derived from animals is expected to increase by 70% from 2005 to 2050, with poultry meat seeing the highest growth at 121%. In 2016, worldwide over 106 million tons of chicken meat were produced, and around 66 billion chickens were slaughtered, which will average to nine birds per person (Al-Nasser et al., 2020; FAOSTAT, 2016). By 2018, poultry meat production increased by over 8%, with almost 70 billion chickens slaughtered (FAOSTAT, 2018). Some factors that are collectively contributing to this projected rise in poultry meat consumption over the next decade are:

Affordability: Poultry is generally less expensive compared to other meats like beef and pork, making it more accessible to a broader population.
Health Perceptions: Poultry is often perceived as a healthier option due to its lower fat content compared to red meats.
Cultural and Dietary Preferences: Poultry is widely accepted across various cultures and religious practices, increasing its global consumption.
Production Efficiency: Poultry farming has a relatively shorter production cycle and higher feed conversion efficiency, making it more sustainable and easier to scale up.
Urbanization and Income Growth: As urban populations grow and incomes rise, especially in developing countries, the demand for affordable and convenient protein sources like poultry increases.

Over the past 50 years, changes in chicken meat consumption have led to genetic selection for fast-growing birds to meet increasing demand. Currently, broiler chickens are 35% heavier at slaughter and are processed 16 days earlier than in the 1960s (NCC, 2019). Selection has focused on increasing the size of the pectoralis major muscle, which now makes up to 25% of the bird's weight (Aviagen, 2022). However, this has also caused muscular defects like deep pectoral myopathy, where the pectoralis minor muscle suffers from ischemic necrosis due to inadequate blood supply (Baldi et al., 2020).

It was reported earlier that for the world's escalating population, chicken meat is estimated to become the major source of meat protein (Petracci et al., 2019). This extensive appeal has led to increased dependence on poultry to accomplish the dietary requirements of the rising population around the world. The production process for chicken varies based on factors such as economic conditions, environmental factors, and cultural practices, which are unique to different regions of the world. Some factors such as global population growth, climate change, loss of agricultural lands, disease outbreaks and health crisis has greatly affected the production process of chickens. Additionally, export restrictions, currency depreciation, world oil prices also do have negative impacts on this process (Al-Khalaifah et al., 2020a, 2022b; Al-Nasser, 2006; Al-Nasser et al., 2015). Chicken production has a growing appeal because of its lower environmental impact on perceived health benefits and wide cultural acceptance compared to other livestock. Henceforth, the efficiency of chicken production, considering its high feed conversion rate and its shorter production cycles, permits for the increased production to meet the increasing demand.

17.3 Nutritional Importance of Poultry Products

Poultry products, including meat and eggs, hold significant global appeal, valued for their versatility and nutritional richness. Meat and eggs are significant contributors to human nourishment due to their abundance of essential nutrients, high bioavailability of proteins, and relative affordability. As essential sources of proteins and vital micronutrients, comprising of vitamin A, vitamin B12, riboflavin, calcium, iron, and zinc, they contribute to the overall health and well-being of individuals worldwide. Chicken breast meat, renowned for its lean protein content and niacin (vitamin B3), plays a crucial role in supporting metabolic functions, nervous system health, and resistance to oxidation (Al Salihi et al., 2023; Al-Khalaifah, 2018; Al-Khalaifah et al., 2020b; Al-Surrayai & Al-Khalaifah, 2022; Attia et al., 2017).

Eggs, another prized component of poultry, boast not only high-quality proteins but also contain choline, which aids in digestion and supports optimal brain function. Poultry, often cited alongside fish, is acknowledged as one of the healthiest meat options due to its nutritional density, abundant minerals, and relatively low calorie count. Importantly, the affordability of poultry products makes them accessible to individuals across diverse income levels, contributing to their widespread consumption. Recognizing the pivotal role of eggs, especially in promoting the health of children, the Global Alliance for Improved Nutrition (GAIN) has launched campaigns to encourage increased egg consumption (Henchion et al., 2021). Eggs are a vital constituent in human nutrition and well-being because these are nutritionally rich and affordable, offering a substantial amount of essential nutrients relative to their calorie content. Since it offers significant health and nutritional benefits, this positions eggs as an essential and universal healthy food and are considered one of the most nutrient-dense foods available. Nutrients density in food means the provision of high levels of vitamins, minerals, and high-quality proteins with relatively low calorie content. These characteristics make them an excellent dietary choice for peoples of all ages. Their rich nutritional value supports various bodily functions, from growth and development in children to maintaining health and vitality in adults (Al-Khalaifah et al., 2022a; Mine et al., 2023).

Poultry meat and eggs include all nine vital amino acids that the body cannot produce on its own. These amino acids and the protein profile of poultry products are vital for muscle growth, tissue repair, maintaining lean body mass, promoting overall health and the proper functioning of various bodily processes (Alagawany et al., 2020). The highly digestible property of protein and its high bioavailability found in poultry products are efficiently utilized by the body for maintaining muscle mass, particularly in populations with increased protein needs, such as children during growth phases, athletes who require protein for muscle recovery and performance, and for the elderly people, who need sufficient protein to prevent muscle loss and maintain strength (Pratama et al., 2021). Table 17.1 summarizes the importance of vitamins which are rich in poultry products such as niacin (B3), riboflavin (B2), pantothenic acid (B5), pyridoxine (B6), biotin (B7), vitamin D, vitamin A, and vitamin E (Surana et al., 2019).

Moreover, poultry products are valuable sources of heme iron (iron which are found in meat, fish and poultry), which are easily absorbed by our body compared to non-heme iron, found in plant-based foods, where absorption is very less. This formulates poultry products very important in preventing anemia (deficiency in iron), a common nutritional deficiency that can lead to exhaustion, weakness, and impaired cognitive function. Adequate zinc intake is essential for preserving a healthy immune system, fostering proper growth and development, and supporting reproductive health. Zinc is also present in poultry meat and eggs, which is critical for immune function, healing of wounds, and DNA synthesis. In addition, both poultry meat and eggs contain selenium, an important antioxidant which helps the damage to protect cells and supports thyroid hormone metabolism. Selenium helps to lessen the risk of certain chronic diseases, and maintains the immune function by protecting against oxidative stress (Weyh et al., 2022).

Poultry meat and eggs, especially those from birds fed omega-3-enriched diets, are a valuable source of these fundamental fatty acids. Omega-3 fatty acids support the health of heart by decreasing the inflammation, improving cholesterol levels, and lowering blood pressure. They are also important for cognitive function, contributing to brain health and potentially reducing the risk of cognitive decline (Abo El-Maaty

Table 17.1 Importance of vitamins in poultry meat and table eggs (Surana et al., 2019)

Vitamin	Biological importance
Niacin (B3)	Supports energy metabolism and plays a role in maintaining healthy skin and nerves
Riboflavin (B2)	Important for production of energy and the metabolism of fats, drugs, and steroids
Pantothenic acid (B5)	Essential for the amalgamation of coenzyme A
Pyridoxine (B6)	Important for the development and function of brain, and it helps the conversion of food into energy and produce neurotransmitters and red blood cells
Biotin (B7)	Supports healthy hair, skin, and nails, and has a vital role in the absorption of carbohydrates, fats, and proteins
Vitamin D	Critical for calcium absorption, bone health, and immune function
Vitamin A	To maintain healthy vision, immune function, skin health, cell growth and differentiation
Vitamin E	Protects cells damage, immune function support, and boosts healthy skin and eyes

et al., 2020; Al-Khalaifah, 2020; Al-Khalaifah & Uddin, 2022; Al-Khalaifah et al., 2020b; Syan et al., 2024).

Poultry meat is ideal for individuals who are managing their weight or aiming to include more lean proteins in their diet because it is low in calories and fat content. Chicken breast, in particular, provides a substantial amount of high-quality protein with minimal fat, supporting muscle maintenance and overall health without contributing excess calories or unhealthy fats (Caplan, 2023).

17.4 Effect of Extreme Temperature and Humidity on Poultry Production

Elevated temperatures can negatively impact the performance, efficiency, and profitability of the operation. This issue is particularly pertinent to poultry production, as the production output (meat or eggs) increases, it leads to heightened metabolic activity and subsequent heat production. For broiler chickens, which grow rapidly, heat production rises while the ability to dissipate heat declines because of reduction in body surface area relative to body weight (Nawaz et al., 2021).

Maintaining an ideal temperature range of approximately 20–27 °C is essential for poultry, as it helps to support the overall health by considering the effects on growth, and feed adeptness. The Feed Conversion Ratio (FCR), which calculates the efficiency of converting feed into product (meat or eggs), is particularly sensitive to environmental conditions, including temperature (Fouad et al., 2016). Heat stress in poultry can lead to lowered feed intake, increased water consumption, changes in behavior, and even mortality in extreme cases. Heat exhaustion happens when there is a disparity among the flow of total energy from the body of chicken to the environment relative to the heat produced by the chicken. This imbalance is a result of factors like sunlight, thermal radiation, temperature, humidity, and air movement, which are mainly environmental related, along-with the characteristics of chicken, such as species, metabolic rate, and thermoregulatory mechanisms. Extreme temperatures and high humidity can have adverse effects on poultry, leading to decreased feed consumption, nutrient deficiencies, and increased body temperature (Lara & Rostagno, 2013). These environmental factors can trigger behavioral, physiological, and immunological responses in chickens, which can negatively impact their productivities. High environmental temperature can result in decreased growth rate, body weight, meat quality, egg production, egg weight, and egg quality, semen quality, fertility, and hatchability, leading to significant financial losses for the poultry industry (Gouda et al., 2024; Oke et al., 2021).

During the early and post-hatching periods and during hot and humid weather conditions, chickens are particularly susceptible to overheating and are at risk of dying if they are unable to control their body temperature efficiently (Saeed et al., 2019). It is crucial that chicks are maintained at the optimal temperature conditions with artificial heat immediately after hatching, as they are unable to maintain the temperature of their body by their own. The standard temperature range for newborn chicks is between 32 and 35 °C, with a gradual reduction of 3°C per week until the normal temperature of approximately 23–26 °C is reached. If the chicks' temperature is not managed properly during transportation from hatcheries, they may experience overheating or cooling, which can significantly reduce their post-hatch survival rate. Due to high ambient temperatures the rate at which the chicks grow and the amount of feed consumed are reduced to great extent, which will further reduce the performance in growth, reduced feeding time, and lethargy. By preparing and formulating a chick feed with high nutrient density, and a good supply of nutrients, the problem of reduced feed intake can be overcome to a great extent (Sohail et al., 2012).

The optimal range of temperature for poultry is between 15 and 20 °C (59–68 °F), but at 30 °C (86 °F) birds experiencing mild heat stress. A drastic decline in the egg laying rate is seen as the temperature increases above 37 °C (98 °F), where chickens experience severe heat stress. When hens are exposed to a temperature of 30 °C and above, hens will consume less feed, lowering egg production and smaller egg size. As the temperature increases from 30 to 38 °C, shell quality is likely to deteriorate, as signified by an increase in the percentage of cracked eggs.

Since chickens have no sweat glands, and have a high body temperature, they rely on panting and rapid breathing to release heat from their body and cool down. When chickens pant excessively, it increases their respiratory rate which shows a disruption in acid-base balance in the blood. In addition, heat stress can instigate heat exhaustion in birds, leading to a loss of appetite, reduced activity, increased water intake, and lethargy, which can negatively affect the quality of meat and egg production (Saeed et al., 2019).

Chickens subjected to heat exhaustion conditions spend most of their time panting, drinking water and resting, comparatively reduces the movement and feeding (Ranjan et al., 2019). To regulate their body temperature and maintain homeostasis, animals use various mechanisms, like vasodilation including increased radiant, convective and evaporative heat loss and sweating (Mota-Rojas et al., 2021). Birds have air sacs which is an another supplementary mechanism that accelerates the exchange of heat between their body and the environment. During panting, more air is circulated on the surface providing an increased exchange of gas with the air, thus more heat is being evaporated. Therefore, under heat stress conditions, air sacs play a crucial role in maintaining the body temperature but as panting increases beyond a level, it disturbs the metabolic acid–base balance (Ranjan et al., 2019).

Heat stress significantly impacts the performance especially when environmental temperatures rise and the Thermal Heat Index (THI) surpasses critical thresholds. The physiological response to heat stress varies across poultry types. For instance, heat stress induces a greater increase in muscle and rectal temperature in broiler chickens compared to layer chickens, leading to a rise in superoxide production. The above result shows a more substantial decrease in the body weight of the broilers under heat stress conditions, highlighting their varied susceptibility to environmental stressors (Nawaz et al., 2021).

Chronic heat stress adversely affects feed digestibility of proteins, fats, and starch, thereby lowering broiler production (Cardoso et al., 2022). Agitation and increased respiratory rate are some other evident indicators of heat stress which lead to increase in the loss of body fluids and cause dehydration (Gicheha, 2021). In tropical climates, birds cool their bodies by drinking more water by which the feed intake will reduce considerably, which negatively affects meat production efficiency and quality, particularly in birds not adapted to tropical climates (Zhang et al., 2012).

Because of the absence of sweat glands in birds and since they have feathers covering their entire body, which impairs their ability to regulate their body temperature. Consequently, they rely on effective mechanisms, such as panting, to release heat during high relative temperatures. As the birds pant, they open the beak to increase the rate of respiration and promote cooling through the respiratory tract by evaporation. This process results in an increased rate excretion of CO_2, which reduces the strength and intensity of carbonic acid and hydrogen ions in the blood. In response, the body increases the concentration of bicarbonate ions and raises the blood pH, making it more alkaline. To restore the acid–base balance and maintain normal blood pH, birds expel more bicarbonate and preserve hydrogen ions in the kidney. This modification in acid–base balance leads to respiratory alkalosis and metabolic acidosis, which can decrease the availability of blood bicarbonate for eggshell mineralization and increase the availability of organic acids. As a result of this modified status, free calcium levels in the blood decreases, which could have been utilized by the shell glands to produce high-quality eggs. This deficit cannot be corrected through diet, leading to poor egg quality in breeders and laying hens (Ruchi et al., 2022; Wasti et al., 2020).

Moreover, it is essential to recognize the close relationship between humidity and ambient temperature in poultry environments. Regardless of the temperature, higher humidity levels can intensify bird discomfort and compromise the efficacy of evaporative cooling mechanisms. As the humidity increases, it lowers the evaporative heat loss, which increases with temperature. Maintaining an optimal humidity level within the chicken coop is crucial for the well-being of the birds. Ideally, the humidity should be maintained between 40 and 60% (Apalowo et al., 2024). Coops that are excessively humid can contribute to an elevated risk of respiratory issues and mold growth. On the contrary, coops with insufficient humidity may result in dry sinuses and an increased susceptibility to dust-borne respiratory problems (Wang et al., 2023).

Quality of the produced meat is also affected by high temperatures and humidity. When broilers are subjected to high temperatures during the growing stage, it often encounters meager meat characteristics, loss of quality, decreased fat deposition, and lower protein content. This negatively affects the ratio of breast muscle and increases the ratio of thigh muscle in broilers (Zhang et al., 2020).

In chickens, Pale, Soft, Exudative (PSE) type of meat is often produced by a fast decrease in post-slaughter pH level. Birds with an increased metabolic activity and effective rate at which it grows, are more sensitive and vulnerable to heat stress during the growing period, which can lead to poor meat quality. Apart from that the acid–base imbalance and muscle atrophy is also greatly affected (Petracci et al., 2015). Since chicken muscles are primarily composed of fast-twitch fibers, HS can significantly impact meat quality (Zhang et al., 2017). Prior to slaughter, subjecting birds to high stocking density can accelerate the anaerobic glycolysis process in muscles, since the body temperature of the birds is high. This further leads to a decrease in pH which happens during the transformation of muscles into meat. This combination of increased carcass temperature and low pH results in protein degradation and the development of PSE meat (Allan et al., 2010). The PSE meat exhibits poor processing capabilities, resulting in dry and brittle meat with low protein extractability and water-holding capacity (Barbut et al., 2008). The poultry industry often accounts significant losses in meat production during hot weather due to decreased capacity of water-holding capacity, poor meat texture, and pale color.

When cells encounter stressful conditions, such as UV light exposure, extreme temperatures, or tissue repair demands, they initiate a defensive response by integrating proteins known as heat shock proteins (HSPs). During elevated chemical, physiological, or temperature-related stress, these proteins are preserved within the organisms and are produced in increased amounts. The HSPs display a major role in survival mechanisms of the cells and in the intricate processes of protein, particularly in the investigations that results in protein denaturation and folding within the cell. The formulation of heat shock proteins is induced not only by heat shock itself but also by various stressors such as anoxia, hypoxia, and heavy metal toxicity, all of which can disrupt proper protein structure. Prominent members of this protein family include Hsp25, Hsp27, Hsp60, Hsp70, and Hsp90. These proteins perform diverse roles within the cell and have different molecular weights, to protect from stress-induced damage protein, mainly protein folding. The dysregulation of the protein highlights the critical role in maintaining cellular homeostasis associated in various diseases (Balakrishnan et al., 2023).

In the poultry industry, exposure to extreme levels of solar radiation, environmental factors and elevated temperatures are some major factors which affects the decline in the chemical composition and meat quality of broilers which will further mitigate broiler health and meat characteristics. Before slaughtering, high heat stress can increase muscle metabolic rate and rigor mortis, leading to protein denaturation. This will mainly affect the water-holding capacity of meat, as proteins get damaged due to high carcass temperature which inhibits its ability to bind water, resulting in an increase in drip loss and cooking loss (Woelfel et al., 2002).

Study of Nawaz et al. (2021) has shown that high temperatures damage the water-holding capacity of the broiler, resulting in increased drip loss in poultry meat. Drip loss is a common issue where meat loses its original texture and taste once the conversion from muscle to meat happens. Frozen meat when thawed, the texture and taste of the meat will change due to loss of water content which results in the loss of other nutritional substances through the water. Overall meat quality is affected by drip loss because it reduces meat tastiness, juiciness, and acceptability. It results from the loss of water, myofibers, iron, and proteins, which negatively impact meat quality (Ponsuksili et al., 2008). This issue is particularly prevalent in pork and chicken, where experts are actively seeking solutions.

Chronic heat stress encountered in broilers leads to a notable shift in muscle composition, with a decrease in the ratio of breast muscle to an increase in the ratio of thigh muscle (Lara & Rostagno, 2013). Additionally, chronic heat exposure can adversely impact fat deposition, exhibits lower protein content and quality of the meat in broilers, where these effects vary according to the breed. The broiler production efficiency and meat quality, is totally dependent on the absorption of nutrients, the reliability of the gut wall, and the bird's immune system. These changes highlight the multifaceted impression of heat stress on broiler physiology, affecting immune response, muscle composition, and nutritional content.

17.5 Effect on the Overall Health Status and Immune System

The health of broiler and laying chickens directly influences the productivity and profitability of poultry businesses. Temperature, a crucial climatic element, can affect a bird's ability to combat infections. Birds must experience minimal heat stress to avoid struggle for nutrients between fighting infections and maintaining heat balance. The thermal neutrality in chicken and the intake of nutrients are both negatively affected by the increase in temperature which will expose the birds to many diseases.

The heat stress is proven to weaken the immune system of chickens, creating them more vulnerable to infectious and contagious poultry diseases. This is particularly evident during the summer season in tropical countries, where diseases such as Infectious Bursal Disease Virus (IBDV), Newcastle Disease Virus (NDV), infectious bronchitis disease virus, and Gumboro disease are more prevalent (Gicheha, 2021). Additionally, heat stress impacts the levels of Reactive Oxygen Species (ROS) in cells, potentially prominent to oxidative stress if not properly managed. Extreme temperatures force the body to maintain normal ROS concentrations, but heat stress can cause an overproduction of ROS that exceeds the body's capacity to process them (Altan et al., 2003). This state is known as oxidative stress (Pizzino et al., 2017). Oxidative stress caused by heat stress has several undesirable effects on poultry, involving reduced antioxidant status and fertility. Additionally, in poultry a decrease in metabolic oxidative activity of mitochondria is reported due to heat and oxidative stress which will affect the dysfunction of mitochondria and alter the antioxidant enzyme activity (Oke et al., 2024).

The balance of poultry birds' homeostasis can also be disturbed due to high levels of humidity, which will create a favorable environment to pathogenic agents such as bacteria, viruses, and fungi. In hotter temperatures this imbalance will further increase the occurrence and frequency of bacterial diseases like salmonellosis, coccidiosis, and *E. coli* infections. Heat stress is one of the main aspect that weaken the immune system of broilers, making them more inclined to conditions like coccidiosis, which can further lead to necrotic enteritis (Tsiouris et al., 2018). Under heat stress birds will decrease the amount of the total white blood cell counts (WBC) in broilers, and an increase in the heterophils to lymphocytes (H/L) ratio, which indicates that there is a possibility of a change in immune response (Mashaly et al., 2004).

Due to heat stress, there happens a decrease in the size of spleen, thymus, and lymphoid organs which are immune-related organs in poultry (Ghazi et al., 2012; Quinteiro-Filho et al., 2010). The decrease in the weight of immune organ indicates that there will be a drop in the production of antibodies of these affected birds which will increase their chances to be exposed to infectious diseases. Wasti et al. (2020) observed a fact that in poultry production, in the summer months in tropical regions there is a rise in the occurrence and frequency of infectious and contagious diseases, which they characterize as the higher occurrences of heat stress in poultry flocks. Studies have shown that the thymus, Bursa of Fabricius, and spleen all experience a reduction in weight due to heat stress, and these organs may also experience oxidative damage in exposed flocks (Apalowo et al., 2024). These findings highlight the importance of managing heat stress to maintain the health and productivity of poultry, particularly in tropical climates where high temperatures are common.

To maintain poultry health and to improve the effects of heat stress, it is very important to develop effective and operational strategies and approaches. The main strategies basically involve managing the environmental conditions, optimizing nutrition, and applying and executing biosecurity measures to protect against pathogens. This will eventually enhance the productivity and profitability of poultry production. Other strategies which are very effective to reduce heat stress are providing sufficient ventilation, ensuring proper hydration, improving the nutrition to enhance the immune function, and implementing ways to decrease heat load on the poultry birds. Generally, the health and productivity of the poultry birds will be improved by the above interventions where it preserves immune organ integrity, enhances antibody production, and reduces occurrence of infectious diseases (Al-Khalaifah & Al-Nasser, 2018; Al-Surrayai & Al-Khalaifah, 2022; Attia et al., 2019).

17.6 Effect of Productive and Reproductive Performance

The factors affecting environmental and the endocrine system are very important in normalizing the reproductive patterns of chickens (Gicheha, 2021). The main aspects that affect the environment which has a direct or indirect impact on reproductive system of the birds are the length of each day, temperature, and the patterns of rainfall. Additionally, other factors such as fodder and feeding management systems, the number of birds per cage, population density, socialization, and health and nutritional status also affect the productive and reproductive performances (Giwercman & Giwercman, 2011). By changing one of the factors or a combination of these factors can disrupt the reproductive function, either partially or completely. Environmental factors affecting reproductive process can be either ultimate or proximate, depending on the time of their impact on breeding activity. Ultimate influencers such as the availability of nutrient rich food is the most significant environmental aspect that affects breeding, as it is necessary to synchronize animal feed demand and supply do have a long-term effect. Whereas proximate influencers such as ambient temperature is an environmental factor which has short-term effects (Gicheha, 2021).

Heat stress imposes a significant challenge to poultry birds by stimulating oxidative stress that adversely affects the fertility and the antioxidant status. Scientific studies have associated heat stress to mitochondrial dysfunction in poultry. Mitochondria, the cellular organelles responsible for energy production, are particularly vulnerable to heat stress. Normal cellular production of energy is disrupted because of this stress caused by temperature which marks in a decrease in the metabolic oxidative activity of mitochondria. Mitochondrial dysfunction is proved by reduced oxidative activity and alterations in the antioxidant enzymes in birds. These changes can effect the overall health and reproductive capabilities, giving rise an imbalance in cellular redox status, which will further increase the susceptibility to oxidative damage (Algothmi et al., 2024).

Acute heat stress can significantly impact the fertility and reproductive performance of breeder flocks which has a negative effect on the acid–base balance and ion exchange mechanisms (Ronique et al., 2020). These interferences affect the physiological processes and the environment and reproductive performances of the breeders. Heat stress also impacts a crucial role in changes in receptor signaling pathways which affects the gonad sensitivity to metabolic hormones. As hens experience heat stress, the normal functioning of receptor mechanisms is disrupted, which affects how gonads respond to essential metabolic signals. This disruption can lead to a cascade of events that compromise reproductive processes (Oluwagbenga & Fraley, 2023). Long-term consequences of the fertility of breeder flocks extend to the gonadal function of the flocks. The challenge of maintaining fertility in flocks exposed to heat stress is directly related to the altered gonad sensitivity. Stress-induced changes and impacts in receptor signaling may persist, creating a continuous imbalance that obstructs the optimal reproductive system function. Studies have presented that atmospheric high temperatures adversely affect the biological processes in chicken. Temperature between 12 and 26 °C is found to be ideal range for poultry. High temperatures significantly impact reproduction (Ebeid et al., 2012). The heat stress affects reproductive performance by disrupting the thermoregulatory mechanisms, the hypothalamus and pituitary glands (Wasti et al., 2020).

Male breeders' fertility is also significantly affected by the heat stress, where males are being more susceptible to heat stress compared to female counterparts. Increased environmental temperatures adversely affect the functions of different cell types within the testes across various animal species. There is increased lipid peroxidation because of the increase in the production of reactive oxygen species. This oxidative damage has negative effects on seminal parameters, compromising the overall fertility in male breeders (Bonato et al., 2014). The seminal parameters, such as volume of semen, weight of the testes, concentration of the sperm, motility and viability of the sperm, presence of spermatids, spermatocytes, and spermatogonia are vital indicators of male reproductive health. In the beginning, a rise in temperature can promote testicular growth and increase semen quantity and concentration, but extreme temperature increase can suppress the reproductive capabilities in broilers (Chen et al., 2021). As temperature rises beyond the comfort zone level, the quality and quantity of the sperm deteriorates. Volume of semen, weight of the testes, concentration and motility of the sperm, viability of the sperm, as well as the presence of spermatids, spermatocytes, and spermatogonia, are important factors to access the fertility level. Turk et al. (2015) highlighted the susceptibility of the parameters that affect the fertility with respect to environmental factors such as temperature, pH, and ion concentration. Optimal circumstances are established within the thermo-neutral range, where all these parameters are maximized. Beyond this range, excessive ambient temperatures can have detrimental effects on almost all major parameters which affects male fertility.

Fertility is disrupted by elevated temperatures which can significantly compromise spermatozoa quality. Factors like heat stress are known to decrease testosterone synthesis, which in turn reduces testicular weight, volume of semen, concentration of the sperm, the number of sperm cells which are alive, and motility of the sperm (Fouad et al., 2016; McDaniel et al., 2004). The characteristics and the quality of sperm and seminal parameters are linked to each other, and any disruption can have cascading effects on fertility. Temperature, oxidative stress, and seminal parameters are closely related in managing and modifying the impact of heat stress on male fertility. The importance of environmental temperature on semen production are complex and multifaceted. Moderate temperature increases in the early stages of semen production can stimulate testicular growth and improve semen quality. Nevertheless, during the phases of mid and late semen production, increased temperatures lead to decreased seminiferous epithelial cell differentiation, adversely reduces semen concentration, which will reduce the quality and volume, and thereby affecting overall flock fertility (Fouad et al., 2016).

The semen collected from roosters throughout the summer frequently shows irregularities such as deformed and distorted heads, split in the mid-pieces, and cytoplasmic droplets (Ayo et al., 2022). Henceforth, hens impregnated with sperm from roosters which are gone through heat stress may produce a lower number of fertile eggs because of reduced egg production. Additionally, semen collected from rooster that has been exposed to heat for a long time, has a shorter life span in the utero-vaginal body part of hens, and semen collected in the morning shows greater success rates in artificial insemination (Ayo et al., 2022). A study on quails exposed to heat stress conditions (34 °C, 8 h/day, 60–65% RH) for 28 days documented a reduced level of spermatocyte cells, lowered spermatogonia, spermatocytes, spermatids, and spermatozoa (Türk et al., 2015).

Female reproductive system is significantly affected by heat stress by disrupting hormonal balance in both the hypothalamus and the ovary. The hypothalamus regularizes the reproductive system by discharging gonadotropin-releasing hormone (GnRH), which in turn accelerates the pituitary gland to produce luteinizing hormone (LH) and follicle-stimulating hormone (FSH) (Khan et al., 2023). Hormonal balance in the hypothalamus which is very sensitive is significantly interrupted by heat stress. Additionally, ovary the vital female reproductive organ is also affected by heat stress. High temperatures will affect and disrupts the development and maturation of ovarian follicles, leading to impaired oocyte quality and decreased fertilization capacity. This, sequentially affects the overall reproductive success rate to a declining level (Yüzen et al., 2023).

17.7 Strategies to Challenge High Temperature and Humidity in Poultry Production

Understanding the challenges of high temperature and humidity in poultry is crucial for the poultry production, in order to implement effective strategies and plans to alleviate heat stress and maintain optimal broiler performance and meat quality. Major approaches to overcome these challenges are improving the environmental conditions, changing and adjusting nutritional feed, and choosing breeds which are more resistant to heat stress. These measures, procedures and actions are very important and essential to ensure the safety and security of broilers and the quality of meat produced.

In tropical areas, where high temperatures and humidity levels are a threat to poultry industry, it is important to implement a wide range of management practices to reduce those adverse effects (Onagbesan et al., 2023). The best and effective method to reduce heat stress in poultry is through a wide-ranging, multidisciplinary strategy and modifications like ventilation systems, fogger fans, cooling pads, static pressure controllers, and thermostats. Moreover, it is found that providing intermittent light has a positive effect on lowering heat production in poultry (Alaqil, 2022). Additionally, the orientation of the building, insulation, and roof overhang can significantly influence the temperature inside the poultry house. Other factors, such as bird density, nutritional management, monitoring feed intake, adjusting feed composition, providing cool and effervescent water, using sprinkling system, and selecting lightweight breeds of birds, can also be effective in addressing the problem of heat stress in tropical countries, particularly in the poorest regions.

The principal purpose of handling heat stress in poultry is to release nutrients that would otherwise be utilized for conserving thermal impartiality, allowing them to be focused towards growth, good health maintenance, production, and reproduction. The deleterious effects of heat stress caused by elevated temperatures are widely recognized, and implementing effective measures to mitigate the impact of high temperatures increases the production of eggs and chicken meat to fulfill the growing demand for poultry products. Implementing these approaches would increase the poultry productivity and profitability for poultry production enterprises. These measures can be focused on general farm management or dietary manipulation. Even the temperate zones are also experiencing prolonged increase in ambient temperatures. High temperatures and continued heat waves result in increased electricity consumption as farmers mainly rely on ventilation, cooling systems, and artificial lighting to conserve and maintain best possible conditions for chickens. Consequently, energy cost increases, which negatively impacts the profitability of poultry farms.

17.7.1 Nutritional Solutions

Throughout the period of heat stress, poultry performance and productivity have negative impact due to reduced feed consumption and nutrient intake. To find a solution to this issue, it is crucial to modify the nutrient composition of feed to improve the feed intake. Effective feeding strategies include controlling the feed during hot periods of the day to essentially minimize the heat load. Selecting a feed which is rich in protein or energy-rich ingredients to meet nutritional needs, varying the particle sizes of feeds to slow down the digestion process and to reduce metabolic heat, and feeding diets with more water content to promote increased consumption of water are some of the tactics used. As these strategies are implemented, they helped to maximize the consumption of feed, reduce heat load, and mitigate the adverse consequences of heat stress, that ultimately leads to an improved management of poultry nutrition under high temperatures (Attia et al., 2019; Oke et al., 2021).

Limiting feeding during the hot hours like noon time and shifting to the cooler hours like morning, evening, or night time can decrease the impacts of heat stress that occurs during the peak hours (Mohamed et al., 2019). This limited feeding method shows a reduced body heat by 23%, since limiting the feed during peak hours reduces the thermal resistance which indirectly reduces the heat production (Mohamed et al., 2019). Once heat production is reduced, it regulates the body temperature, and therefore minimizes the rate of mortality in poultry. The final outcome is the lowered rectal temperature, decreased mortality rate, and the fat deposition in abdomen is considerably reduced (Uzum & Toplu, 2013). Moreover, restricting feed availability to 8 h per day can improve feed efficiency and reduce anxiety, fear and alarm in broilers, as indicated by studies on tonic immobility. Even though, considering these approaches, the unfavorable outcome of this strategy is the delayed age at which the chicken can be marketed. To counteract this undesirable effect, an alternative strategy is the dual feeding regime. This will provide continuous feed access throughout the day, by which the rush at re-feeding time can be reduced as well as minimizing the overcrowding during the feeding time (Wasti et al. 2020).

The metabolic heat produced by proteins is much higher than the heat produced from carbohydrates. As per Westerterp (2004), this difference in production of heat is a very important implication with regard to the diet given for to heat-stressed animals. Therefore, to reduce body temperature and mortality in heat-stressed broilers, it is recommended to provide a diet rich in protein during the cooler period of the day, from 4 pm to 9 am, and a diet rich in energy during the warmer period, from 9 am to 4 pm (Lozano et al., 2006).

Diets with higher-energy have been found to partially reduce the impacts of heat stress in poultry. Production of heat in poultry during metabolism is comparatively low in fat feed compared to feed more with proteins and carbohydrates (Musharaf & Latshaw, 1999). As an end result, in places where temperatures are too high, it is a common practice to supplement the level of fat in diet to increase the energy levels. In the gastrointestinal tract, supplementing the diet with more fat not only improves the utilization of nutrients but also slows down the rate at which food passes and it also increases the energy value of other feed constituents (Attia & Hassan, 2017). Addition of 5% fat to the diet of laying hens increased the feed consumption by 17% (Daghir, 2008), and providing the same fat content to broilers improved their performance significantly. Escalating the amount of oil supplementation in protein high concentrated diets lessened the undesirable effects of chronic heat stress on broiler performance (Attia et al., 2017). Moreover, broilers supplemented with more fat diet showed visible variations in abdominal fat, meat lipids, and physiological and immunological traits.

Birds loose an enormous quantity of water through their respiratory tract when they are stressed due to heat. Henceforth there is a significant increase in the water consumption to restore thermoregulatory balance. When more water is added to the feed, it helps to decrease viscosity in the gut, resulting in quicker passage of the feed. Under heat-stress circumstances, the benefits of feed with more amount of water are substantially greater than those of conventional dry mash or pellet feed, particularly in situations where birds are unable to devour adequate quantities of the latter. Interestingly, wet feeding accelerates the normal digestion and pre-digestion, which helps to boost up the absorption of nutrients from the gut, by accelerating the digestive enzymes on the feed (Syafwan et al., 2011). During summer, laying hens which are fed with wet feed showed improved yolk index, increased shell weight, increased percentage of yolk, moisture percentage, and efficiency of conversion of feed (Waiz et al., 2016a, 2016b). These improvements resulted in enhanced performance and productivity of laying hens. It is recommended for broilers raised in the tropics, to add one portion of water to one portion of feed (50:50) along with the drinking water. Wet feeding in broilers led to increased intake of feed, increased weight of the hen, and weight of the GI tract, in addition to improved ratio of feed conversion and growth potential of the birds (Khoa, 2007). In laying hens, wet feeding increases the intake of dry matter, weight and production of egg (Lin et al., 2006). Even though, these results have shown valuable effects in heat-stressed birds, it is not commonly used by poulterer because of the threat of the growth of fungus in the feed, which leads to mycotoxicosis in the birds (Raj et al., 2023).

During the periods of high temperatures, at each degree of increase in temperature, the demand for water increases by roughly around 7%. Therefore, it is recommended that an additional 25% of drinking space be provided to meet the increased demand. Providing water at a temperature slightly cooler than the birds' body temperature can also help with heat dissipation (Czenze et al., 2020). It is very important to maintain the water system in a clean and functional state to prevent heat stress in layers, and regular checks of all drinkers to ensure they are functioning properly.

Promising results have emerged from dietary interventions, specifically the inclusion of flavonoids in the birds' diets has shown to benefit chronically heat-stressed poultry. Flavonoids, found in various plant sources and known for their antioxidant properties can significantly enhance poultry's resilience to heat stress. By scavenging free radicals and reducing oxidative stress, flavonoids help protect cellular structures and maintain overall health, which is crucial for poultry under chronic heat stress conditions. This intervention not only supports better physiological and metabolic functions but also potentially improves reproductive parameters and overall productivity (Zmrhal et al., 2023).

Antioxidant vitamins and their supplements are required for the maintenance, support and performance of the growth of the body and the efficiency of immune functions. During heat stress, birds cannot synthesize vitamin C, so supplementing it at 150–500 mg/kg of diet can improve resistance of heat, enhance the quality of the carcass, reduction in the mortality rate, and act as an antioxidant, protecting against the damage that occurs due to oxidation. Supplementing with vitamin A (9000–15,000 IU/kg diet) can help alleviate the negative impact that happens because of heat on the production of eggs, boost immunity, and protects the oxidative damage. Supplementing with Vitamin E (150–500 mg/kg diet) is also useful since it enhances the production and quality of eggs, increasing the intake of feed and supporting immunity, while acting as an antioxidant to protect against oxidative damage (Al-Khalifa, 2015, 2017; Al-Khalifa et al., 2016; Du et al., 2022; Horváth & Babinszky, 2018). Table 17.2 summarizes some of the studies that used nutritional supplements to improve poultry production.

Table 17.2 Beneficial effects of supplements on heat stressed birds

Supplements	Favorable effects on heat-stressed birds	References
Vitamin E	Avoids the harm of liver, increase egg production in laying hen	Yardibi and Hoştürk (2008)
Vitamin E	Lowers the concentration of MDA in liver and serum; increases the concentration of vitamin E and A in the liver and serum in broilers	Sahin and Kucuk (2001)
Vitamin A	Increases the weight of egg	Lin et al. (2006)
Vitamin A	Increases gain in weight, improved efficiency of the feed and reduces serum concentration in broilers	Kucuk et al. (2003)
Vitamin C	Enhanced rate in growth, utilization of the nutrients, production of more and quality eggs, immune response, and the level of antioxidants	Khan et al. (2012)
Vitamin C	Improved body weight gain and FCR in broilers	Njoku (1986)
Zinc	Improved body mass growth and feed intake, increases the level of lipid peroxide, and activity of SOD in broilers	Rao et al. (2016)
Zinc	Mitigate the eggshell defects	Qiu et al. (2020)
Chromium	Improves weight of the body, intake of feed, quality of carcass and level of serum insulin in broilers; lowers the level and concentration of serum corticosterone, concentration of glucose level in serum, concentration of cholesterol	Sahin et al. (2002)
Chromium	Improves cellular and humonal immune responses in broilers	Jahanian and Rasouli (2015)
Sodium Bicarbonate	Improves eggshell quality in hens.	Balnave and Muheereza (1997)

Along with the previously mentioned measures, the inclusion of specific minerals and vitamins can improve water consumption, boost tolerance to heat stress, and enhance overall production performance. Sodium chloride (1%) and potassium chloride (0.15–0.6%) help maintain electrolyte balance and improve water intake, which is essential for thermoregulation. Sodium bicarbonate (0.2%) can act as a buffer, aiding to mitigate the effects of heat stress by maintaining acid–base balance. The inclusion of zinc (30/60 mg/kg diet) and selenium (0.1–1 mg/kg diet) is important for reducing oxidative stress by swapping as cofactors for antioxidant enzymes like superoxide dismutase and glutathione peroxidase, while micro minerals like zinc and selenium cannot directly prevent or reduce reactive oxygen species formation, they are indispensable cofactors for enzymes that have the ability to react with free radicals. Therefore, supplementation is essential to prevent the negative effects of heat stress on the birds. Vitamins A, E, and C are potent antioxidants that help counter balance free radicals, reducing oxidative damage and preventing lipid peroxidation in cell membranes. Adjusting protein intake to boost body heat generation and adopting lower-protein diets for sick birds can be considered (Horváth & Babinszky, 2018). Implementing the above dietary strategies can significantly improve the resilience of poultry to heat stress, ensuring better health and performance during hot periods.

17.7.2 Poultry Housing Systems

A poultry building have to be planned and designed by taking into consideration the inclusion of ventilation system and its construction materials which significantly impacts the solar heat gain. Features such as roof color, reflectivity, pitch, orientation, and the building's location, need to be considered in relation to the position of sun during the day. Even during the early design stage the building and ventilation system should be completed effectively with the expert guidance and assistance of professionals. Open house system is a widely recognized method of poultry housing in hot climate which is very effective and economical. It basically manages the heat within the building by natural ventilation (Oloyo, 2018). In the case of rearing of laying birds, the heat stress related problems are very much reduced by taking in to consideration the architectural design, the materials used for construction, the mechanical amenities and facilities installed, and the location and orientation of the building. Even in temperate climates, it is important to plan and consider appropriate measures for extreme summer temperatures and take necessary actions to minimize heat stress in the design of a poultry building. The first and the most important step to avoid heat stress is a planned and well-designed ventilated poultry house. Natural ventilation or fan-powered systems where the outside air is circulated is the primary methods used to remove heat from the building. Tunnel and conventional ventilation or a combined form of these two can be used for the effective design of ventilation. The movement of air is the primary goal of a safe and sound ventilation system, where it eliminates moisture-laden air which is inside the poultry house and substitutes it with fresh air from outside. Moreover, this air movement will considerably help to reduce the moisture, ammonia and carbon dioxide from the poultry house, and bringing in fresh air from outside (Bist et al., 2023).

Well planned and designed ventilated poultry houses provides a continuous and uninterrupted air flow. Tunnel ventilation are very effective since it creates an air flow of high speed which moves the air from the inlets to exhaust fans of the building. This fast air movement enhances the heat loss by convection, which facilitates to reduce the body temperature of the layers. Installing circulation fans at 1–1.5 m above the floor level and tilting it downward at a 5° angle helps a maximum circulation of air over the birds. The use of air conditioning can also be considered as an alternative method to control and manage heat (Czenze et al., 2020).

Exhaust fans are essential to create and maintain a healthy environment of chicken in a poultry house. Regular and proper cleaning and maintenance of these fans are essential to maintain the efficiency which in-turn helps and maintain their ability to regulate the indoor climate effectively and ideal performance. In the perspective of the well-being and comfort of the birds, it is suggested to run the fans not only when temperatures are too high, rather run it all throughout the day and also during night hours. By using fans in such a way, the optimal temperature can be maintained, which will reduce the harmful effects of heat on the chickens (Hajiyev et al., 2024). Adjusting air inlets helps to maintain ventilation efficiency and to control the temperature inside the poultry house. Regular and thorough maintenance, along with adjusting air inlets for uniform airflow, fulfills the comprehensive methodology to regulate the temperature within the building (Hajiyev et al., 2024).

Standby generators and alarm systems are a very important factors in the proper functioning of the poultry buildings. These systems should be thoroughly checked frequently, at least once a month to ensure their proper functioning. Settings the alarm with a broad span both at high and low temperatures within a sensitive range is necessary for detecting the issue well in advance before they escalate and cause significant damage. In addition to monitoring generators and alarms, foggers or misters should be examined regularly. When humidity levels are low, to lower the threat of stress due to heat and to contribute a more comfortable environment, they should be operated for two minutes out of every ten to maintain the desired humidity levels. In extreme summer temperatures, installation and proper functioning of roof sprinklers are necessary. These sprinklers can effectively cool the surroundings and lower the overall temperature within the housing. Apart from providing insulation to the building, it is very important to make sure that the insulation is sufficient to reduce heat transfer through the walls and roof. Even on a normal summer day, a lack of proper insulation can allow a significant amount of solar heat to penetrate the roof (Trokhaniak et al., 2023).

Planting more trees in and around the chicken shelter and shielding it with grass creates an ecological cooling environment, which is an effective method for decreasing the heat on birds (Saeed et al., 2019). Along with this environmental methodology, the usage of a reflective roof surfaces helps to minimize heat from solar radiation, which promotes the birds' well-being. During hot weather, providing appropriate shelter, shade, sprinkling systems, and ventilation are some of the essential management strategies that prioritize the birds' comfort and welfare, making sure they remain healthy and stress-free. Additionally, it is suggested to plant other vegetation and grass on the air-inlet side of the building, ensuring that these areas are well-maintained without obstructing the inlet airflow and prevents any sorts of interference with the inlets. These humane practices facilitate to relieve the negative effects of stress due to heat on poultry flocks. When the vegetation is properly managed, it contributes to the efficient operation of ventilation systems, enhancing overall heat dissipation and maintaining a conducive environment.

Another major concern to prevent stress due to heat in poultry is to minimize the disturbances to the birds, particularly during the hottest part of the day. Already the birds are stressed and due to further excessive heat and additional disturbances can worsen their discomfort. Therefore, it is advisable to schedule essential tasks and activities within the poultry house when temperatures are less intense, such as early mornings or evenings. This reduces the impact of human and environmental impact which helps to reduce the stress for the birds. Working during cooler period of the day helps to create a calmer and stress-free routine for the birds (Munonye et al., 2023).

A major aspect to consider for the well-being and overall health of birds is the proper housing conditions by avoiding overstocking of cages, as this can have a detrimental effect on their comfort and benefit. Birds require enough and sufficient space to move around and spread their wings, which improves their physical activity and increases airflow around their bodies which will facilitate and maintain the health of the poultry. It is important to follow the proposed floor space in preventing overstocking and should be allocated according to the size and weight of the birds. For example, a floor space of 0.06 m^2 per bird is recommended for birds weighing 1.7 kg, and 0.13 m^2 per bird for those weighing 3.5 kg. These measurements correspond to a density of 27.8 kg/m^2. Adhering to these guidelines helps prevent overcrowding, ensuring each bird has adequate room to move, perch, and engage in natural behaviors (Aviagen, 2018).

In addition to making available the proper floor space, providing and maintaining appropriate environmental conditions is crucial for successful poultry farming. This includes setting sufficient brooders in the facility based on the flock size. Brooders are crucial for regulating temperature, especially during different production stages, and it is important to monitor and optimize the temperature to mitigate the specific needs of the birds for ideal growth and development.

17.7.3 Management and Husbandry

Effective management and husbandry practices are crucial for preventing heat stress in poultry, regardless of what system is being used. Ensuring the well-being of the birds requires, establishing clearly written instructions, leaflets, handouts and communication details on all aspects of poultry operations. The poultry industry should ensure that their staff is well-trained to identify the early signs of stress due to heat. Designating senior stockmen, explicit and experienced staff members who are knowledgeable about the ventilation systems in the poultry house is essential for maintaining optimal conditions.

Well advance preparation is important for effectively managing hot weather emergencies. Written plans and procedures should be established to outline the actions required from specialized individuals during regular operations, as well as during the clearance and culling, and transportation of birds. These plans should also include contingency measures for equipment failures, such as backup systems, and assign different roles to staff members as needed (Oluwagbenga & Fraley, 2023). Emergency instructions should be clearly displayed in areas accessible to all staff, and should also include contact information for veterinary surgeons and equipment engineers. It is vital to have a responsible person with authority to take immediate action during emergencies to safe guard and protect the well-being of the birds available at all times. Another major factor that compromises the critical aspect of immune system efficiency in poultry management during extreme temperatures is disease prevention. Implementing vigorous and healthy biosecurity measures is essential to minimize the risk of disease spread within the flock. This incorporates restricting access to the public to farm premises, employing proper sanitation practices, and controlling the movement of personnel equipment and vehicles. By limiting exposure to potential disease vectors, such as wild birds or contaminated materials, biosecurity measures create and act as a protective barrier against infectious agents (Onagbesan et al., 2023).

Vaccination is definitely a vital component of disease prevention in poultry. Developing vaccination programs in collaboration and consultation with veterinarians who understand and are familiar with the local diseases environment ensure that the program addresses the most relevant threats prevailing in that region. Common vaccines include those for Newcastle disease, infectious bronchitis, and avian influenza. Maintaining a strict vaccination schedule is especially important during extreme temperatures, as stress can weaken the immune response and increase susceptibility to infections. Following strict biosecurity protocols and implementing a well-planned vaccination strategy are critical for strengthening the overall health and resilience of the poultry flock. Regular monitoring, surveillance, and adjustments to the disease prevention plan in response to seasonal variations or emerging threats further enhance the effectiveness of these protective measures (Onagbesan et al., 2023).

17.7.4 Perinatal Heat Acclimation and Genetic Selection

Researchers are intensively focusing on advanced approaches to manage heat stress in poultry. Perinatal heat acclimation is an assuring method where the chicks are exposed to elevated temperatures during their early developmental stages, which enhances their ability to manage heat stress substantially as they mature. Broilers are basically vulnerable to heat stress because of their high metabolic rates, henceforth more importance and focus are to be given to improve the broiler breeds so that they exhibit heightened heat tolerance. Poultry lines with genetic traits are developed mostly in hot and arid regions to cope up with the high temperature and to facilitate and boost productivity (Juiputta et al., 2023).

In high temperatures and hot environments, high prominence is given to the genetic selection breeding programs which is a cutting-edge strategy in the field of heat tolerance. The main aspect is selecting some specific kind of genes such as the frizzle gene, naked neck gene, dwarf gene, HSP70, and HSP90 which are associated with thermal regulation. These genetic markers are helpful in creating new strains with improved thermo-tolerant abilities. The studies of Wasti et al. (2020) propose incorporating biological markers in breeding programs, do exhibit a more practical methodology for developing a better heat-tolerant poultry strains.

The Naked Neck (Na) gene, shows a significant role in enhancing heat tolerance in poultry with regard to their body structure. The Na gene is a single dominant autosomal gene, which has a reduced feather coverage in their neck region. In Na/na (heterozygous) gene birds, the feather coverage is reduced by 20% whereas in Na/Na (homozygous) it is reduced by 40%. This feather reduction facilitates better heat dissipation compared to normal feathered birds (Fernandes et al., 2023; Merat, 1986). Additionally, Na gene in broilers is associated with several benefits such as increased breast muscle and increased body weight (Azoulay et al., 2011), reduced abdominal fat (Rajkumar et al., 2010), and lower body temperature. The Naked-necked birds also show lower total plasma cholesterol levels and H/L ratios compared to their fully feathered counterparts during high temperatures (Rajkumar et al., 2010). In laying hens, even in high temperatures, the Naked Neck gene produces improved eggs which are more in their number, mass, and quality (Fathi et al., 2013). These findings highlight the facts to incorporate the Naked Neck gene into breeding programs to develop chicken breeds that are better equipped to sustain and cope up well with the heat stress, ultimately enhancing productivity and welfare in hot environments.

The Frizzle (F) gene, significantly influences heat stress management in poultry by curling of their feathers. The Frizzle gene releases and radiates heat more easily from the body by insulating properties and by reducing feather weight (Fathi et al., 2022). In laying hens, the homozygous frizzle genotype (FF) shows improvement in the production and quality of eggs by improving heat dissipation more effectively than heterozygous carriers and normal-feathered hens. Researchers have found that there is a notable interaction between feathering genotype and environmental temperature under heat stress which effects reproductive characteristics, such as egg production, hatchability, and chick production. This interaction is significant for all characteristics except sexual maturity (Sharifi et al., 2010). These results highlight the ability of the Frizzle gene to better manage stress due to heat in developing poultry breeds while maintaining the optimal productivity and reproductive performance.

The comparative characteristics of the Frizzle (F) gene and Naked Neck (Na) gene on poultry and their roles in heat stress management under high temperature is detailed above. Even though sexual maturity remains unaffected, feathered hens experience a significant decrease in reproductive characteristics compared to frizzle-feathered hens. On the contrary, the FF genotype shows reduced egg production and a delay in sexual maturity at lower temperatures (around 19 °C). Even though under high temperature, Frizzle genes shows less prominent characteristics in broilers compared to the Naked Neck gene, an additive effect is noticed in double heterozygous broilers (Na/Na F/f) (Yunis & Cahaner, 1999). This suggests that combining both genes together could enhance heat tolerance, making the Frizzle gene an important target for developing heat-tolerant poultry breeds.

The Dwarf (dw) gene, which is a sex-linked recessive gene, leads to a decrease in body weight by approximately 40% in homozygous males and 30% in homozygous females (Daghir, 2008). Research shows that in fast-growing broiler chickens under persistent heat stress conditions, the dw gene does not develop heat tolerance (Deeb & Cahaner, 2001). The characteristics of thermo-tolerant genes like the frizzle gene, naked neck gene, and polymorphisms in heat shock proteins (HSP) genes indicate a promising targets for genetic selection. Using these genes, marker-assisted selective breeding is exceptionally effective in broiler lines, which are susceptible to stress due to heat because of their high metabolic rates (Butcher & Miles, 2012). This approach shows a promising outcome for developing poultry breeds with enhanced heat tolerance.

Both perinatal heat acclimation and genetic selection approach assures promising results for enhancing poultry production in hot climates. Apparently, these are still in the early stages of investigation, therefore, further research and developments are needed to fully understand their efficiency and effectiveness to optimize their implementation. Further scientific research and practical application is needed towards the effectiveness of these approaches and ensuring their cost-effectiveness within the extensive study of poultry farming, especially as climate conditions continue to change.

Table 17.3 summarizes key strategies to overcome the challenges of heat stress in poultry. It explains and shows an outline, how to manage extreme temperatures by taking into consideration various aspects of environment, nutrition, and overall management. By regularly monitoring and correcting these strategies according to the specific needs of the flock, and proper consultation with poultry health professionals, the health and performance of the poultry can be standardized under extreme temperature conditions.

17.8 Conclusions and Recommendations

As the change in climatic conditions are persistent, a comprehensive approach need to be considered to ensure optimal poultry productivity and reproduction. Proper and efficient management strategies like effective nutritional techniques, advanced housing systems, and disease management are essential. A cutting edge to overcome the unfavorable effects of heat stress in poultry, researchers and practitioners are coming up with various intervention strategies. These approaches aim to create a more comfortable environment for poultry and reduce the impact of extreme temperatures. By regularly monitoring and adjusting these strategies according to the specific needs of the flock, and consulting with poultry health professionals, poultry health

Table 17.3 The strategies to overcome the extreme temperature in poultry

Category	Strategies
Housing and environment	– Provide well-ventilated housing with proper insulation
	– Install fans and misting systems for cooling in hot weather
	– Ensure proper heating in cold weather with heating systems
Water management	– Ensure access to cool, clean water at all times
	– Adjust water delivery systems to prevent freezing in cold weather
Nutrition	– Formulate balanced diets to meet specific temperature needs
	– Adjust feed composition for heat-stressed conditions
	– Provide supplements like electrolytes during extreme heat
Adaptation techniques	– Gradual acclimatization to temperature changes
	– Implement shade structures in outdoor areas
Health monitoring	– Regular health checks by veterinarians
	– Monitor for signs of heat or cold stress
Ventilation management	– Improve ventilation systems for temperature regulation
	– Install curtains or baffles to control drafts
Bedding and litter	– Provide appropriate bedding materials for insulation
	– Adjust litter depth based on temperature conditions
Lighting management	– Adjust lighting schedules to manage heat production
	– Ensure adequate darkness during hot periods
Behavioral observation	– Monitor flock behavior for signs of distress
	– Separate and treat stressed individuals promptly
Emergency protocols	– Develop protocols for extreme temperature events
	– Have contingency plans for power outages and equipment failures
Disease prevention	– Implement biosecurity measures to prevent disease spread
	– Vaccinate birds appropriately based on local disease risks

and performance can be optimized under extreme temperature conditions. Recent research has focused on two innovative approaches that are perinatal heat acclimation and the genetic selection of breeds with increased heat tolerance. However, both approaches require further research and development to fully realize their potential and ensure their effectiveness in mitigating heat stress.

References

Abo El-Maaty, H. A., El-Khateeb, A. Y., Al-Khalaifah, H., El-Hamed, E. A., Hamed, S., El-Said, E. A., Mahrose, K. M., Metwally, K., & Mansour, A. M. (2020). Effects of ecofriendly synthesized calcium nanoparticles with biocompatible Sargassum latifolium algae extract supplementation on egg quality and scanning electron microscopy images of the eggshell of aged laying hens. *Poultry Science, 100*(2), 675–684. https://doi.org/10.1016/j.psj.2020.10.043. PMID: 33518121.

Alagawany, M., Elnesr, S. S., Farag, M. R., Tiwari, R., Yatoo, M. I., Karthik, K., & Dhama, K. (2020). Nutritional significance of amino acids, vitamins and minerals as nutraceuticals in poultry production and health—A comprehensive review. *Veterinary Quarterly, 41*(1), 1–29.

Alaqil, A. A., Abd El-Atty, H. K., & Abbas, A. O. (2022). Intermittent lighting program relieves the deleterious effect of heat stress on growth, stress biomarkers, physiological status, and immune response of broiler chickens. *Animals, 12*(14), 1834. https://doi.org/10.3390/ani12141834

Algothmi, K. M., Mahasneh, Z. M. H., Abdelnour, S. A., Khalaf, Q. A. W., Noreldin, A. E., Barkat, R. A., Khalifa, N. E., Khafaga, A. F., Tellez-Isaias, G., Alqhtani, A. H., Swelum, A. A., & Abd El-Hack, M. E. (2024). Protective impacts of mitochondria enhancers against thermal stress in poultry. *Poultry Science, 103*(1), Article 103218.

Al-Khalaifah, H. S. (2018). Benefits of probiotics and/or prebiotics for antibiotic-reduced poultry. *Poultry Science, 97*, 3807–3815.

Al-Khalaifah, H., & Al-Nasser, A. (2018). Cytokines as effective elements of the avian immune system. *Journal of Microbiology and Genetics: JMGE-119, 3*(1), 1–7. https://doi.org/10.29011/2574-7371.00019

Al-Khalaifah, H. (2020). Modulatory effect of dietary polyunsaturated fatty acids on immunity, represented by phagocytic activity. *Frontiers in Veterinary Science*, 7. https://doi.org/10.3389/fvets.2020.569939

Al-Khalaifah, H., & Uddin, S. (2022). Assessment of sargassum sp., spirulina sp., and gracilaria sp. as poultry feed supplements: Feasibility and environmental implications. *Sustainability, 14*(14), 8968. https://doi.org/10.3390/su14148968

Al-Khalifa, H., Al-Nasser, A., Al-Bahouh, M., Ragheb, G., Al-Qalaf, S., Al-Omani, N., & Ahmad, A. (2016). The effect of polyunsaturated fatty acids on avian immune cell subpopulations in peripheral blood, spleen, and thymus. *World's Poultry Science Journal, 1*(03), 1–4. https://doi.org/10.1017/S0043933916000428

Al-Khalaifah, H., Al-Nasser, A., Abdulmalek, N., Al-Mansour, H., Ahmed, A., & Ragheb, G. (2020a). Impact of SARS-Con-V2 on the poultry industry in Kuwait: A case study. *Frontiers in Veterinary Science, 7*, Article 577178. https://doi.org/10.3389/fvets.2020.577178

Al-Khalaifah, H., Al-Nasser, A., Givens, D. I., Rymer, C., & Yaqoob, P. (2020b). Comparison of different dietary sources of n-3 polyunsaturated fatty acids on immune response in broiler chickens. *Heliyon, 6*(1), e03326. https://doi.org/10.1016/j.heliyon.2020.e03326

Al-Khalaifah, H., Al-Nasser, A., Al-Surrayai, T., Sultan, H., Al-Attal, D., Al-Kandari, R., Al-Saleem, H., Al-Holi, A., & Dashti, F. (2022a). Effect of ginger powder on production performance, antioxidant status, hematological parameters, digestibility, and plasma cholesterol content in broiler chickens. *Animals, 12*(7), 901. https://doi.org/10.3390/ani12070901

Al-Khalaifah, H., Alotaib, M., & Al-Nasser, A. (2022b). The relation between avian coronaviruses and SARS-CoV-2 coronavirus. *Frontiers in Microbiology, 13*, Article 976462. https://doi.org/10.3389/fmicb.2022.976462

Al-Khalifa, H. (2015). Production of added-value poultry meat: Enrichment with n-3 polyunsaturated fatty acids. *World's Poultry Science Journal, 71*, 319–326.

Al-Khalifa, H. (2017). Enrichment of poultry diets with polyunsaturated fatty acids (PUFA) for human consumption. *Approaches in Poultry, Dairy & Veterinary Sciences, 1*, 95–96.

Allan, E. W., Magali, B. M., Francisco, J. H., Elza, I. I., & Massami, S. (2010). Protease activity and the ultrastructure of broiler chicken PSE (pale, soft, exudative) meat. *Food Chemistry, 119*, 1201–1204.

Al-Nasser, A. (2006). Poultry industry in Kuwait. *World's Poultry Science Journal, 62*, 702–708.

Al-Nasser, A., Al-Khlaifa, H., Al-Bahouh, M., Khalil, F., Boareki, M., & Ragheb, G. (2015). Challenges facing poultry production in Kuwait. *World's Poultry Science Journal, 71*, 339–348.

Al-Nasser, A., Al-Khalaifah, H., Khalil, F., & Al-Mansour, H. (2020). Poultry industry in the Gulf Cooperation Council with emphasis on Kuwait. *World's Poultry Science Journal, 76*, 577–589.

Al Salihi, K. A., AL-Mhsenawi, Z. K. F., Awad, Z. H., & Ali, R. M. (2023). A comparative study between red and white meat: a literature review. *Mirror of Research in Veterinary Sciences and Animals, 12*(2), 1–9. https://doi.org/10.22428/mrvsa2023-00122-01

Al-Surrayai, T., & Al-Khalaifah, H. (2022). Dietary supplementation of fructooligosaccharides enhanced antioxidant activity and cellular immune response in broiler chickens. *Frontiers in Veterinary Science, 9*, Article 857294.

Altan, O., Pabuçcuoğlu, A., Altan, A., Konyalioğlu, S., & Bayraktar, H. (2003). Effect of heat stress on oxidative stress, lipid peroxidation and some stress parameters in broilers. *British Poultry Science, 44*(4), 545–550.

Apalowo, O. O., Ekunseitan, D. A., & Fasina, Y. O. (2024). Impact of heat stress on broiler chicken production. *Poultry, 3*, 107–128.

Attia, Y.A., & Hassan, S. S. (2017). Broiler tolerance to heat stress at various dietary protein/energy levels. *European Poultry Science, 81.* https://doi.org/10.1399/eps.2017.171

Attia, Y. A., Al-Harthi, M. A., & Elnaggar, A. S. (2017). Productive, physiological and immunological responses of two broiler strains fed different dietary regimens and exposed to heat stress. *Italian Journal of Animal Science, 17*(3), 686–697

Attia, Y., Al-Khalaifah, H. S., Abd El-Hamid, H., Al-Harthi, M., & El-Shafey, A. (2019). Multi-enzymes and immune response effect of different levels of multi-enzymes on immune response, blood hematology and biochemistry, antioxidants status and organs histology of broiler chicks fed standard and low-density diets. *Frontiers in Veterinary Science, 6*, 510.

Aviagen. (2018). *Ross-Broiler management handbook.*

Aviagen. (2022). *Ross 308 and Ross 308 FF performance objectives.* http://eu.aviagen.com/tech-center/download/1339/Ross308-308FF-BroilerPO2019-EN.pdf

Ayo, J. O., Obidi, J. A., Rekwot, P. I., & Onyeanusi, B. I. (2022). Modulatory effects of lycopene and vitamin E on cloacal temperature, thyroid hormonal and reproductive performance responses in laying hens during the hot-dry season. *Journal of Thermal Biology, 104*, Article 103105. https://doi.org/10.1016/j.jtherbio.2021.103105

Azoulay, Y., Druyan, S., Yadgary, L., Hadad, Y., & Cahaner, A. (2011). The viability and performance under hot conditions of featherless broilers versus fully feathered broilers. *Poultry Science, 90*(1), 19–29.

Balakrishnan, K. N., Ramiah, S. K., & Zulkifli, I. (2023). Heat shock protein response to stress in poultry: A review. *Animals, 13*(2), 317.

Baldi, G., Soglia, F., & Petracci, M. (2020). Current status of poultry meat abnormalities. Meat and Muscle *Biology, 4*(2). https://doi.org/10.22175/mmb.9503

Balnave, D., & Muheereza, S. K. (1997). Improving eggshell quality at high temperatures with dietary sodium bicarbonate. *Poultry Science, 76*(4), 588–593.

Barbut, S., Sosnicki, A. A., Lonergan, S. M., Knapp, T., Ciobanu, D. C., Gatcliffe, L. J., Huff-Lonergan, E., & Wilson, E. W. (2008). Progress in reducing the pale, soft and exudative (PSE) problem in pork and poultry meat. *Meat Science, 79*(1), 46–63.

Bist, R. B., Subedi, S., Chai, L., & Yang, X. (2023). Ammonia emissions, impacts, and mitigation strategies for poultry production: A critical review. *Journal of Environmental Management*. 328. https://doi.org/10.1016/j.jenvman.2022.116919

Bonato, M., Malecki, I. A., Rybnik-Trzaskowska, P. K., Cornwallis, C. K., & Cloete, S. W. (2014). Predicting ejaculate quality and libido in male ostriches: Effect of season and age. *Animal Reproduction Science, 151*(1–2), 49–55.

Butcher, G. D., & Miles, R. (2012). *Heat stress management in broilers.* VM65 series of the Veterinary Medicine Large Animal Clinical Sciences Department, Florida Cooperative Extension Service, Institute of Food and Agricultural Sciences. University of Florida. https://edis.ifas.ufl.edu/publication/VM019

Caplan, R. M. (2023). Diet, weight loss. *Long life strategy* (pp. 125–130). Springer.

Cardoso, D. M., Cardeal, P. C., Soares, K. R., Sousa, L. S., Castro, F. L. S., Araújo, I. C. S., & Lara, L. J. C. (2022). Feed form and nutritional level for rearing growing broilers in thermoneutral or heat stress environments. *Journal of Thermal Biology, 103*, 103159.

Carlin, D., Arshad, M., & Baker, K. (2023). *Climate risks in the agriculture sector*. UN Environment Programme, Report.

Chen, S., Yong, Y., & Ju, X. (2021). Effect of heat stress on growth and production performance of livestock and poultry: Mechanism to prevention. *Journal of Thermal Biology, 99*, Article 103019. https://doi.org/10.1016/j.jtherbio.2021.103019

Czenze, Z. J., Kemp, R., Jaarsveld, B. Y., Freeman, M. T., Smit, B., Wolf, B. O., & McKechnie, A. E. (2020). Regularly drinking desert birds have greater evaporative cooling capacity and higher heat tolerance limits than non-drinking species. *Functional Ecology, 34*(8), 1589–1600.

Daghir, N. J. (ed.). (2008). *Poultry production in hot climates* (2nd ed.). CABI.

Deeb, N., & Cahaner, A. (2001). Genotype-by-environment interaction with broiler genotypes differing in growth rate: 2. The effects of high ambient temperature on dwarf versus normal broilers. *Poultry Science, 80*(5), 541–548.

Du, J., Shi, Y., Zhou, C., Guo, L., Hu, R., Huang, C., Hu, G., Gao, X., & Guo, X. (2022) Antioxidative and anti-inflammatory effects of vitamin C on the liver of laying hens under chronic heat stress. *Frontiers in Veterinary Science, 9*, 1052553.

Ebeid, T. A., Suzuki, T., & Sugiyama, T. (2012). High ambient temperature influences eggshell quality and calbindin-D28k localization of eggshell gland and all intestinal segments of laying hens. *Poultry Science, 91*(9), 2282–2287.

FAOSTAT. (2016). *Food and Agriculture Organization of the United Nations*, Rome.

FAOSTAT. (2018). *FAOSTAT Statistical Database*. Food and Agriculture Organization, Rome.

FAO Statistics. (2022). *Statistical yearbook world food and agriculture*. Food and Agriculture Organization, Rome.

Fathi, M. M., Galal, A. A., Salah, E., & Yousef, M. M. (2013). Naked neck and frizzle genes for improving chickens raised under high ambient temperature: I. Growth performance and egg production. *World's Poultry Science Journal, 69*, 813–832.

Fathi, M. M., Gala, A., Radwan, L. M., Abou-Emera, O. K., & Al-Homidan, I. H. (2022). Using major genes to mitigate the deleterious effects of heat stress in poultry: An updated review. *Poultry Science, 101*(11), Article 102157.

Fernandes, E. A., Raymundo, A., Martins, L. L., Lordelo, M., & De Almeida, A. M. (2023). The naked neck gene in the domestic chicken: A genetic strategy to mitigate the impact of heat stress in poultry production—A review. *Animals, 13*(6), 1007.

Fouad, A. M., Chen, W., Ruan, D., Wang, S., Xia, W. G., & Zheng, C. T. (2016). Impact of heat stress on meat, egg quality, immunity and fertility in poultry and nutritional factors that overcome these effects: A Review. *International Journal of Poultry Science, 15*, 81–95.

Ghazi, S. H., Habibian, M., Moeini, M. M., & Abdolmohammadi, A. R. (2012). Effects of different levels of organic and inorganic chromium on growth performance and immunocompetence of broilers under heat stress. *Biological Trace Element Research, 146*(3), 309–317.

Gicheha, G. M. (2021). The effects of heat stress on production, reproduction, health in chicken and its dietary amelioration. https://doi.org/10.5772/intechopen.97284

Giwercman, A., & Giwercman, Y. L. (2011). Environmental factors and testicular function. Best practice and research. *Clinical Endocrinology and Metabolism, 25*(2), 391–402.

Godde, C. M., Mason-D'Croz, D., Mayberry, D. E., Thornton, P. K., & Herrero, M. (2021). Impacts of climate change on the livestock food supply chain; A review of the evidence. *Global Food Security, 28*, Article 100488.

Gouda, A., Al-Khalaifah, H., Al-Nasser, A., Kamel, N. N., Gabr, S., & Eid, K. (2024). Early feeding strategy mitigates major physiological dynamics altered by heat stress in broilers. *Animals, 14*, 1485.

Hajiyev, R., Huseynova, M., Taghiyev, U., Mammadov, G., Allahverdiyeva, G. (2024). The study of the efficiency evaluation of the ventilation system of the poultry house in the summer. *EUREKA Physics and Engineering*, (1), 82. https://doi.org/10.21303/2461-4262.2024.003248

Henchion, M., Moloney, A. P., Hyland, J., Zimmermann, J., & McCarthy, S. (2021). Review: Trends for meat, milk and egg consumption for the next decades and the role played by livestock systems in the global production of proteins. *Animal: The International Journal of Animal Biosciences, 15*(1). 100287. https://doi.org/10.1016/j.animal.2021.100287

Horváth, M., & Babinszky, L. (2018). Impact of selected antioxidant vitamins (Vitamin A, E and C) and micro minerals (Zn, Se) on the antioxidant status and performance under high environmental temperature in poultry. A review. *Acta Agriculturae Scandinavica, Section A—Animal Science, 68*(3), 152–160. https://doi.org/10.1080/09064702.2019.1611913

Jahanian, R., & Rasouli, E. (2015). Dietary chromium methionine supplementation could alleviate immunosuppressive effects of heat stress in broiler chicks. *Journal of Animal Science, 93*(7), 3355–3363.

Juiputta, J., Chankitisakul, V., & Boonkum, W. (2023). Appropriate genetic approaches for heat tolerance and maintaining good productivity in tropical poultry production: A review. *Veterinary Sciences, 10*(10), 591. https://doi.org/10.3390/vetsci10100591

Khan, R. U., Naz, S., Nikousefat, Z., Selvaggi, M., Laudadio, V., & Tufarelli, V. (2012). Effect of ascorbic acid in heat-stressed poultry. *World's Poultry Science Journal, 68*, 477–490.

Khan, I., Mesalam, A., Heo, Y. S., Lee, S. H., Nabi, G., & Kong, I. K. (2023). Heat stress as a barrier to successful reproduction and potential alleviation strategies in cattle. *Animals, 13*(14), 2359. https://doi.org/10.3390/ani13142359

Khoa, M. A. (2007). *Wet and coarse diets in broiler nutrition: Development of the GI tract and performance*. Ph.D. Thesis. Wageningen University, Wageningen, The Netherlands.

Kucuk, O., Sahin, N., & Sahin, K. (2003). Supplemental zinc and vitamin A can alleviate negative effects of heat stress in broiler chickens. *Biological Trace Element Research, 94*, 225–235.

Lacetera, N. (2019). Impact of climate change on animal health and welfare. *Animal Frontiers, 9*(1), 26–31.

Lara, L. J., & Rostagno, M. H. (2013). Impact of heat stress on poultry production. *Animals, 3*, 356–369. https://doi.org/10.3390/ani3020356

Lin, H., Jiao, H. C., Buyse, J., & Decuypere, E. (2006). Strategies for preventing heat stress in poultry. *World's Poultry Science Journal, 62*(1), 71–86.

Lozano, C., De Basilio, V., Oliveros, I., Alvarez, R., Colina, I., Bastianelli, D., Yahav, S., & Picard, M. (2006). Is sequential feeding a suitable technique to compensate for the negative effects of a tropical climate in finishing broilers? *Animal Research, 55*(1), 71–76.

Mashaly, M. M., Hendricks, G. L., Kalama, M. A., Gehad, A. E., Abbas, A. O., & Patterson, P. H. (2004). Effect of heat stress on production parameters and immune responses of commercial laying hens. *Poultry Science, 83*(6), 889–894.

McDaniel, C. D., Hood, J. E., & Parker, H. M. (2004). An attempt at alleviating heat stress infertility in male broiler breeder chickens with dietary ascorbic acid. *International Journal of Poultry Science, 3*, 593–602.

Merat, P. (1986). Potential usefulness of the Na (naked neck) gene in poultry production. *World's Poultry Science Journal, 42*, 124–142.

Mine, Y., Guyonnet, V., Hatta, H., Nau, F., & Qiu, N. (eds.). (2023). *Handbook of egg science and technology* (1st ed.). CRC Press.

Mohamed, A. S. A., Lozovskiy, A. R., & Ali, A. M. A. (2019). Strategies to combat the deleterious impacts of heat stress through feed restrictions and dietary supplementation (vitamins, minerals) in broilers. *Journal of the Indonesian Tropical Animal Agriculture, 44*(2), 155–166. https://doi.org/10.14710/jitaa.44.2.155-166

Mota-Rojas, D., Titto, C. G., de Mira, G. A., Martínez-Burnes, J., Gómez, J., Hernández-Ávalos, I., Casas, A., Domínguez, A., José, N., Bertoni, A., Reyes, B., & Pereira, A. M. F. (2021). Efficacy and function of feathers, hair, and glabrous skin in the thermoregulation strategies of domestic animals. *Animals, 11*(12), 3472.

Munonye, J., Munonye, C., & Esiegwu, A. (2023). Assessment of thermal comfort of broiler birds in warm-humid climate. *Tropical Agricultural Research & Extension, 26*(3), 231–240. https://doi.org/10.4038/tara.v26i3.5639

Musharaf, N. A., & Latshaw, J. D. (1999). Heat increment as affected by protein and amino acid nutrition. *World's Poultry Science Journal, 55*(3), 233–240.

Nawaz, A. H., Amoah, K., Leng, Q. Y., Zheng, J. H., Zhang, W. L., & Zhang, L. (2021). Poultry response to heat stress: Its physiological, metabolic, and genetic implications on meat production and quality including strategies to improve broiler production in a warming world. *Frontiers in Veterinary Science, 8*, Article 699081. https://doi.org/10.3389/fvets.2021.699081

NCC. (2019). *National chicken council*. Retrieved December 20, 2019, from https://www.nationalchickencouncil.org/

Njoku, P. C. (1986). Effect of dietary ascorbic acid (vitamin C) supplementation on the performance of broiler chickens in a tropical environment. *Animal Feed Science and Technology, 16*, 17–24.

Oke, O. E., Uyanga, V. A., Iyasere, O. S., Oke, F. O., Majekodunmi, B. C., Logunleko, M. O., Abiona, J. A., Nwosu, E. U., Abioja, M. O., Daramola, J. O., & Onagbesan, O. M. (2021). Environmental stress and livestock productivity in hot-humid tropics: Alleviation and future perspectives. *Journal of Thermal Biology, 100*, Article 103077. https://doi.org/10.1016/j.jtherbio.2021.103077

Oke, O. E., Akosile, O. A., Oni, A. I., Opowoye, I. O., Ishola, C. A., Adebiyi, J. O., Odeyemi, A. J., Adjei-Mensah, B., Uyanga, V. A., & Abioja, M. O. (2024). Oxidative stress in poultry production. *Poultry Science, 103*(9), Article 104003.

Oloyo, A. (2018). The use of housing system in the management of heat stress in poultry production in hot and humid climate: A review. *Poultry Science Journal, 6*(1), 1–9. https://doi.org/10.22069/psj.2018.13880.1284

Oluwagbenga, E. M., & Fraley, G. S. (2023). Heat stress and poultry production: A comprehensive review. *Poultry Science, 102*(12), Article 103141. https://doi.org/10.1016/j.psj.2023.103141

Onagbesan, O. M., Uyanga, V. A., Oso, O., Tona, K., & Oke, O. E. (2023). Alleviating heat stress effects in poultry: Updates on methods and mechanisms of actions. *Frontiers in Veterinary Science, 10*, 1255520. https://doi.org/10.3389/fvets.2023.1255520

Petracci, M., Mudalal, S., Soglia, F., & Cavani, C. (2015). Meat quality in fast-growing broiler chickens. *Worlds Poultry Science Journal, 71*, 363–374.

Petracci, M., Soglia, F., Madruga, M., Carvalho, L., Ida, E., & Estévez, M. (2019). Wooden-breast, white striping, and spaghetti meat: Causes, consequences and consumer perception of emerging broiler meat abnormalities. *Comprehensive Reviews in Food Science and Food Safety, 18*(2), 565–583.

Pizzino, G., Irrera, N., Cucinotta, M., Pallio, G., Mannino, F., Arcoraci, V., Squadrito, F., Altavilla, D., & Bitto, A. (2017). Oxidative stress: harms and benefits for human health. *Oxidative Medicine and Cellular Longevity*, 8416763. https://doi.org/10.1155/2017/8416763

Ponsuksili, S., Jonas, E., Murani, E., Phatsara, C., Srikanchai, T., Walz, C., Schwerin, M., Schellander, K., & Wimmers, K. (2008). Trait correlated expression combined with expression QTL analysis reveals biological pathways and candidate genes affecting water holding capacity of muscle. *BMC Genomics, 9*, 367. https://doi.org/10.1186/1471-2164-9-367

Pratama, R. I., Rostini, I., Andriani, Y., & Dewanti, L. P. (2021). A review on protein quality and bioavailability of amino acids. *International Journal of All Research Writings, 3*(1), 90–97.

Qiu, J. L., Zhou, Q., Zhu, J. M., Lu, X. T., Liu, B., Yu, D. Y., Lin, G., Ao, T., & Xu, J. M. (2020). Organic trace minerals improve eggshell quality by improving the eggshell ultrastructure of laying hens during the late laying period. *Poultry Science, 99*(3), 1483–1490.

Quinteiro-Filho, W. M., Ribeiro, A., Ferraz-de-Paula, V., Pinheiro, M. L., Sakai, M., Sá, L. R., Ferreira, A. J., & Palermo-Neto, J. (2010). Heat stress impairs performance parameters, induces intestinal injury, and decreases macrophage activity in broiler chickens. *Poultry Science, 89*(9), 1905–1914.

Raj, J., Farkaš, H., Jakovčević, Z., Vasiljević, M., Kumar, R., & Asrani, R. K. J. P. S. (2023). Effects of supplemented multicomponent mycotoxin detoxifying agent in laying hens fed aflatoxin B1 and T2-toxin contaminated feeds. *Poultry Science, 102*(8), Article 102795. https://doi.org/10.1016/j.psj.2023.102795

Rajkumar, U., Reddy, B. L. N., Rajaravindra, K. S., Niranjan, M., Bhattacharya, T. K., Chatterjee, R. N., Panda, A. K., Reddy, M. R., & Sharma, R. P. (2010). Effect of naked neck gene on immune competence, serum biochemical and carcass traits in chickens under a tropical climate. *Asian-Australasian Journal of Animal Science, 23*, 867–872.

Ranjan, A., Sinha, R., Devi, I., Rahim, A., & Tiwari, S. (2019). Effect of heat stress on poultry production and their managemental approaches. *International Journal of Current Microbiology and Applied Sciences, 8*(2), 1548–1555. https://doi.org/10.20546/ijcmas.2019.802.181

Rao, S. V. R., Prakash, B., Raju, M. V. L. N., Panda, A. K., Kumari, R. K., & Reddy, E. P. K. (2016). Effect of supplementing organic forms of zinc, selenium and chromium on performance, antioxidant and immune responses in broiler chicken reared in tropical summer. *Biological Trace Element Research, 172*(2), 511–520.

Renaudeau, D., Collin, A., Yahav, S., de Basilio, V., Gourdine, J. L., & Collier, R. J. (2012). Adaptation to hot climate and strategies to alleviate heat stress in livestock production. *Animal, 6*(5), 707–728. https://doi.org/10.1017/S1751731111002448

Ronique, C. B., Ellestad, L. E., Proszkowiec-Weglarz, M., Farley, L., Brady, K., Angel, R., Liu, H. C., & Porter, T. E. (2020). Effects of heat stress on performance, blood chemistry, and hypothalamic and pituitary mRNA expression in broiler chickens. *Poultry Science, 99*(12), 6317–6325.

Ruchi, T., Rathour, K., Tyagi, L. K., & Tiwari, G. (2022). Egg shell: An essential waste product to improve dietary calcium uptake. *Pharmacophore, 13*(4), 32–40.

Saeed, M., Abbas, G., Alagawany, M., Kamboh, A. A., Abd El-Hack, M. E., Khafaga, A. F., & Chao, S. (2019). Heat stress management in poultry farms: A comprehensive overview. *Journal of Thermal Biology, 84*, 414–425.

Safi, A., Imtiaz, M., Ahsan, H. M. T., Riaz, A., Ali, S., & Uzair, M. D. (2022). Effect of heat stress in poultry and its mitigation. *Current Research in Agriculture and Farming, 3*(6), 9–18.

Sahin, K., & Kucuk, O. (2001). Effects of vitamin E and selenium on performance, digestibility of nutrients, and carcass characteristics of Japanese quails reared under heat stress (34 °C). *Journal of Animal Physiology and Animal Nutrition, 85*(11–12), 342–348. https://doi.org/10.1046/j.1439-0396.2001.00340.x

Sahin, K., Sahin, N., Onderci, M., Gursu, F., & Cikim, G. (2002). Optimal dietary concentration of chromium for alleviating the effect of heat stress on growth, carcass qualities, and some serum metabolites of broiler chickens. *Biological Trace Element Research, 89*, 53–64.

Sharifi, A. R., Horst, P., & Simianer, H. (2010). The effect of frizzle gene and dwarf gene on reproductive performance of broiler breeder dams under high and normal ambient temperatures. *Poultry Science, 89*(11), 2356–2369.

Shrestha, S., & Mahat, J. (2022, April). Sustainable food security: How to feed an increasing population? A Review. In *INWASCON Technology Magazine (i-TECH MAG)* (Vol. 4, pp. 15–18). Zibeline International Publishing.

Sohail, M. U., Hume, M. E., Byrd, J. A., Nisbet, D. J., Ijaz, A., Sohail, A., Shabbir, M. Z., & Rehman, H. (2012). Effect of supplementation of prebiotic mannan-oligosaccharides and probiotic mixture on growth performance of broilers subjected to chronic heat stress. *Poultry Science, 91*(9), 2235–2240.

Surana, K. R., Ahire, E. D., Keservani, R. K., Gupta, A. K., Tripathi, M., Yadav, A., Bharti, S., Singh, B. K., & Shukla, N. (2019). Nutritional advantages and effects of vitamins in human health. *Chhattisgarh Journal of Science and Technology, 16*(3), 67–70.

Syafwan, S., Kwakkel, R. P., & Verstegen, M. W. A. (2011). Heat stress and feeding strategies in meat-type chickens. *World's Poultry Science Journal, 67*(4), 653–674.

Syan, V., Kaur, J., Sharma, K., Patni, M., Rasane, P., Singh, J., & Bhadariya, V. (2024). An overview on the types, applications and health implications of fat replacers. *Journal of Food Science and Technology, 61*(1), 27–38.

Trokhaniak, V., Spodyniuk, N., Lendiel, T., Luzan, P., Mishchenko, A., Tarasenko, S., Popa, L., & Ionita, C. (2023). Investigation of an improved side ventilation system in a poultry house using CFD. *INMATEH—Agricultural Engineering, 69*(1), 121–130. https://doi.org/10.35633/inmateh-69-11

Tsiouris, V., Georgopoulou, I., Batzios, C., Pappaioannou, N., Ducatelle, R., & Fortomaris, P. (2018). Heat stress as a predisposing factor for necrotic enteritis in broiler chicks. *Avian Pathology, 47*(6), 616–624.

Türk, G., Şimşek, Ü. G., Çeribaşı, A. O., Çeribaşı, S., Kaya, ŞÖ., Güvenç, M., Çiftçi, M., Sönmez, M., Yüce, A., Bayrakdar, A., Yaman, M., & Tonbak, F. (2015). Effect of cinnamon (Cinnamomum zeylanicum) bark oil on heat stress-induced changes in sperm production, testicular lipid peroxidation, testicular apoptosis, and androgenic receptor density in developing Japanese quails. *Theriogenology, 84*(3), 365–376. https://doi.org/10.1016/j.theriogenology.2015.03.035

UNDESA (United Nations Department of Economic and Social Affairs). (2017). World population projected to reach 9.8 billion in 2050, and 11.2 billion in 2100. 2017 Revision of World Population Prospects, 21 June 2017.

Uzum, M. H., & Toplu, H. D. O. (2013). Effects of stocking density and feed restriction on performance, carcass, meat quality characteristics and some stress parameters in broilers under heat stress. *Revue De Medecine Veterinaire, 164*(12), 546–554.

Waiz, H. A., Gautam, L., Nagda, R. K., & Bhat, G. A. (2016a). Effect of wet feeding on feed conversion efficiency in laying hens during summer season. *Iranian Journal of Applied Animal Science, 6*, 383–387.

Waiz, H. A., Gautam, L. K., Nisar, N. A., Rathore, N. S., & Nagda, R. K. (2016b). Effect of wet feeding on egg quality parameters in laying hens. *Veterinary Practitioner, 17*(1), 142–144.

Wang, K., Shen, D., Dai, P., & Li, C. (2023). Particulate matter in poultry house on poultry respiratory disease: A systematic review. *Poultry Science, 102*(4), Article 102556. https://doi.org/10.1016/j.psj.2023.102556

Wasti, S., Sah, N., & Mishra, B. (2020). Impact of heat stress on poultry health and performances, and potential mitigation strategies. *Animals, 10*(8), 1266. https://doi.org/10.3390/ani10081266

Westerterp, K. R. (2004). Diet induced thermogenesis. *Nutrition and Metabolism, 1*(1), 5. https://doi.org/10.1186/1743-7075-1-5

Weyh, C., Krüger, K., Peeling, P., & Castell, L. (2022). The role of minerals in the optimal functioning of the immune system. *Nutrients, 14*(3), 644. https://doi.org/10.3390/nu14030644

Woelfel, R. L., Owens, C. M., Hirschler, E. M., Martinez-Dawson, R., & Sams, A. R. (2002). The characterization and incidence of pale, soft, and exudative broiler meat in a commercial processing plant. *Poultry Science, 81*(4), 579–584.

Yardibi, H., & Hoştürk, G. T. (2008). The effects of vitamin E on the antioxidant system, egg production, and egg quality in heat stressed laying hens. *Turkish Journal of Veterinary & Animal Sciences, 32*(5), 319–325.

Yunis, R., & Cahaner, A. (1999). The effects of the naked neck (Na) and frizzle (F) genes on growth and meat yield of broilers and their interactions with ambient temperatures and potential growth rate. *Poultry Science, 78*, 1347–1352. https://doi.org/10.1093/ps/78.10.1347

Yüzen, D., Graf, I., Diemert, A., & Arck, P. C. (2023). Climate change and pregnancy complications: From hormones to the immune response. *Frontiers in Endocrinology, 14*, 1149284. https://doi.org/10.3389/fendo.2023.1149284

Zhang, X., Owens, C. M., & Schilling, M. W. (2017). Meat: The edible flesh from mammals only or does it include poultry, fish, and seafood? *Animal Frontiers, 7*, 12–18. https://doi.org/10.1007/s00484-020-01929-6

Zhang, Z. Y., Jia, G. Q., Zuo, J. J., Zhang, Y., Lei, J., Ren, L., & Feng, D. Y. (2012). Effects of constant and cyclic heat stress on muscle metabolism and meat quality of broiler breast fillet and thigh meat. *Poultry Science, 91*(11), 2931–2937. https://doi.org/10.3382/ps.2012-02255

Zhang, M., Dunshea, F. R., Warner, R. D., DiGiacomo, K., Osei-Amponsah, R., & Chauhan, S. S. (2020). Impacts of heat stress on meat quality and strategies for amelioration: A review. *International Journal of Biometeorology, 64*(9), 1613–1628.

Zmrhal, V., Svoradova, A., Venusova, E., & Slama, P. (2023). The influence of heat stress on chicken immune system and mitigation of negative impacts by Baicalin and Baicalein. *Animals, 13*(16), 2564. https://doi.org/10.3390/ani13162564

Chapter 18
Impact of Climate Change on Shrimp Farming in Arid Region Countries: Challenges, Opportunities and Threats

Sherain N. Al-Subiai

Abstract Aquaculture is an important source of food and contributes to national food security at large. It also contributes to meet the growing food demand of escalating and rapidly growing population and addresses nutritional deficiencies. Being the most diversified food product in the world, aquaculture is the fastest growing food production system. However, scientific community has witnessed that the climate change is a serious threat confronting the aquaculture sector in the world, specifically hot spots in the arid region, as a result of scarcity of the precipitation and elevation of sea surface temperature which have profound antagonistic effects on shrimp production. Shrimp farming practices in the arid region have indicated that changes in rainfall have caused a variety of changes in water availability. Moreover, rising temperatures similarly has an adverse effect on water quality parameters affecting shrimp metabolic rate and its immunophysiological process, leading to increase the physiological stress and a decline in production, while also increasing the risk of disease outbreaks and mortality. In order to overcome these challenges, it is required to develop and apply sustainable indoor aquaculture technologies such as recirculating aquaculture system (RAS) and biofloc technology (BFT) to cope with unpredictable weather changes. These systems are based on shrimp production with limited water exchange resulting in reduced water usage, year-round production, higher yields, minimizing transmission of infectious diseases, great potential for integrated aquaponic farming, enhancing immunity, and lower feed requirements. Above all, its advantage in arid environments is to minimize high evapotranspiration and temperature rise by blocking strong sunlight. This gives a superior opportunity in the short term to face the gradual climate change, additionally could prepare the shrimp farming sector for rapid future climate change scenarios.

Keywords Aquaculture · Low-salinity groundwater · Aquaponics · Arid regions · Biofloc · Climate change · Rising temperature

---

S. N. Al-Subiai (✉)
Coastal and Marine Resources Program, Environment and Life Sciences Research Center, Kuwait Institute for Scientific Research, Safat, Kuwait
e-mail: ssubiai@kisr.edu.kw

M. K. Suleiman and S. A. Shahid (eds.), *Fostering Arid Lands Agriculture in the Face of Climate Change*, Advances in Global Change Research 81,
https://doi.org/10.1007/978-3-031-98490-7_18

18.1 Introduction

Kuwait is located in the western edge of the Mesopotamian shallow shelf of the northern Arabian Gulf, with a coastal line of 195 km. The northern coastal waters are influenced by the discharges of Shatt al Arab River (Fig. 18.1). This area is estuarine where the depth is about 10 m, whereas average maximum depth in Kuwait waters is 35 m. The extensive mudflats near the Shatt Al Arab provide breeding and nursery grounds for a wide variety of fauna including shrimp. A considerable amount of the shrimp catch in Kuwait originates from the marshlands associated with Shatt Al Arab (Bishop, 1988). The clear, shallow waters, warm temperatures and an inflow of nutrients from the Tigris and Euphrates Rivers make the Arabian Gulf one of the most productive water bodies in the world.

Shrimp fisheries in Kuwait mainly consist of three species. According to Al-Husaini et al. (2015), the species and their contribution to the total catch are *Penaeus semisulcatus* (60%), *Metapenaeus affinis* (30%) and *Parapenaeopsis stylifera* (10%). *P. semisulcatus* is distributed towards the south; whereas the other two species are mainly concentrated in the north western Arabian Gulf. It is believed that these species undertake shoreward migrations for spawning in shallow waters, where food supply for the larvae will be abundant (Yimin & Mohammad, 1999). The peak catch occurs during the spawning season from December to January when the shrimps undertake shoreward migration. The movement of shrimp from Kuwait Bay to offshore waters has been reported by Mathews (1986).

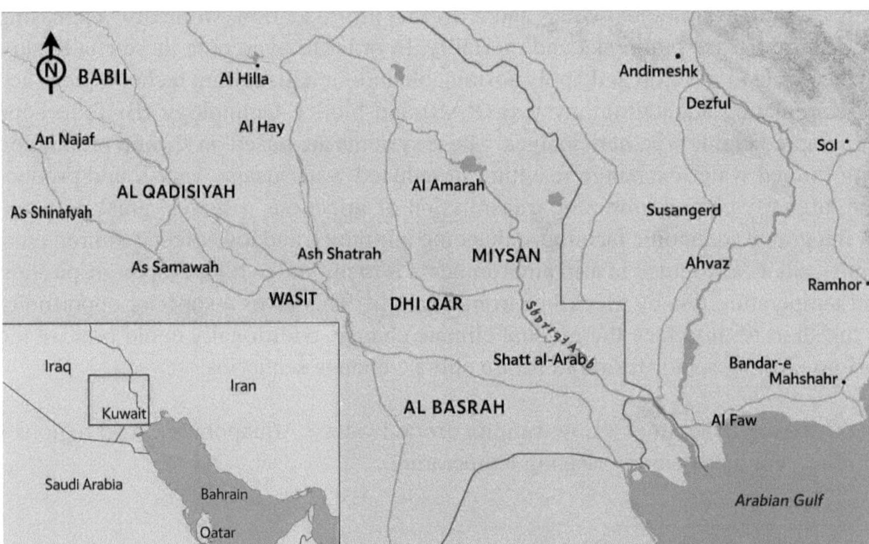

Fig. 18.1 Arabian Gulf showing Shatt al-Arab discharge

The shrimp landing in Kuwait decreased from 5,125 tons in 1990–1991 to 1,010 tons in 2023 (CSB, 2024). The reasons attributed include an increase in fishing effort resulting in reduction in recruitment to the fishery due to exploitation of spawners and environmental degradation as a result of anthropogenic influencers such as oil spills, reduced discharge from Shatt al-Arab, sewage/industrial pollution and climate changes (Castilho et al., 2008).

The assumption regarding the environmental degradation in Kuwait is true, considering the sewage discharge and oil spills (Alqattan & Gray, 2021). According to Al-Sarawi et al. (2018), samples collected from different locations of Kuwaiti waters were recorded high counts of *E.coil* exceeding the international standards that indicate sewage discharge. Moreover, oil leaks are unavoidable in a very active small body of water like the Arabian Gulf, where over 25,000 oil tankers are shipping (Al-Saad & Salman, 2012). Shrimps are physiologically strong at adult stage, but their eggs and larvae are highly delicate and readily harmed by environmental changes. They can come into direct touch with floating oil since they are pelagic. The larvae consume both phytoplankton and zooplankton as food. Because the floating oil blocks sunlight, less plankton is proliferated. A plankton survey in these waters has revealed a considerable reduction in the abundance of eggs and larvae since 1992 (Price et al., 1993). Reduction in the plankton productivity directly affects the benthic productivity, which forms food for adult shrimp. Oil can persist in the environment for many years, and therefore, the chances of revival of shrimp fishery in recent years is quite impossible.

18.2 Nutritional Facts and Importance of Shrimp Food for Human

Shrimp is a popular type of seafood that is high in protein content (24g of protein/100g shrimp), low in calories (99 kcal/100g), rich in several beneficial nutrients including n-3 to n-6 polyunsaturated fatty acid, as well as macro and micro minerals such as vitamin B12, selenium, choline, zinc, iron, copper and phosphorus (Liu et al., 2024). It provides a range of health benefits for humans, from enhancing immunity and maintaining the health of bones and joints to supporting cardiovascular and brain function. According to several studies, astaxanthin, the primary carotenoid, can improve skin health by scavenging free radicals and preventing the effects of oxidative damage (Liu et al., 2021; Niamnuy et al., 2008; Pu et al., 2010). Shrimp is also known of its anticancer, antidiabetic and fertility enhancing effect (Fakhri et al., 2018). However, shrimp has high cholesterol content, which could be harmful. De Oliveira e Silva et al. (1996) have verified that eating shrimp led to significantly lower triacylglycerol concentrations as well as ratios of total to high-density lipoprotein (HDL) and low-density lipoprotein (LDL) cholesterol to HDL cholesterol when compared to a high cholesterol diet like eggs. They conclude that eating shrimp has no detrimental impacts on a person's total plasma lipoprotein profile and has no appreciable effect on a person's risk of heart disease.

Besides its nutritional values, an important indicator of animal production efficiency is the Feed Conversion Ratio (FCR), which shows how much feed is needed to achieve a specific level of weight growth in the animal. In general, the FCR of aquaculture is lower than that of poultry and beef cattle, which offers considerable benefits in terms of cost, sustainability, and efficiency (Davison et al., 2023; Galkanda-Arachchige et al., 2020). For every kilogram of body mass gained, aquatic species like fish and shrimp need between 1.1 to 1.6 kg of feed. As opposed to this, the FCR of beef cattle usually falls between 6 and 8. A lower FCR increases overall profitability for aquaculture producers since they use less feed per kilogram of meat produced. Therefore, aquaculture has become more resource-friendly and feed-efficient.

Shrimp is also one of the most delicious seafood, part of each nations' traditional dish. Because of its versatility, it can be served in different ways either cooked or raw. Shrimp can be cooked within 2–5 min, which makes it energy efficient. The shrimp's small size makes it simple to serve for meals, fitting perfectly with today's fast-paced lifestyle.

18.3 Global, Regional Climate and Climate Change Trend

The scientific community has observed that the aquaculture industry globally, particularly hot spots in arid regions, is facing a severe threat from climate change due to factors such as rising global temperatures, limited precipitation, and human activity. The world is on track to set new temperature records in 2024 and 2028 that will surpass 1.5 °C (WMO, 2024). It suggests that further adverse climatic consequences are anticipated, such as heatwaves, severe weather, and increasing sea levels. These factors will undoubtedly have a significant negative impact on shrimp output. Shrimp farming practices in the arid region have revealed that variations in rainfall have led to a range of adjustments in the amount of water available. Shrimp can tolerate temperatures between 15 and 30 °C on average, although 28 to 29 °C is ideal for optimal growth (Araneda et al., 2020). When temperatures rise over acceptable ranges, shrimp become more susceptible to disease outbreaks and mortality due to thermal stress (Zheng et al., 2019). Likewise, it negatively impacts the dissolved oxygen (DO) in water, which impacts the growth performance and metabolic rate of shrimp. This results in inefficient feed consumption and higher FCR. The environmental impact is increased when FCR is higher, since more land will be needed to cultivate crops for animal feed. In addition to the effect of the frequent usage of the freshwater resources, there will be an increase in gas emissions from the transportation, feed production and processing.

18.4 Global, Regional and National Productions of Shrimp

Protein scarcity is the world's major issue today, and aquaculture practice is considered a solution for this, since aquaculture can produce more protein per unit area compared to agricultural or animal husbandry practice. Increasing aquaculture production is also essential to meet the huge demand for fish and shrimp products in the global market, augmented by the decreasing production from the capture fisheries sector. Global production of aquatic animals reached a record of 185 Million Tons (MT) in 2022, more than 50% are from the aquaculture sector (FAO, 2022). Asian countries are contributing 70% of the total production of aquatic animals, followed by Latin America. White leg shrimp, *Litopenaeus vannamei*, was the top cultured species, with 6.8 MT produced in 2022. Shrimp is the largest single seafood commodity, constituting about 20% of the total value of internationally traded fishery products. Since 2022, the significant proportion of the shrimp has been from the aquaculture sector (FAO, 2022). The contribution by major shrimp producing countries in aquaculture production is given in Fig. 18.2. According to the recent report issued by Jory, 2023, "*Ecuador is the fastest growing aquaculture industry in the world*".

The Gulf Cooperation Council (GCC) countries, including Saudi Arabia, the United Arab Emirates (UAE), Oman, Kuwait, Qatar, and Bahrain, have recently shown significant interest in shrimp farming, driven by the need for economic diversification, expanding seafood consumption, and food security. Saudi Arabia is the key player and the front-runner in shrimp farming in the GCC region. The National Aquaculture Group (NAQUA), situated in Saudi Arabia near the shore of the Red Sea,

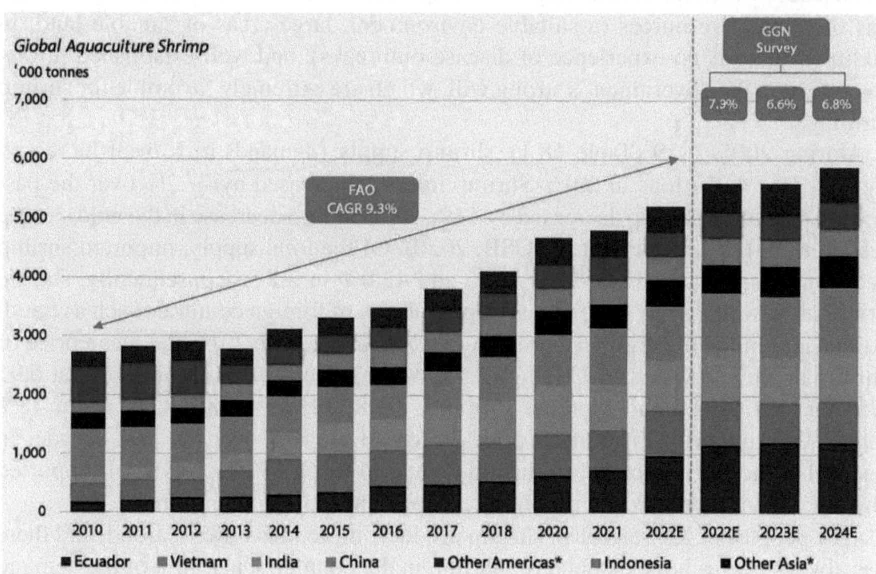

Fig. 18.2 Shrimp production by region. *Sources* Jory (2023); FAO GOAL survey

is one of the biggest shrimp farms in the world. Its main output is whiteleg shrimp and it produces around 90,000 Metric Tons (MT) annually (NAQUA, 2024). Although there are still no commercial shrimp farms in the other GCC countries, there has been a rise in investment in shrimp farming. In Qatar, there are three different-sized private shrimp farms: a small shrimp farm (within an agricultural farm) with an annual production of about 10 tons, a medium-sized (within an agricultural farm) farm with an annual production of approximately 200 tons, and a coastal farm in the Al-Arish region with an annual production capacity of 1,000 tons, which is under construction (MME, State of Qatar). Shrimp farming has been growing in Oman as well. As a component of its Vision 2040 strategy to diversify the Omani economy and improve food security, the government has made investments in aquaculture projects. In South Al Sharqiyah Governorate, a test operation of the shrimp farming project was started in 2021. The aspects of this mega project's production started to emerge, and Oman is trying to increase this step until the yearly production of shrimp reaches 4,600 MT (Personal communication, MAFWR, Sultanate of Oman). The UAE has been concentrating more on sustainable shrimp farming projects, particularly in Ras Al Khaimah and other coastal areas. Despite this, shrimp production in the UAE is still very limited, with an average production of 168 tons, due to the high production cost and the inability to compete with low-cost imported shrimp in the markets (Personal communication, UAE Ministry of Climate Change and Environment-MOCCAE). Kuwait possesses over forty years of research experience in aquaculture, and the country is expected to see future expansion in the shrimp farming industry. Despite the industry being in its early stages of development, shrimp farming is expected to grow in the future. The demand for shrimp has been increasing in Kuwait during the last decade, which has had an impact on imports and raised shrimp annually. Kuwait has the natural resources (a suitable environment, large areas of suitable land; in addition, there is no experience of disease outbreaks), and well-established infrastructure, and the government's strong will, which are extremely favorable for shrimp farming.

During 2005–2019 (Table 18.1), shrimp supply (demand) in Kuwait increased by 24.4% to 4,054 tons in 2016. Shrimp imports increased by 77.2% over the past decade. Captured shrimp decreased by 14%, suggesting a decrease in the wild shrimp resources in the Kuwaiti waters (CSB, 2020). Of the total supply, imported shrimp accounted for occupied 59.9% in 2016 and 42.0% in 2005. Consequently, shrimp prices in Kuwait largely depend on the production of foreign countries such as Saudi Arabia and India, resulting in a great fluctuation of price. In 2016, the mean price of shrimp in Kuwait was 8.10 US$/kg (2.456 KD). Imported shrimp was sold at 6.92 US$/kg (2.079 KD) and captured shrimp is 9.98 US$/kg (3.023 KD), about 45% higher than imported. The mean price increased by 40% over the past decade. In particular, the price of captured shrimp increased by 61.8%. However, all imported shrimp in Kuwait are frozen products, except captured shrimp are fresh.

The peoples of Kuwait relish shrimp products more than other seafood, and there has always been a huge demand for shrimp in the country. Only 38% of the demand is satisfied by the catch from natural waters (CSB, 2024). The remaining is imported from countries like Iran, Saudi Arabia and India. If the decline in shrimp stocks

Table 18.1 Shrimp supply (imported and captured) and price in Kuwait (2005–2019)

	Imported			Local captured			Total value (KD) (imported + local captured)
Year	Price/ kg (KD)	Quantity (kg)	Value (KD)	Price/ kg (KD)	Quantity (kg)	Value (KD)	Value (KD)
2005	1.588	1,370,080	2,175,687	1.868	1,890,255	3,530,996	5,706,683
2007	1.502	1,265,750	1,901,157	1.766	1,539,857	2,719,387	4,620,544
2009	1.447	1,275,715	1,845,960	1.871	1,749,307	3,272,953	5,118,913
2010	1.543	1,118,700	1,726,154	1.926	2,102,721	4,049,841	5,775,995
2011	1.807	939,360	1,697,424	2.400	1,439,142	3,453,941	5,151,365
2012	1.759	1,546,270	2,719,889	2.119	1,468,478	3,111,705	5,831,594
2013	2.073	1,303,550	2,702,259	2.417	1,510,080	3,649,863	6,352,122
2014	2.234	1,588,060	3,547,726	2.854	1,487,551	4,245,471	7,793,197
2015	2.119	2,122,200	4,496,942	2.831	1,444,309	4,088,839	8,585,781
2016	2.079	2,427,760	5,047,313	3.023	1,626,319	4,916,362	9,963,675
2017	1.796	1,710,730	3,072,471	2.736	1,100,830	3,011,871	6,084,342
2018	1.650	1,957,896	3,230,528	2.985	617,280	1,842,581	5,073,109
2019	1.813	1,461,740	2,650,135	2.765	917,920	2,538,049	5,188,184

Source CSB, (2020)

continues in the same momentum, in the future, the country may require higher percentages of shrimp imports. In this context, the development and commercialization of shrimp aquaculture is the only solution to meet the demand for shrimp. Since shrimp is a commodity with great demand in the global market, surplus shrimp can be exported to strengthen the economy of the country.

18.5 Potential Arid Regions Countries for Shrimp Production

The aquaculture production in the Middle East is expected to increase as more and more, especially if they have seriously taken up the practice not only to satisfy the demand of nationals, but also to gain export earnings in the global market. The total aquaculture production of shrimp in Saudi Arabia was 56,000 tons in 2018 (Miao & Wang, 2020), while Iran produced 22,310 tons (Abdolhay, 2016).

The techniques for raising shrimp in desert conditions have been developed in countries and regions like Australia, Algeria, Egypt, Israel, Mexico, Southern Africa, the United States of America and Central Asia (Crespi & Lovatelli, 2011). Desert fish farming was formulated in 1963–1965 and tested experimentally, showing that it was possible to use desert salt or brackish waters to rear fish successfully (Fishelson & Loya, 1969). In these areas, modern technology and alternative sources of energy have helped to achieve this in places where the climate is highly extreme. Aquaculture in desert and arid regions has been defined as "Aquaculture activities practiced in desert and arid lands characterized by low precipitation (<250 mm/year), high solar radiation, high rate of evaporation, using subsurface and surface water" (Crespi & Lovatelli, 2011). In the Sahara Desert, Algeria, white leg shrimp *L. vannamei* is being cultured in low saline groundwater (2–4 parts per thousand) with ionic modification. From the indoor tanks and outdoor ponds using biofloc technology, a production of 500 kg was obtained within four months of culture (Jang & Gong, 2016).

Kuwait has a total area of 17,818 km^2 and features a typical Middle Eastern desert and arid climate. The total population is about 4.934 million and most of them are concentrated in few areas (World Population Review, 2024). There are vast areas uninhabited, so there is no land scarcity for aquaculture in Kuwait. In aquifers with low saline waters can be found in most areas of southwest Kuwait are utilized as small-scale agricultural resources. Fresh and brackish waters are available in Kuwait group and Dammam formations (Alsharhan et al., 2001). These aquifers are regularly recharged naturally and water moves northeast and joins the Arabian Gulf, as a result the salinity increases towards north and east. A large percentage of the southwest area of Kuwait may be converted into shrimp farms where culture is feasible using groundwater which is pristine quality, without the presence of any pathogens. The limiting factors include high temperatures, solar radiation, scarce precipitation, low water exchange rates and environmental issues to be foreseen such as salinization of aquifers, and land subsidence (Crespi & Lovatelli, 2011). The coastal areas of Kuwait as well as islands can also be utilized for culture of the shrimp at higher salinity, with proper monitoring of waste water disposal, if any.

Among the species of shrimp cultured worldwide, *L. vannamei* is the major species of shrimp because of its hardy nature, less vulnerability of disease, compatibility in high density culture, adaptation to wide range of salinities and consumer preference for its quality (Mc Graw et al., 2002). Its production on a global basis was 5,812,180 tons in 2020 (FAO, 2022). Specific pathogen free strains have been developed in countries like the USA and are used for culture in most of the shrimp producing countries. The species has proved to be ideal for culture in low saline waters of desert lands (Jang & Gong, 2016).

Biofloc technology is one of the recent advances in aquaculture, which facilitates culture with zero water exchange, and therefore ideal for areas where the main limitation is the availability of water (Avnimelech, 2015). Biofloc is a conglomeration of organisms such as bacteria, algae, fungi, ciliates, rotifers, metazoans and other microorganisms that grow on waste produced by the cultured organisms, which in turn forms food for these organisms (Ray et al., 2011). Biofloc has also reported to increase expression of immune-related genes enhancing resistance to disease in cultured species (Kim et al., 2014). Shrimp species such as *L. vannamei* (Mishra et al., 2008; Samocha et al., 2007) *Penaeus monodon* (Arnold et al., 2009), *Farfantepenaeus paulensis* (Ballester et al., 2010), *Farfantepenaeus brasiliensis* and *Litopenaeus stylirostris* (Emerenciano et al., 2012) have been successfully cultured using this technology. The *L. vannamei* has been found to be especially good in utilizing biofloc as its nutrition due to specified filtering apparatus on the maxillipeds of mouthparts, which makes capturing and utilizing bioflocs more efficient compared to other species (Kim et al., 2015), thus making it the best species for high intensive, zero water exchange biofloc culture, excluding all diseases.

Kuwait has already taken up a few projects to initiate shrimp culture. The Sabbiyah shrimp project of the Public Authority for Agriculture Affairs and Fish Resources (PAAFR), which is intended to produce 3,000 tons of shrimp annually is one of its kind (Fig. 18.3). However, the project is still in its infancy (personal communication). Another important project is the Government Initiative project of the Kuwait Institute for Scientific Research (KISR), namely, 'Establishment of an Economic Production Complex for Fish and Shrimp' (Fig. 18.4). This project aims to commercialize shrimp aquaculture production in Kuwait through advanced technologies that have been developed on local finfish species and shrimp at KISR i.e. biofloc technology (BFT) and recirculating aquaculture system (RAS) technology. The project intends to produce 1,000 MT of finfish *(Epinephelus coioides, Acanthopagrus cuvieri, Acanthopagrus bifasciatus)* and 200 MT of shrimp (*L. vannamei*) annually from less than a hectare area, from a facility that is expected to be constructed in Bar-Ghudai, in the next 3–4 years. The success of these projects determines the future of aquaculture in Kuwait. The project will also include other operation units such as fish/shrimp hatcheries, feed mill, processing plant and other research facilities for quality assurance and development (Fig. 18.5).

Thus, although the Kuwait government has initiated a few aquaculture ventures, there is tremendous potential for the country to be exploited with advanced and proven technologies. Exploring and using these potentials can not only satisfy the nation but also prevent overexploitation of natural resources, reduce imports, increase exports, strengthen the national economy, and contribute to the world's food supply.

18.5.1 Coastal

The practice of coastal shrimp farming is well-established long time ago and has widespread applications in many tropical and subtropical regions, where environment condition is suitable for shrimp production. Shrimp farming has gained importance due to the increase of the global demand for shrimp, which has led to high export prices, generation of job opportunities, and increase in foreign exchange earnings.

Shrimp culture systems vary in their intensity of operation, from most extensive pond systems where the stocked shrimp are not fed up to the super-intensive recirculating systems, which are supported by feed, aeration and recycled water treatment. The selection of the optimal shrimp culture system mainly depends on the condition of the site (coastal or inland), water resources (fresh, brackish or seawater) and

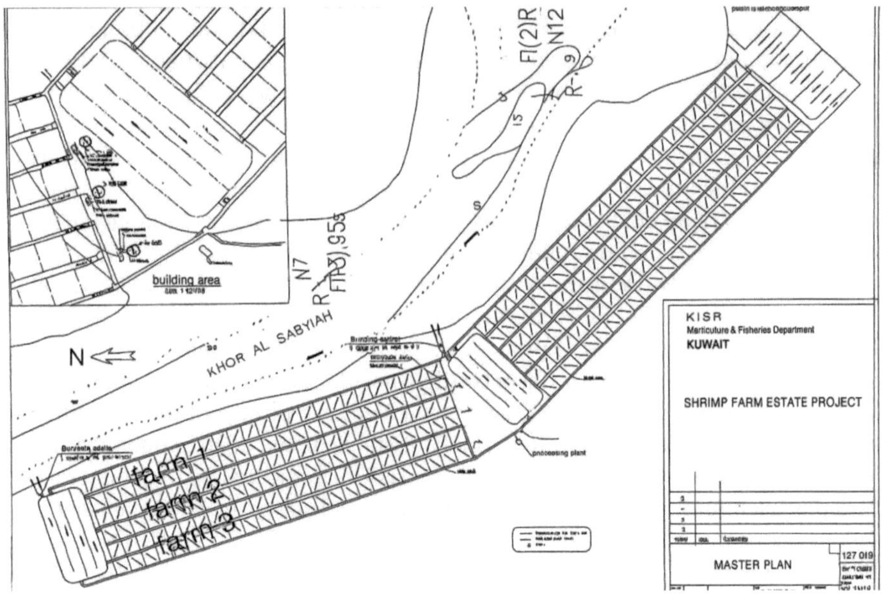

Fig. 18.3 Proposed plan of Sabbiyah shrimp project, Kuwait (Al-Ameeri et al., 2002)

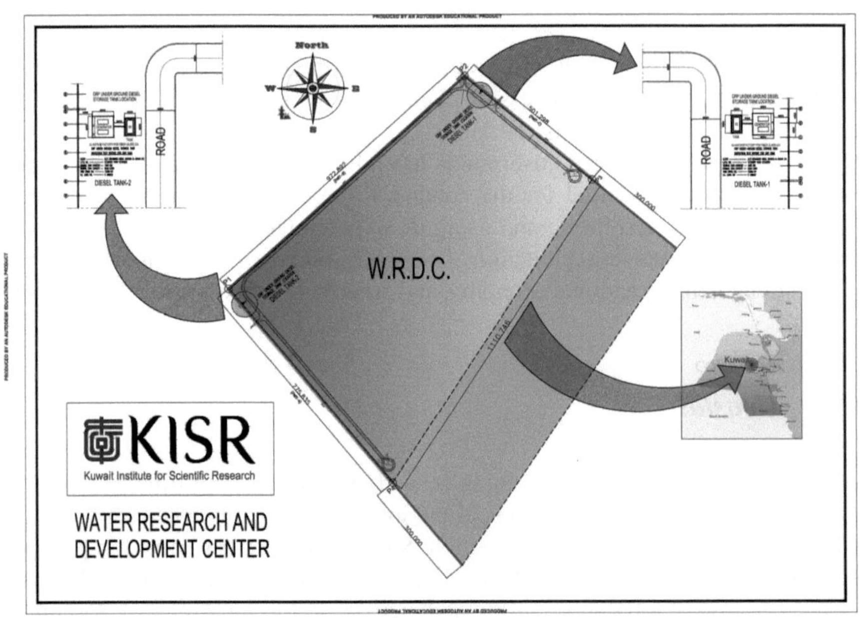

Fig. 18.4 Suggested area for an economic production complex for fish and shrimp at Bar-Ghudai site

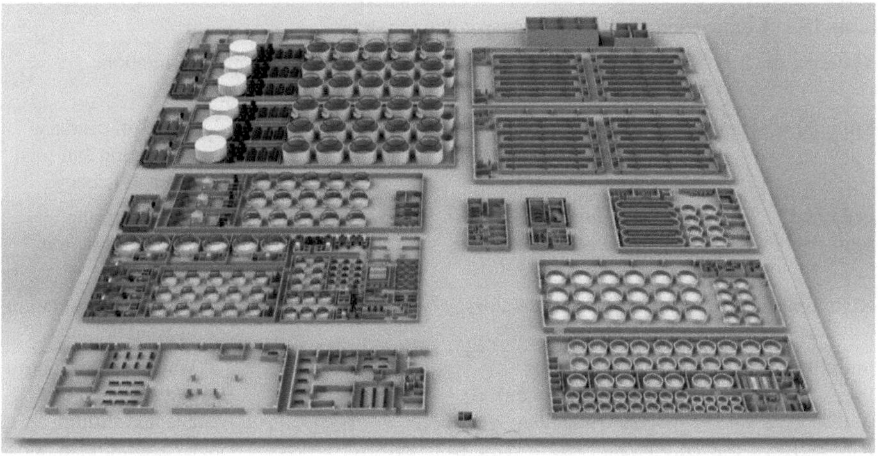

Fig. 18.5 A 3D floor plan for a commercial production complex for fish and shrimp at Bar-Ghudai site, KISR under designing and construction. *Credit* Wonjin & FNS Co, used with permission

production target. An overview of these systems is presented in Table 18.2 showing the ranking of the culture intensity with its production and limitations. In order to achieve super-intensity culture, RAS has been successfully used in the aquaculture industry. This system is based on a flow-through system, but the disposed water is treated using mechanical solids separation and biofilters before returning to the production tank. The RAS allows production at high intensity which goes up to 10 kg/m^3 using very little water in small area. However, water treatment in the system has added substantial costs in terms of energy and material; and makes it expensive for investment. Another eco-friendly culture system known as Biofloc technology has been recently developed to facilitate super-intensive culture and discussed in detail in the next section.

Almost all commercial shrimp farms use seawater, which provides the natural habitat for the healthy growth and reproduction of shrimp. The favorable salinity condition and the ionic composition of the seawater match shrimp physiological needs. Seawater is also rich in natural food sources for the shrimp like plankton. The studies have shown that shrimp grow faster and tend to have the high survival rate in the seawater (Samocha et al., 1998; Emanuell et al., 2010; Sumara et al., 2015). However, although shrimp culture in seawater fulfilled all biological needs, disease outbreaks can be major concerns in coastal shrimp farming. Therefore, adopting proper water management with advanced biosecurity help alleviate these risks.

Table 18.2 Comparison of different shrimp culture systems

System type	Intervention	Approximate shrimp yields (kg/m^3)	Limitations
Minimal feed extensive	Minimal feeding with grains, farm and home residues	<0.01–0.05	Limits of primary production and food chain efficiency
Fed pond semi-intensive	Feeding by commercial diet pellets	0.05–0.25	Early morning oxygen depletion
Night time aeration	Night time aerators ~ 1–5 hp/ha	0.15–0.8	Sludge accumulation in pond bottom
Intensive mixed aerated	24 h aeration (completely mixed) 20–150 hp/ha	0.8–2.0	Water quality control and selection of adapted shrimp
Super intensive: it can be circular, square, RW	24 h aeration. Highly mixed aerated (often pure oxygen added)	5.0–10	Energy utilization

Source Modified fromBallester et al. (2010)

18.5.2 Inland

The use of underground saline water of desert areas for shrimp culture is a new concept, and a move has been made in this direction in various areas in the world, starting from Middle Eastern Africa to Latin America for the culture of *L. vannamei*. However, there is often a deficiency of potassium, magnesium, and sulfate ions in the rearing medium when shrimp culture is executed in the brackish water, which usually results in the reduction of growth and survival of the species (Al-Subiai et al., 2024). Therefore, it is necessary to evaluate the quality and ion concentration of the water source to utilize it for inland shrimp culture.

Ion imbalance, super-saturated gas or iron in groundwater, and algal blooms by blue-green algae are major problems of inland shrimp farming. However, the most important limiting factor for inland shrimp farming is the ionic imbalance of the groundwater, specifically ions of potassium, magnesium and sulfate (Van Wyk et al., 1999). Most groundwater sources are deficient in magnesium and potassium, which cause reversed Mg/Ca or high or low Na/K compared to those in the standard seawater. The ionic imbalance of the water has been thought as principal causes for molt-related mortality, slow growth, and cramping or white muscle. These symptoms can result in continuous or mass mortality of the crop (Han et al., 2015).

18.6 Case Study—Kuwait

Kuwait is one of the arid countries in the Middle East that features a desert climate with average 105 mm of annual precipitation and high annual evapotranspiration of 2270 mm (Alsharhan et al., 2001). Most of the groundwater sources in Kuwait are brackish, and have a wide range of salinity, and it is divided into three classes, below 1,000 parts per million (ppm), between 1,000 and 10,000 ppm, and exceeding 10,000 ppm. Natural water resources in Kuwait are very limited, and the groundwater is used for landscaping and agriculture for saving the high production cost of quality freshwater by desalination. Furthermore, aquaculture using brackish groundwater may be a better option as a water source than using seawater in Kuwait, which has high temperature and salinity (Misak & Hussain, 2022).

Shrimp aquaculture has taken its roots in Kuwait through the farming research activities initiated by Kuwait Institute for Scientific Research (KISR) in 2015. From an experimental tank-based trials carried out till now, and also after analyzing the scope for upscaling this important food production activity based on the experiences gained so far, it is time to translate shrimp culture from "lab to land" by selecting the most appropriate farming technique to reap maximum returns. These important steps require careful selection of the most suitable culture system for efficient operation in an arid region using local resources. The decision of ELSRC/KISR management to productively use the brackish water available at KISR Station for Research and Innovation (KSRI), Kabd was a bold decision and the shrimp team has accepted it as a challenging task to use the underground low saline water for shrimp production in suitable systems. As such, Inland Shrimp Research Facility (ISRF) has been established at the Kabd site using biofloc system (Fig. 18.6). The farm has indoor and outdoor shrimp culture facilities, including ponds, tanks and raceways.

A series of trials have been conducted to cultivate Pacific white shrimp using low salinity water drawn from 60 m underground at the Kabd site in Kuwait. The water samples from the existing bore wells at the Kabd site were analysed for their suitability for shrimp culture. The biggest obstacles for shrimp culture at Kabd site consisted of the following: first, the extreme weather conditions, which can be solved by operating functional indoor biofloc systems and installing shading sheets on the outdoor ponds; the second obstacle was the inadequate ionic composition of the groundwater in Kabd site, which is required to be appropriately balanced.

The salinity of the groundwater (4 parts per thousand) was acceptable for *L. vannamei* survival and growth. However, the ionic composition of the groundwater was seen to be greatly unbalanced, compared with the natural seawater (Table 18.3). Ionic composition is extremely important because marine shrimp, such as *L. vannamei* requires specific ratios of some ions for normal osmoregulation process during molting and optimal growth, particularly in inland low-salinity water environments. The ratio of the Na/K of Kabd groundwater is 29.6, almost close to that of the normal seawater (27.6) (Table 18.3). However, Mg/Ca ratio is 0.44, which is much lower than 3.18 of normal seawater. Therefore, additional magnesium sources, such as magnesium chloride ($MgCl_2$) or magnesium sulfate ($MgSO_4$)

Fig. 18.6 Inland Shrimp Research Facility (ISRF) at Kabd site, KISR, Kuwait. *L. vannamei* has been produced in indoor and outdoor systems since 2020

should be provided to the culture systems to adjust the required ratio before stocking post-larvae into the tanks.

Based on the higher survival and growth rate observed after four weeks of nursery culture at the experimental and semi-commercial levels, Al-Subiai et al. (2024)

Table 18.3 Chemical analysis of the groundwater at Kabd site and Kuwait Bay seawater

Parameters (mg/L)	Unit	Groundwater[1]	Kuwait Bay[2]	Open sea[3]
pH	–	7.91	NA	NA
Salinity	ppt	4.08	45	34.5
Sodium (Na)	mg/l	478	15,850	10,500
Potassium (K)	mg/l	15.9	460	380
Magnesium (Mg)	mg/l	119	1,765	1,350
Calcium (Ca)	mg/l	271	500	400
Chloride (Cl)	mg/l	561	23,000	19,000
Total Alkalinity (as $CaCO_3$)	mg/l	67.4	NA	NA
Bicarbonate (as $CaCO_3$)	mg/l	<0.10	NA	NA
Carbonate (as $CaCO_3$)	mg/l	<0.10	NA	NA
Ma/Ca ratio	–	0.44	3.53	3.38
Na/K ratio	–	29.6	34.46	27.6
Cl:Na:Mg	–	4.7:4:1	1.45:8.98	1.81:7.78

[1] Water Research Center, KISR
[2] www.lenntech.com/composition-seawater.htm
[3] Boyd et al. (2009)

suggested that Mg/Ca of 1.0 might be the lower limit for *L. vannamei* nursery in a zero-exchange biofloc culture system using low-salinity groundwater in Kuwait (Fig. 18.7). On the other hand, the grow-out trials of eight weeks in different Mg/Ca conditions showed that the performance of experimental groups, especially in 2.12 (Mg2-G) and 3.28 (Mg3-G) of Mg/Ca, was significantly better than that of control in terms of survival, growth, and feed efficiency (Fig. 18.8). This finding has been confirmed in the pond trial, and it is safe to say that adding Mg salt has contributed to the better shrimp growth performance (Al-Subiai et al., 2022) (Fig. 18.9). Thus, it can be concluded that compensation of deficient Mg in the Kabd groundwater can improve the survival and the production of inland low-salinity shrimp culture. Further research is required to determine if other affordable mineral sources or salt sources can be used to adjust ion imbalance of Kabd groundwater, instead of expensive magnesium chloride, and if there are other mineral factors that can affect shrimp performance.

Outdoor ponds at Inland Shrimp Research Facility, KISR were installed with shading sheets and aerators controlling sunlight and water temperature. About 13 tons/ha/cycle of *L. vannamei* has been produced since 2021. The outcomes of the nursery and grow-out of BFT shrimp farming in low salinity groundwater of Kabd site are expected to be the foundation stone for improving aquaculture technology, developing new industry and, moreover, ensuring food security and safety in Kuwait.

Fig. 18.7 Healthy biofloc water and postlarvae in nursery tanks at Inland Shrimp Research Facility at Kabd site, KISR

Fig. 18.8 Well-managed biofloc raceway tanks (top left), harvest (top right) and sampled *L. vannamei* sub-adults (bottom left) at Inland Shrimp Research Facility at Kabd site, KISR. Mean shrimp production (bottom right) reaches 3–4 kg/m^2

18.7 Conclusions and Recommendations

Shrimp farming accounts for around 50% of all shrimp produced worldwide, and this percentage is rising. Shrimp farming has significantly been improved in the arid region. Shrimp aquaculture is predicted to play an even larger role in the future because of the persistently high demand for quality seafood and the stagnating or declining levels of wild capture. Similar to most human activity, shrimp farming has had a wide range of environmental effects. The degree and importance

Fig. 18.9 Outdoor pond management and harvest at Kabd site, Kuwait

of these effects vary greatly within and between regions, depending on farm operation, water management, disease outbreaks, and other physical/biological factors of shrimp culture. As the challenges brought about by climate change intensify, the shrimp aquaculture industry must embrace a holistic strategy that acknowledges the mutual reliance of environmental sustainability and climatic resilience. By investigating ways that simultaneously meet both goals, the industry may support global efforts to preserve long-term security and build a more resilient and sustainable future. Sufficient technology and managerial experience now exist to mitigate these key problems, but it definitely needs incentives and/or government involvement to be implemented.

Acknowledgements Research funding for Kuwait case study was provided by the Kuwait Foundation for the Advancement of Sciences (KFAS) (Grant no. PR19-12SL-08) and Kuwait Institute for Scientific Research (KISR) (FM105C), State of Kuwait. We are particularly grateful to Dr. Darryl Jory for helping us and sharing the shrimp production survey. Also, we would like to acknowledge Wonjin & FNS Co for providing assistance in sharing 3D-plan. Additionally, we would like to express our special thanks to Ms. Hebah Jaber Kamal from Systems and Software Development Department, Science and Technology Sector, Kuwait Institute for Scientific Research for her assistance in the preparation of the map.

References

Abdolhay, H. (2016). *National aquaculture sector overview*. National Aquaculture Iran Sector. Overview Fact Sheets.

Al-Ameeri, A., Cruz, E., Almatar, S., AlHazeem, S., & Al-Nafisi, R. (2002). Study on the suitability of the Sabiyah area for the development of shrimp farming in Kuwait. Kuwait Institute for Scientific Research, Final Report No. 6336, Kuwait.

Al-Husaini, M., Bishop, J. M., Al Foudari, H. M., & Al-Baz, A. M. (2015). A review of the status and development of Kuwait's fisheries. *Marine Pollution Bulletin, 100*(2),597–606. https://doi.org/10.1016/j.marpolbul.2015.07.053.

Alqattan, M. E. A., & Gray, T. S. (2021). Marine pollution in Kuwait and its impacts on fish-stock decline in Kuwaiti waters: Reviewing the Kuwaiti Government's Policies and Practices. *Frontiers in Sustainability, 2*. Art. No. 667822. https://doi.org/10.3389/frsus.2021.667822.

Al-Saad, H. T., Salman, N. A. (2012). Status of oil pollution in the Arabian Gulf and Shatt Al-Arab Estuary: a review. In V. Subramanian (Ed.), *Coastal Environments: Focus on Asian Regions*, (Vol. 7, pp. 36–47). Springer.

Al-Sarawi, H. A., Jha, A. N., Baker-Austin, C., Al-Sarawi, M. A., Lyons, B. P. (2018). Baseline screening for the presence of antimicrobial resistance in *E. coli* isolated from Kuwait's marine environment. *Marine Pollution Bulletin, 129*, 893–898. https://doi.org/10.1016/j.marpolbul.2017.10.044.

Alsharhan, A. S., Rizk, Z. A., Nairn, A. E. M., Bakhit, D. W., & Alhajari, S. A. (2001). *Hydrogeology of an Arid Region: The Arabian Gulf and Adjoining Areas* (p. 325). Elsevier Publishing.

Al-Subiai, S. N., Hussain, S., Al-Foudari, M., Jutaily, S., Al-Hasan, E., AlNuaimi, S., & Ali, Z. (2022). Phase I: Inland biofloc shrimp farming of the white leg shrimp *L.vannamei* in Kuwait. Kuwait Institute for Scientific Research, Final Report No. 17623, Kuwait.

Al-Subiai, S. N., Jang, I. K., Bae, S., Yoon, H., Hussain, S., AlNuaimi, S., Al-Foudari, M., & Al-Hasan, E. (2024). Enhancing the performance of *Litopenaeus vannamei* nursery and grow-out by modifying Mg/Ca ratios in biofloc systems using low-salinity groundwater of Kuwait Desert. *Aquaculture, 591*, Article 741405. https://doi.org/10.1016/j.aquaculture.2024.74140

Araneda, M., Gasca-Leyva, E., Vela, M. A., & Domínguez-May, R. (2020). Effects of temperature and stocking density on intensive culture of Pacific white shrimp in freshwater. *Journal of Thermal Biology, 94*, Article 102756.

Arnold, S. J., Coman, F. E., Jackson, C. J., & Groves, S. A. (2009). High-intensity, zero water exchange production of juvenile tiger shrimp, *Penaeus monodon*: An evaluation of artificial substrates and stocking density. *Aquaculture, 293*, 42–48.

Avnimelech, Y. (2015). *Biofloc technology—a practical guide book* (3rd ed.). The World Aquaculture Society.

Ballester, E. L. C., Abreu, P. C., Cavalli, R. O., Emerenciano, M., Abreu, L., & Wasielesky, W. (2010). Effect of practical diets with different protein levels on the performance of *Farfantepenaeus paulensis* juveniles nursed in a zero exchange suspended microbial flocs intensive system. *Aquaculture Nutrition, 16*, 163–172.

Bishop, J. M. (1988). *A survey of Kuwait shrimp nursery grounds and establishment of a shrimp recruitment index*. Kuwait Institute of Scientific Research, Final Report No 2622, Kuwait.

Boyd, C. A., Chaney, P. L., Boyd, C. E., & Rouse, D. B. (2009). Distribution of groundwater suitable for use in saline-water aquaculture in centeral and west-central Alabama. *Journal of Applied Aquaculture, 21*, 228–240.

Castilho, A. L., Pie, M. R., Fransozo, A., Pinheiro, A. P., & Costa, R. C. (2008). The relationship between environmental variation and species abundance in shrimp community (Crustacea: Decapoda: Penaeoidea) in south-eastern Brazil. *Journal of the Biological Association of the UK, 88*, 119–123. https://doi.org/10.1017/S0025315408000313

Crespi, V., & Lovatelli, A. (2011). Aquaculture in desert and arid lands: Development constraints and opportunities. In *FAO Technical Workshop. 6–9 July 2010, Hermosillo, Mexico. FAO Fisheries and Aquaculture Proceedings*, (No. 20, p. 202). FAO.

CSB. (2020). *Annual Bulletin of Statistics of Fish Wealth.* Central Statistical Bureau of Kuwait.

CSB. (2024). *Annual Bulletin of Statistics of Fish Wealth.* Central Statistical Bureau of Kuwait.

Davison, C., Michie, C., Tachtatzis, C., Andonovic, I., Bowen, J., & Duthie, C. A. (2023). Feed conversion ratio (FCR) and performance group estimation based on predicted feed intake for the optimization of beef production. *Sensors, 23*, 4621. https://doi.org/10.3390/s23104621

De Oliveira e Silva, E. R., Seidman, C. E., Tian, J. J., Hudgins, L. C., Sacks, F. M., & Breslow, J. L. (1996). Effects of shrimp consumption on plasma lipoproteins. *The American Journal of Clinical Nutrition, 64*, 712–717.

Emanuell, S., Calazans, N., Soares, M., Soares, R., & Peixoto, S. (2010). Effect of salinity on survival, growth, food consumption and haemolymph osmolality of the pink shrimp *Farfantepenaeus subtilis* (Pérez-Farfante, 1967). *Aquaculture, 306*, 352–356.

Emerenciano, M., Ballester, E. L. C., Cavalli, R. O., & Wasielesky, W. (2012). Biofloc technology application as a food source in a limited water exchange nursery system for pink shrimp *Farfantepenaeus brasiliensis* (Latreille, 1817). *Aquaculture Research, 43*, 447–457.

Fakhri, S., Abbaszadeh, F., Dargahi, L., & Jorjani, M. (2018). A mechanistic review on its biological activities and health benefits. *Pharmacological Research, 136*, 1–20. https://doi.org/10.1016/j.phrs.2018.08.012.

FAO. (2022). *The state of world fisheries and aquaculture in 2024* (p. 7). Food and Agriculture Organization of the United Nations.

Fishelson, L., & Loya, Y. (1969). Experiments on rearing Tilapia-hybrids in brackish water pond near the Dead Sea. *Limnology, 17*, 602–610.

Galkanda-Arachchige, H. S. C., Wilson, A. E., & Davis, D. A. (2020). Success of fishmeal replacement through poultry by-product meal in aquaculture feed formulations: A meta-analysis. *Reviews in Aquaculture, 12*(3), 1624–1636.

Han, J. E., Tang, K. F. J., Pantoja, C. R., White, B. L., & Lighter, D. V. (2015). QPCR Assay for detecting and quantifying a Virulence Plasmid in Acute Hepatopancreatic Necrosis Disease (AHPND) due to Pathogenic *Vibrio parahaemolyticus. Aquaculture, 442*, 12–15.

Jang, I. K., & Gong, H. (2016). Growing marine shrimp in the Sahara Desert. *Aquaculture Magazine, 42*(2).

Jory, D. (2023). Annual farmed shrimp production survey: A slight decrease in production reduction in 2023 with hopes for renewed growth in 2024. Global Seafood Advocate, Portsmouth, NH 03801 USA.

Kim, S. K., Pang, Z., Seo, H., Cho, Y., Samocha, T., & Jang, I. K. (2014). Effect of bioflocs on growth and immune activity of Pacific white shrimp, *Litopenaeus vannamei* postlarvae. *Aquaculture Research, 45*, 362–371.

Kim, S. K., Guo, Q., & Jang, I. K. (2015). Effect of biofloc on the survival and growth of the postlarvae of three penaeids (*Litopenaeus vannamei*, *Fenneropenaeus chinensis*, and *Marsupenaeus japonicus*) and their biofloc feeding efficiencies, as related to the morphological structure of the third maxilliped. *Journal of Crustaceous Biology, 35*(1), 41–50.

Liu, Z., Liu, Q., Zhang, D., Wei, S., Sun, Q., Xia, Q., Shi, W., Ji, H., & Liu, S. (2021). Comparison of the proximate composition and nutritional profile of byproducts and edible parts of five species of shrimp. *Foods, 10*, 2603.

Liu, A., To, V., Dumas, A., Hernandez, J. M., & Santigosa, E. (2024). Vitamin nutrition in shrimp aquaculture: A review focusing on the last decade. *Aquaculture, 578*, Article 740004.

Mathews, C. P. (1986). Shrimp fisheries management project. Phase III Final report (MB-47), Kuwait Institute for Scientific Research, KISR Report No 2180, Kuwait.

Mc Graw, W. J., Davis, D. A., Teichert- Coddington, D., & Rouse, D. B. (2002). Acclimation of *Litopenaeus vannamei* Postlarvae to low salinity: Influence of age, salinity endpoint, and rate of salinity reduction. *Journal of the World Aquaculture Society, 33*, 78–84.

Miao, W., & Wang, W. (2020). Trends of aquaculture production and trade: Carp, Tilapia, and Shrimp. *Asian Fisheries Society, 33*(S1), 1–10.

Misak, R., & Hussain, W. (2022). Groundwater in Kuwait. In: The Geology of Kuwait (pp. 199–214). Springer International Publishing.

Mishra, J. K., Samocha, T. M., Patnaik, S., Speed, M., Gandy, R. L., & Ali, A. (2008). Performance of an intensive nursery system for the Pacific white shrimp, *Litopenaeus vannamei*, under limited discharge condition. *Aquaculture Engineering, 38*, 2–15.

NAQUA. (2024). National Aquaculture Group. https://www.naqua.com.sa

Niamnuy, N., Devahastin, S., Soponronnarit, S., & Raghavan, G. S. V. (2008). Kinetics of astaxanthin degradation and color changes of dried shrimp during storage. *Journal of Food Engineering, 87*, 591–600.

Price, A. R. G., Mathews, C. P., Ingle, R. W., & Al-Rasheed, K. (1993). Abundance of zooplankton and penaeid shrimp larvae in the western Gulf: Analysis of pre-war (1991) and post-war data. *Marine Pollution Bulletin, 27*, 273–278.

Pu, J., Bechtel, P. J., & Sathivel, S. (2010). Extraction of shrimp astaxanthin with flaxseed oil: Effects on lipid oxidation and astaxanthin degradation rates. *Biosystems Engineering, 107*(364), 371. https://doi.org/10.1016/j.biosystemseng.2010.10.001

Ray, A. J., Dillon, K. S., & Lotz, J. M. (2011). Water quality dynamics and shrimp (*Litopenacus vannamei*) production in intensive, mesohaline culture systems with two levels of biofloc management. *Aquaculture Engineering, 45*, 127–136.

Samocha, T. M., Lawrence, A. L., & Pooser, D. (1998). Growth and survival of juvenile *Penaeus vannamei* in low salinity water in a semi-closed recirculating system. *Israeli Journal of. Aquaculture—Bamidgeh, 50*(2), 55–59.

Samocha, T. M., Patnaik, S., Speed, M., Ali, A., Burger, J. M., Almeida, R. V., Ayub, Z., Harisanto, M., Horowitz, A., & Brock, D. L. (2007). Use of molasses as carbon source in limited discharge nursery and grow-out systems for *Litopenaeus vannamei. Aquacultural Engineering, 36*, 184–191.

Sumara, M., Vadher, K. H., & Husni, A. (2015). Effect of higher salinities on growth and survival of pacific white shrimp, *Litopenaeus vannamei* (Boone, 1931). *Journal of Applied and Natural Science, 7*(2), 885–888.

Van Wyk, P., Davis-Hodgkins, M., Laramore, R., Main, K. L., & Scarpa, J. (1999). Farming marine shrimp in recirculating fresh water systems. FDACS contract M520. Florida Department of Agriculture and Consumer Services, Tallahassee, Florida, USA.

World Meteorological Organization (WMO). (2024). WMO global annual to decadal climate update 2024–2028. https://library.wmo.int/records/item/68910-wmo-global-annual-to-decadal-climate-update. Last accessed date 27 Nov 2024.

World Population Review (2024). https://worldpopulationreview.com/countries/kuwait. Last accessed date 27 Nov 2024.

Yimin, Y., & Mohammad, H. M. A. (1999). An analysis of variation in catchability of green tiger prawn, *Penaeus semisulcatus*, in waters off Kuwait. *Fishery Bull US, 97*, 702–712.

Zheng, J., Cao, J., Mao, Y., Su, Y., & Wang, J. (2019). Effects of thermal stress on oxidative stress and antioxidant response, heat shock proteins expression profiles and histological changes in *Marsupenaeus japonicas. Ecological Indicators, 101*, 780–791. https://doi.org/10.1016/j.ecolind.2018.11.044

Part V
Agricultural Pests and Soil Microbial Population

Chapter 19
Soil Microbial Population Dynamics and Shifts in Arid Lands Due to Global Warming

Ali M. Quoreshi and Faizah Al-Enezi

Abstract Soil microorganisms make considerable contributions to soil properties through multiple actions to support plant growth for food security and are greatly affected by a changing climate. The microorganisms require optimum conditions to thrive and perform their functions including but not necessarily limited to nutrient cycling, decomposition of organic matter to compost and organic fertilizers etc. It is not clear how climate change (global warming) is going to impact the dynamics of microbes (risks of extinction or irreversible loss of microorganisms) and impact on arid land agriculture. In order to fully comprehend and enhance our knowledge of the community composition and functional attributes of soil microbial population, it is essential to record changes in their dynamics and functions under global warming scenarios. It is believed that without effective and equitable mitigation and adaptation actions, climate change is likely to increasingly threaten the soil microbiology. Scientists are concerned about the potential shift to soil microbial community structures and functions where they dwell and how they interact with one another due to long-term environmental changes. However, these conditions and microbial population dynamics may shift when the environment is facing the global warming condition. In this review, we explored the existing state of knowledge about the impacts of global warming on soil microbiomes, which already exist under extreme weather-sensitive arid soil environments associating with ecosystem services including crop production for food security. The chapter discusses and reviews knowledge in understanding how global warming may influence biogeochemical cycles by the soil microbiome. It also provides insights into the potential ways that soil microbiomes can be utilized to help mitigate the consequences of long-term global warming through their inputs to soil health and resilience. The book chapter offers valuable insights for researchers and students interested in further understanding in microbial population dynamic

A. M. Quoreshi (✉) · F. Al-Enezi
Food Security Program, Environment and Life Sciences Research Center, Kuwait Institute for Scientific Research, Safat, Kuwait
e-mail: aquoreshi@kisr.edu.kw

F. Al-Enezi
e-mail: fnenezi@kisr.edu.kw

M. K. Suleiman and S. A. Shahid (eds.), *Fostering Arid Lands Agriculture in the Face of Climate Change*, Advances in Global Change Research 81,
https://doi.org/10.1007/978-3-031-98490-7_19

research and focus their studies especially on the arid region soil microbiomes with respect to improving arid land agriculture productivity for food security.

Keywords Soil ecosystems · Soil microbiome · Microbial diversity · Community structure · Nutrient cycling

19.1 Introduction

Arid lands harbor a wide range of resilient microbial communities, which can remain free-living groups, mutual associations with plants, or assemblages within soil biocrusts (Coleine et al., 2024). These soil systems comprise a complex community of algae, cyanobacteria, lichens, and associated bacteria, fungi, microfungi, mosses, and archaea. Prevailing low precipitation, high temperatures, significant diurnal temperature variations, and sporadic rainfall patterns characterize arid lands. They are also considered to have limited vegetation and specific adaptations of flora and fauna. These microbial communities are crucial in organic matter decomposition, nutrient cycling, carbon storage, defending plants from abiotic and biotic pressures, soil health, and overall ecosystem functionality (Bardgett & Van Der Putten, 2014; Delgado-Baquerizo et al., 2016). The soil microbiome is accountable for essential ecosystem processes; nevertheless, it is inadequately known about its sensitivity and alternations to global warming (Quoreshi & Madouh, 2023). Climate change, mediated by anthropogenic activities, is influencing these microbial populations, leading to significant shifts in their dynamics, diversity, and community structure.

Arid and semiarid ecosystems encompass nearly 45% of the Earth's land surface and are found on every continent characterized by some of the most severe environments on the planet (Coleine et al., 2024). Over the past two centuries, the Earth's biosphere has undergone unprecedented changes. Climate change, largely driven by anthropogenic activities, increasing industrialization, and air pollution, is widely acknowledged as the primary factor influencing global alterations in rainfall patterns (IPCC, 2023). The arid regions have expanded due to global warming. They are primarily located in the Middle East, Africa, North America, Asia, Africa and Australia, where they depend profoundly on agricultural systems for food supply (Ma & Yang, 2017). However, these areas are confronted with significant abiotic challenges, including drought, salinity, and intense temperatures, which severely influenced plant productivity, growth (Shahid & Rahman, 2011; Yadav et al., 2020; Zaman et al., 2018). The arid and hyper-arid ecosystems are considered to have extreme evaporation rates, very elevated ultraviolet radiation, and extreme temperatures (Coleine, 2024). These stresses are further intensified by global warming and environmental degradation, alteration in carbon balance, and water scarcity caused by anthropogenic activities (Martinez-Valderrama et al., 2020), leading to widespread consequences on crop yields and soil nutrient richness. Global warming further complicates the sustainability of agricultural systems and ecosystem processes (Dubey et al., 2019). A recent synthesis report (IPCC, 2023) acknowledges that continued greenhouse gases

(GHGs) release will progress to escalating global warming, with the best approximation of attaining 1.5 °C soon based on numerous situations and modeled pathways believed. The predictable upsurge is up to 4.4 °C for an extremely high GHGs emissions situation. Climbing atmospheric CO_2 and warming of the ambiance is anticipated to alter environmental properties and processes, declined vegetation cover, and variations in existence and development rates (Fernández-Martínez et al., 2019; Maurer et al., 2020). The influence of climate warming on soil microbiomes is a serious issue that confronts ecosystem health, agricultural productivity, and climate stability. Since microbial communities respond to increasing temperatures, comprehending these shifts is vital for developing effective management practices that promote resilience and sustainability. This chapter explores the consequence of global warming on soil microbial populations in arid ecosystems, the consequent shifts in microbial diversity, and the implications for ecosystem resilience.

19.2 Arid Lands

Arid regions are characterized by infrequent and reduced precipitation, typically less than 250 mm per year, as well as extreme temperatures, significant diurnal temperature variations, low nutrient availability, limited vegetation, and specific adaptations of flora and fauna. These regions support unique soil microbiomes that have adapted to harsh conditions (Middleton & Thomas, 1997). Arid lands are recognized as substantially sensitive to global warming and climate variability and expected to alter the amount, incidence, and intensity of expected rainfall pulses, consequently, influencing soil moisture (McHugh et al., 2015). Nevertheless, a profound limitation of water availability is the key condition of arid lands. The aridity of an ecosystem can be designated by its aridity index (AI). AI is the numerical indicator that can be used to express a degree of dryness in a region's climate or soil's water availability. It is just opposite to the humidity index. There are few calculations of AIs, and it has been suggested that the best accepted utilized metrics described the variation between median yearly precipitation (P) and mean annual potential evapotranspiration (PET), in which evapotranspiration is the sum of evaporation and plant transpiration (UNEP, 1992). Recently, UNEP adopted this index and defined it as AI = P/PET (UNEP, 1992).

19.3 Climate and Climate Change Aspects

Climate change represents the utmost alterations in the increase in global GHGs, temperatures, and weather patterns over recent times. In general, climate refers to average weather patterns in a particular territory over an extended period and performs a vital role in shaping our planet's ecosystem. Though usual environmental variations in climate have always occurred throughout Earth's record. The recent challenges of climate change are essentially driven by human caused activities. The present catastrophe in climate change is mainly attributed to the increase in discharges of GHGs, (CO_2, CH_4, and N_2O) in the atmosphere due to human-induced actions, the burning of fossil fuels, massive deforestation, growing industrial processes that have enhanced, consequently, leading to the greenhouse effect (GHE), and resulted in an uprising global temperature, and heat events (IPCC, 2023). Recently, the climate change committee recognized that continuous GHGs discharge will progress to escalating global warming, with the best estimation of reaching 1.5° C in the future based on different modeling trials, and further predictable upsurge up to 4.4° C in case of a very high emissions scenario (IPCC, 2023). Continued global warming projections will further exacerbate the global water cycle, rainy season precipitation, and the harshness of wet and dry events. Scientists have recognized the harmful effects of climate change that have already triggered huge impacts and associated damage to ecosystems, human systems, and economic damages (AFED, 2009; IPCC, 2023).

Arid regions are anticipated to experience substantial changes under global warming situations, however, there is great inconsistency and uncertainty in the predictions of possible influences of the increasing CO_2 and warming of the globe in arid ecosystems (Lioubimtseva, 2004). The loss of soil carbon is expected to escalate in response to climate warming, however, microbiomes and microenvironments may render less intense of this effect (Liu et al., 2024). The inherent complexity of microbial communities existing in soil systems and other various manners they interact with make it extremely challenging to determine the various positive and negative responsiveness of soil communities to climate change. Nevertheless, global warming and/or climate change can have intense influences on soil biodiversity and related processes (Khursheed, 2016). Aridity is considered the foremost climatic driver of ecosystem composition and functioning of drylands (Berdugo et al., 2020). Studies have reported that aridity has intensified during the contemporary periods and is anticipated to continue to increase in the coming years due to elevated CO_2, amount, and variability of precipitation, combined with global warming (Chai et al., 2021). The accurate predictions of carbon cycling greatly depend on precise moisture response functions, nevertheless, there is surprising deficiency in agreement in the form these functions, suggesting more integrated modelling approach necessary for understanding microbial moisture responses and refining models.

The association between the soil microbiome and climate change has become a focus of interest because of its possible implications for storage of carbon across the globe caused by rising temperatures, changing precipitation patterns, and increased atmospheric CO_2 levels. These drivers play a role in influencing the performance and structure of microbial communities in the soil and can impact how much carbon is accumulated in the soil, thus impacting the overall carbon balance.

19.4 Soil Microbiomes and Biodiversity

Soils are highly dynamic, diverse, and extremely complex. It is known that 1.0 g of soil comprises up to 1 billion bacterial populations containing tens of thousands of taxa, about 20 million fungal hyphae, and a wide variety of mites, earthworms, arthropods, and nematodes (Bardgett & Van Der Putten, 2014). The soil microbiome is a major determinant of ecosystem composition and function as it mediates key biogeochemical processes, e.g., nutrient cycling, decomposition of organic matter, and plant-soil interactions contributing to vegetation growth and development moreover the maintenance of healthy ecosystems (Anthony et al., 2023; Hartman and Six, 2023). The soil is a reservoir for varied organisms that contains bacteria, fungi, cyanobacteria, algae, archaea, and other microbes. These microbes are crucial for soil ecosystem processes and essential for maintaining biodiversity and the structure of natural soils. The accessibility of nutrients such as carbon (C), nitrogen (N) and phosphorus (P) are necessary for all the living organisms existing on Earth, which are primarily regulated by microorganisms. Despite the immense significance of soil system biodiversity, adequate knowledge is deficient in the quantitative assessment of soil biodiversity (Anthony et al., 2023). These challenges make scientists stipulate the importance of safeguarding, managing, and restoring soil life and how soil biodiversity might influence ecosystem sustainability (Anthony et al., 2023). Our knowledge of the function that soil organisms perform, their distribution, and biogeography is quickly developing to some extent because of the increasing acknowledgement of its substantial role in ecosystem processes and recent concern about global climate change.

Soil microbiomes regulate global biogeochemical fluxes and precisely contribute to rates of climate change and human well-being (Banerjee and van der Heijden, 2023). Climate change patterns are recognized to influence both micro and macroorganisms and are believed to be a main global issue affecting life on the planet (Sergaki et al., 2018). Global average temperature is increasing and predicted to continue to enhance and alter precipitation patterns, with even greater increases in arid regions. Consequently, leading to prolonged drought and erratic rainfall events, which can be detrimental for microbial communities, which can shift microbial population structure and function. It is described that alterations in structure, richness, constitution, and function of plant-related microbial taxa due to altered climate change (Dubey et al., 2019) as well as influence both direct and indirect effects on plant-soil-microbe relations and microbial communities (Bojko & Kabala, 2017).

Global warming uses abiotic stress and instigate modifications in soil microbial community diversity, composition, and their processes (Shade et al., 2012). The shifting of microbial communities may occur in support of the species which are well adapted to increased temperature and higher growth rates. The alterations in soil biodiversity and community composition may also affect processes associated with nutrient cycling. It is commonly agreed that steady alterations in climatic properties affect the performance and relative richness of the microbial populations existing in the soil, consequently, modifying the distribution and behavior of species, and greatly impacting vegetation and crop production (Dubey et al., 2019). Many findings have specified that elevated CO_2 conditions may have positive effects on plant growth-promoting fungi (PGPF) such as arbuscular mycorrhizal fungi, ectomycorrhizal fungi, and other endophytic fungi and bacteria (Das & Verma, 2009). Other studies reported that plant growth stimulating microorganisms both bacteria and fungi induce microbial community composition changes influenced by warming and drought (Compant et al., 2010).

19.5 Extinction or Irreversible Loss Due to Global Warming and Safeguarding

Soil ecosystem harbors a massive quantity of biodiversity. However, our ecological knowledge of soil microbiome is inadequate as it initially believed (Maestre et al., 2012). Although soil microbiota plays an essential function in sustaining the performance of ecosystems, most of the extinction research emphasis is on aboveground organisms (Veresoglou et al., 2015). The reason for alarmingly sparse extinction study reports related to soil organisms partially because of poorly noticeable organisms and partially soil biota being dominated by microorganisms as well as due to the unique features of the soil habitat and its heterogeneity (Veresoglou et al., 2015). Alterations in soil properties due to global warming are anticipated to influence biodiversity. Human-induced and climate change-related drivers are recognized to alter soil physiochemical properties, which in turn influence soil biota (Kovacs and Kovacs, 2023). The microbial communities and their functions are complex and alter in both temperature and moisture availability.

Physiological adaptations of single species and shifts within the community could be the mechanisms of the temperature adaptations. Consequently, soil biological properties could change and may result in exclusion or introduction of organisms that could influence this soil functioning, and loss or alterations of ecosystem services (Kovacs and Kovacs, 2023). The soil microorganisms are considered the main component of soil biodiversity owing to their contribution in numerous processes and interactions in terrestrial ecosystems. Such components of soil biota directly arbitrate and effect the constancy and cycling of related elements of climate change (Kovacs and Kovacs, 2023). Since climate change modifies global temperatures, precipitation, soil moisture, and land use patterns, the stability of the microbiome is threatening, directing to possible declines in soil health and nutritional status.

However, microbiota components are native to the environment and expected to adapt to changed environmental conditions and can exist under different adverse conditions. Even though microbiotas are important and believed in regulating climate change, microbes are infrequently considered for climate change studies and not reflected in climate change policy development (Cavicchioli et al., 2019). The recent Consensus Statement represents a cautionary to humanity to raise awareness of the microbial world and demand more engagement for microbiological research and integration for addressing climate change impacts as climate change will depend greatly on response of microorganisms (Cavicchioli et al., 2019). This awareness is indispensable for obtaining an environmentally sustainable future.

Several strategies can be considered for safeguarding soil microbiome to mitigate global warming concern. Practicing sustainable land management strategies in agriculture, such as no-till farming, decreasing soil disturbance (conservation agriculture), use of mulching, regular crop rotations can help increase microbial diversity and consequently enhance soil health (Kumar et al., 2016). Other strategies could be managing or preserving microbial environment by reducing soil disturbance and strategic land use can help sequester further soil carbon storage, managing salinity, increasing more organic matter in soils, growing more mycorrhizal host plants for preserving mycorrhizal fungi may provide a stable ecosystem for effective microbial communities. Ecological monitoring of microbiological communities and soil health should be practiced comprehending changes and adopt suitable management practices, such as restoration and revegetation of degraded lands. These approaches can assist restore microbial communities and ecosystem functions. Safeguarding of soil microbiota against global warming is crucial for conserving soil health, ecosystem processes and services for food security under climate change scenarios.

19.6 Nutrient Cycling, Nitrification and Carbon Cycle

Microorganisms are crucial components for most of the ecosystem's functions and services. The soil microbiome performs an indispensable role in the biogeochemical cycling of soil systems (Naylor et al., 2020; Quoreshi et al., 2019). The major nutrients that act a fundamental role in functioning are carbon, nitrogen, and phosphorus, and cycled across the ecosystem by several processes, such as decomposition, mineralization, immobilization and nutrient uptake by plants. Microorganisms are accountable for the decomposition of organic matter, which regulates the fundamental ecological processes, the maintenance of soil structure and sustainability that help in the sustaining of life on Earth. The nutrient cycling in the soil responds in a way where energy and matter are relocated between living and non-living fractions of the environment.

The nitrogen (N) cycle encompasses soil, plant and animal pools that hold comparatively small amounts of biologically active N, in contrast to the huge pools of fairly inert N in the lithosphere and atmosphere. Nitrogen is an important constituent of living organisms, and main limiting nutrient throughout terrestrial ecosystems. Currently, we are confronting pressures from climate changes, involving global warming, atmospheric nitrogen accumulation, agricultural fertilization, which have now changed soil biodiversity and nutrient cycling. Our perception of N-cycling mediated by microorganisms has considerably progressed over the past decades due to new discovery of new microorganisms involved in nutrient cycling (Wooliver et al., 2019). These new microorganisms have opened the new approach and have essentially changed our knowledge of global N cycling and amplified the complication of fully examining N-cycling microorganisms (Li et al., 2023; Lin et al., 2023). There is no doubt that our understanding of terrestrial N cycling is altering because new developments are being discovered related to the resources, and losses of N from ecosystems (Wooliver et al., 2019). However, the recent progress in our perception of N-cycling microorganisms, their relationships and feedback related to global change situations are yet predominantly unknown. Recent research (Gong et al., 2024) demonstrated that increased nitrogen emission from anthropogenic fossil fuel burning and agricultural fertilizers have net cooling effect on the climate. However, researchers cautioned that elevated atmospheric nitrogen may not be an explanation for combating the overall climate change scenario. Because nitrogen fertilizers pollute water bodies and fossil fuels burning produces nitrogen oxide that pollute air. Therefore, an increased rate of atmospheric nitrogen to tackle climate change is not a suitable approach to consider (Gong et al., 2024). A recent comprehensive global synthesis study of warming experiment demonstrated that increasing warming under experimental conditions mostly encourages carbon cycling, comprising both productivity and respiration (Cui et al., 2024).

Increased global warming mediated mainly by anthropogenic activities is changing ecosystems in various ways. One of the most unseen shifts is occurring, which involve microbes that dwell in terrestrial and aquatic environments. These microbial communities are crucial for various ecosystem services, such as nutrient cycling, ecosystem stability, and overall health. The soil microbiomes are the basis for ecosystem, agricultural productivity, and health, govern biogeochemical cycles and sustain plant growth and development (Naylor et al., 2020). Global warming as well as increased atmospheric CO_2 levels are a great concern that influences and have both direct and indirect effects on soil microbiomes and their stability (Cavicchioli et al., 2019). However, a recent meta-analysis of the influence of global climate change factors (GCFs) related to microbial diversity and functionality revealed that the anticipated effect on microbial diversity is uncertain and found that climate change may not always be the reason for microbial diversity loss and have small negative effect compared to aboveground communities (Hooper et al., 2012; Zhou et al., 2020).

Recent research emphasizes the soil microbe's crucial function in regulating global carbon and nitrogen cycles, which are essential to the global climate system. Soil microorganisms are responsible for decomposition of organic matters, a process vital to carbon cycling. Microorganisms stimulate the decomposition of organic materials, such as dead plant and animal materials, releasing nutrients back into the soil and stabilizing carbon within the soil organic material (SOM). This fundamental relationship between soil microbes and carbon flow is now seen as a vital process in attempts to entrap carbon and slow the rise of CO_2 in the atmosphere, an essential driver of climate change.

Mild increase in temperatures have the potential to encourage biological reactions rates and enhance the plant growing season, consequently increasing plant productivity (Zhang et al., 2022). On the other hand, higher heat situations can enhance plant respiration rates, loss of water and resulting in reduced plant productivity (Quan et al., 2019). Fluctuations of temperatures may affect soil denitrification processes and possible increase in nitrogen losses into the environment with increase in temperatures (Cui et al., 2024), but responses of various variables of biogeochemical cycles to warming can be different and shows spatial and temporal diversity (Reich et al., 2015). A study of warming experiments demonstrated that increasing warming under experimental conditions mostly encourages carbon cycling, comprising both productivity and respiration (Cui et al., 2024). Despite all this evidence, there is a deficiency in inclusive research exploring the consequence of global warming on the complete approach of the nitrogen cycle and interactions between carbon and nitrogen within forest ecosystems.

19.7 Biological Nitrogen Fixation

Biological nitrogen fixation (BNF) is a constituent of nitrogen cycle and activity performed out by a specific group of prokaryotes that transform atmospheric dinitrogen (N_2) into ammonia (NH_3) (Einsle & Rees, 2020). The enzyme complex called nitrogenase is utilized by these microorganisms to convert the conversion process into a usable nitrogenous compound that can be utilized through biological pathways. Dinitrogen is an abundant gas molecule and found about 78% in the atmosphere on earth. Although abundant nitrogen is present in the atmosphere, nitrogen is a crucial limiting element for plant growth and production. Nitrogen is a main component of chlorophyll required for photosynthesis, amino acids, and essential for other important biomolecules. Atmospheric nitrogen fixation to ammonia by prokaryotes is fundamental process for plant productivity and about 60% of fixed nitrogen on globe consequences from biological nitrogen fixation (Mart'inez-Hidalgo & Hirsch, 2017). Plants can freely uptake NH_3 to produce the above mentioned nitrogenous essential biomolecules. The ability of nitrogen fixation is limited to particular bacteria and few actinomycetes and referred to as diazotrophs. These microorganisms include free-living soil bacteria, cyanobacteria, most importantly symbiotic bacteria, such as *Rhizobium* spp., and *Bradyrhizobium* spp. (Xu et al., 2018).

Some free-living heterotrophic bacteria inhabit the soil and are able to fix nitrogen besides the direct involvement with other organisms. Free-living bacteria can form associative and/or endophytic relationships with plants and reduced nitrogen and make available form of nitrogen to plants. These bacteria include *Azotobacter*, *Bacillus*, *Clostridium*, and *Klebsiella*. Another group of bacteria performed associative nitrogen fixation by the species *Azospirillum;* and was capable of forming intimate associations with different members of the *Poaceae* family, including important cereal crops. These bacteria can reduce nitrogen within the rhizosphere of the host plants. Most important nitrogen fixation is symbiotic nitrogen fixation mediated by many microorganisms within the legumes host plants. The host plant provides carbohydrates that are used by the nitrogen fixing bacteria (*Rhizobium* and *Bradyrhizobium*) for the energy required for the nitrogen fixation for the host for growth and development. Another example of symbiotic nitrogen fixation through symbiotic association between actinorhizal trees and shrubs and *Frankia*. The BNF process is crucial for enriching soil nitrogen situations, particularly nitrogen-poor arid environments. The process is not only improving the soil conditions, but it becomes essential for plant growth and resilience in an arid environment.

The levels of BNF may differ strongly with fluctuations of temperatures, causing it susceptible to climate change, but inclusive prediction across the terrestrial and oceanic is deficient (Deutsch et al., 2024). The influence of climate change on BNF is arbitrated by a broad array of environmental and ecological conditions, and all are influenced by temperature (Deutsch et al., 2024). Recent research by Deutsch et al. (2024), projected global BNF under global warming across terrestrial and oceanic ecosystems. In this publication, authors concluded that the rate of nitrogen fixation intensely depends on temperature irrespective of the habitation. The projected increase in BNF at higher latitudes is larger on land than in the ocean. Global warming will probably have a considerable positive influence on nitrogen fixation in the higher latitudes, particularly on lands (Peng et al., 2020). In a warmer climate, a shift of thermal niche of N_2 fixation to higher latitudes has been projected. However, the mechanism affects the N cycling and inventories of N are still unclear, as it is mainly based on the countering reactions, or denitrification (Deutsch et al., 2024). Current global datasets emphasize new research is required to elucidate how N_2 fixation may swing in the future in relation to changing climate.

19.8 Current Status and Knowledge Gaps on Arid Lands Soil Microbial Communities

Forty percent of the earth's terrestrial surface is composed of arid lands. The arid regions have extreme characteristics such as significant temperature fluctuations, intense solar irradiation, high evaporation rates, and low precipitation. These stress environments significantly influence the roles of soil microbial communities, which are essential for nutrient cycling, soil structure maintenance, and overall ecosystem productivity. The microorganisms' communities in the soil of arid regions consist of fungi, bacteria, archaea, and other microorganisms that have adapted to these extreme conditions. Advanced techniques such as whole genome sequencing have revealed unique microbial populations in arid environments, including cyanobacteria, actinobacteria, and extremophiles such as halophiles and acidophiles (Peng et al., 2024).

The microbial communities in both semiarid and arid lands have an important effect on organic material degradation, nutrient turnover process, and soil stability. In addition, they interfere with the composition and structure of the biological soil crusts by preventing soil erosion, enhancing seed germination, and improving water infiltration. Extreme environments, like those affected by global warming, enhance various microorganisms to employ several strategies to tolerate both immediate and prolonged stress that are triggered by climate change. Climate change affects both the structure and the diversity of soil microbial communities (microbiome) in both direct and indirect ways (Mariappan et al., 2022). Therefore, the research gaps for improving the relationship between the soil microbiome and the plant in arid areas are crucial, and therefore future studies should focus on studying the effect of climate change on soil microbial communities at both small and large scales is crucial. Even with the major progress, the arid soil microbiome remains not well-studied, and the ecosystem services provided by these microbial communities, such as biogeochemical cycling, are critical in regions with limited macrofaunal and plant biodiversity (Neilson et al., 2017).

19.9 Direct Impacts of Global Warming on the Soil Microbiome

The soil microbiome composition and activities have been directly influenced by climate change. Many research studies have described the changes in the temperature effect on microbial communities in different areas for example, distribution, phenology, and geographical range. The influence on soil microbiome can be categorized into many factors. For Example, fluctuation in the temperature changes the soil microbiome function and structure by allowing specific groups, such as thermophilic bacteria, to overgrow over others. In addition, this can be occupied with enhancing microbial activity directly. High-temperature leads to an increase in microbial respiration, which causes the soil to lose carbon (Bond-Lamberty & Thomson, 2010). These temperature-elevations cause a disruption in soil carbon mechanisms and nutrient cycling processes that end up affecting microbial communities.

In addition, both the soil pH and structure are affected by temperature changes that cause considerable alterations in microbial communities' activities and composition. Many studies reported that there is a correlation between soil microbiome diversity and environmental temperature. As temperatures increase, the microbial growth rate increases, and species numbers increase. On the other hand, decreasing temperatures can cause a reduction in both bacterial and fungal diversity by extending population doubling times, increasing mutation rates, and altering relations, such as competition, parasitism, and predation (Zhou et al., 2016).

In semi-arid and arid lands, changes in both moisture and temperature lead to a significant effect on soil microbiomes. In addition, there is a direct effect on gene copy numbers, microbial productivity, and respiration (Armstrong et al., 2016), as well as bacterial community composition (Stovicek et al., 2017). Moreover, increased atmospheric CO_2 levels have been shown to stimulate rhizosphere-colonizing bacteria and plant-growth-promoting fungi, while non-rhizospheric species remain unaffected (Mekala & Polepongu, 2019).

19.10 Indirect Impacts of Global Warming on the Soil Microbiome

The soil microbiome in arid lands has also been affected indirectly by global warming. There are indirect effects of climate change on soil microbiomes specifically in arid lands. High temperatures can impact moisture levels and nutrient cycling, which also disrupt both microbial structure and function. Nevertheless, the relations between plant and soil microbiomes is not clear due to the differing rates at which soil microbiomes and plants respond to increased temperature (Van Der Putten et al., 2013). Increased temperatures can impact soil microbial diversity by reducing soil moisture and altering the carbon-to-nitrogen ratio, which affects nutrient availability. The changes in soil pH will affect the fungal communities in an indirect way (Liu et al., 2024). The drought stress caused by prolonged high temperatures and low precipitation, can lead to soil organic matter decomposition and plant growth reduction (Seleiman et al., 2021). Also, Wang et al. (2014) show that climate changes cause a reduction in agricultural productivity and affect plant photosynthesis and soil microbiomes. Additionally, elevated CO_2 levels and soil warming, or drought conditions can negatively affect arbuscular mycorrhizal fungi (AMF) which are essential for nutrient uptake and carbohydrate exchange between plants and soil.

19.11 Biological Responses to Increased Temperature

Agricultural and marine ecosystems are affected by climate change. For example, in marine ecosystems increases in temperatures interfere with a range of issues like; the proliferation of waterborne diseases, challenges in food security, increased outbreaks of food poisoning and declines in marine biodiversity. While in the agricultural ecosystem, a rise in temperature affects the agricultural ecosystem by elevating the level of crop diseases or reduction in soil biodiversity. In agricultural ecosystems the increase in temperatures causes crop diseases or enhanced CO_2 sequestration and a reduction in soil biodiversity (Riebesell et al., 2018). Furthermore, climate change can influence humans, where climate change impacts the microorganisms that cause more plant and animal pathogens to appear. A key factor in this is the microorganisms' ability to acclimatize and mutate to the fluctuations in the climate that cause remarkable effects on animal health (Cavicchioli et al., 2019).

In response to the high temperature bacteria show adaptive actions such as transitioning from planktonic states to biofilm formation. This change is enabled by reduced oxygen solubility at elevated temperatures that promotes biofilm formation. Biofilms play a vital role in enhancing the bacterial capacity to survive and persist underneath heat stress. Furthermore, these microorganisms go through mutations that allow them to survive under elevated temperatures and microbial resistance (Rodríguez-Verdugo et al., 2020). Numerous pathogens in hot conditions have been identified by many researchers. To date, 25 viruses, 33 bacteria, 21 metazoans, and

23 protists have been reported as major causes of marine diseases affecting corals, turtles, plants, and mammals.

Despite all these research studies the marine microbiome variety remains mainly unexplored (Abirami et al., 2021). These pathogens can develop a thermotolerance mechanism to survive through climate change. This thermotolerance mechanism contains several molecular and cellular mechanisms including the activation of heat shock response genes that encode heat shock proteins (HSPs). Metabolic adaptations, such as Deoxyribonucleic acid (DNA) repair mechanisms, are employed to maximize energy production at high temperatures. Additionally, gene regulatory networks and signalling pathways are modulated to express heat-responsive genes (Richter et al., 2010).

19.12 Impact of Global Warming on Arid Lands Agricultural Farms

Global warming is considered a major threat to the human society in various aspects, and has greatly affected human life, agricultural production is particularly vulnerable to climate change that has transformed environmental elements such as temperature, precipitation, and affected crop growth periods, erratic extreme weather events, the patterns of pests and diseases, ultimately affecting crop yield and quality (Yuan et al., 2024). The distribution of pathogens around the world caused by global warming has a direct and indirect effect on human health by influencing the food sector including agriculture, animals, and fishing (Ibáñez et al., 2023). The agriculture sector is a critical sector to ensure food security. The plants' growth rates, sensitivity to the temperature and physiological mechanisms have been disrupted by the changes in soil microbiome changes. These disruptions threaten the stability and productivity of agricultural systems.

In arid regions, rising temperatures exacerbate water resource limitations, placing significant pressure on efforts to achieve self-sufficiency and ensure food security (Mostafa et al., 2021). Tian et al. (2023) show that the water stress in both arid and semi-arid lands caused the increases in water evaporation due to global warming. Also, warmer temperatures host several pests and diseases further threatening agricultural productivity. High temperatures caused by global warming end up with a severe drought in arid and semi-arid regions like Algeria and Tunisia (Schilling et al., 2020). The increase in temperature often correlates with decreased precipitation which contributes to desertification through soil erosion and salinization (Shahid, 2023; Shahid et al., 1999). Climate change has also triggered soil degradation and shortage of land resources, thus affecting the sustainable advancement of agricultural production (Eekhout & De Vente, 2022). These processes result in the loss of essential nutrients, water and carbon from the soil. There is a significant reduction in crop production due to the soil erosion and the land quality degradation has been appearing in several regions (Peláez et al., 1987).

19.13 Soil Microbiota

The soil microbiome plays an important role in controlling biogeochemical cycles of micronutrients, macronutrients and other essential elements necessary for the growth of plants and animals. Soil microbiotas are engaged in almost every biochemical reaction that happens in soil systems (Rakshit et al., 2020). A significant consequence of increased temperatures is the prevalence of drought, with desertification expected to become more common in semi-arid and arid regions in the future (Huang et al., 2016). There is a well-documented inverse relationship between drought and microbial functions; as drought intensifies, microbial functions crucial for soil sustainability tend to decline (McHugh et al. 2015).

In arid lands that experience extreme variations from hot to cold deserts and limited rainfall, understanding microbial responses to increasing drought conditions poses a challenge. The desert soil environment is frequently characterized by limited availability of carbon and moisture. Pointing & Belnap (2012) reported that this limitation in carbon and moisture results in the dominance of surface-dwelling photoautotrophs which play a key role in performing photosynthesis under extreme conditions. This leads to the formation of soil crusts, or biocrusts which are major producers of nutrients (Garcia-Pichel et al., 2003). A pressing concern is whether these biocrusts will be able to tolerate or adapt to even harsher and more extreme conditions in the future (Steven et al., 2015).

Different strategies have been used by soil microbiomes to adapt to drought conditions, including the synthesis of extracellular enzymes, osmoregulation and entering a dormant state. Some microorganisms can tolerate lower water availability by transitioning into a dehydrated state and recovering when moisture becomes available again (Meisner et al., 2017). Additionally, certain microorganisms produce extracellular polymeric substances to retain water, while bacterial taxa such as bacilli and cyanobacteria can persist in drought impacted soils by becoming dormant under dry conditions (Naylor et al., 2017).

19.14 Plant-Soil-Microbes' Interactions

A wide range of microorganisms can be found in plant microbiota such as bacteria, fungi, and archaea. These microorganisms interact with the plant and influence their growth and health (Chialva et al., 2022). In addition, their ability to engage with the plants through different relationships like commensalism and parasitism mechanisms can substantially affect both plant's development and resilience (Hardoim et al., 2015). The salinity and drought stressors can affect the composition of the plant microbiome in parallel to soil, water, and nutrients as the primary source for the plant microbiome (Wang et al., 2020; Yadav et al., 2020). Both Genomic and metabolic techniques prove the important role of plant microbiota in plant health. The plant microbiomes, especially the plant growth promoting bacteria enhance plant resilience through different mechanisms such as nitrogen fixation phytohormone production and mineral solubilization. These mechanisms are vital for improving stress tolerance and overall plant performance (Orozco-Mosqueda et al., 2021).

Genomic techniques such as metagenomics, metaproteomics, and metatranscriptomics play an important role in understanding both the functional and diversity of rhizosphere microbiome for sustainable agriculture. For example, molecular methods and automation of Sanger sequencing were used to identify the microorganisms in the rhizosphere to be able to culture and their potential use in improving sustainable agriculture (Priya et al., 2021). To fully comprehend how functional traits of microbes fluctuate in their response to climate warming across masses, complete metagenomics analysis on the soil aggregates from contrasting soil conditions required. Recent research findings indicate that microbes have the inherent ability to utilize diverse approaches to adapt to warming and heat stresses through altering in the functional gene abundance and structures within variable microenvironments, prejudiced by soil's physical protection (Liu et al., 2024).

Plant soil microbes' interactions are crucial for the ecosystem functions with biophysical and biogeochemical cycles that influence plant, soil, and microbial structures. These interactions have been impacted by global warming by influencing the dynamics between plants, soil, and microbes. For example, climate change may drive plant species to migrate to more favourable environments and adapt through altered gene expression and earlier flowering in response to warming, drought and disease pressures (Cavicchioli et al., 2019; Nolan et al., 2019).

Evidence suggests that climate change affects both plant and soil microbiomes due to their interconnected nature; for example, the root-associated microbes support the plant in flowering time, surviving and other functional traits under climate changes (Wagner et al., 2014). Any increase in the temperature can lead to an increased plant species variety that will affect the nutrient availability and modifying plant–microbe interactions, particularly influencing populations of growth-promoting rhizobacteria (PGPR) that depends on rhizodeposition (Sharma et al., 2022). Additionally, changes in precipitation patterns and drought stress can negatively impact the development of mycorrhizal mycelium in plant roots (Dong et al., 2023). Numerous studies demonstrated the useful interactions of arbuscular mycorrhizal fungi (AMF) in enhancing plant growth and development under adverse conditions. It is well known that AMF facilitates in plant nutrient and water uptake, decreases soil erosion, and improves plant stress tolerance. Plants successfully colonized with AFM can survive and handle well with ecological extremes.

19.15 Ecological Resilience and Resistance in Response to Global Warming

To manage the effect of global warming on ecosystems both understanding, and enhancement of ecological resilience is crucial. Ecological resilience refers to the capacity of a natural system to maintain its original functions, structure, and biodiversity despite disturbances, either through resistance or adaptation (Ma et al., 2021). Van Nes & Scheffer (2007) reported that it is essential to quantify the ecological resilience of natural systems for forecasting critical transitions and gaining deeper insights into the fundamental mechanisms underlying these systems. In the context of global warming, ecological resilience encompasses a combination of resistance to progressively harsh disturbances, the ability to recover and reorganize and the capacity to adapt to new conditions. Three primary ecological attributes contribute to resilience: diversity, connectivity, and adaptive capacity. Ecosystem-based management that emphasizes resilience acknowledges the crucial role of these attributes in maintaining stability across different spatial and temporal scales (Leslie, 2005).

Assessing resilience presents challenges because the factors influencing it vary across different scales of time and space. Coral reefs and mangroves (Alongi, 2008; Woodroffe, 2007) exhibit resilience because of their persistence and stability in the geological record. Therefore, this resilience is attributed to their ability to sustain desirable ecosystem conditions over extended periods demonstrating the importance of understanding and managing ecological resilience in response to global environmental changes.

19.16 Conclusions and Recommendations

Comprehending the complex interactions between soil microbiomes and alterations of climatic variables is crucial for predicting and mitigating the effects of global warming on arid ecosystems. The sensitivity of soil microorganisms to these stresses highlights the necessity of progressing further research and expanding adaptive management strategies. The management practices should target to safeguard microbial diversity, encourage ecosystem resilience, and make sure the sustainable utilization of arid lands, which are essential for global food security and climate regulations.

In arid lands, microbes play complex functions in nutrient cycling, organic matter decomposition, and biological nitrogen fixation. These fundamental activities are crucial for maintaining soil health, sustaining plant growth, and maintaining ecosystem resilience in relation to environmental stresses. As a result, such variations in the climate for example temperature and moisture will interrupt the soil microbiomes and consequently the ecosystem balance and crop production may be affected. A latest meta-analysis revealed that the anticipated effects on microbial diversity and functionality under global climate change factors (GCFs) may be uncertain and observed that climate change may not always be the reason for microbial diversity loss and have small negative effects. The rate of biological nitrogen fixation may alter significantly with fluctuations of temperatures, making it susceptible to global warming. However, further in-depth research is required in order to project shifts in nitrogen and carbon cycles in the terrestrial ecosystems.

To confront these issues several steps can be recommended. Initially, it is important to improve and monitor research studies in arid regions. The comprehensive investigations should place more effort on the microbial variety and their functions under diverse environmental conditions like temperature and salinity. Therefore, focusing on long term findings will help to understand how microbial communities acclimatize to different environmental changes and this kind of study or research will provide valuable insights into microbial resilience and ecosystem health. Next, it is necessary to apply a resilience strategy. Moreover, sustainable land management practices should prefer the protection and restoration of biological soil crusts, increasing organic natter in soils, reducing tillage, adopting agroforestry approaches, and aim to enhance microbial variety to sustain ecosystem functions, be able to build resilient microbial communities. Integrating climate adaptation measures like enhancing watering methods and soil moisture retention can mitigate water scarcity in agriculture fields. Also, selecting drought and heat resistant crop varieties can assist lessen the impacts of climate change.

Additionally, promoting integrative research is essential for filling gaps between soil science, microbiology, climate science, agronomy and other allied disciplines, this integration will enrich our perception of the interactions between microbial diversity, nutrient cycling, soil health, and climate change. Adopting these recommendations, we can enhance our understanding and management of climate change's effect on soil microbiomes, thereby ensuring the continued health and functionality of arid land ecosystems and withstand the challenges by climate change. Finally, integrating soil microbiological research into greater environmental policies is vital, and integrating microbiome research results into climate change strategies and policymaking will be essential in addressing the challenges of climate change. These efforts can help rebuild degraded soil systems and stabilize microbial functions in response to warming stresses.

Acknowledgements We wish to thank Ms. M.T. Sivadasan, Research Associate, Food Security Program, ELSRC-KISR for her help in formatting, editing of reference section and the manuscript.

References

Abirami, B., Radhakrishnan, M., Kumaran, S., & Wilson, A. (2021). Impacts of global warming on marine microbial communities. *Science of the Total Environment, 791*, Article 147905.

AFED. (2009). Arab environment: Climate change. Impact of Climate Change on Arab Countries. In M. K. Tolba, & N. W. Saab (Eds.), *Report of the arab forum for environment and development*.

Alongi, D. M. (2008). Mangrove forests: Resilience, protection from tsunamis, and responses to global climate change. *Estuarine, Coastal and Shelf Science, 76*(1), 1–13.

Anthony, M. A., Bender, S. F., & Heijden, M. G. A., Van der. (2023). Enumerating soil biodiversity. *Ecology, 120*(33).

Armstrong, A., Valverde, A., Ramond, J. B., Makhalanyane, T. P., Jansson, J. K., & Hopkins, D. W. (2016). Temporal dynamics of hot desert microbial communities reveal structural and functional responses to water input. *Scientific Reports, 6*, 1–8.

Banerjee, S., & van der Heijden, M. G. A. (2023). Soil microbes and one health. *Nature Reviews Microbiology, 21*, 6–10.

Bardgett, R. D., & Van der Putten, W. H. (2014). Belowground biodiversity and ecosystem functioning. *Nature, 515*(7528), 505–511.

Berdugo, M., Delgado, B. M., Soliveres, S., & Clemente, R. H. (2020). Global ecosystem thresholds driven by aridity. *Science, 367*(6479), 787–790.

Bojko, O., & Kabala, C. (2017). Organic carbon pools in mountain soils-sources of variability and predicted changes in relation to climate and land use changes. *CATENA, 149*, 209–220. https://doi.org/10.1016/j.catena

Bond-Lamberty, B., & Thomson, A. (2010). Temperature-associated increases in the global soil respiration record. *Nature, 464*(7288), 579–582.

Cavicchioli, R., Ripple, W. J., Timmis, K. N., Azam, F., Bakken, L. R., Baylis, M., Behrenfeld, M. J., Boetius, A., Boyd, P. W., Classen, A. T., & Crowther, T. W. (2019). Scientists' warning to humanity: Microorganisms and climate change. *Nature Reviews Microbiology, 17*(9), 569–586.

Chai, K. C., Li, Q., Bao, X. L., Zhu, J., & He, X. X. (2021). An empirical study of economic cycle, air quality, and national health since reform and opening up. *Frontier. Public Health, 9*, Article 706955. https://doi.org/10.3389/fpubh.2021.706955

Chialva, M., Lanfranco, L., & Bonfante, P. (2022). The plant microbiota: Composition, functions, and engineering. *Current Opinion in Biotechnology, 73*, 135–142.

Coleine, C., Delgado-Baquerizo, M., DiRuggiero, J., Guirado, E., Harfouche, A. L., Perez-Fernandez, C., Singh Br, K., Selbmann, L., & Egidi, E. (2024). Dryland microbiomes reveal community adaptations to desertification and climate change. *The ISME Journal, 18*(1), wrae056. https://doi.org/10.1093/ismejo/wrae056.

Compant, S., Clemant, C., & Sessitch, A. (2010). Plant growth promoting bacteria in the rhizo and endosphere of plants: Their role, colonization, mechanism involved and prospects for utilization. *Soil Biology and Biochemistry*. https://doi.org/10.1016/j.soilbio.2009.11.024

Cui, J., Deng, O., Zheng, M., Zhang, X., Bian, Z., Pan, N., Tian, H., Xu, J., & Gu, B. (2024). Warming exacerbates global inequality in forest carbon and nitrogen cycles. *Nature Communications, 15*, 9185.

Das, A., & Verma, A. (2009). The art of living. In *Symbiotic Fungi*, pp. 1–28.

Delgado-Baquerizo, M., Maestre, F. T., Reich, P. B., Jeffries, T. C., Gaitan, J. J., Encinar, D., Berdugo, M., Campbell, C. D., & Singh, B. K. (2016). Microbial diversity drives multifunctionality in terrestrial ecosystems. *Nature Communications, 7*, 10541. https://doi.org/10.1038/ncomms10541

Deutsch, C., Inomura, K., Luo, Y. W., & Wang, Y. P. (2024). Projecting global biological N_2 fixation under climate warming across land and ocean. *Trends in Microbiology, 32*, 6.

Dong, J., Jiang, Y., Lyu, M., Cao, C., Li, X., Xiong, X., Lin, W., Yang, Z., Chen, G., Yang, Y., & Xie, J. (2023). Drought changes the trade-off strategy of root and Arbuscular Mycorrhizal Fungi growth in a subtropical Chinese Fir Plantation. *Forests, 14*(1), 114. https://doi.org/10.3390/f14010114

Dubey, A., Malla, M. A., Khan, F., Chowdhary, K., Yadav, S., Kumar, A., Sharma, S., Khare, P. K., & Khan, M. L. (2019). Soil microbiome: A key player for conservation of soil health under changing climate. *Biodiversity and Conservation, 28*, 2405–2429. https://doi.org/10.1007/s10531-019-01760-5

Eekhout, J. P. C., & De Vente, J. (2022). Global impact of climate change on soil erosion and potential for adaptation through soil conservation. *Earth-Science. Reviews, 226*, Article 103921.

Einsle, O., & Rees, D. C. (2020). Structural enzymology of nitrogenese enzymes. *Chemical Reviews, 12*, 4969–5004.

Fernández Martínez, M., Sardans, J., Chevallier, F., Ciais, P., Obersteiner, M., Vicca, S., Canadell, J. G., Bastos, A., Friedlingstein, P., Sitch, S., Piao, S. L., Janssens, I. A., & Peñuelas, J. (2019). Global trends in carbon sinks and their relationships with CO_2 and temperature. *Nature Climate Change*, (9), 73–79. https://doi.org/10.1038/s41558-018-0367-7.

Garcia-Pichel, F., Johnson, S. L., Youngkin, D., & Belnap, J. (2003). Small-scale vertical distribution of bacterial biomass and diversity in biological soil crusts from arid lands in the Colorado Plateau. *Microbial Ecology, 46*, 312–321.

Gong, C., Tian, H., Liao, H., Pan, N., Pan, S., Ito, A., Jain, A. K., Kou-Giesbrecht, J. F., Sun, Q., Shi, H., Vuichard, N., Zhu, Q., Peng, C., Maggi, F., Tang, F. H. M., & Zaehle, S. (2024). Global net climate effects of anthropogenic reactive nitrogen. *Nature, 632*, 557–563.

Hardoim, P. R., Van Overbeek, L. S., Berg, G., Pirttilä, A. M., Compant, S., Campisano, A., Döring, M., & Sessitsch, A. (2015). The hidden world within plants: Ecological and evolutionary considerations for defining functioning of microbial endophytes. *Microbiology and Molecular Biology Reviews, 79*(3), 293–320.

Hartmann, M., & Six, J. (2023). Soil structure and microbiome functions in agroecosystems. *Nature Reviews Earth & Environment, 4*(1), 4–18.

Hooper, D., Adair, E., & Cardinale, B. (2012). A global synthesis reveals biodiversity loss as a major driver of ecosystem change. *Nature, 486*, 105–108. https://doi.org/10.1038/nature11118

Huang, J. P., Yu, H. P., Guan, X. D., Wang, G. Y., & Guo, R. X. (2016). Accelerated dryland expansion under climate change. *Nature Climate Change, 6*, 166–171.

Ibáñez, A., Garrido-Chamorro, S., & Barreiro, C. (2023). Microorganisms and climate change: A not so invisible effect. *Microbiology Research, 14*(3), 918–947.

IPCC. (2023). Summary for policymakers. In Core Writing Team, H. Lee, & J. Romero (Eds.), *Climate change 2023: Synthesis report. A Report of the Intergovernmental Panel on Climate Change. Contribution of Working Groups I, II and III to the Sixth Assessment Report of the Intergovernmental Panel on Climate Change*. Geneva, Switzerland: IPCC.

Khursheed, S. (2016). Soil biodiversity and climate change. *Advances in Plants & Agriculture Research, 3*, 166–168. https://doi.org/10.15406/apar.2016.03.00113.

Kovacs, E. D., & Kovacs, M. H. (2023). Global change drivers impact on soil microbiota: Challenges for maintaining soil ecosystem services. In L. Hufnagel, M. A. El-Esawi (eds) *Vegetation Dynamics, Changing Ecosystems and Human Responsibility*. IntechOpen. https://doi.org/10.5772/intechopen.111585.

Kumar, A., Sharma, S., & Mishra, S. (2016). Evaluating effect of arbuscular mycorrhizal fungal consortia and *Azotobacter chroococcum* in improving biomass yield of *Jatropha curcas*. *Plant Biosystems, 150*, 1056–1064.

Leslie, H. M. (2005). A synthesis of marine conservation planning approaches. *Conservation Biology, 19*, 1701–1713.

Li, Z. X., Liu, Y., & Ernst, R. (2023). A dynamic 2000–540 Ma Earth history: From carbonic amalgamation to the age of supercontinent cycle. *Earth-Science RevIews, 238*(104336), 1–45. https://doi.org/10.1016/j.earscirev.2023.104336

Lin, Y., Hu, H. W., Gao, G. F., & Cai, Y. (2023). Editorial: Nitrogen cycling micro-organisms under global change: Response and Feedback. *Frontiers in Microbiology, 14*, 1166306.

Lioubimtseva, E. (2004). Climate change in arid environments: Revisiting the past to understand the future. *Progress in Physical Geography: Earth and Environment, 28*, 502–530. https://doi.org/10.1191/0309133304pp422oa

Liu, Y., Delgado-Baquerizo, M., Bing, H., Wang, Y., Wang, J., Chen, J., Qiu, S., Zhu, H., Wu, Y., Fang, L., & Chang, R. (2024). Warming-induced shifts in alpine soil microbiome: An ecosystem-scale study with environmental context-dependent insights. *Environmental Research, 255*, Article 119206.

Ma, Z., & Yang, Q. (2017). Global patterns of aridity trends and time regimes in transition. In *Aridity Trend in Northern China*, pp. 67–90. Singapore: World Scientific.

Ma, S., Liu, D., Tian, Y., Fu, C., Li, J., Ju, P., Sun, P., Ye, Z., Liu, Y., & Watanabe, Y. (2021). Critical transitions and ecological resilience of large marine ecosystems in the Northwestern Pacific in response to global warming. *Global Change Biology, 27*(20), 5310–5328.

Maestre, F. T., Quero, J. L., Gotelli, N. J., Escudero, A., Ochoa, V., Delgado-Baquerizo, M., García-Gómez, M., Bowker, M. A., Soliveres, S., Escolar, C., García-Palacios, P., Berdugo, M., Valencia, E., Gozalo, B., Gallardo, A., Aguilera, L., Arredondo, T., Blones, J., Boeken, B., & Zaady, E. (2012). Plant species richness and ecosystem multifunctionality in global drylands. *Science, 335*(6065), 214–218.

Mariappan, S., Hartley, I. P., Cressey, E. L., Dungait, J. A., & Quine, T. A. (2022). Soil burial reduces decomposition and offsets erosion-induced soil carbon losses in the Indian Himalaya. *Global Change Biology, 28*(4), 1643–1658.

Martínez-Hidalgo, P., & Hirsch, A. M. (2017). The nodule microbiome: N_2 fixing rhizobia do not live alone. The American pathological society. *Phytobiome, 1*, 70–82.

Martinez-Valderrama, J., Guirado, E., & Maestre, F. T. (2020). Desertifying deserts. *Nature Sustainability, 3*, 572–575.

Maurer, G. E., Hallmark, A. J., Brown, R. F., Sala, O. E., & Collins, S. L. (2020). Sensitivity of primary production to precipitation across the United States. *Ecology Letters, 23*(3), 527–536. https://doi.org/10.1111/ele.13455Mc.

McHugh, T. A., Morrissey, E. M., Reed, S. C., Hungate, B. A., & Schwartz, E. (2015). Water from air: An overlooked source of moisture in arid and semiarid regions. *Scientific Reports, 5*, 1–6.

Meisner, A., Leizeaga, A., Rousk, J., & Bååth, E. (2017). Partial drying accelerates bacterial growth recovery to rewetting. *Soil Biology and Biochemistry, 112*, 269–276.

Mekala, S., & Polepongu, S. (2019). Impact of climate change on soil microbial community. In *Plant Biotic Interactions* (pp. 31–41). Springer International Publishing. ISBN 978–3–030–26657–8.

Middleton, N., & Thomas, D. (1997). World atlas of desertification, 2nd Ed. UNEP (017)/W927, pp. 171–180.

Mostafa, S. M., Wahed, O., El-Nashar, W. Y., El-Marsafawy, S. M., & Abd-Elhamid, H. F. (2021). Impact of climate change on water resources and crop yield in the Middle Egypt region. AQUA—Water Infrastructure. *Ecosystems and Society, 70*(7), 1066–1084.

Naylor, D., DeGraaf, S., Purdom, E., & Coleman-Derr, D. (2017). Drought and host selection influence bacterial community dynamics in the grass root microbiome. *The ISME Journal, 11*(12), 2691–2704.

Naylor, D., Sadler, N., Bhattacharjee, A., Graham, E. B., Anderton, C. R., Lipton, M., Hofmockel, K. S., & Jansson, J. K. (2020). Soil Microbiomes under climate change and implications for carbon cycling. *Annual Review of Environment and Resources, 45*, 29–59.

Neilson, J. W., Califf, K., Cardona, C., Copeland, A., Van Treuren, W., Josephson, K. L., Knight, R., Gilbert, J. A., Quade, J., Caporaso, J. G., & Maier, R. M. (2017). Significant impacts of increasing aridity on the arid soil microbiome. *mSystems, 2*(3), 10–1128.

Nolan, R. H., Sinclair, J., Waters, C. M., Mitchell, P. J., Eldridge, D. J., Paul, K. I., Roxburgh, S., Butler, D. W., & Ramp, D. (2019). Risks to carbon dynamics in semi-arid woodlands of eastern Australia under current and future climates. *Journal of Environmental Management, 235*, 500–510.

Orozco-Mosqueda, M. D. C., Flores, A., Rojas-Sánchez, B., Urtis-Flores, C. A., Morales-Cedeño, L. R., Valencia-Marin, M. F., Chávez-Avila, S., Rojas-Solis, D., & Santoyo, G. (2021). Plant growth-promoting bacteria as bioinoculants: Attributes and challenges for sustainable crop improvement. *Agronomy, 11*(6), 1167.

Peláez, D. P., Allen, J., & Beers, A. (1987). World agriculture and soil erosion. *BioScience, 37*(4), 277–283.

Peng, J., Wang, Y. P., Houlton, B. Z., Dan, L., Park, B., & Tang, X. (2020). Global carbon sequestration is highly sensitive to model-based formulations of nitrogen fixation. *Global Biogeochemical Cycles*. https://doi.org/10.1029/2019GB006296

Peng, Y., Xu, H., Shi, J., Wang, Z., Lv, J., Li, L., & Wang, X. (2024). Soil microbial composition, diversity, and network stability in intercropping versus monoculture responded differently to drought. *Agriculture, Ecosystems and Environment, 365*, Article 108915.

Pointing, S. B., & Belnap, J. (2012). Microbial colonization and controls in dryland systems. *Nature Reviews Microbiology, 10*, 551–562.

Priya, P., Aneesh, B., & Harikrishnan, K. (2021). Genomics as a potential tool to unravel the rhizosphere microbiome interactions on plant health. *Journal of Microbiological Methods, 185*, Article 106215.

Quan, Q., Tian, D., Zhang, Y. L., Crowther, T. W., Zhu, K., Chen, H. Y., Zhou, Q., & Niu, S. (2019). Water scaling of ecosystem carbon cycle feedback to climate warming. *Science Advances, 5*, 1–8.

Quoreshi, A. M., & Madouh, T. (2023). Kuwait deserts and ecosystems in the context of changing climate. In M. K. Suleiman, S. A. Shahid (Eds.) *Terrestrial Environment and Ecosystems of Kuwait*, (Ch. 14, pp. 341–359), Springer International Publisher.

Quoreshi, A. M., Suleiman, M., Kumar, V., Jasmine, A., Sivadasan, M. T., Islam, A., & Khasa, D. P. (2019). Untangling the bacterial community composition and structure in selected Kuwait desert soils. *Applied Soil Ecology, 138.*

Rakshit, A., Singh, H. B., Singh, A. K., Singh, U. S., & Fraceto, L. (2020). New frontiers in stress management for durable agriculture. In A. Rakshit, H. B. Singh, A. K. Singh, & U. S. Singh (Eds.) *New Frontiers in Stress Management for Durable Agriculture*. Springer. https://doi.org/10.1007/978-981-15-1322-0.

Reich, P. B., Sendall, K. M., Rice, K., Rich, R. L., Stefanski, A., Hobbie, S. E., & Montgomery, R. A. (2015). Geographic range predicts photosynthetic and growth response to warming in co-occurring tree species. *Nature Climate Change, 5*, 148–152.

Richter, K., Haslbeck, M., & Buchner, J. (2010). The heat shock response: Life on the verge of death. *Molecular Cell, 40*, 253–266.

Riebesell, U., Aberle-Malzahn, N., Achterberg, E. P., Algueró-Muñiz, M., Alvarez-Fernandez, S., Arístegui, J., Bach, L. T., Boersma, M., Boxhammer, T., Guan, W., Haunost, M., Horn, H. G., Löscher, C. R., Ludwig, A., Spisla, C., Sswat, M., Stange, P., & Taucher, J. (2018). Toxic algal bloom induced by ocean acidification disrupts the pelagic food web. *Nature Climate Change, 8*(12), 1082–1086. https://doi.org/10.1038/s41558-018-0344-1

Rodríguez-Verdugo, A., Lozano-Huntelman, N., Cruz-Loya, M., Savage, V., & Yeh, P. (2020). Compounding effects of climate warming and antibiotic resistance. *iScience, 23*, 101024.

Schilling, J., Hertig, E., Tramblay, Y., & Scheffran, J. (2020). Climate change vulnerability, water resources and social implications in North Africa. *Regional Environmental Change, 20*(1).https://doi.org/10.1007/s10113-020-01597-7.

Seleiman, M. F., Al-Suhaibani, N., Ali, N., Akmal, M., Alotaibi, M., Refay, Y., Dindaroglu, T., Abdul-Wajid, H. H., & Battaglia, M. L. (2021). Drought stress impacts on plants and different approaches to alleviate its adverse effects. *Plants, 10*(2), 259.

Sergaki, C., Lagunas, B., Lidbury, I., Gifford, M. L., & Schafer, P. (2018). Challenges and approaches in microbiome research: From fundamental to applied. *Frontier in Plant Sciences, 9*, 1205. https://doi.org/10.3389/fpls.2018.01205

Shade, A., Peter, H., Allison, S. D., Baho, D. L., Berga, M., Bürgmann, H., Huber, D. H., Langenheder, S., Lennon, J. T., Martiny, J. B., Matulich, K. L., Schmidt, T. M., & Handelsman, J. (2012). Fundamentals of microbial community resistance and resilience. *Frontier in Microbiology, 3*, 417. https://doi.org/10.3389/fmicb.2012.00417

Shahid, S. A. (2023). Desertification: A central problem to restoring ecosystems. In M. K. Suleiman, S. A. Shahid (Eds.) *Terrestrial Environment and Ecosystems of Kuwait* (Ch. 5, pp. 119–137). Springer. https://doi.org/10.1007/978-3-031-46262-7_5.

Shahid, S. A., & Rahman, K. (2011). Soil salinity development, classification, assessment and management in irrigated agriculture. In M. Pessarakli (Ed.), *Handbook of plant and crop stress* (3rd ed., Ch. 2, pp. 23–39). CRC Press.

Shahid, S. A., Omar, S. A. S., & Al-Ghawas, S. (1999). Indicators of desertification in Kuwait and their possible management. *Desertification Control Bulletin, 34*, 61–66.

Sharma, K., Sharma, S., Vaishnav, A., Jain, R., Singh, D., Singh, H. B., Goel, A., & Singh, S. (2022). Salt-tolerant PGPR strain Priestia endophytica SK1 promotes fenugreek growth under salt stress by inducing nitrogen assimilation and secondary metabolites. *Journal of Applied Microbiology, 133*, 2802–2813.

Steven, B., Kuske, C. R., Gallegos-Graves, L. V., Reed, S. C., & Belnap, J. (2015). Climate change and physical disturbance manipulations result in distinct biological soil crust communities. *Applied and Environmental Microbiology, 81*, 7448–7459.

Stovicek, A., Kim, M., Or, D., & Gillor, O. (2017). Microbial community response to hydration-desiccation cycles in desert soil. *Scientific Reports, 7*, 45735. https://doi.org/10.1038/srep45735

Tian, X., Dong, J., Jin, S., He, H., Yin, H., & Chen, X. (2023). Climate change impacts on regional agricultural irrigation water use in semi-arid environments. *Agricultural Water Management, 281*, Article 108239.

UNEP. (1992). *World Atlas of Desertification* (1st Edn.). UNEP (017)/W927.

Van der Putten, W. H., Bardgett, R. D., Bever, J. D., Bezemer, T. M., Casper, B. B., Fukami, T., Kardol, P., Klironomos, J. N., Kulmatiski, A., Schweitzer, J. A., & Suding, K. N. (2013). Plant–soil feedbacks: The past, the present and future challenges. *Journal of Ecology, 101*(2), 265–276.

Van Nes, E. H., & Scheffer, M. (2007). Slow recovery from perturbations as a generic indicator of a nearby catastrophic shift. *The American Naturalist, 169*, 738–747. https://doi.org/10.1086/516845

Veresoglou, S., Halley, J., & Rillig, M. (2015). Extinction risk of soil biota. *Nature Communications, 6*, 8862. https://doi.org/10.1038/ncomms9862

Wagner, M. R., Lundberg, D. S., Coleman-Derr, D., Tringe, S. G., Dangl, J. L., & Mitchell-Olds, T. (2014). Natural soil microbes alter flowering phenology and the intensity of selection on flowering time in a wild Arabidopsis relative. *Ecology Letters, 17*(6), 717–726.

Wang, Y., Hao, Y., Cui, X. Y., Zhao, H., Xu, C., Zhou, X., & Xu, Z. (2014). Responses of soil respiration and its components to drought stress. *Journal of Soils and Sediments, 14*, 99–109.

Wang, X., Wang, M., Xie, X., Guo, S., Zhou, Y., Zhang, X., Yu, N., & Wang, E. (2020). An amplification-selection model for quantified rhizosphere microbiota assembly. *Science Bulletin, 65*(12), 983–986.

Woodroffe, C. D. (2007). The natural resilience of coastal systems: Primary concepts. In L. McFadden, E. Penning-Rowsell, & R. J. Nicholls (Eds.), *Managing coastal vulnerability* (pp. 45–60). Elsevier.

Wooliver, R., Pellegrini, A. F. A., Waring, B., Houlton, B. Z., Averilll, C., Schimel, J., Hedin, L. O., Bailey, J. K., & Schweitzer, J. A. (2019). Changing perspectives on terrestrial nitrogen cycling: The importance of weathering and evolved resource-use traits for understanding ecosystem responses to global change. *Functional Ecology, 33*, 1818–1829.

Xu, J., Kloepper, J. W., Huang, P., McInroy, J. A., & Hu, C. H. (2018). Isolation and characterization of N_2 fixing bacteria from giant reed and switch grass for plant growth promotion and nutrient uptake. *Journal of Basic Microbiology, 58*, 459–471.

Yadav, S., Payal, M., Akanksha, D., Akdasbanu, V., Patel, D., & Patel, M. (2020). Effects of abiotic stress on crops. Section 1 Crop production and farming system. In M. C. M. Teixeira Filho, T. A. R. Nogueira, & F. S. Galindo (Eds.) *Sustainable crop production*, p. 83521. https://doi.org/10.5772/Intechopen.

Yuan, X., Chen, X., Ochega, F. U., Hamdi, R., Tabari, H., Li, B., He, B., Zhang, C., Maeyer, P. D., & Luo, G. (2024). Weakening of global terrestrial carbon sequestration capacity under increasing intensity of warm extreme. *Nature Ecology and Evolution, 2174*, 103. https://doi.org/10.1038/s41559-024-02576-5.

Zaman, M., Shahid, S. A., & Heng, L. (2018). *Guideline in salinity assessment, mitigation and adaptation using nuclear and related techniques* (p. 164). Springer.

Zhang, H., Zhu, J., Gong, Z., & Zhu, J. K. (2022). Abiotic stress responses in plants. *Nature Reviews Genetics, 23*, 104–119. https://doi.org/10.1038/s41576-021-00413-0

Zhou, J., Deng, Y. E., Shen, L., Wen, C., Yan, Q., Ning, D., Qin, Y., Xue, K., Wu, L., He, Z., & Voordeckers, J. W. (2016). Temperature mediates continental-scale diversity of microbes in forest soils. *Nature Communications, 7*(1), 12083.

Zhou, Z., Wang, C., & Luo, Y. (2020). Meta- analysis of the impacts of global change factors on soil microbial diversity and functionality. *Nature Communications, 11*, 3072.

Chapter 20
Dynamics of Agricultural Pests in a Changing Climate and Implications for Pesticide Use: A Review

Mustapha F. A. Jallow

Abstract The Impact of climate change is a major concern for agriculture and global food security. Changes in agro-ecological conditions due to climate change are feared to create favorable conditions for agricultural pests, which could affect crops and disrupt food systems. Climatic factors, including increased temperatures, elevated carbon dioxide levels, changing rainfall patterns and distribution, have significant impacts on crop production as well as on agricultural pests and beneficial insects such as natural enemies and pollinators. As a consequence of climate change, the diversity and abundance of agricultural pests have changed, and their distribution have extended far beyond their geographical range. New invasive insect pests and pathogens have emerged, and the effectiveness of biological control agents, especially natural enemies of insect pests, has been significantly impacted. These changes have led to reduced efficacy of crop protection technologies and the increase dependence on chemical pesticides, posing a serious threat to global agriculture, food security systems, environmental sustainability, and human health. This chapter reviews the impact of climate change on agricultural insect pests and pathogens, particularly invasive species, and examines the effects of climate change on beneficial insects and their interactions with insect pests and crop plants. Additionally, the review explored pesticide overuse as a consequence of climate change, and potential effects on human health and the environment. Finally, strategies to prevent, mitigate and adapt to the effects of climate change on pests and pesticide use is discussed. This is of crucial importance and a key element of sustainable agriculture and future food security, and is integral to the long-term viability of the environment and human health.

Keywords Climate change · Food security · Agricultural pests · Invasive species · Pest management · Pesticide overuse · Environmental and human exposure

---

M. F. A. Jallow (✉)
Food Security Program, Environment and Life Sciences Research Center, Kuwait Institute for Scientific Research, Safat, Kuwait
e-mail: mjallow@kisr.edu.kw

M. K. Suleiman and S. A. Shahid (eds.), *Fostering Arid Lands Agriculture in the Face of Climate Change*, Advances in Global Change Research 81,
https://doi.org/10.1007/978-3-031-98490-7_20

20.1 Introduction

Natural resources are being heavily stressed globally due to increased variability and changing climate (Prakash et al., 2014). Many of these changes, which have been linked to anthropogenic influences, include rising global temperatures, increased atmospheric greenhouse gases (e.g. carbon dioxide (CO_2), variable rainfall patterns, heavy precipitation events, droughts and other extreme weather events (NAS, 2020). Climate change affects biological systems at multiple levels, from genes to entire ecosystems, including cultivated lands (Rawat et al., 2024). It can have significant effects on crops, pests, and their interactions with natural enemies, pollinators and other organisms. It is estimated that 15 to 20 percent of global crop production is lost due to different insect pests, pathogens, weeds and other animals (Tilman et al., 2011). This issue is further exacerbated by post-harvest loss caused by different insect pests and pathogens. Furthermore, it is postulated that pest problems are likely to become more unpredictable and numerous with changing climate, which will have a significant impact on global food security (Bale et al., 2002; Godfray et al., 2010). Climate change can profoundly affect the biology and status of pests. These effects could either be direct, through the influence of climatic variability on the pests' physiology and behavior, or indirectly, mediated through their host plants, natural enemies, competitors, mutualist, and vectors (Bale et al., 2002; Shrestha, 2019). For example, rising temperatures and atmospheric CO_2 levels, and changing rainfall patterns can directly impact the overwintering and feeding rates of insects, number of generations, survival, population size and fertility, dispersal, and geographical distribution (Shrestha, 2019). Similarly, increased temperatures, $CO_{2,}$ and changes in precipitation patterns can alter the biochemistry of plants, and consequently plant defense responses to insects (Streck, 2005). These changes can have important implications for the interactions between insect pests, plants and natural enemies.

Like crop pests, climate change has severe direct negative impacts on the abundance, distribution, and seasonal phenology of beneficial insects (natural enemies and pollinators), making vital ecosystem services impossible and increasing the vulnerability of crops to pest damage (Adebesin et al., 2017; Eigenbrode & Adhikari, 2023). Similarly, climate change can enhance the virulence of pathogens, their frequent outbreaks, emergence of new strains, and cause range shifts and spread into new agricultural regions. Climate change also significantly affects the environmental fate, behavior and bioavailability of pesticides through faster photodegradation and volatilization rendering pesticides less effective (Bloomfield et al., 2006). The impact of climate change on the efficacy of pest management technologies such as pesticides and biological control, along with the global increase in pest pressure due to a changing climate, will likely increase pesticide use (both in volume and frequency) across various agricultural ecosystems. These combined factors pose a significant threat to global food security, environmental sustainability, and human health. This chapter reviewed the impact of climate change on agricultural insect pests and diseases, with a particular focus on invasive species that pose major challenges in crop production, as well as effects of climate change on beneficial insects and their interactions with insect pests and crop plants. In addition, the review examines the overuse and toxicity of pesticides as a consequence of climate change, and their potential effects on human health and the environment. The review also explored the preventative, mitigation, and adaptation measures necessary to alleviate the projected increase in pest risk to global agricultural production due to climate change.

20.2 Changing Climatic Patterns

Climate change is the alteration of environmental factors including temperature, humidity, and precipitation over many years. It is associated with an increased likelihood of extreme climate events such as drought, flooding, heat waves, and rise in sea level (IPCC, 2013). Heat waves are getting more frequent and lasting longer; drier regions are experiencing reduced precipitation and rising temperatures, while other areas are witnessing heavy precipitation events. For example, the countries in the Middle East and Northern Africa (MENA) region are experiencing a notable increase in aridity, whereas the Northern Hemisphere appears to have an increase in annual precipitation (Henry et al., 2022). One of the key drivers of climate change is extreme temperatures. Global warming is a serious problem facing the world today, as evidenced by unprecedented rates of increase in atmospheric temperatures (IPCC, 2013). Global near-surface temperature has increased by about 1.7 °C from 1901 to 2016, and is expected to rise to 2.4 °C by the end of the century (Rawat et al., 2024). This could pose a significant challenge to global food production and food security.

The rise in global climate temperature is the results of increased emissions of greenhouse gases (GHGs) (IPCC, 2014). Atmospheric concentrations of GHGs have been at unprecedented levels, with CO_2 the most important and most prevalent in the atmosphere (IPCC, 2014). Atmospheric CO_2 concentrations is estimated to have increased from 280 parts per million (ppm) in pre-industrial periods to 416 ppm in 2020, and is likely to double by the end of this century (Rawat et al., 2024). Global warming is attributed to natural variability and anthropogenic causes. However, most of the global warming observed over the last 50 years is attributed to anthropogenic activities, such as burning of fossil fuels, intensification of industrial and agricultural activities, and changes in land use patterns (NAS, 2020). Human contribution to the global average temperature increase from 1951–2020 is about 0.2 °C per decade, and this is likely to increase further if global initiatives are not put in place to combat these effects (NAS, 2020).

20.3 Climate Change Impact on Agricultural Insect Pests

Insects represent more than half of all known animal species and are among the most susceptible organisms to climate change (Bale et al., 2002). Temperature is the primary climate change variable affecting insects, as they are ectothermic, meaning their body temperature depends on the surrounding environment. Therefore, their physiology and behavior are strongly influenced by temperature (Parmesan & Yohe, 2003). As a result, all biological parameters of insects are sensitive to changes in temperature. For example, research has demonstrated that the metabolic rate of insects increases with rising temperature up to a certain temperature threshold (Bale et al., 2002). An increase in metabolic rate accelerates insect food consumption, consequently, influencing several biological parameters including fecundity, survival, population size and geographical range distribution (Kocmánková et al., 2010). The effect of increased temperature is greater on aboveground insects than for soil dwelling insects, because soil act as a thermal insulating medium, moderating temperature changes and reducing its impact (Kocmánková et al., 2010). Insect species that can withstand extreme climates and adapt to rising temperatures are generally expected to thrive, reproduce, and expand their population and geographical range. Conversely, species that cannot adapt to increasing temperatures are likely to struggle to maintain their populations.

Insect populations in tropical and subtropical regions are predicted to decline because of warming temperatures reaching beyond their maximum developmental threshold, while insects in temperate regions will experience an increase in population (Schweiger, 2008). The general prediction is that rising temperature in temperate regions will enhance the ability of insects to overwinter at higher altitudes and increase their population, as they are no longer constrained by low temperatures. This will likely change the dynamics of insect pest risk across climatic zones, presumably resulting in decrease in insect pest risk in tropical regions while rising in temperate regions. However, on a global scale, rising temperatures are expected to increase insect pest abundance and severity through the expansion of their geographic range; increased overwintering survival; changes in their population growth rates and number of generations; increased risk of introduction of invasive alien species; increased incidence of insect vectored plant diseases; and the reduced effectiveness of natural enemies (Bale et al., 2002). This will have significant consequences on global food production.

Similar to temperature, elevated CO_2 levels can indirectly affect the food consumption rates, fecundity, survival, population size and other biological parameters of insect pests, with these indirect effects mediated through their host plants. Increasing CO_2 levels has direct effects on plant biochemistry through changes in photosynthetic activity and the synthesis of secondary plant metabolites (Chen et al., 2005). However, C4 plants (e.g. maize and sugarcane) and C3 plants (e.g. wheat and soybean) respond to elevated CO_2 differently, with C3 plants more sensitive to elevated CO_2 levels (Lincoln, 1993). Consequently, the indirect effects of elevated CO_2 levels on insect pests may depend on the plants the insect feeds on, with insects feeding on C3 plants most likely to be negatively affected. A rise in atmospheric CO_2 generally increases the synthesis of plant secondary metabolites and alters the carbon (C) to nitrogen (N) ratio in plants, with increased ratio of carbon to nitrogen (Cotrufo et al., 1998). This in turn reduces the palatability and nutritional quality of these plants. Nitrogen is the key element necessary for insect development. Consequently, to compensate for the low nitrogen availability, insects tend to feed more on these plants in order to achieve sufficient dietary nitrogen leading to increased levels of plant damage, although this is not always the case (Stiling & Cornelissen, 2007). While some authors have demonstrated that the biological parameters of insects feeding on plants grown under elevated CO_2 are negatively affected due to reduced nutritional quality of the plants, other authors have demonstrated the contrary. Several studies have demonstrated an increase in fecundity, population size, survival and growth rate of insects feeding on plants under elevated CO_2 levels. The response of insects to rising CO_2 also varies between species and feeding guild. Elevated CO_2 tends to negatively affect leaf-chewing insects (e.g. cutworms, flea beetles) more than insects with other specialized feeding strategies, such as phloem feeders (e.g. aphids, leafhoppers) and leafminers (e.g. tomato leafminer).

Variable rainfall patterns may also affect agricultural insect pests directly as well as indirectly through changes in humidity levels. Climate change has resulted in extreme droughts and flooding in many regions. Heavy precipitation and flooding are limiting factors in the survival of adults and neonates of small-bodied insect pests such as aphids, whiteflies, mites, etc., as these pests get washed away by heavy rains (Staley et al., 2007). On the other hand, changing rainfall patterns can create favorable conditions for insect pests to thrive and expand their populations (Staley et al., 2007). For example, crops suffering from water stress due to drought are more vulnerable to insect attack because of a decrease in the synthesis of chemical defense compounds in these plants (Bale et al., 2002). Most plant-feeding insects tend to survive well under drought conditions. However, this is not always the case. For example, poor aphid survival and performance on mildly-stressed winter wheat (*Triticum aestivum* L.) was correlated with increased plant defenses, whereas poor performance on highly-stressed plants was correlated with stronger plant defense induction and reduced plant nutritional quality, thus reduced insect damage (Kansman et al., 2022). Similarly, humid conditions favor insect pest outbreaks and population increase.

20.4 Climate Change Impact on Beneficial Insects

Beneficial insects, including pollinators and natural enemies of insect pests play a significant role in providing various ecosystem services. For example, approximately 90 percent of the world's flowering plants are pollinated by bees, wasps, ants and other important pollinator species, thus, playing a vital role in global food production and the sustainability of natural ecosystems (Adebesin, 2017). These pollinators, in turn, benefit by obtaining floral resources such as nectar, and this mutualistic relationship has evolved over centuries. However, these important pollinators and the services they provide in various ecosystems are under severe threat due to climate change. Rising temperatures, elevated CO_2 levels, and changing precipitation patterns can alter the phenology, morphology, abundance and geographical range of plants and insects, which can directly impact the interaction between plants and pollinators (Gérard et al., 2020). For example, pollinators have evolved to synchronize their life cycle with plant phenological events such as flowering, and this synchrony is expected to be disrupted by changing climate; consequently, affecting pollinator efficiency. Most plants are flowering earlier than they did decades ago due to changing climate, and this shift in flowering periods is creating mismatches between when plants flower and the life cycles of many pollinators (Kambrekar et al., 2015). Similarly, climate change can influence plant-pollinator interactions through the modification of plant secondary metabolites (e.g. nectar and volatile organic compounds) necessary for pollinator attraction (Kambrekar et al., 2015).

Climate change affects insect predators and parasitoids directly through the influence of climatic variables on the natural enemies' biology, or indirectly through trophic interactions between herbivores, plants, and natural enemies. For example, the population size, fertility, and geographical distribution of insect natural enemies are directly affected by drought and temperature extremes (Hance et al., 2007). Similarly, elevated CO_2 concentration, temperature increase, as well as decreased precipitation has direct effects on plant phenology and biochemistry, which influences the nutritional resources available to insect herbivores; consequently, indirectly altering the fitness of natural enemies that feed on these herbivores (Chen et al., 2005). Climate change can also influence the synchrony between the phenology of herbivores and the appearance of their natural enemies, as well as the ability of natural enemies to track shifts in geographic range of herbivores. This can decrease the effectiveness of insect natural enemies as biological control agents (Legrand et al., 2004).

20.5 Climate Change Impact on Plant Pathogens

Plant pathogens are significantly influenced by the environment in which they thrive, and consequently, climate change will facilitate crop infection across various agricultural ecosystems and climatic regions. Many climatic variables can affect plant pathogens, with rising temperatures, elevated CO_2 levels, and changing rainfall patterns/water availability being among the most significant. These variables are predicted to change considerably under different climate change scenarios (Whitfield et al., 2015). Rising temperatures can significantly impact the population dynamics of pathogens, their survival, population growth, and interaction with host plants and transmitting vectors. For example, *Xanthomonas oryzae,* a major bacterial pathogen affecting rice was found to be more virulent at higher temperatures (Whitfield et al., 2015). Warming temperatures shorten the pathogen's incubation period, resulting in an increased population of the pathogen. Similarly, high temperatures combined with increased humidity are linked to enhanced virulence of the fungus *Sclerotinia Sclerotiorum* which affects lettuce crop (Lorang et al., 2012). It is also anticipated that the expected increase in global temperatures will result in the range shift of plant pathogens. A number of important plant pathogens (e.g. *Fusarium graminearum* and *Puccinia graminis*) are enlarging their range to higher latitudes as a consequence of rising temperature, increasing their potential survival in areas outside their typical ranges (Bebber et al., 2013). Warming temperatures can also result in the prevalence of new pathogen strains that are better adapted to their environment and more virulent

(Melotto et al., 2006). The effects of rising temperatures on plant pathogens can also be indirect, mediated through their host plants and transmitting vectors. For example, rising temperatures can positively affect virus transmitting vectors and/or suppress plant immunity leading to increased pathogen infection and their severity (Lorang et al., 2012).

There is evidence that elevated CO_2 levels can increase disease severity in plants due to the reduction in the effectiveness of plant defensive pathways under this condition (Ciliberti et al., 2015). For example, elevated CO_2 levels increased the severity of *Fusarium graminearun* as well as *Septoria tritici* on wheat, *Shaerotheca fuliginea* on curcurbits and *Phytophthora infestans* on potato (Atkinson, 2012). While rising CO_2 levels can enhance the susceptibility of plants to pathogens, there is also evidence to the contrary. Soybean plants grown under elevated CO_2 levels have shown greater tolerance to the downy mildew pathogen *Peronospora manshurica*, resulting in less disease severity (Ciliberti et al., 2015). Similarly, plant pathogens tend to strive better under high air humidity and soil moisture levels. For example, the severity and virulence of some pathogens that infect aerial plant tissues (e.g. *Magnaporthe oryzae* in rice and *Sclerotinia sclerotiorum* in lettuce) and soil-inhabiting pathogens (*Ralstonia solanacearum* in tomato) is correlated with increased rain and high humidity (Charaya et al., 2021). However, drought can also have a positive impact on the severity of some pathogens. The tomato mosaic virus (ToMV), onion white rot (*Sclerotium cepivorum*) and potato bacteria scab (*Streptomyces scabies*) can increase in severity with drought condition (Trebicki, 2020).

20.6 Climate Change, Invasive Species and Future Pest Risk

Invasive pest species are identified as the greatest threat to agricultural production and global food security (Shrestha, 2019). Apart from their threat to agricultural production, invasive pests also disrupt natural ecosystems by competing with native species, leading to a decline in biodiversity (Masters & Norgrove, 2010). Weather events (e.g. hurricanes, heavy rainstorms, etc.) can directly spread pests into new regions where they may find environmental conditions favorable for their establishment. Many scientists have used climate-based simulation models to forecast pest risk under different climatic scenarios and have concluded that there will be a global shift in the geographical range of pests, at least in part due to climate change. For example, rising temperatures may facilitate the survival and establishment of pest in regions where they would otherwise be unable to establish, especially in the absence of natural enemies to regulate their populations (Hance et al., 2007; Subedi et al., 2023).

20.6.1 Invasive Insect Pests

Invasions involving insect pests are perhaps the most pervasive among all invasive species, and many are emerging as a threat to global food production with annual impact in billions of dollars because of crop damage and cost of pest control. Some of the most destructive invasive agricultural insect pests include the South American tomato pinworm (*Tuta absoluta* or *Phthorimaea absoluta*); fall armyworm (*Spodoptera frugiperda*); cotton bollworm (*Helicoverpa armigera*); oriental fruit fly (*Bactrocera dorsalis*); desert locust (*Schistocera gregaria*); red palm weevil (*Rhynchophorus ferrugineus*); spotted wing drosophila (*Drosophila suzukii*); and the brown marmorated stink bug (*Halyomorpha haly*).

The South American tomato pinworm (*Tuta absoluta* or *Phthorimaea absoluta*) is polyphagous moth that attacks mainly solanaceae crops including tomatoes, potato, eggplant, and pepper (Desneux et al., 2011). Young larvae penetrate the leaves, aerial fruits or stems, where they feed and develop. Feeding within the mesophyll of leaves creates leaf mines, which reduce the photosynthetic activity of the plant and eventually cause death of leaf tissues (Sayed & Saad, 2014). Additionally, larval entry into fruits creates entry points for secondary pathogens leading to fruit rot. *Tuta absoluta* is native to South America and was first reported as a major pest in the early 1980s; and since then, it has rapidly invaded Europe, Africa, the Middle East, and parts of Asia (EPPO, 2024). Current simulations and estimates all indicate the potential spread of *T. absoluta* to all tomato growing regions of the world including Australia, China and Japan (Azrag et al., 2023; Desneux et al., 2011). The success of *T. absoluta* as an invasive species is attributed to its innate dispersal ability, wide temperature tolerance and phenotypic plasticity (Biondi et al., 2018). Furthermore, the pest is polyphagous with a short developmental life cycle and a high reproductive capacity (Desneux et al., 2011).

The fall armyworm (*Spodoptera frugiperda*) is a highly polyphagous moth that feeds on over 80 different crop species including maize and sorghum, being its preferred hosts (Goergen et al., 2016). Larvae cause damage by consuming foliage and other plant parts resulting in extensive defoliation and significant crop damage. *Spodoptera frugiperda* is native to tropical and subtropical areas of the American continent, and has rapidly spread to Africa, the Middle East, southern and eastern Asia, Australia, New Zealand and the Pacific Island countries (EPPO, 2020). There are indications based on climate models that *S. frugiperda* will likely spread to Europe, especially in the warmer southern regions of Spain, Italy and Greece (Jeger et al., 2018). The success of *S. frugiperda* as an invasive species is attributed to its polyphagy, high dispersal capacity, and tolerance to a wide bioclimatic range (Jeger et al., 2018).

The cotton bollworm (*Helicoverpa armigera*) is one of the most destructive agricultural pest in the world (Zhao et al., 2022). The pest is highly polyphagous and has been recorded feeding on more than 200 plant species, including economically important agricultural crops such as cotton, tobacco, tomato, potato, maize, soyabean, lucerne, and chickpea (Jallow et al., 2004; Mastrangelo et al., 2014). *Helicoverpa armigera* is native to Asia and Australasia and has expanded its range into Europe, Africa, South and Central America (Mastrangelo et al., 2014). Climate models predict that the natural spread of *H. armigera* into Northern America, especially southern United States, is a matter of time due to the existence of suitable climate and extensive host crops capable of supporting the establishment of the pest (Zhao et al., 2022). The behavioral, ecological, and physiological characteristics of *H. armigera* (polyphagy, high mobility, high fecundity, wide thermal tolerance and phenotypic plasticity) enable this moth to invade, survive and establish in various climatic regions (Zhao et al., 2022).

The Desert Locust (*Spodoptera frugiperda*) is a migratory pest with the ability to fly across great distances and invade new areas. The pest is a voracious feeder, and can cause extensive damage to food and forage crops and any green vegetation. *Spodoptera frugiperda* is native to Africa and has become established over much of the Arabian Peninsula, western and southern Asia and south-west Europe (Kimathi et al., 2020). Climate models have predicted that South America, Australasia are at a high risk of invasion. Changing weather and abiotic factors (warmer temperatures, heavy precipitation, soil properties, wind direction and speed, and food availability) create favorable conditions for the outbreak, breeding, survival and migration of the desert locust (Kimathi et al., 2020). It is suggested that the more frequent outbreak of desert locust in Africa and the Middle East over the last decade are a consequence of changing climate conditions (Meynard et al., 2017).

The red palm weevil (*Rhynchophorus ferrugineus*) is one of the most economically damaging insect pest of date palm trees and several other palm species including coconuts (El-Shafie & Faleiro, 2020). The larvae of the weevil feed within the apical growing points of the palms, causing extensive damage to palm tissues and weakening the structure of the palm trunk resulting in tree death. *Rhynchophorus ferrugineus*, native to southern Asia and the Melanesia region, has rapidly expanded its geographical range into Southern Europe, North Africa, the Middle East, Australia, the Caribbean, Central America and Southwestern United States (Hoddle et al., 2015). There are indications based on climate models that the number of areas highly favorable for the establishment of *R. ferrugineus* will increase as a consequence of climate change, resulting in the expansion of the pest geographical range (El-Shafie & Faleiro, 2020).

Spotted-wing drosophila (*Drosophila suzukii*) is one of the most important invasive pest of soft fruits worldwide. The species is polyphagous and infest a wide range of soft fruits such as cherry, raspberry, blackberry, blueberry, strawberry, peach, plums and grape (Cini et al., 2012). *Drosophila suzukii* oviposit and feeds on healthy fruits resulting in significant fruit damage. The pest is native to East Asia and has rapidly spread into Europe, North Africa, North and South America, and the Middle East (Iran) (Kiss et al., 2016). *Drosophila suzukii* has been highly successful as an invasive species largely due to international trade, climate change and its ability to rapidly adapt to changing environmental conditions (Kiss et al., 2016). Furthermore, the behavioral, ecological and physiological characteristics of *D. suzukii* (polyphagy, high mobility, high reproductive rate, short generation time and other traits) has contributed towards its spread into new regions (Andreazza et al., 2017).

The Brown Marmorated Stink Bug *(Halyomorpha halys) is an invasive polyphagous pest that feeds on a wide range of vegetable and fruit crops including* tomato, pepper, corn, apple, pear, plums, apricot, and cherry (Leskey & Nielsen, 2018). Both adults and nymphs feed on leaf and fruit tissue, resulting in significant crop damage. *Halyomorpha halys is native to East Asia,* and has become established in Europe and North America. Climatic models indicate the potential for *H. halys* to invade Central and South America, Africa, Australia and New Zealand (Inkley, 2012; Stoeckli et al., 2020).

20.6.2 Invasive Plant Pathogens

Root-knot nematodes are sedentary endoparasitic nematodes that feed on roots of plants. Among root-knot nematodes, *Meloidogyne spp.* are one of the most destructive nematode genera, and infects a variety of vegetable, fruit, and ornamental crops (Barker & Koenning, 1998). There are more than 100 reported species of *Meloidogyne* globally, among them *M. incognita, M. javanica and M. javanica are the most prevalent. Plants infested* with root-knot nematodes show an abnormal development of the root system characterized by the formation of typical galls, which alter the uptake of water and nutrients (De Waele & Elsen, 2007). As a result, aboveground deficiency symptoms appear which may lead to severe yield decreases, depending on the severity of the infestation. Infested plants also loss their ability to withstand other stresses, like drought and other plant pathogens. The distribution and infection rates of *Meloidogyne spp.* are higher in tropical regions than in temperate regions due to more favorable environmental conditions for the pathogens to colonize, reproduce, develop, and disperse (Barker & Koenning, 1998). However, climate change is expected to enhance the spread and dispersal of *Meloidogyne spp.* across temperate regions (Dutta & Phani, 2023).

Many plant viruses depend on insect vectors for their transmission and dissemination. The invasive whitefly, *Bemisia tabaci,* is one of the most important virus vectors worldwide, responsible for the transmission of more than four hundred virus species, the majority belonging to genus Begomovirus, family Geminiviridae (Ghosh & Ghanim, 2021). Whitefly transmitted geminiviruses (WTGs) and their biotypes have emerged as a major threat to global vegetable and fruit crop production, particularly in the tropics and semi-tropics, where environmental conditions are favorable for both whiteflies and the viruses they vector to strive and spread (Taravati et al., 2014). The emergence and spread of whitefly-transmitted geminiviruses (WTGs) in temperate regions can be linked to the northward expansion of whiteflies, potentially a consequence of global warming and the international transport of infected plant materials (Ghosh & Ghanim, 2021). Major diseases transmitted by whiteflies include the tomato mottle virus, tomato yellow leaf curl virus, cassava mosaic and brown streak virus (Martin & Mound, 2007).

Fusarium wilt of banana is an invasive fungal disease caused by the soil-born fungus *Fusarium oxysporum* f. sp. *cubense* tropical Race 4 (TR4) (Butler, 2013).

The pathogen has emerged as a threat to banana production, especially in countries producing Cavendish bananas in monoculture plantations. The fungus enters the plant through the roots and colonizes the xylem vessels, thereby blocking the flow of water and nutrients; consequently, resulting in plant death. *Fusarium oxysporum* (TR4) was first reported in Australia early 1990s attacking Cavendish bananas; and since then, it has rapidly spread to many countries in Asia, the Middle East, and parts of Africa and South America (Dita et al., 2018). Changing weather and abiotic factors such as warmer temperatures, heavy precipitation and waterlogged soils, increase the risk of the disease outbreak (Martyn, 2014).

Banana bunchy top virus (genus *Babuvirus*, family *Nanoviridae*) is the most devastating virus disease of bananas, causing severe crop losses in many banana-growing regions (Dale, 1987; Murhububa et al., 2024). The virus is vectored by the banana aphid (*Pentalonia nigronervosa*), and can be spread through trans-boundary movement of infected planting materials (Allen, 1978). Banana bunchy top virus likely originated in the South Pacific region in Fiji, and has since been spreading to other banana producing countries in the South pacific, African, Asia, and the Middle East (Stainton et al., 2015). Climate change, especially variable rainfall patterns and rising global temperatures, complemented by the international trade in banana has increased the risk of banana bunchy top virus expanding its global geographical range (Kumar et al., 2011).

20.7 Global Pesticide Use and Trends

Pesticides are considered a vital component of modern farming, playing a major role in maintaining high agricultural productivity. Consequently, in high-input, intensive systems of agricultural production, the widespread use of pesticides to manage pests has emerged as a dominant feature (Sharma et al., 2019). For most farmers, crop production is impossible without intensive use of chemical pesticides. Global pesticide use in agriculture has increased significantly in the last three decades, with over 3.5 million metric tons applied in 2021 (FAO, 2023, Table 20.1), most of which are herbicides (50%), insecticides (30%) and fungicides (15%) (Shattuck et al., 2023). Average application rates ranged from 6.5 to 25 kg active ingredients per hectare depending on the crop (FAO, 2023; Sharma et al., 2019; Table 20.2). In 2021, Asia, North, South and Central America represented more than seventy percent of the agricultural pesticide consumption worldwide, with over 250,000, 457,000 and 720,000 metric tons consumed in China, the United States and Brazil, respectively (FAO, 2023). Although Africa and Oceania applied small amount of pesticides over time compared to the rest of the world, nonetheless these two regions had one of the highest growth in pesticides application (over 175% and 86%, respectively) (Table 20.1). The demand for pesticides is predicted to surge further in many regions, driven, among other factors, by climate change and growth in agricultural activity to meet the food

Table 20.1 Global pesticide use in 2000 and 2021 and percentage change in use (Metric Tons) [a]

Region	2000	2021	Increased change (%)
World	2,178,696	3,535,375	62.27
Africa	73,856	203,580	175.64
Americas	892,215	1,775,009	98.94
Asia including Middle East	727,564	979,816	34.7
Europe	446,630	505,157	13.1
Oceania	38,431	71,813	86.90
Middle East	27,563	36,719	33.22

Source FAO (2023), [a] The totals include herbicides, insecticides, fungicides, bactericides, mineral oils and rodenticides

Table 20.2 Global pesticide use per cropland area in 2000 and 2021 and percentage change in use (kg active ingredients per hectare)[a]

Region	2000	2021	Increased change (%)
World	1.47	2.26	53.74
Africa	0.32	0.70	118.75
Americas	2.41	4.70	95.02
Asia including Middle East	1.33	1.71	28.57
Europe	1.47	1.75	19.05
Oceania	1.49	2.10	40.93
Middle East	2.01	3.31	64.67

Source FAO (2023), [a] The totals include herbicides, insecticides, fungicides, bactericides

demands of an increasing global population. By 2050, global pesticide use is forecasted to increase by 5.5 million tones, with significant demands in Africa, South and Central America, and Asia (Shattuck et al., 2023).

20.8 Climate Change Impact on Pesticide Use and Toxicology

As a consequence of climate change, the diversity and abundance of agricultural pests are expected to increase globally, with pests spreading beyond their native distribution area and become more prevalent in various climatic regions (Shrestha, 2019). Climate change-related increases in temperature are also expected to affect pesticide toxicity. Some pesticide chemistries will breakdown faster under certain climate change scenarios, and become less effective over time (Bloomfield et al., 2006). For example, increasing temperatures are linked to the reduce efficacy of the organophosphate insecticide chlorpyrifos (Bailey, 2004). This is attributed to the fact that organophosphate insecticides breakdown faster at higher temperatures. These combined factors (pest prevalence and reduce efficacy of pesticide chemistries) are expected to contribute to an increase in pesticide use across different regions, especially in regions with increasing temperatures and variable precipitation. The rise in pesticides use may occur in both quantity and frequency of application due to increased pest pressure. This is supported by the fact that, despite measures put in place for global pesticide use reduction, the demand for pesticides is increasing in many regions (Delcour et al., 2015).

20.8.1 Environmental and Human Health Impacts of Pesticides

Pesticide use has considerable repercussions, because of the unintended long-term adverse effects on the environment and human health (Carvalho, 2017; Dahaba et al., 2017). Problems like pesticide resistance and secondary pest outbreaks (Guedes, 2017; Jallow et al., 2018a); destruction of non-target species (Zaller & Brühl, 2019); soil, water, and air contamination (Syafrudin et al., 2021); and pesticide residues in primary and derived agricultural products that endanger both the environment and human-health are a consequence of improper and excessive use of pesticides (Jallow et al., 2017c; Ramadan et al., 2020). Pesticides are associated with a range of human health concerns, from acute impacts (short-term exposure) such as headaches and nausea, to chronic impacts (long-term exposure) including various cancers, birth defects, infertility, endocrine and immune system disruption, diabetes, Parkinson's and Alzheimer's disease (Asghar et al., 2016). Children, in particular, are more endangered by short-term and chronic exposure to pesticides (Lozowicka, 2015). There is mounting evidence that pesticide exposure from various sources can induce neurodevelopment and neurobehavioral effects (delayed cognitive and intellectual development, attention deficit and hyperactivity, etc.) and various cancers in children (Lanphear, 2015).

Exposure to pesticides can occur through the ingestion of residue in food and drinking water, inhalation of pesticide-contaminated air, and dermal contact with contaminated soil and other surfaces (Gangemi et al., 2016). The widest exposure to pesticides, however, is through residues in primary and derived agricultural products, especially fruits and vegetables consumed either raw or semi-processed (Quijano et al., 2016). Farmers, particular, those directly involved in the handling of pesticides, face a high risk of exposure to pesticides through contact with pesticide residues on treated crops, unsafe handling, storage and disposal practices, poor maintenance of spraying equipment, and the lack of protective equipment or failure to use it properly (Jallow et al., 2017a). The World Health Organization estimates 15,000–20,000 fatalities of agricultural workers yearly through occupational exposure to pesticides, mostly in developing countries (WHO & FAO, 2019).

Pesticides can enter the environment through contamination of water, air and soil, and the problem of pesticides in aquatic environments, in particular, is a serious challenge worldwide. Increasing evidence indicate that agricultural pesticides are entering surface waters in many regions with high intensive systems of agricultural production, where pesticides are the dominant pest management strategy. High levels of pesticide residues in surface and ground water resources have been detected in intensive agricultural regions of Iran (Shakerkhatibi et al., 2014), Saudi Arabia (Al-Hatim et al., 2015), and India (Syafrudin et al., 2021), posing a threat to human health directly, and via the food chain through bioaccumulation. Similarly, soil contamination by pesticides, particularly *organochlorine* pesticides (OCPs), has become an issue of increasing concern because of the long-term persistence of these chemicals in the environment and the possibility of entering the food chain via plant uptake (Alshemmari et al., 2021). Although most OCPs, such as DDT, aldrin and endosulfan, have long been banned in majority of countries because of their extreme harmful effects on human health and the environment, these pesticides remain the most frequently detected in contaminated soils (Sun et al., 2018). Pesticides can be toxic to soil microbial communities, including nitrogen-fixing microbes, leading to higher fertilizer requirements and crop vulnerability to insect pests and diseases (Feld et al., 2015). Similarly, pesticides are known to inhibit the growth of symbiotic mycorrhizal fungi that help in plant nutrient uptake (Siviter & Muth, 2020).

There is substantial published literature on the impact of pesticides on wildlife and biodiversity, including beneficial insects. For example, pesticide overuse has significantly contributed to the decline in the populations of beneficial insects (natural enemies and pollinators) rendering vital ecosystem services impossible (Siviter & Muth, 2020). Systemic insecticides such as neonicotinoids, have been shown to impair the foraging behavior of honeybees, consequently reducing their function as pollinators (Lundin et al., 2015). In addition, pesticides can negatively affect the behavior and life-history parameters of natural enemies of insect pests in agroecosystems, aggravating subsequent incidence of pest outbreaks (Guedes, 2017). The potential drivers of excessive pesticide use, apart from climate change, is linked to lack of alternatives to synthetic pesticides, stringent market requirements for crop aesthetic perfection unwillingness of farmers to accept the risk of crop loss, lack of education and training in pesticide use, weak enforcement of pesticide laws, lack of access to extension services, *farmers*' perceptions of pesticide risk and the inadequate information on related hazards (Jallow et al., 2017b).

20.9 Prevention, Mitigation and Adaptation Strategies for Pest Management in a Changing Climate

Implementing integrated preventative, mitigation and adaptation strategies to reduce the potential risks of climate change impacts on agricultural pests is fundamental to reducing the risks of spreading new pests and managing existing pests in agricultural ecosystems. Preventative measures are the most effective approaches to addressing evolving climate change-induced pest threats, and include: 1) regulating the movement of agricultural produce through phytosanitary measures to reduce the spread of pests; 2) surveillance and monitoring of pests to allow detection, and consequently, implementing immediate control and eradication measures; and 3) pest risk modelling and forecasting to project the possible range of pest expansion in new regions (Munkvold & Gullino, 2020; Skendžić et al., 2021). Effective phytosanitary measures, as well as surveillance and monitoring, are essential to prevent invasive species from spreading and becoming established pests in new geographical regions. This requires cooperation among countries, national and regional organizations, research institutes, farmers, extension workers and other stakeholders to manage pest threats more effectively. Farmers and extension agents need to be sensitized and trained through capacity building programs on climate change and its implications for pest incidence, especially invasive species, in order to manage different pest risk scenarios.

In addition to the preventive measures, there is a need to implement mitigation and adaptation strategies to reduce the negative impacts of climate change. Climate change mitigation and adaptation is viewed as an ongoing process of implementing risk management strategies as the climate changes, to reduce the potential risk from climate change. A range of adaptation and mitigation strategies has been identified that farmers and other stakeholders can implement to mitigate the negative impacts of agricultural insect pests and diseases as a consequence of climate change. Some of these strategies include: 1) developing crop varieties better suited to future climate conditions and resilient to pests, diseases, drought and other abiotic stress factors; 2) crop diversification; 3) crop rotation and adjustment of planting time to minimize exposure to pest outbreaks; 4) conservation of beneficial insects including natural enemies and pollinator and; 5) adjustment of agronomic practices such as irrigation, fertilization and improving soil management to reduce erosion (Almulla et al., 2024; Jallow et al., 2022; Juroszek & Von-Tiedemann, 2011; Skendžić et al., 2021). Other strategies include developing new formulations of biopesticides, pheromones and reduced-risk insecticides resilient to changing climate (Jallow et al., 2018b, 2020). Most currently available pest control products, especially biopesticides based on entomopathogens (viruses, nematodes, fungi, and bacteria), are high sensitivity to environmental factors, such as temperature and relative humidity (Williams & Walters, 2000). Increasing temperatures as a consequence of climate change may cause some of these management techniques to be less effective. To maintain and/or enhance the efficacy of current pest control products in the phase of climate change may require developing new formulations and application technologies resilient to climate change.

20.10 Conclusions and Remarks

Climate change, driven by both natural and anthropogenic activities, has accelerated over the past decades, and is predicted to continue increasing if measures to mitigate it are not implemented. Climate parameters, such as rising temperatures, can directly or indirectly affect the population dynamics and distribution of agricultural pests and beneficial organisms such as pollinators and insect natural enemies. While the consequence of climate change can be both positive and negative for pest species and beneficial organisms, most climate scenarios tend to negatively impact beneficial organisms and favoring pest proliferation. Consequently, climate change is generally expected to pose a significant threat to global food security due to potential global increase in current pest pressure, the emergence of new pests and the reduced efficacy of crop protection technologies such as biological and chemical control. This will in turn contribute to a likely increase in pesticide use across different regions, thus, impacting environmental sustainability and human health. Preventative, mitigation and adaptation strategies are strongly needed to mitigate the projected increases in pest risk and pesticide use in agriculture as a consequence of climate change. Priority must be given to developing and promoting climate-smart pest management strategies and non-synthetic methods of pest control that strengthen the resilience of agricultural systems in the face of climate change. Farmers, extension workers, scientists, public and private stakeholders should be sensitized about the impacts of climate change and trained to address the evolving climate change-induced threats to global food security, environmental sustainability and human health. Finally, interventions are also needed to strengthen national and regional enforcement mechanisms concerning greenhouse gases emissions and developing climate-smart agricultural policies and strategies.

References

Adebesin, F., Widhalm, J. R., Boachon, B., Lefèvre, F., Pierman, B., Lynch, J. H., & Porter, J. A. (2017). Emission of volatile organic compounds from petunia flowers is facilitated by an ABC transporter. *Science, 356*, 1386–1388.

Al-Hatim, H. Y., Alrajhi, D., & Al-Rajab, A. J. (2015). Detection of pesticide residue in dams and well water in Jazan Area, Saudi Arabia. *American Journal of Environmental Science, 11*, 358–365.

Allen, R. N. (1978). Spread of bunchy top disease in established banana plantations. *Australian Journal of Agricultural Research, 29*, 1223–1233.

Almulla, L., Thomas, B. M., Jallow, M. F. A., Al-Roumi, A., Devi, V. Y., & Jacob, J. (2024). Rotary drum composting of organic school wastes and compost valorization. *Sustainability, 16*, 2428. https://doi.org/10.3390/su16062428

Alshemmari, H., Al-Shareedah, A. E., Rajagopalan, S., Lina, L. A., Talebi, A., & Hajeyah, H. (2021). Pesticides driven pollution in Kuwait: The first evidence of environmental exposure to pesticides in soils and human health risk assessment. *Chemosphere, 273*, 129688. https://doi.org/10.1016/j.chemosphere.

Andreazza, F., Bernardi, D., DosSantos, R. S. S., Garcia, F. R. M., Oliveira, E. E., Botton, M., & Nava, D. E. (2017). *Drosophila suzukii* in Southern Neotropical Region: Current status and future perspectives. *Neotropical Entomology, 46*, 591–605. https://doi.org/10.1007/s13744-017-0554-7

Asghar, U., Malik, M. F., & Javed, A. (2016). Pesticide exposure and human health: A review. *Journal of Ecosystem and Ecography S, 5*, 005. https://doi.org/10.4172/2157-7625.S5-005

Atkinson, N. J., & Urwin, P. E. (2012). The interaction of plant biotic and abiotic stresses: From genes to the field. *Journal of Experimental Botany, 63*, 3523–3543.

Azrag, A. G. A., Obala, F., Tonnang, H. E. Z., Hogg, B. N., Ndlela, S., & Mohamed, S. A. (2023). Predicting the impact of climate change on the potential distribution of the invasive tomato pinworm *Phthorimaea absoluta* (Meyrick) (Lepidoptera: Gelechiidae). *Scientific Reports, 13*, 16477. https://doi.org/10.1038/s41598-023-43564-2

Bailey, S. W. (2004). Climate change and decreasing herbicide persistence. *Pest Management Science, 60*, 158–162.

Bale, J. S., Masters, G. J., Hodkinson, I. D., Awmack, C., Bezemer, T. M., Brown, V. K., Butterfield, J., Buse, A., Coulson, J. C., Farrar, J., Good, J., Harrington, R., Hartley, S., Jones, T., Lindroth, R., Press, M., Symrnioudis, L., Watt, A., & Whittaker, J. (2002). Herbivory in global climate change research: Direct effects of rising temperature on insect herbivores. *Global Change Biology, 8*, 1–16. https://doi.org/10.1046/j.1365-2486.2002.00451.x

Barker, K. R., & Koenning, S. R. (1998). Developing sustainable systems for nematode management. *Annual Review of Phytopathology, 36*, 165–205.

Bebber, D., Ramotowski, M., & Gurr, S. (2013). Crop pests and pathogens move poleward in a warming world. *Nature Climate Change, 3*, 985–988.

Biondi, A., Guedes, R. N. C., Wan, F., & Desneux, N. (2018). Ecology, worldwide spread, and management of the invasive South American tomato pinworm, *Tuta absoluta*: Past, present, and future. *Annual Review of Entomology, 63*, 239–258.

Bloomfield, J. P., Williams, R. J., Gooddy, D. C., Cape, J. N., & Guha, P. (2006). Impacts of climate change on the fate and behaviour of pesticides in surface and groundwater—A UK perspective. *Science of the Total Environment, 369*, 163–177.

Butler, D. (2013). Fungus threatens top banana. *Nature, 504*, 195–196.

Carvalho, F. P. (2017). Pesticides, environment and food safety. *Food and Energy Security, 6*, 48–60.

Charaya, M. U., Upadhyay, A., Bhati, H. P., & Kumar, A. (2021). Plant disease forecasting: Past practices to emerging technologies. In S. Nehra (Ed.), *Plant Dis Management Strategy* (pp. 1–30). Agrobios Research Rajasthan.

Chen, F. J., Wu, G., Lu, J., & Ge, F. (2005). Effects of elevated CO_2 on the foraging behaviour of cotton bollworm, *Helicoverpa armigera*. *Insect Science, 12*, 359–365.

Ciliberti, N., Fermaud, M., Roudet, J., & Rossi, V. (2015). Environmental conditions affect *Botrytis cinerea* infection of mature grape berries more than the strain or transposon genotype. *Phytopathology, 105*, 1090–1096.

Cini, A., Ioriatti, C., & Anfora, G. (2012). A review of the invasion of *Drosophila suzukii* in Europe and a draft research agenda for integrated pest management. *Bulletin of Insectology, 65*, 149–160.

Cotrufo, M. F., Ineson, P., & Scott, A. (1998). Elevated CO_2 reduces the nitrogen concentration of plant tissues. *Global Change Biology, 4*, 43–54. https://doi.org/10.1046/j.1365-2486.1998.00101.x

Dahaba, A. A., Jallow, M. F. A., & Albaho, M. S. (2017). Environmental and human health impacts of pesticide use in agriculture. In *Pesticides* (pp. 1–15). Avid Science.

Dale, J. L. (1987). Banana bunchy top: An economically important tropical plant virus disease. *Advances in Virus Research, 33*, 301–325.

De Waele, D., & Elsen, A. (2007). Challenges in tropical plant nematology. *Annual Review of Phytopathology, 45*, 457–485.

Delcour, I., Spanoghe, P., & Uyttendaele, M. (2015). Literature review: Impact of climate change on pesticide use. *Food Research International, 68*, 7–15.

Desneux, N., Luna, N. G., Guillemaud, T., & Urbaneja, A. (2011). The invasive South American tomato pinworm, *Tuta absoluta*, continues to spread in Afro-Eurasia and beyond: The new threat to tomato world production. *Journal of Pest Science, 84*, 403–408.

Dita, M., Barquero, M., Heck, D., Mizubuti, E. S. G., & Staver, C. P. (2018). Fusarium wilt of Banana: Current knowledge on epidemiology and research needs toward sustainable disease management. *Frontiers in Plant Science, 9*, 1468. https://doi.org/10.3389/fpls.2018.01468

Dutta, T. K., & Phani, V. (2023). The pervasive impact of global climate change on plant-nematode interaction continuum. *Frontiers in Plant Science, 6*(14), 1143889. https://doi.org/10.3389/fpls.2023.1143889

Eigenbrode, S. D., & Adhikari, S. (2023). Climate change and managing insect pests and beneficials in agricultural systems. *Agronomy Journal, 115*. https://doi.org/10.1002/agj2.21399.

El-Shafie, H. A. F., & Faleiro, J. R. (2020). Red palm weevil *Rhynchophorus ferrugineus* (Coleoptera: Curculionidae): Global invasion, current management options, challenges, and future prospects. In H. A. F. El-Shafie (Ed.), *Invasive species-introduction pathways, economic impact, and possible management options* (pp. 3–31). Intechopen.

EPPO (European and Mediterranean Plant Protection Organization). (2020). List of pests recommended for regulation as quarantine pests, version 2020–09. In *European and mediterranean plant protection organization.* https://www.eppo.int/ACTIVITIES/plant_quarantine/alert_list. Accessed on 18 July 2024.

EPPO (European and Mediterranean Plant Protection Organization). (2024). *Tuta absoluta* global distribution map. In *European and mediterranean plant protection organization* Online Database. https://gd.eppo.int/taxon/GNORAB/distribution. Accessed on 6 Jun 2024.

FAO. (2023). Pesticides use. In *World food and agriculture—statistical yearbook 2023, Rome.* https://doi.org/10.4060/cc8166en.

Feld, L., Hjelmsø, M. H., Nielsen, M. S., Jacobsen, A. D., Rønn, R., Ekelund, F., Krogh, P. H., Strobel, B. W., & Jacobsen, C. S. (2015). Pesticide side effects in an agricultural soil ecosystem as measured by amoA expression quantification and bacterial diversity changes. *PLoS ONE, 10*, Article e0126080.

Gangemi, S., Gofita, E., Costa, C., Teodoro, M., Briguglio, G., Nikitovic, D., Tzanakakis, G., Tsatsakis, A. M., Wilks, M. F., Spandidos, D. A., & Fenga, C. (2016). Occupational and environmental exposure to pesticides and cytokine pathways in chronic diseases (Review). *International Journal of Molecular Medicine, 38*, 1012–1020.

Gérard, M., Vanderplanck, M., Wood, T., & Michez, D. (2020). Global warming and plant-pollinator mismatches. *Emerging Topics in Life Sciences, 4*, 77–86. https://doi.org/10.1042/ETLS20190139

Ghosh, S., & Ghanim, M. (2021). Factors determining transmission of persistent viruses by *Bemisia tabaci* and emergence of new virus-vector relationships. *Viruses, 1113*(9), 1808. https://doi.org/10.3390/v13091808

Godfray, H. C. J., Beddington, J. R., Crute, I., Haddad, L., Lawrence, D., Muir, J. F., Pretty, J., Robinson, S., Thomas, S. M., & Toulmin, C. (2010). Food security: The challenge of feeding 9 billion people. *Science, 327*, 812–818. https://doi.org/10.1126/science.1185383

Goergen, G., Kumar, P. L., Sankung, S. B., Togola, A., & Tamò, M. (2016). First report of outbreaks of the fall armyworm Spodoptera frugiperda (J E Smith) (Lepidoptera, Noctuidae), a new alien invasive pest in West and Central Africa. *PLoS ONE, 11*(10), Article e0165632. https://doi.org/10.1371/journal.pone.0165632

Guedes, R. N. C. (2017). Insecticide resistance, control failure likelihood and the first law of geography. *Pest Management Science, 73*, 479–484.

Hance, T., Van Baaren, J., Vernon, P., & Boivin, G. (2007). Impact of extreme temperatures on parasitoids in a climate change perspective. *Annual Review of Entomology, 52*, 107–126. https://doi.org/10.1146/annurev.ento.52.110405.091333

Henry, J., Hesham, M., & Aatadal, A. S. (2022). Climate smart agriculture and mitigation techniques for sustainable resilient farming in Middle East Region. *Global Business and Management Research, 14*, 12.

Hoddle, M. S., Hoddle, C. D., Faleiro, J. R., El-Shafie, H. A. F., Jeske, D. R., & Sallam, A. A. (2015). How far can the red palm weevil (Coleoptera: Curculionidae) fly?: Computerized flight mill studies with field-captured weevils. *Journal of Economic Entomology, 108*, 2599–2609.

Inkley, D. B. (2012). Characteristics of home invasion by the brown marmorated stink bug (Hemiptera: Pentatomidae). *Journal of Entomological Science, 47*, 125–130.

IPCC (Intergovernmental Panel on Climate Change). (2013). Climate change 2013: The physical science basis. In T. F. Stocker, D. Qin, G. K. Lattner, M. Tignor, S. K. Allen, J. Boschung, A. Nauels, Y. Xia, V. Bex, & P. M. Midgley (Eds.),*Contribution of Working Group I to the Fifth Assessment Report of the Intergovernmental Panel on Climate Change* (p. 1535). Cambridge University Press.

IPCC (Intergovernmental Panel on Climate Change). (2014). Climate change 2014: Synthesis report. In R. K. Pachauri, L. A.Meyer (Eds.), *Contribution of Working Groups I, II and III to the Fifth Assessment Report of the Intergovernmental Panel on Climate Change* (p. 151). IPCC.

Jallow, M. F. A., Cunningham, J. P., & Zalucki, M. P. (2004). Intra-specific variation for host plant use in *Helicoverpa armigera* (Hübner) (lepidoptera: Noctuidae): Implications for management. *Crop Protection, 23*, 955–964.

Jallow, M. F. A., Awadh, D. G., Albaho, M. S., Devi, V. Y., & Thomas, B. M. (2017a). Pesticide risk behaviors and factors influencing pesticide use among farmers in Kuwait. *Science of the Total Environment, 574*, 490–498.

Jallow, M. F. A., Awadh, D. G., Albaho, M. S., Devi, V. Y., & Thomas, B. M. (2017b). Pesticide knowledge and safety practices among farm workers in Kuwait: Results of a survey. *International Journal of Environmental Research and Public Health, 14*, 330. https://doi.org/10.3390/ijerph14040340.

Jallow, M. F. A., Awadh, D. G., Albaho, M. S., Devi, V. Y., & Ahmad, N. (2017c). Monitoring of pesticide residues in commonly used fruits and vegetables in Kuwait. *International Journal of Environmental Research and Public Health, 14*, 833. https://doi.org/10.3390/ijerph14080833.

Jallow, M. F. A., Dahaba, A. A., Albaho, M. S., Devi, V. Y., Awadh, D. G., & Thomas, B. M. (2018a). Baseline susceptibility and assessment of resistance risk to flubendiamide and chlorantraniliprole in *Tuta absoluta* (Lepidoptera: Gelechiidae) populations from Kuwait. *Applied Entomology and Zoology, 54*. https://doi.org/10.1007/s13355-018-0598-0.

Jallow, M. F. A., Dahaba, A. A., Albaho, M. S., & Devi, V. Y. (2018b). Efficacy of some biorational insecticides against *Tuta absoluta* (Meyrick) (Lepidoptera: Gelechiidae) under laboratory and greenhouse conditions in Kuwait: *Journal of Applied Entomology, 143*. https://doi.org/10.1111/jen.

Jallow, M. F. A., Dahaba, A. A., Albaho, M. S., Devi, V. Y., Jacob, J., & Al-Saeed, O. (2020). Efficacy of mating disruption compared with chemical insecticides for controlling *Tuta absoluta* (Lepidoptera: Gelechiidae) in Kuwait. *Applied Entomology and Zoology, 55*. https://doi.org/10.1007/s13355-020-00673-y.

Jallow, M. F. A., Dahab, A. A., Albaho, M. S., & Devi, V. Y. (2022). Efficacy of *Nesidiocoris tenuis* (Hemiptera: Miridae) and Bacillus thuringiensis (Berliner) for controlling *Tuta absoluta* (Lepidoptera: Gelechiidae) in greenhouse tomato crops under Kuwait hot desert climate. *International Journal of Pest Management,* 1–11. https://doi.org/10.1080/09670874.2022.2162998.

Jeger, M., Bragard, C., Caffier, D., Candresse, T., Chatzivassiliou, E., Dehnen-Schmutz, K., Gilioli, G., Gregoire, J., Miret, J. A. J., Navarro, M. N., Niere, B., Parnell, S., Potting, R., Rafoss, T., Rossi, V., Urek, G., Bruggen, A. V., Werf, W. V., West, J., & MacLeod, A. (2018). Pest risk assessment of *Spodoptera frugiperda* for the European Union. *European Food Safety Authority Journal, 16*, 5351. https://doi.org/10.2903/j.efsa.2018.5351

Juroszek, P., & Von-Tiedemann, A. (2011). Potential strategies and future requirements for plant disease management under a changing climate. *Plant Pathology, 60*, 100–112.

Kambrekar, D. N., Guledgudda, S. S., & Anand, K. (2015). Impact of climate change on insect pests and their natural enemies. *Karnataka Journal of Agricultural Science, 28*, 814–816.

Kansman, J. T., Basu, S., Casteel, C. L., Crowder, D. V., Lee, B. W., Nihranz, C. T., & Finke, D. L. (2022). Water stress reduces aphid performance: Exploring mechanisms driven by water stress intensity. *Frontiers Ecology and Evolution, 10*, 846. https://doi.org/10.3389/fevo.2022.846908.

Kimathi, E., Tonnang, H. E. Z., Subramanina, K., Abdel-Rahman, E. M., Tesfayohannes, M., Niassy, S., Torto, B., Dubois, T., Tanga, C. M., Kassie, M., Ekesi, S., Mwangi, D., & Kelemu, S. (2020). Prediction of breeding regions for the desert locust *Schistocerca gregaria* in East Africa. *Scientific Reports, 10*, 11937. https://doi.org/10.1038/s41598-020-68895-2

Kiss, B., Kis, A., & Kákai, Á. (2016). The rapid invasion of spotted wing drosophila, *Drosophila suzukii* (Matsumura) (Diptera: Drosophilidae), in Hungary. *Phytoparasitica, 6*(44), 429–433. https://doi.org/10.1007/s12600-016-0520-7

Kocmánková, E., Trnka, M., Juroch, J., Dubrovský, M., Semerádová, D., Možný, M., Žalud, Z., Pokorný, R., & Lebeda, A. (2010). Impact of climate change on the occurrence and activity of harmful organisms. *Plant Protection Science, 45*, S48–S52. https://doi.org/10.17221/2835-PPS

Kumar, P. L., Hanna, R., Alabi, O. J., Soko, M. M., Oben, T. T., Vangu, G. H. P., & Naidu, R. A. (2011). Banana bunchy top virus in sub-Saharan Africa: Investigations on virus distribution and diversity. *Virus Research, 159*, 171–182.

Lanphear, B. P. (2015). The impact of toxins on the developing brain. *Annual Review of Public Health, 36*, 211–230.

Legrand, M. A., Colinet, H., Vernon, P., & Hance, T. (2004). Autumn, winter and spring dynamics of aphid *Sitobion avenae* and parasitoid *Aphidius rhopalosiphi* interactions. *Annals of Applied Biology, 145*, 139–144.

Leskey, T. C., & Nielsen, A. L. (2018). Impact of the invasive brown marmorated stink bug in North America and Europe: History, biology, ecology, and management. *Annual Review of Entomology, 63*, 599–618.

Lincoln, D. E. (1993). The influence of plant carbon dioxide and nutrient supply on susceptibility to insect herbivores. *Vegetation, 104*, 273–280. https://doi.org/10.1007/BF00048158

Lorang, T., Kidarsa, C. S., Bradford, B., Gilbert, M., Curtis, S. C., Tzeng, C. S., Maier, T. J., & Wolpet, T. J. (2012). Tricking the guard: Exploiting plant defense for disease susceptibility. *Science, 338*, 659–662.

Lozowicka, B. (2015). Health risk for children and adults consuming apples with pesticide residue. *Science of the Total Environment*, 502184–502198.

Lundin, O., Rundlöf, M., Smith, H. G., Fries, I., & Bommarco, B. (2015). Neonicotinoid insecticides and their impacts on bees: A systematic review of research approaches and identification of knowledge gaps. *PLoS ONE, 10*(8), Article e0136928.

Martin, J. H., & Mound, L. A. (2007). An annotated check list of the world's whiteflies (Insecta: Hemiptera: Aleyrodidae). *Zootaxa* 1–84. https://doi.org/10.11646/zootaxa.1492.1.1.

Martyn, R. D. (2014). Fusarium wilt of watermelon: 120 years of research. *Horticultural Reviews, 42*, 349–442. https://doi.org/10.1002/9781118916827.ch07

Masters, G., & Norgrove, L. (2010). Climate change and invasive alien species (p. 30). CABI. CABI Working Paper 1.

Mastrangelo, T. D. F., Paulo, L. W., Bergam, E. G. F., Morais, M., Silva, G., Bezerra-Silva, A. M. L., & Azeredo-Espin, A. (2014). Detection and genetic diversity of a Heliothine invader (Lepidoptera: Noctuidae) from north and northeast of Brazil. *Journal of Economic Entomology, 107*, 970–980. https://doi.org/10.1603/EC13403

Melotto, W., Underwood, W., Koczan, J., Nomura, K., & He, S. Y. (2006). Plant stomata function in innate immunity against bacterial invasion. *Cell, 126*, 969–980.

Meynard, C. N., Gay, P. E., Lecoq, M., Foucart, A., Piou, C., & Chapuis, M. P. (2017). Climate-driven geographic distribution of the desert locust during recession periods: Subspecies' niche differentiation and relative risks under scenarios of climate change. *Global Change Biology, 23*, 4739–4749.

Munkvold, G. P., & Gullino, M. L. (2020). Seed and propagative material. In M. L. Gullino, R. Albajes, & P. C. Nicot (Eds.), *Integrated pest and disease management in greenhouse crops* (pp. 331–354). Dordrecht.

Murhububa, I. S., Bragard, C., Tougeron, K., & Hance, T. (2024). Preference of *Pentalonia nigronervosa* for infected banana plants tends to reverse after *Banana bunchy top virus* acquisition. *Scientific Reports, 14*, 2993. https://doi.org/10.1038/s41598-024-53205-x

NAS (National Academy of Sciences). (2020). *Climate change: Evidence and causes: Update 2020* (p. 9). The National Academies Press. https://doi.org/10.17226/25733

Parmesan, C., & Yohe, G. (2003). A globally coherent fingerprint of climate change impacts across natural systems. *Nature Cell Biology, 421*, 37–42. https://doi.org/10.1038/nature01286

Prakash, A., Rao, J., Mukherjee, A. K., Berliner, J., Pokhare, S. S., Adak, T., Munda, S., & Shashank, P. R. (2014). *Climate change: Impact on crop pests.* Applied Zoologists Research Association (AZRA), Central Rice Research Institute.

Quijano, L., Yusá, V., Font, G., & Pardo, O. (2016). Chronic cumulative risk assessment of the exposure to organophosphorus, carbamate and pyrethroid and pyrethrin pesticides through fruit and vegetables consumption in the region of Valencia (Spain). *Food and Chemical Toxicology, 89*, 39–46.

Ramadan, M. F. A., Abdel-Hamid, M. M. A., Altorgoman, M. M. F., AlGaramah, H. A., Alawi, M. A., Shati, A. A., Shweeta, H. A., & Awwad, N. S. (2020). Evaluation of pesticide residues in vegetables from the Asir Region, Saudi Arabia. *Molecules, 25*, 205. https://doi.org/10.3390/molecules25010205.

Rawat, A., Kumar, D., & Khati, B. S. (2024). A review on climate change impacts, models, and its consequences on different sectors: A systematic approach. *Journal of Water and Climate Change, 1*, 104–126. https://doi.org/10.2166/wcc.2023.536

Sayed, A., & Saad, A. K. (2014). Integrated pest management for *Tuta absoluta* (Lepidoptera:Gelechiidae) infesting tomato plants. *Egyptian International Journal of Agricultural Research, 19*, 756–767.

Schweiger, O., Settele, J., Kudrna, O., Klotz, S., & Kühn, I. (2008). Climate change can cause spatial mismatch of trophically interacting species. *Ecology, 89*, 3472–3479.

Shakerkhatibi, M., Mosaferi, M., Asghari-Jafarabadi, M., Lotfi, E., & Belvasi, M. (2014). Pesticides residue in drinking groundwater resources of rural areas in the northwest of Iran. *Health Promotion Perspectives, 4*, 195–205.

Sharma, A., Kumar, V., Shahzad, B., Tanveer, M., Sidhu, J. P. S., Handa, N., Kohli, S. K., Yadav, P., Bali, A. S., Parihar, R. D., Dar, O. I., Sing, K., Jasrotia, S., Bakshi, P., Ramakrishnan, M., Kumar, S., Bhardwaj, R., & Thukral, A. S. (2019). Worldwide pesticide usage and its impacts on ecosystem. *SN Applied Sciences, 1*, 1446. https://doi.org/10.1007/s42452-019-1485-1

Shattuck, A., Werner, M., Mempel, F., Dunivin, Z., & Galt, R. (2023). Global pesticide use and trade database (GloPUT): New estimates show pesticide use trends in low-income countries substantially underestimated. *Global Environmental Change, 81*, Article 102693. https://doi.org/10.1016/j.gloenvcha.2023.102693

Shrestha. (2019). Effects of climate change in agricultural insect pest. *Acta Scientific Agriculture, 3*, 74–80. https://doi.org/10.31080/ASAG.2019.03.0727

Siviter, H., & Muth, F. (2020). Do novel insecticides pose a threat to beneficial insects? *Proceedings of the Royal Society B: Biological Sciences 287*, 20201265. https://doi.org/10.1098/rspb.2020.1265.

Skendžić, S., Zovko, M., Živković, I. P., Lešić, V., & Lemić, D. (2021). The impact of climate change on agricultural insect pests. *Insects, 12*, 440. PMID: 34066138; PMCID: PMC8150874. https://doi.org/10.3390/insects12050440.

Stainton, D., Martin, D. P., Muhire, B. M., Lolohea, S., Halafihi, M., Lepoint, P., Blomme, G., Crew, K. S., Sharman, M., Kraberger, S., Dayaram, A., Walters, M., Collings, D. A., Mabvakure, B., Lemey, P., Harkins, G. W., Thomas, J. E., & Varsani, A. (2015). The global distribution of banana bunchy top virus reveals little evidence for frequent recent, human-mediated long distance dispersal events. *Virus, 10*, 1. https://doi.org/10.1093/ve/vev009

Staley, J. T., Hodgson, C. J., Mortimer, S. R., Morecroft, M. D., Masters, G. J., Brown, V. K., & Taylor, M. E. (2007). Effects of summer rainfall manipulations on the abundance and vertical distribution of herbivorous soil macro-invertebrates. *European Journal of Soil Biology, 43*, 189–198. https://doi.org/10.1016/j.ejsobi.2007.02.010

Stiling, P., & Cornelissen, T. (2007). How does elevated carbon dioxide (CO_2) affect plant–herbivore interactions. A field experiment and meta-analysis of CO_2-mediated changes on plant chemistry and herbivore performance. *Global Change Biology, 13*, 1823–1842.

Stoeckli, S., Felber, R., & Haye, T. (2020). Current distribution and voltinism of the brown marmorated stink bug, *Halyomorpha halys*, in Switzerland and its response to climate change using a high-resolution CLIMEX model. *International Journal of Biometeorology, 64*, 2019–2032. https://doi.org/10.1007/s00484-020-01992-z

Streck, N. A. (2005). Climate change and agroecosystems: The effect of elevated atmospheric CO_2 and temperature on crop growth, development, and yield. *Ciência Rural, 35*, 730–740. https://doi.org/10.1590/S0103-84782005000300041

Subedi, B., Poudel, A., & Aryal, S. (2023). The impact of climate change on insect pest biology and ecology: Implications for pest management strategies, crop production, and food security. *Journal of Agriculture and Food Research, 14*(6), Article 100733. https://doi.org/10.1016/j.jafr.2023.100733

Sun, S., Virinder, S., Rong, Y., & Zheng, Y. (2018). Pesticide pollution in agricultural soils and sustainable remediation methods: A review. *Current Pollution Reportd, 4*(3), 1–11.

Syafrudin, M., Kristani, R. A., Yuniarto, A., Hadibarata, T., Rhee, J., Al-Onazi, W. A., Algarni, T. S., Almarri, A. H., & Al-Mohaimeed, A. M. (2021). Pesticides in drinking water—A review. *International Journal of Environmental Research and Public Health, 18*, 468. https://doi.org/10.3390/ijerph18020468.

Taravati, S., Glenn, H., & Mannion, C. (2014). Daily flight activity of the rugose spiraling whitefly (Hymiptera: Aleyrodidae). *Florida Entomologist, 97*, 1842–1844.

Tilman, D., Balzer, C., Hill, J., & Befort, B. L. (2011). Global food demand and the sustainable intensification of agriculture. *Proceedings of National Academy of Sciences USA, 108*, 20260–20264. https://doi.org/10.1073/pnas.1116437108

Trebicki, P. (2020). Climate change and plant virus epidemiology. *Virus Research, 286*, Article 198059. https://doi.org/10.1016/j.virusres.2020.198059

Whitfield, A. E., Falk, B. W., & Rotenberg, D. (2015). Insect vector-mediated transmission of plant viruses. *Virology, 479*, 278–289.

WHO and FAO. (2019). Global situation of pesticide management in agriculture and public health: report of a 2018 WHO-FAO survey. World Health Organization. Available online: https://apps.who.int/iris/handle/10665/329971. Accessed Jul 2024.

Williams, E., & Walters, K. (2000). Foliar application of the entomopathogenic nematode *Steinernema feltiae* against leafminers on vegetables. *Biocontrol Science and Technology, 10*, 61–70.

Zaller, J. G., & Brühl, C. A. (2019). Editorial: non-target effects of pesticides on organisms inhabiting agroecosystems. *Frontiers in Environmental Science, 7*, 75. https://doi.org/10.3389/fenvs.2019.00075.

Zhao, H., Xian, X., Zhao, Z., Zhang, G., Liu, W., & Wan, F. (2022). Climate change increases the expansion risk of *Helicoverpa zea* in China according to potential geographical distribution estimation. *Insects, 13*, 79. https://doi.org/10.3390/insects13010079

Index

A

Adaptation, 30, 34, 75, 76, 78, 111, 112, 122, 148–150, 183, 185, 186, 199, 204, 207, 234, 238, 239, 273–277, 282, 283, 287, 289, 291, 293, 296–302, 307, 308, 310, 314, 317, 318, 320, 324, 325, 327, 340, 354, 361, 372, 410, 425–427, 430, 437, 440, 441, 451, 462–464
Agricultural pests, 449, 460, 462, 464
Agricultural policies, 251–253, 255, 260, 308, 464
Agricultural production, 32, 54, 58, 68, 110, 111, 132, 133, 158, 161, 163, 183, 191, 198, 231, 232, 236, 239, 256–258, 261, 262, 312, 437, 451, 455, 459, 462
Agriculture, 25, 27, 30, 35, 36, 38, 45, 54, 57–62, 72, 88–90, 92, 93, 95, 97, 100–103, 109–112, 118, 123, 124, 131, 133–135, 142, 145, 147–150, 155, 157, 158, 160–163, 165–167, 172, 173, 185–192, 196, 198–200, 204, 231–233, 236, 237, 240, 242, 251–256, 259, 260, 262, 263, 265, 266, 274, 276–283, 287–290, 292, 294, 296–302, 307–317, 321, 323, 325–328, 340, 350, 371, 410, 411, 415, 425, 426, 431, 437, 439, 441, 449, 459, 464
Agriculture economic policies, 252
Antarctica, 4
Aquaculture, 54, 157, 165, 403, 406–408, 410, 411, 413, 415, 417, 418, 419
Aquaponics, 170, 263
Arable land, 25, 26, 37, 38, 87, 91, 93, 96, 100, 112, 126, 132, 157, 159, 160, 169, 195, 232, 236, 275, 279, 287, 288, 293, 294, 309, 313
Arid, 25, 26, 37, 45–49, 51, 52, 54, 57–62, 67, 68, 70, 71, 78, 87, 95, 96, 98, 103, 109–111, 115, 117, 118, 122–124, 126, 131–133, 148–150, 155–158, 160, 164–168, 170, 172, 181–183, 187, 189–193, 195–198, 202, 204, 206–208, 210–212, 216–219, 229–233, 235–237, 239–242, 244, 251, 256–262, 266, 273, 274, 276–278, 281, 283, 287–290, 294, 296–298, 307, 309–313, 315, 317–320, 327, 340, 342–344, 347–355, 358–364, 392, 403, 406, 410, 415, 418, 425–429, 434–438, 440, 441
Aridity, 26, 45, 68, 70, 96, 111, 149, 156, 196, 197, 200, 207, 236, 274, 287, 288, 296, 309, 339, 355, 427, 428, 451
Aridity index, 111, 274, 288, 427
Arid lands, 25, 26, 37, 45, 54, 57, 68, 70, 109, 110, 117, 122, 131–133, 148, 150, 156, 182, 196–199, 201, 202, 240, 251, 252, 273, 274, 276–278, 287–290, 294, 296, 297, 309, 312, 318, 340, 342–344, 348, 352, 410, 426, 427, 434–438, 440
Arid regions, 45–47, 49–52, 54, 60–62, 67, 68, 75, 78, 109, 131, 132, 148, 156, 157, 164–166, 173, 187, 195–198, 206, 211, 216, 229–231, 233,

M. K. Suleiman and S. A. Shahid (eds.), *Fostering Arid Lands Agriculture in the Face of Climate Change*, Advances in Global Change Research 81,
https://doi.org/10.1007/978-3-031-98490-7

235–237, 239, 241–243, 256, 258–262, 266, 287, 309, 311–313, 315, 317, 319, 320, 327, 340, 342, 347–355, 358, 360, 363, 364, 392, 406, 410, 426–429, 434, 437, 438, 441

B

Biofloc, 403, 410, 411, 413, 415, 417, 418
Biome, 196, 348, 349, 359
Business as usual, 28, 89, 114, 123, 126, 301

C

Capacity building, 181–184, 316, 326, 328, 463
Climate action, 276–278, 282, 283
Climate change, 29, 30, 33–35, 38, 45, 54, 61, 67–69, 71–79, 89, 93, 104, 109–112, 114, 118, 121–123, 131, 133, 147–150, 155, 157, 158, 160, 162, 172, 181, 182, 187, 190, 195, 196, 198–201, 204, 211, 212, 217, 219, 229–232, 234, 236, 238, 239, 242, 251–254, 256–259, 261, 262, 264, 266, 273–280, 282, 283, 287–291, 293, 294, 297, 300–302, 307–318, 320–325, 327, 328, 339–342, 347–350, 354, 356–364, 371, 373, 403, 405, 406, 408, 419, 425–432, 434–437, 439, 441, 449–451, 453–455, 457–460, 462–464
Climate constraints, 182
Climate impact, 187, 259, 359
Climate literacy, 276, 278
Climate resilient, 31, 131, 239, 240, 242, 308, 324–326
Climate smart, 61, 62, 101, 110, 204, 219, 262, 263, 280, 282, 283, 297, 302, 316, 318, 321, 327, 328
Climate-Smart Agriculture (CSA), 110, 160, 186, 263, 280, 282, 283, 297, 300, 321
Community structure, 425, 426
Conservation agriculture, 157, 186, 259, 266, 301, 310, 316, 326, 431
Crop production, 25, 26, 37, 38, 45, 46, 57, 61, 62, 71, 88, 91–93, 100–104, 109–112, 115, 118, 121, 123–126, 131, 132, 134, 137, 142, 143, 146–149, 157–159, 161, 164–167, 170, 172, 182, 183, 185, 188, 190, 192, 196, 198, 200, 229, 230, 232, 241–243, 262, 274, 277, 280, 282, 283, 287, 288, 293, 294, 296, 297, 302, 308, 313, 317–320, 325, 326, 328, 341, 359, 425, 430, 437, 440, 449, 450, 458, 459

D

Desertification, 32, 33, 35, 54, 58, 96, 156, 197, 234, 280, 289, 309, 310, 313, 314, 316, 339, 342, 347, 349–351, 354, 355, 437, 438
Desert plants, 215, 218, 219, 350
Developed countries, 172, 251, 252, 254, 256, 281
Developing countries, 32, 68, 87, 103, 109, 111, 157, 200, 202, 211, 233, 235, 243, 251–258, 260–262, 266, 276, 277, 279, 287, 289, 290, 297, 302, 373, 461
Drought, 30, 37, 50, 51, 61, 67, 68, 75, 77, 78, 95, 97, 109, 111, 121, 122, 124, 131, 148, 150, 156–158, 161, 181, 182, 185, 189, 198, 200, 203, 204, 206, 207, 210, 229, 232, 234, 236, 240, 242, 264, 273, 274, 280, 282, 288–290, 293, 296, 298, 300, 307, 312, 313, 316, 317, 320, 322–325, 327, 340, 342, 347, 350, 354, 357, 358, 426, 429, 430, 436–439, 441, 450, 451, 453–455, 458, 463
Drylands, 46, 62, 93, 126, 132, 156, 196, 233, 274, 277, 287, 288, 290, 311, 339, 350, 364, 428
Dynamic Agricultural Economic Policy (DAEP), 252, 253, 257–262, 264–266

E

Economic development, 100, 103, 280, 315
Ecosystem services, 25, 26, 31–35, 182, 234, 236–238, 273, 276, 277, 289, 294, 302, 315, 316, 320, 348, 356, 359, 361, 425, 430, 432, 435, 450, 453, 462
Eggs, 372–378, 383, 384, 387, 392, 405
Environmental and human exposure, 450
Escalating population, 26, 28, 38, 90, 97, 104, 109, 111, 123, 135, 155, 181, 183, 189, 192, 193, 195, 288, 290, 293, 302, 373

Extension, 29, 36, 58, 61, 62, 116, 146, 150, 181, 241–244, 280, 281, 283, 300–302, 463, 464
Extension services, 36, 58, 61, 62, 116, 146, 181, 280, 281, 283, 300, 301, 326, 462

F
Feed, 28, 30, 67, 69, 71–74, 76–78, 88, 90, 91, 97, 102, 109, 110, 112, 123, 125, 126, 158, 181–183, 189, 191, 192, 202, 216, 217, 219, 231, 241, 262, 279, 288, 302, 308, 373, 374, 376–378, 381, 384–387, 394, 403, 406, 411, 414, 417, 452, 454, 456–458
Fodder, 50, 70, 73, 79, 149, 189, 215, 216, 234, 235, 237, 240, 241, 243, 381
Food losses and waste, 109, 125, 279
Food policies, 88
Food security, 25, 26, 31–35, 37, 38, 45, 58, 59, 62, 78, 88, 93, 100, 101, 103, 109, 111, 112, 123, 125, 126, 131, 132, 134, 136, 147–150, 155, 157, 158, 182, 183, 187, 189, 192, 196, 198–202, 206, 212, 216, 229–232, 235, 237, 242, 244, 251–257, 259–262, 264–266, 273, 274, 276, 277, 279, 280, 282, 283, 287–290, 294–298, 307–309, 312–316, 318, 324, 326–328, 341, 355, 359, 372, 403, 407, 408, 417, 425, 431, 436, 437, 440, 441, 449–451, 455, 464
Food trade, 94
Forage, 69, 72, 73, 75–78, 92, 142, 216, 229, 231–233, 238, 239, 241, 243, 322, 457
Foreign investor, 87, 100, 101, 104, 105

G
Gaps, 26, 110, 125, 187, 188, 218, 261, 273–277, 280–283, 327, 435, 441
Global warming, 30, 114, 121, 133, 148, 162, 232, 275, 276, 280, 288, 298, 308, 310, 341, 359, 360, 425–434, 436, 437, 439–441, 451, 458
Greenhouse gas mitigation, 68

H
Hunger, 30, 35, 88, 90, 110, 114, 147, 157, 172, 182, 195, 201, 202, 211, 278, 290, 302, 307, 314, 315, 326–328
Hydrological cycle, 356

I
Innovations, 58, 109, 121, 123, 124, 149, 362
Invasive species, 350, 354, 449, 450, 456, 457, 463

L
Land uses, 35, 239
Livelihood, 67, 68, 71, 90, 101, 123, 126, 156, 159, 182, 198, 204, 206, 219, 234, 237, 238, 243, 254, 255, 276, 280, 290, 294, 300, 308, 310, 312
Livestock, 67–79, 89, 96, 111, 156, 165, 185, 196, 200, 216, 229, 230, 232, 233, 240–243, 264, 290, 308, 315, 322–324, 347, 351, 372–374
Low-salinity groundwater, 417

M
Marginal, 25, 37, 38, 60, 62, 88, 89, 101–104, 123, 126, 131, 149, 183, 287, 300, 301
Marginal lands, 38, 88, 89, 101–103, 105, 126, 190, 239, 279, 300
Meat, 67, 76, 160, 371–380, 384, 386
Meteorology, 183
Microbial diversity, 239, 427, 431, 432, 436, 440, 441
Mitigation, 30, 33, 69, 75, 76, 78, 79, 109, 111, 114, 150, 157, 159, 230, 238, 239, 243, 273–275, 277, 282, 283, 287, 289, 293, 294, 297, 298, 301, 302, 307–310, 314, 317, 318, 320, 322–325, 327, 328, 359, 363, 425, 451, 462–464
Mitigation options, 301

N
National Agriculture Research System (NARS), 110, 126, 132, 281
Nutrient cycling, 322, 323, 353, 354, 358, 425, 426, 429, 431, 432, 434–436, 440, 441
Nutritional facts, 195, 405

Nutrition insecurity, 196

P
Paradigm shift, 89, 110, 114, 126
Pesticide overuse, 449, 462
Pest management, 161, 164, 230, 232, 316, 318, 321, 450, 462, 464
Policy development, 36, 89, 125, 298, 326, 364, 430
Population, 26, 28–31, 37, 38, 58, 67–69, 88, 90, 91, 93, 95, 97, 102–104, 110, 123, 125, 126, 131, 133, 135, 156, 158–160, 172, 181, 182, 189, 198–200, 202, 213, 217, 219, 229–232, 235, 236, 238, 242, 251, 256, 275, 276, 278–280, 282, 283, 287–291, 294, 297, 300, 301, 312, 315, 323, 324, 326, 328, 347, 348, 356, 357, 359, 360, 364, 372, 373, 375, 381, 403, 410, 425, 427, 429, 435, 439, 450, 452–455, 460, 462, 464
Poultry, 58, 59, 62, 165, 371–382, 384–386, 388–393, 406
Precision Agriculture (PA), 110, 145, 158, 161, 163, 192, 300, 308, 316, 362
Protected areas, 355, 364

R
Rationale uses, 26
Regenerative, 61, 62, 101, 110, 133, 155, 162, 181, 274, 282, 297, 300, 302, 320–322
Regenerative practices, 274
Research, 34, 36, 38, 54, 58, 61, 63, 79, 109, 110, 112, 114, 115, 118, 121–126, 132, 163, 165, 182, 183, 187–193, 206, 211, 213, 217–219, 229–231, 234, 237, 241, 262, 264, 266, 274, 278, 281, 283, 299, 300, 302, 316, 317, 325, 327, 328, 339, 359, 361, 364, 393, 394, 408, 411, 415–419, 426, 430, 432–435, 437, 439–441, 452, 463
Resilient, 31, 109–111, 148, 155, 160, 185, 187, 203, 206, 278, 307, 316, 320, 321, 323, 328, 354, 355, 358, 363, 419, 426, 441, 463
Rising temperature, 30, 273, 276, 283, 325, 358, 360, 403, 428, 450–452, 454, 455, 464

S
Saline agriculture, 131, 145, 157
Salinity dynamics, 132, 133, 142
Semiarid, 46, 68, 71, 73, 93, 123, 124, 126, 156, 196, 204, 207, 210, 233, 236, 274, 277, 296, 309, 311, 343, 426, 434
Shrinking, 28, 37, 155, 158, 159, 172, 196, 279, 290, 312
Smallholder, 88, 90, 97, 101, 125, 234, 239, 274, 277, 280, 283, 291–293, 326
Soil classification, 27, 36, 294
Soil ecosystems, 426
Soil governance, 25, 26, 33–38
Soil improvement, 262
Soilless agriculture, 156
Soil microbiome, 425–427, 429–432, 435–441
Soil protection, 26, 259
Soil science societies, 25, 26, 37, 38, 88, 89, 103, 105, 279
Sustainability, 33, 76, 78, 111, 133, 149, 156, 161, 168, 172, 185, 186, 211, 230, 234, 241, 254, 255, 258, 259, 261, 266, 279, 315, 316, 320, 323, 327, 342, 343, 347, 350, 355, 360, 362–364, 406, 419, 426, 427, 429, 431, 438, 449, 450, 453, 464
Sustainable Development Goal (SDG), 25, 30, 33, 88, 133, 147, 200, 219, 289, 307, 314
Sustainable farming systems, 165, 239
Synergy, 287–289, 293, 294, 297–302

T
Temperature, 25, 30, 37, 46, 54, 57, 59, 67, 71–75, 78, 92, 96, 109, 113, 121, 122, 124, 148, 149, 156–158, 161–163, 165–168, 172, 181–183, 185, 188, 191, 196, 198, 199, 204, 207, 232, 238, 273, 275–277, 282, 283, 289, 293, 302, 307–313, 318, 325, 327, 340–344, 347–351, 354, 356–360, 363, 371, 372, 376–386, 389–394, 403, 404, 406, 410, 415, 417, 426–430, 432–441, 449–457, 459–461, 463, 464

V
Ventilation, 371, 381, 384, 388–391, 394
Virtual water, 91

W

Water footprint, 91, 160, 189
Water management, 30, 45, 61, 133, 134, 149, 161, 164, 273, 282, 296, 310, 317, 320, 321, 363, 364, 394, 413, 419
Water scarcity, 30, 37, 54, 60, 62, 77, 92, 109, 118, 122, 123, 126, 148, 149, 155, 158, 172, 186, 187, 189, 198, 273, 274, 281–283, 288, 293, 296, 298, 307, 312, 313, 315, 339, 340, 342, 358, 359, 363, 426, 441
Wildfire, 358, 359

MIX
Papier aus verantwortungsvollen Quellen
Paper from responsible sources
FSC® C105338

If you have any concerns about our products,
you can contact us on
ProductSafety@springernature.com

In case Publisher is established outside the EU,
the EU authorized representative is:
Springer Nature Customer Service Center GmbH
Europaplatz 3, 69115 Heidelberg, Germany

Printed by Libri Plureos GmbH
in Hamburg, Germany